Biology of *Salmonella*

NATO ASI Series

Advanced Science Institutes Series

A series presenting the results of activities sponsored by the NATO Science Committee, which aims at the dissemination of advanced scientific and technological knowledge, with a view to strengthening links between scientific communities.

The series is published by an international board of publishers in conjunction with the NATO Scientific Affairs Division

A	**Life Sciences**	Plenum Publishing Corporation
B	**Physics**	New York and London
C	**Mathematical and Physical Sciences**	Kluwer Academic Publishers
D	**Behavioral and Social Sciences**	Dordrecht, Boston, and London
E	**Applied Sciences**	
F	**Computer and Systems Sciences**	Springer-Verlag
G	**Ecological Sciences**	Berlin, Heidelberg, New York, London,
H	**Cell Biology**	Paris, Tokyo, Hong Kong, and Barcelona
I	**Global Environmental Change**	

Recent Volumes in this Series

Volume 243A —Biological Effects and Physics of Solar and Galactic Cosmic Radiation, Part A
edited by Charles E. Swenberg, Gerda Horneck, and E. G. Stassinopoulos

Volume 243B —Biological Effects and Physics of Solar and Galactic Cosmic Radiation, Part B
edited by Charles E. Swenberg, Gerda Horneck, and E. G. Stassinopoulos

Volume 244 —Forest Development in Cold Climates
edited by John Alden, J. Louise Mastrantonio, and Søren Ødum

Volume 245 —Biology of *Salmonella*
edited by Felipe Cabello, Carlos Hormaeche, Pasquale Mastroeni, and Letterio Bonina

Volume 246 —New Developments in Lipid–Protein Interactions and Receptor Function
edited by K. W. A. Wirtz, L. Packer, J. Å. Gustafsson, A. E. Evangelopoulos, and J. P. Changeux

Volume 247 —Bone Circulation and Vascularization in Normal and Pathological Conditions
edited by A. Schoutens, J. Arlet, J. W. M. Gardeniers, and S. P. F. Hughes

Series A: Life Sciences

Biology of *Salmonella*

Edited by

Felipe Cabello

New York Medical College
Valhalla, New York

Carlos Hormaeche

Cambridge University
Cambridge, United Kingdom

Pasquale Mastroeni and Letterio Bonina

University of Messina
Messina, Italy

Springer Science+Business Media, LLC

Proceedings of a NATO Advanced Studies Institute on
Biology of *Salmonella*,
held May 11–15, 1992,
in Portorosa, Italy

NATO-PCO-DATA BASE

The electronic index to the NATO ASI Series provides full bibliographical references (with keywords and/or abstracts) to more than 30,000 contributions from international scientists published in all sections of the NATO ASI Series. Access to the NATO-PCO-DATA BASE is possible in two ways:

—via online FILE 128 (NATO-PCO-DATA BASE) hosted by ESRIN, Via Galileo Galilei, I-00044 Frascati, Italy

Additional material to this book can be downloaded from http://extra.springer.com.

Library of Congress Cataloging-in-Publication Data

Biology of salmonella / edited by Felipe Cabello ... [et al.].
p. cm. -- (NATO ASI series. Series A, Life sciences ; vol. 245)
"Proceedings of a NATO Advanced Studies Institute on Biology of Salmonella, held May 11-15, 1992, in Portorosa, Italy"--T.p. verso.
Includes bibliographical references and index.
ISBN 978-1-4613-6236-4 ISBN 978-1-4615-2854-8 (eBook)
DOI 10.1007/978-1-4615-2854-8
1. Salmonellosis--Congresses. 2. Salmonella--Congresses.
I. Cabello, Felipe C., 1942- . II. NATO Advanced Studies Institute on Biology of Salmonella (1992 : Messina, Italy)
III. Series: NATO ASI series. Series A, Life sciences ; v. 245.
[DNLM: 1. Salmonella--genetics--congresses. 2. Salmonella--pathogenicity--congresses. 3. Drug Resistance, Microbial--congresses. 4. Salmonella Infections--prevention & control--congresses. 5. Vaccines--congresses. QW 138.5.S2 B615 1993]
QR201.S25B56 1993
616'.0145--dc20
DNLM/DLC
for Library of Congress 93-4806
CIP

ISBN 978-1-4613-6236-4

Originally published by Plenum Press, New York in 1993
Softcover reprint of the hardcover 1st edition 1993

PREFACE

Salmonella infections of man and animals continue to be a distressing health problem worldwide. Far from disappearing, the incidence of typhoid fever in developing countries may be far higher than we had imagined. *Salmonella* food poisoning has increased to one of the major causes of gastroenteritis in the developed world, in itself also an indication that animal salmonellosis is still a major cause for concern. The situation requires a concerted multidisciplinary research effort in order to generate the new information and technology needed to assist in the control of these diseases.

This concept was the driving force behind the NATO Advanced Research Workshop on "Biology of *Salmonella*" held at Portorosa, Messina, Italy, May 11-15, 1992. With additional support from the University of Messina, Medeva Group Research (UK) and the Swiss Serum and Vaccine Research Institute, the meeting brought together epidemiologists, microbiologists, molecular biologists, immunologists and clinicians. All the participants were actively working on different but related aspects of *Salmonella* and salmonellosis, with most of the leading laboratories worldwide being represented.

The workshop provided an excellent opportunity for interdisciplinary consultation; it is not often that the topic of *Salmonella* and salmonellosis is covered to such breadth and depth in one extended meeting. Keynote addresses by invited speakers were interspersed with offered papers, many by younger members of the scientific community, and this volume presents the collated manuscripts of the lectures and extended summaries of the offered papers.

Reports from the epidemiologists emphasized the continuing extent of salmonellosis worldwide, the increasing problem of drug resistance and the use of molecular approaches to epidemiology, while the section on cellular and molecular biology included papers on genetics, structure, gene regulation and new approaches to diagnosis. The section on pathogenesis covered a wide range of cellular and molecular aspects of the host-parasite relationship, while the presentations on immunity addressed new developments in both fundamental and practical aspects of mechanisms of acquired resistance and *Salmonella* vaccines.

We would like to thank the many people involved in helping to organize the meetings, the numerous members of the University of Messina, who gave their time so generously to assist in its success and Mrs. Harriett V. Harrison who helped in the final preparation and arrangement of the manuscripts for this volume. We would like to also thank Dr. Henry Godfrey and Ms. Gail Price for their help.

It is the hope of the organizers that we will be able to schedule similar meetings every two or three years.

Felipe Cabello
Carlos Hormaeche
Pasquale Mastroeni
Letterio Bonina

CONTENTS

SHORT COMMUNICATIONS

CHANGING TRENDS IN ANTIBIOTIC RESISTANCE IN SALMONELLA ISOLATED FROM HUMANS IN ENGLAND AND WALES

E. John Threlfall, Bernard Rowe
and Linda R. Ward

Division of Enteric Pathogens
Central Public Health Laboratory
London NW9 5HT, UK

INTRODUCTION

The Division of Enteric Pathogens (DEP) of the Public Health Laboratory Service (PHLS) is the national reference centre for salmonellas isolated from humans in England and Wales. The DEP also phage types strains of *Salmonella typhimurium*, *S enteritidis* and *S virchow* from food animals and referred by laboratories of the Veterinary Investigation Service. The majority of isolates from both humans and food animals are screened for resistance to a range of commonly-used antimicrobial drugs including ampicillin (A), chloramphenicol (C), gentamicin (G), kanamycin (K), streptomycin (S), sulphonamides (Su), tetracyclines (T), trimethoprim (Tm), furazolidone (Fu) and nalidixic acid (Nx). This provides a framework for observing trends in the occurence of resistance and for investigating factors which have contributed to the acquisition of resistance.

INCIDENCE OF RESISTANCE IN HUMAN ISOLATES

In the ten-year period 1981 - 1990, of a total of 186244 non-typhoidal salmonellas isolated from humans in England and Wales and referred to the DEP, the three most common serotypes were *S enteritidis*, *S typhimurium* and *S virchow*. These three serotypes respectively comprised 38%, 35% and 6% of the total isolates. The remaining 21% were comprised of isolates of a further 302 serotypes, none of which comprised more than 1% of the total (Threlfall *et al*, 1992a). From 1981 to 1987 *S typhimurium* was the most common serotype but since 1988 *S enteritidis* has predominated as a result of a national epidemic of *S enteritidis* phage type

Biology of Salmonella, Edited by F. Cabello *et al.*.
Plenum Press, New York, 1993

4 (=PT4) associated with poultry meat and shell eggs from chickens (Anonymous 1992).

Table 1 shows that since 1981 both drug resistance and multiple resistance (= 4 drugs or more) has remained uncommon and essentially unchanged in human isolates of *S enteritidis* but has doubled in *S typhimurium* and increased more than ten-fold in *S virchow*. In the remaining serotypes both resistance and muliple resistance has increased slightly but these increases are difficult to evaluate because of the large number of serotypes involved and the relatively small numbers of strains.

Table 1. Drug resistance in *S enteritidis*, *S typhimurium* and *S virchow* isolated from humans in England and Wales in 1981 and 1990

Serotype	Year	No.	% Sens.	% Resistant to 1	2 drugs	3	4+
S enteritidis	1981	1087	85	13	1	<1	<1
	1990	17794*	89	5	2	3	1
S typhimurium	1981	3992	64	7	18	6	6
	1990	5451	47	14	8	13	18
S virchow	1981	663	84	13	2	<1	1
	1990	1216	25	52	6	6	11

* 17794 of 18840 strains referred were tested
Source: Strains referred to Division of Enteric Pathogens

RESISTANCE TO INDIVIDUAL DRUGS

Table 2 shows that for *S enteritidis* the only resistance to increase substantially was that to ampicillin, from 1% to 4%. In contrast the incidence of resistance to high level furazolidone (MIC: >20mg/L) has fallen, from 13% to 6%. For *S typhimurium* the most significant increases were for ampicillin, tetracyclines and trimethoprim, although resistance to kanamycin has declined, from 6% to 2%. For *S virchow*, there have been substantial increases in the incidence of resistance to ampicillin, chloramphenicol, kanamycin, streptomycin, sulphonamides, tetracyclines, trimethoprim and furazolidone.

Table 2. Resistance to individual drugs in *S enteritidis*, *S typhimurium* and *S virchow* isolated from humans in 1981 and 1990

Antibiotic	% resistant					
	S enteritidis		*S typhimurium*		*S virchow*	
	1981	1990	1981	1990	1981	1990
Ampicillin (A)	1	4	5	17	1	9
Chloramphenicol (C)	<1	<1	5	6	<1	5
Gentamicin (G)	0	<<1	1	1	<1	1
Kanamycin (K)	1	<<1	6	2	<1	4
Streptomycin (S)	2	3	16	19	3	8
Sulphonamides (Su)	2	2	30	40	6	13
Tetracyclines (T)	1	2	22	45	1	10
Trimethoprim (Tm)	<1	1	8	21	0	12
Furazolidone (Fu)	13	6	1	2	9	71
Nalidixic Acid (Nx)	0	<1	<<1	<1	0	3

MULTIPLE RESISTANCE

S enteritidis

Although multiple resistance is uncommon in *S enteritidis*, from 1987 to 1990 there was the dissemination of a clone of PT24 which possessed resistance to three antibiotics - ampicillin, streptomycin and tetracyclines (R-type AST). Genetic studies demonstrated that this clone had been derived from a drug-sensitive progenitor strain of *S enteritidis* PT4 following the acquisition of an Inc N plasmid coding for AST and for resistance to some of the *S enteritidis* typing phages. The latter property was responsible for the change in phage type from PT4 to PT24 (Frost *et al*, 1989). In 1990 the most common patterns of multiple resistance in *S enteritidis* were ASTFu and CSSuTTm. Strains of R-type ASTFu belonged to this drug-resistant clone of PT24 and had acquired additional chromosomally-encoded resistance to furazolidone. In contrast strains of R-type CSSuTTm belonged to the epidemic phage type PT4, a phage type in which both resistance and multiple resistance was rarely encountered. These few strains of PT4 R-type CSSuTTm were associated with an outbreak in a sandwich bar in the City of London.

S typhimurium

In *S typhimurium* multiple resistance is relatively common and has increased substantially since 1981. In that year the most common patterns of multiple resistance were ACKSSuTTm and ACGKSSuTTm (Ward *et al*, 1990) and the majority of strains with these R-types belonged to PT204c. *S typhimurium* PT204c has been an important cause of bovine salmonellosis in England and Wales since 1979 and in addition to causing epidemics in calves since 1981, has also entered the food chain and caused numerous infections in humans (Rowe and Threlfall, 1984). An important feature of this phage type has been the sequential acquisition in the bovine host of plasmids and transposons coding for resistance to a variety of antimicrobial drugs, probably in response to the introduction and use of a range of antimicrobials in attempts to combat salmonella infection in calves. For example resistance to gentamicin, an important drug in human medicine, was shown to have arisen in PT204c as one of the consequences of using apramycin to treat bovine salmonellosis. In strains of PT204c with resistance to gentamicin, gentamicin resistance was found to be mediated by the production of the enzyme aminoglycoside 3-N-acetyl-transferase IV [AAC(3)IV] (Threlfall *et al* 1986), which also confers resistance to apramycin. Gentamicin is rarely used in animal husbandry in the United Kingdom but apramycin has been used extensively since it was licenced for veterinary use in 1980. This is is the first documented example of the use of a veterinary antibiotic (apramycin) resulting in the appearance and spread of salmonellae resistant to an antibiotic used almost exclusively in human medicine (gentamicin).

In 1990 the most common patterns of multiple resistance in *S typhimurium* were those of ASSuT and ACSSuTTm. The majority of strains of R-type ASSuT belonged to PT193 and since 1985 isolates of this phage type/R-type combination have increased in incidence in humans to such an extent that by 1988 it had become the most common multiresistant phage type (Ward *et al*, 1990). PT193 ASSuT first appeared in England and Wales in 1985 in bovine animals and although increasing in incidence in bovines has subsequently been associated with pigs and in 1990, with poultry. In strains of PT193 ASSuT the complete spectrum of resistance was encoded by a conjugative plasmid of 80 MDa (E. J. Threlfall & H. Chart, unpublished observations). Strains of R-type ACSSuTTm belonged to an epidemic clone of PT204c. This clone had lost a 36 MDa plasmid coding for AK, present in strains of PT204c of R-type ACKSSuTTm and ACGKSSuTTm, the two most common R-types in this phage type from 1980 to 1986 (Ward *et al*, 1990). Genetic studies demonstrated that in PT204c of R-type ACSSuTTm, the ampicillin resistance genes mediated by the 36 MDa plasmid in PT204c of R-type ACKSSuTTm have been retained by transposition onto other plasmids in the cell (Threlfall *et al*, 1987).

S virchow

In *S virchow* the incidence of multiple drug resistance has increased dramatically since 1980 and in 1990 the most common R-types in multiresistant strains were those of ACSSuTTmFu and CSSuTTmFu. Since 1986 the phage type in which multiple resistance was most common has been PT19, a phage type associated with poultry meat imported from France (Ward *et al*, 1990).

DRUG RESISTANCE IN STRAINS FROM FOOD ANIMALS

Isolates from Cattle

For the last ten years *S typhimurium* has been the serotype most frequently isolated from bovine animals in England and Wales and in particular, has predominated in calves (Anonymous, 1990b). Since 1981 the overall incidence of resistance in bovine isolates of this serotype has increased only slightly, from 71% to 79% but in contrast, the incidence of multiple resistance has quadrupeled, from only 15% in 1981 to 66% in 1990 (Table 3).

Table 3. Drug resistance in *S typhimurium* isolated from bovines in 1981 and 1990

				% Resistant to						
Year	Total	%DR	%MR	A	C	G	K	T	Tm	Fu
1981	1157	71	15	13	15	0	14	62	16	2
1990	1178	79	66	62	45	8	2	76	53	1

Source: Strains referred to the Division of Enteric Pathogens
Resistance symbols: see Table 2
DR = resistant to 1 or more antimicrobial
MR = resistant to 4 or more antimicrobials

The most important factors contributing to this dramatic increase in multiple drug resistance have ben discussed above: - the epidemic spread of strains of *S typhimurium* PT204c of R-types ACKSSuTTm, ACGKSSuTTm and ACSSuTTm in bovines from 1981 to 1990; and more recently, an upsurge in strains of PT193 R-type ASSuT.

Isolates from Poultry

Before 1988 the number of poultry isolates received by DEP were too few for meaningful analysis of drug resistance trends but over the last two years the number referred has increased substantially. In general both resistance and multiple resistance have remained uncommon in poultry isolates of *S enteritidis* and *S typhimurium* although since 1988 a few strains of S enteritidis PT24 of R-types AST and ASTFu have been isolated from poultry (Threlfall *et al*, 1992b). Furthermore, in 1989 and 1990 a few strains of *S typhimurium* of R-types ACGKSSuTTm, ACKSSuTTm and ACSSuTTm were isolated from chickens and turkeys in different parts of the country (Threlfall *et al*, 1990; Threlfall *et al* 1989). The latter strains either belonged to PT204c (Threlfall *et al*, 1990) or were untypable (Threlfall et al, *1989)*.

CONCLUSIONS

In the Swann report, published in 1969, it was concluded that the use of antibiotics as growth promotors had been a major contributory factor in promoting the appearance of multiple anibiotic resistance in zoonotic salmonellas (Anonymous, 1969). However, since 1981 it would appear that the use of antimicrobials for prophylaxis and therapy particularly in calf husbandry has been of major importance in promoting the emergence of multiresistant strains of *S typhimurium*. The spread of such strains would then have been facilitated by practices in the marketing of calves and in this respect in 1984 the British Veterinary Association produced guidelines aimed primarily at reducing the number of occasions a calf may be sent to market during the first 8 weeks of life, when most at risk to salmonella infections. Although changes in marketing techniques have had some impact and resulted in an overall decline in the incidence of *S typhimurium* (Table 4), the results presented above demonstrate that in 1990 multiresistant strains still predominate in bovine isolates of this serotype.

Although the use of antimicrobials in poultry has not been as intensive as that in cattle, there is increasing evidence that the use of nitrofurans in poultry is contributing to the emergence of strains with decreased sensitivity to these compounds (Rampling *et al*, 1991). In addition a small but increasing number of strains have been isolated with resistance to nalidixic acid and it is known that quinolones are already being used in animal husbandry in some European countries (Endtz *et al*, 1991). It now seems imperative that new antibiotics should not be used in veterinary medicine if the use of such compounds can result in cross-resistance to antibiotics used in human medicine.

REFERENCES

Anonymous, 1969. Report of the Joint Committee on the use of antibiotics in Animal Husbandry and Veterinary Medicine. Cmnd.4190: London HMSO. 1969.

Anonymous, 1990. The Microbiological Safety of Food. London: HMSO, 1990.

Anonymous, 1992. Animal Salmonellosis. 1990 Annual Summaries.

Ministry of Agriculture, Fisheries and Food: Welsh Office Agriculture Department: Department of Agriculture and Fisheries for Scotland. 1992.

Endtz, H.Ph., Ruijs, G.J., van Klingeren, B., Jansen, W.H., van der Reyden, T. and Mouton, R.P., 1991. Quinolone resistance in *Campylobacter* isolated from man and poultry following the introduction of fluoroquinolones in veterinary medicine. *J Antimicrob Chemother*. 27: 199.

Frost, J.A., Ward, L.R. and Rowe, B., 1989. Acquisition of a drug-resistance plasmid converts *Salmonella enteritidis* phage type 4 to phage type 24. *Epidemiol Inf*. 103: 243.

Rampling, A., Upson, R. and Brown, D.J.F., 1991. Nitrofurantoin resistance in isolates of *Salmonella enteritidis* phage type 4 from humans. *J Antimicrob Chemother*. 25: 285.

Rowe, B. and Threlfall, E.J., 1984. Drug resistance in Gram-negative aerobic bacilli. *Br Med Bull.* 40: 68.
Threlfall, E.J., Rowe, B., Ferguson, J.L. and Ward, L.R., 1986. Characterization of plasmids conferring resistance to gentamicin and apramycin in strains of *Salmonella typhimurium* phage type 204c isolated in Britain. *J Hyg.* 97: 419.
Threlfall, E.J., Rowe, B. and Ward, L.R., 1987. Increase in prevalence of a neomycin/kanamycin-sensitive variant of *Salmonella typhimurium* DT204c in cattle in Britain. *Vet Rec.* 15: 366.
Threlfall, E.J., Brown, D.J., Rowe, B. and Ward, L.R., 1989. Multiply drug-resistant strains of *Salmonella typhimurium* in poultry. *Vet Rec.* 120: 538.
Threlfall, E.J., Brown, D.J., Rowe, B. and Ward, L.R., 1990. Occurence of *Salmonella typhimurium* DT 204c in poultry in England and Wales. *Vet Rec.* 127: 234.
Threlfall, E.,J., Hall, M.L.M. and Rowe, B., 1992a. Salmonella bacteraemia in England and Wales, 1981-1990. *J Clin Path.* 45: 34.
Threlfall, E.J., Rowe, B. and Ward, L.R., 1992b. Recent changes in the occurence of antibiotic resistance in *Salmonella* isolated in England and Wales. *PHLS Microb Digest.* 9: 69.
Ward, L.R., Threlfall, E.J. and Rowe, B., 1990. Multiple drug resistance in salmonellas isolated from humans in England and Wales: a comparison of 1981 with 1988. *J Clin Path.* 43: 563.

THE PROBLEM OF MULTIRESISTANT SALMONELLA TYPHI - USE OF MOLECULAR MARKERS IN EPIDEMIOLOGY

Bernard Rowe, E. John Threlfall
and Linda R. Ward

Division of Enteric Pathogens
Central Public Health Laboratory
London NW9 5HT, UK

INTRODUCTION

Typhoid fever is endemic in many developing countries, particularly in the Indian sub-continent, South and Central America and Africa. In contrast only 200 - 300 cases occur in the UK each year and the majority of infections are in patients who have returned from countries where *Salmonella typhi* is endemic (Anonymous, 1985). Treatment with an appropriate antibiotic is essential and should commence as soon as the clinical diagnosis is made. Of necessity this may be before the results of laboratory sensitivity are available.

In 1948 chloramphenicol was introduced and in developed countries the use of this antibiotic for typhoid fever resulted in a reduction in the death rate from 10% to less than 2%. However in the 1970's a series of outbreaks caused by chloramphenicol-resistant strains in countries in widely-separated geographical areas (Anonymous, 1972; Paniker and Vilma, 1972) gave rise to fears that the supremacy of this antibiotic might have become jeopardised (Anonymous, 1973).

RESISTANCE TO CHLORAMPHENICOL

British Isolates

Of 2356 strains of *S typhi* isolated in the UK between 1978 and 1985, in only 6 (0.25%) was there clinically-significant resistance to chloramphenicol (MIC: >32mg/L) (Table 1). As a result of these findings, in 1987 it was recommended that chloramphenicol should remain the first-line drug for typhoid fever in the UK and in particular, that it should be used whilst awaiting the results of laboratory sensitivity tests (Rowe *et al*, 1987).

Biology of Salmonella, Edited by F. Cabello *et al.*.
Plenum Press, New York, 1993

Table 1. Chloramphenicol-resistant *Salmonella typhi*: UK, 1978 -1991

Years	Studied	Chloramphenicol-resistant (MIC: >32mg/L) Total	%
1978-85	2356	6	0.25
1986-89	790	12	1.5
1990	248	50	20.2
1991	226	48	21.2

Source: Strains referred to Division of Enteric Pathogens (DEP)

In the succeeding 4-year period 1986-1989 the occurence of chloramphencol-resistant *S typhi* increased slightly and 12 of 790 (1.5%) isolates were resistant to this antibiotic (Rowe *et al*, 1990) (Table 1). However it was considered that the increase, from 0.25% to 1.5%, was not sufficient to warrant changing the recommendation made in 1987 about the use of chloramphenicol.

The situation changed in 1990 when there was a dramatic increase in the number of chloramphenicol-resistant strains isolated in the UK. In that year 50 of 248 (20%) patients with *S typhi* were found to be infected with chloramphenicol-resistant strains (Table 1), the majority of which were also resistant to ampicillin and trimethoprim (Rowe *et al*, 1992). Most patients with chloramphenicol-resistant strains were infected with *S typhi* of Vi-phage type M1 and had either acquired their infections in Pakistan or had been in contact with patients who had recently returned from that country. However 10 patients infected with chloramphenicol-resistant strains had recently returned from India and of these the majority were infected with Vi-phage type E1. The remaining 3 isolates were from patients who had recently returned from Sri Lanka, South Africa or Japan.

There was little change in 1991 when 21% of typhoid patients in the UK were infected with chloramphenicol-resistant strains (Table 1), most of which were also resistant to ampicillin and trimethoprim (Rowe *et al*, 1992). However in contrast to 1990, equal proportions of patients with chloramphenicol-resistant *S typhi* had a history of return from Pakistan or India. Patients infected in Pakistan were infected with Vi-phage type M1 whereas the predominant phage type amongst patients infected in India was E1. In addition, early in 1991 there was a restaurant-associated outbreak in London. The phage type involved was that of M1 and although none of the patients had a history of recent foreign travel, some of the staff had lately returned from Pakistan (Rowe *et al*, 1992).

Isolates from Other Countries

Since 1989 chloramphenicol-resistant *S typhi* have been isolated with increasing frequency in several countries.

Table 2. Multiresistant *Salmonella typhi* in India, 1989 - 1992

City	Year	Phage type	Plasmids
Calcutta	1989	51	H1
Vellore*	1990	E1	H1
Delhi	1990	?	?
Calicut	1990 - 91	?	?
Chandigarh	1990 - 91	?	?
Dindigul*	1990 - 92	E1 (14)	H1
		O (2)	H1

* = strains examined in DEP
? = not known
In parentheses, numbers examined in DEP

In India several outbreaks caused by chloramphenicol-resistant strains have been reported (Prakash and Pillau, 1992) and such strains have been isolated in Chandigarh in the North ((Panigrahi *et al*, 1991), Calcutta in the East (Anand *et al*, 1990), Kerala in the South-west (Kumar, 1991), Dindigul in the South-east (Threlfall *et al*, 1992) and also in Vellore (Jesudasan and John, 1990) and Delhi (Gupta *et al*, 1990) (Table 2). The most common phage type has been that of E1 (Rowe *et al*, 1992) but strains of Vi-phage types O, 28 and 51 have been isolated (Prakash and Pillau, 1992; Anand *et al*, 1990). In Pakistan the predominant multiresistant phage type is M1 and since 1989 there has been an extensive outbreak in Rawalpindi caused by strains of this phage type (Karamat, 1990). Chloramphenicol-resistant strains of Vi-phage types E1 and M1 linked to immigrant workers from the Indian sub-continent have also been isolated in several countries in the Arabian Gulf (Wallace and Yousif, 1990) and in addition, multiresistant strains of different phage types have been isolated in Egypt (E2, C1 and D1-N), South Africa (A) and Kuala Lumpar (E1, B1 and DVS/UVS) (Table 3).

PLASMIDS

In all 98 chloramphenicol-resistant strains isolated in the UK since 1990 resistance to chloramphenicol, ampicillin and trimethoprim has beens mediated by either Inc H1 plasmids with molecular weights (MW's) ranging from approximately 110 megacdaltons (MDa) to 120 MDa (95 strains) or by Inc H2 plasmids of approximately 140 MDa (3 strains). In both 1990 and 1991 a few strains have been isolated which were resistant to ampicillin and trimethoprim but sensitive to chloramphenicol and in these strains resistance to these antimicrobials, presumably previously plasmid-mediated, has become incorporated into the chromosome (Rowe *et al*, 1990). In strains isolated in other countries and examined in the Division of Enteric

Table 3. Multiresistant *Salmonella typhi*: countries other than India, 1989 - 1992

Country	Year	Phage type		Plasmids	
Pakistan	1989 - 91	M1	(57)	H1	
Saudi Arabia	1991	E1	(5)	H1	
		51	(1)	H1	
Bahrain	1990 - 91	E1	(7)	H1	
		51	(1)	H1	
		A	(1)	H1	
Oman	1991	E1	(4)	H1	
Egypt	1991 - 92	E1	(4)	H1	
		E2	(18)	H1	
		D1-N	(2)	H1	
		C1	(2)	H1	
South Africa	1991	A	(9)	H1	
		E1	(1)	H1	
Kuala Lumpar	1991 - 92	E1	(28)	H1	(27)
				B	(1)
		DVS/UVS	(2)	H1	

Source: Strains referred to DEP
In parentheses, numbers examined in DEP

Pathogens, in almost all instances the complete spectrum of resistance has been encoded by Inc H1 plasmids.

RECOMMENDATIONS FOR THERAPY

Because of the increased isolation of chloramphenicol-resistant strains since 1990 and in view of the increasing incidence of such strains in many developing countries visited by travellers to the UK, chloramphenicol should no longer be regarded as the drug of choice for typhoid. Since the great majority of chloramphenicol-resistant strains have also been resistant to trimethoprim and ampicillin, these drugs have also been compromised. However all chloramphenicol-resistant strains isolated in the UK in 1990 and 47/48 of such strains isolated in 1991 were sensitive to ciprofloxacin with an MIC of <0.012 mg/L. In the UK it has therefore been recommended that, with the possible exception of young children, physicians should consider 4-quinolone drugs such as ciprofloxacin as the first-line drug for the treatment of typhoid, particularly in travellers returning to the UK from areas where multiresistant strains are now common (Rowe *et al*, 1992).

REFERENCES

Anand, A.C., Kataria, V.K. *et al.*, 1990. Epidemic multiresistant enteric fever in eastern India. *Lancet*. 335: 352.

Anonymous, 1972. Typhoid fever - Mexico. *Morb Mortality Wkly Rpt*. 21: 177.
Anonymous, 1973. Chloramphenicol resistance in typhoid. *Lancet*. ii: 1008.
Anonymous, 1985. Typhoid and paratyphoid fevers. *OPCS Monitor*. MB2 85/2: 9.
Gupta, B.L., Bhujwala, R.A. and Shrinwas., 1990. Multiresistant *Salmonella typhi* in India. *Lancet*. 336: 252.
Jesudasan, M.V. and John, T.J., 1990. Multiresistant *Salmonella typhi* in India. *Lancet*. 336: 256.
Karamat, K.A., 1990. Multiple drug resistant *Salmonella typhi* and ciprofloxacin. In: Proc 2nd WPCID & Chemother. 1990, p480.
Kumar, P.D., 1991. Ciprofloxacin for typhoid fever. *Lancet*. 338: 1143.
Panigrahi, D., Roy, P. and Sehgal, R., 1991. Ciprofloxacin for typhoid fever. *Lancet*. 338: 1601.
Paniker, C.K.J. and Vilma K.N., 1972. Transferable chloramphenicol resistance in *Salmonella typhi*. *Nature*. 239: 109.
Prakash, K. and Pillai, P.K., 1992. Multidrug-resistant *Salmonella typhi* in India. *APUA Newslett*. 1: 1.
Rowe, B., Threlfall, E.J. and Ward, L.R., 1987. Does chloramphenicol remain the drug of choice for typhoid? *Epidem Inf*. 98: 379.
Rowe, B., Threlfall, E.J. and Ward, L.R., 1990. Spread of multiresistant *Salmonella typhi*. *Lancet*. 336: 1065.
Rowe, B., Ward, L.R. and Threlfall, E.J., 1991. Treatment of multiresistant typhoid fever. *Lancet*. 337: 1422.
Rowe, B., Ward, L.R. and Threlfall, E.J., 1992. Ciprofloxacin and typhoid fever. *Lancet*. 339: 740.
Threlfall, E.J., Ward, L.R. *et al*., 1992. Widespread occurence of multiple drug-resistant *Salmonella typhi* in India. *Eur J Clin Microbiol Infect Dis*. (in press).
Wallace, M. and Yousif, A.A., 1990. Spread of multiresistant *Salmonella typhi*. *Lancet*. 336; 1065.

SALMONELLA TYPHI IN SPAIN: EPIDEMIOLOGICAL MARKERS AND ANTIMICROBIAL SUSCEPTIBILITY (1979-1991)

Miguel A. Usera

Laboratorio de Enterobacterias.
Centro Nacional de Microbiología Virología
e Inmunología Sanitarias, Instituto de Salud Carlos III
Majadahonda, 28220 Madrid, Spain

INTRODUCTION

Typhoid fever continues to be a serious public health problem in developing countries, although in the developed world the likelihood of acquiring the infection is extremely low. Imported cases from endemic regions continue to cause problems in non-endemic regions (Edelman and Levine, 1986).

Surveillance typhoid fever data from Spain (1979-91) indicates that its´ incidence has increased from 6.7 cases per 100,000 population in 1979 to 15.4 cases per 100,000 in 1985, decreasing subsequently to 3.4 cases per 100,000 in 1991 (Usera, 1991) (Figure 1).

Since 1986 multiresistant *Salmonella typhi* strains have been isolated in Asia (Morshed et al., 1986). In 1991 Rowe and colleages (Rowe et al., 1991) reported several multiresistant *S. typhi* isolates in Great Britain. In Spain, *S. typhi* multiresistant strains still represent a low percentage (Usera, 1991).

We have studied a total of 1419 *S. typhi* strains isolated in Spain from 1979-1991, looking for antibiotic susceptibility, phagetypes, biotypes, colicintypes, chloramphenicol resistant strains plasmid profiles and the ampicillin, chloramphenicol and cotrimoxazole minimal inhibitory concentrations (MIC) of all of them.

MATERIAL AND METHODS

On receipt, each isolate was subcultured on Triptic Soy Agar (TSA) plates, confirmed as *S. typhi* by conventional biochemical reactions (Ewing, 1986) and serotyped with antisera prepared in our laboratory (Lindberg and Le Minor, 1984).

Biology of Salmonella, Edited by F. Cabello *et al.*,
Plenum Press, New York, 1993

Phagetyping was performed in our laboratory using the set provided by the WHO International Phagetyping Laboratory for *Salmonella* at Colindale. London (Guinnée and Van Leeuwen, 1978).

Xylose fermentation (Xy) and Tetrationate reductase enzyme (TTR) assays were determined according to Edward and Ewing's Manual (Ewing, 1986). The presence of bacteriocin was detected by Institut Pasteur's method (Vieu, 1983).

Antibiotic susceptibility was carried out by the Kirby-Bauer method (Bauer et al., 1966). Antibiotics tested are listed in Table 1. MIC was performed by the agar dilution technique on ampicillin, chloramphenicol, trimethoprim and sulfamethoxazole and

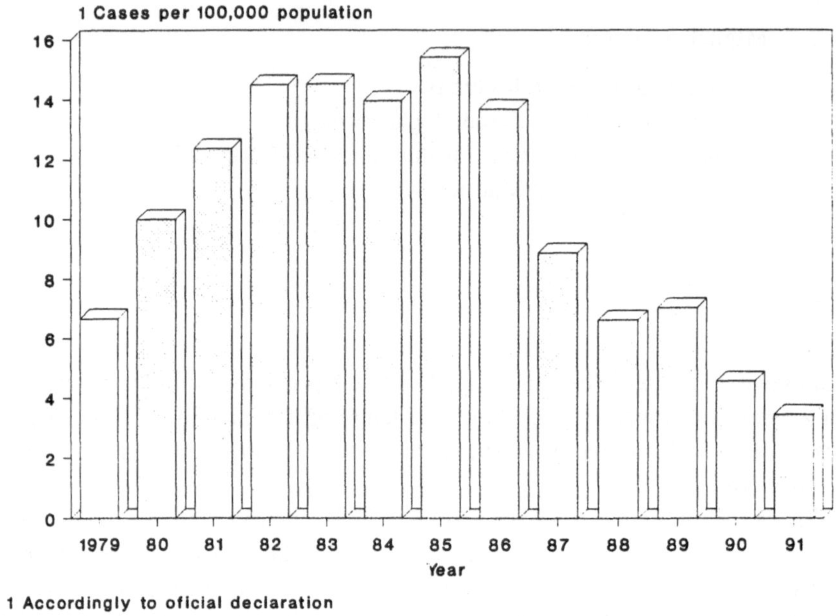

Figure 1. Typhoid fever rates in Spain. 1979-91.

evaluated accordingly to the NCCLS criteria (National Committee for Clinical Laboratory Standards, 1988). Strains resistant to four or more antibiotics were considered multiresistant.

The plasmid profile was obtained by the Birnboim and Doly method (Birnboim and Doly, 1979). Plasmids of *Escherichia coli* K12 strain V517 and RP1 were used as standards for determining plasmid sizes on extraction from *S. typhi*. Purified plasmids were digested with the restriction enzyme *Hind* III (Sambrook et al., 1982) and transferred to an *E. coli* BM21 plasmid free recipient strain according to Karmaker's method (Karmaker et al., 1991). Statistical analysis was made by σ^2 determination (Carrasco, 1986). A p value of < 0.01 was considered to be statistically significant.

Table 1. Resistant *S. typhi* to one or more antimicrobial in Spain (1979 -1991)

Antimicrobial[a]	Strains (%)
C	1 (0.07)
Cfp	1 (0.07)
Gm	2 (0.14)
Sx	1 (0.07)
Te	4 (0.28)
T	2 (0.14)
AmCb	1 (0.07)
GmSx	1 (0.07)
TeT	1 (0.07)
AmCbSx	1 (0.07)
AmCbT	1 (0.07)
CSxTe	1 (0.07)
AmCbCTe	1 (0.07)
AmCbCSxTeT	7 (0.49)
Total	26 (1.83)

[a] Am (Ampicillin), Cb (Carbenicillin), Cfp (Cefoperazone), C (Chloramphenicol), Gm (Gentamicin), Sx (Sulfamethoxazole), Te (Tetracycline), T (Trimethoprim).

RESULTS

Thirty six provinces of the 50 sent 1419 strains. Only two Autonomous Regions (CA) did not participate (Table 2).

Barcelona, Logroño, Madrid and Zaragoza sent strains regularly since 1980. Alicante started sending strains regularly in 1982. Bilbao, Gerona, Navarra and Valladolid sent strains throughout most of the study period and the remaining provinces only sent strains sporadically.

There were no statistically significant sex differences. Most were isolated from patients older than 14 years (61.8%). The monthly distribution is shown in Figure 2. Strains were isolated from blood (67.6%), faeces (28.7%), bile (1.7%) and other human sources (1.8%) except for one from pork.

Most typhoid fever patients were treated with standard antimicrobials therapy such as ampicillin, chloramphenicol or cotrimoxazol, although third generation cephalosporins and quinolones have increased in use over the last two years.

Table 2. Strains distribution by Autonomous Regions (CA)

Autonomous Regions (CA)	Number of Strains
Andalucía	26
Aragón	112
Asturias	9
Baleares	-
Canarias	6
Cantabria	1
Castilla-La Mancha	12
Castilla-León	204
Cataluña	230
Comunidad Valenciana	100
Extremadura	-
Galicia	20
Madrid	454
Murcia	3
Navarra	61
Pais Vasco	121
La Rioja	57
No data available	3
Total	1419

Following the International Federation of Enteric Phage Typing (IFEPT) criteria (Ward, 1990), phagetype distribution was:

a) E1 (30.8%), C1 (14.2%), A (9.7%), Ade (9.6%), 46 (6.6%), D1 (5.6%), NT+ (4.7%), 34 (4.2%), I+IV (3.3%), NT- (2,6%)

b) F1 (1.2%), D1N (1.1%), B2 (1%), D4 (0.7%), C4 (0.6%), 51 (0.5%)

c) D6 (0.4%), G1 (0.4%), D7 (0.3%), N (0.3%), 40 (0.3%), C5 (0.2%), J3 (0.2%), 38 (0.2%), B1, B3, C3, D10, D2, J4, L1, M1, O, 31, 35, 42, 53, 55 since B1 to 55 (0.1%)

Phagetype analysis gave significant differences ($p < 0.01$) between phagetypes E1, C1 and year of isolation. Also gave significant differences between phagetypes E1, C1, D1, 46 and provinces of origin.

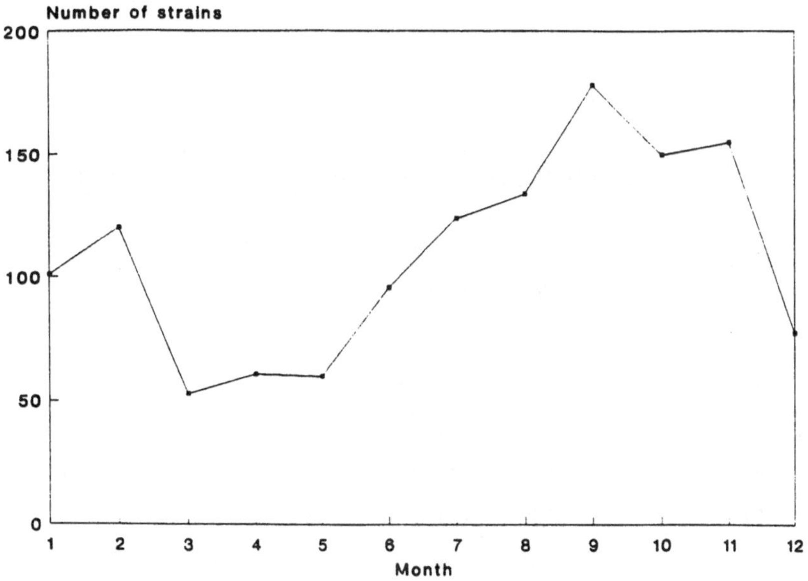

Figure 2. Monthly distribution of *Salmonella typhi* isolates.

Phagetype E1 occurrence was 30.8% with respect to the total. This was isolated significantly more often in 1980 (36.9%) and 1981 (61.1%) and less often in 1983 (19%) and 1990 (14.8%) than the mean. Also phagetype C1 had in 1981 (5.7%) and 1991 (5.8%) lower isolation frequencies and in 1990 (23.4%) higher than the mean (14.2%).

There was a high incidence of phagetype D1 in Alicante (47.5%), E1 in Madrid (41.2%) and Valladolid (48.2%), C1 in Navarra (37.7%); 46 in Barcelona (11.7%) and Bilbao (11.3%). A low incidence of phagetypes E1 in Alicante (8.7%) and Navarra (8.2%) compared with the mean and of C1 in Madrid (9.8%) and Valladolid (7.9%).

Overall of 1419 isolates, 13% were Xy negative and only 1% TTR negative. Xy fermentation showed significant differences with respect to phagetypes and provinces: 32.6% of phagetype A isolates and 17.4% of phagetype E1 were Xy negative along with 28.7% of isolates in Valladolid and 6.8% in Zaragoza.

Twenty-six *S. typhi* isolates (1.83%) were resistant to one or more drugs (Table 1). Fifteen different resistance patterns were found. Eight isolates (0.56%) were multiresistant and isolated in 1990 and 1991.

Table 3 summarizes the epidemiological data and epidemiological marker results of resistant strains.

Minimal inhibitory concentration results with strains resistant to drugs commonly used in treatment were: Ampicillin ≥ 64 μg/ml, chloramphenicol ≥ 256 μg/ml, trimethoprim ≥ 64 μg/ml and sulfamethoxazole ≥ 256 μg/ml.

Chloramphenicol resistant strains produced CAT and harboured a large 79 Mdal plasmid, except for one that harboured three of 75, 5.5 and 3.8 Mdal (Figure 3).

The 79 Mdal plasmid was digested with *Hind* III yielding a similar pattern in all strains tested (3 of 9) (Figure 4). This plasmid was demonstrated by conjugation to be transferrable (Figure 5).

Thirteen outbreaks were reported on during the study period, as summarized in Table 4.

Table 3. Epidemiological data and markers of resistant strains

Strain	Resistance[a]	Year	Origin	Sex[b]	Phagetype[c]	Xy	TTR
48123	C	1983	Faeces	F	E1	+	+
14022	Cfp	1980	Unknown	M	53	+	+
11091	Gm	1980	Unknown	U	C1	+	+
48647	Gm	1983	Blood	F	NT-	+	-
14556	Sx	1991	Faeces	F	34	+	+
47788	Te	1983	Unknown	M	A	+	+
49443	Te	1983	Blood	F	Ade	+	+
56857	Te	1984	Unknown	F	E1	-	+
5177	Te	1987	Faeces	F	C1	+	+
48564 & 8797	T	1983-1989	Faeces	M	NT+	+	+
4698	AmCb	1989	Blood	F	NT-	+	+
82942	GmSx	1986	Faeces	M	Ade	+	+
8706	GmTe	1980	Unknown	U	Ela	+	+
5831	TeT	1988	Unknown	U	NT+	+	+
50902	AmCbSx	1983	Faeces	F	NT+	+	+
6062	AmCbT	1979	Unknown	F	NT	+	+
5131	CSxTe	1987	Blood	M	A	+	+
586	AmCbCTe	1991	Blood	M	Ela	+	+
9044 & 11904	AmCbCSxTeT	1990	Blood	M	M1	-	+
1633 & 1634	AmCbCSxTeT	1991	Blood & Faeces	M	Ela	+	+
5085	AmCbCSxTeT	1991	Faeces	F	Ela	+	+
15627 & 15628	AmCbCSxTeT	1991	Blood & Faeces	M	Ela	+	+

[a]Look table 1; [b]F = female; M = male; U = unknown; [c]NT+ = Non typable, Vi positive; NT- = Non typable, Vi negative.

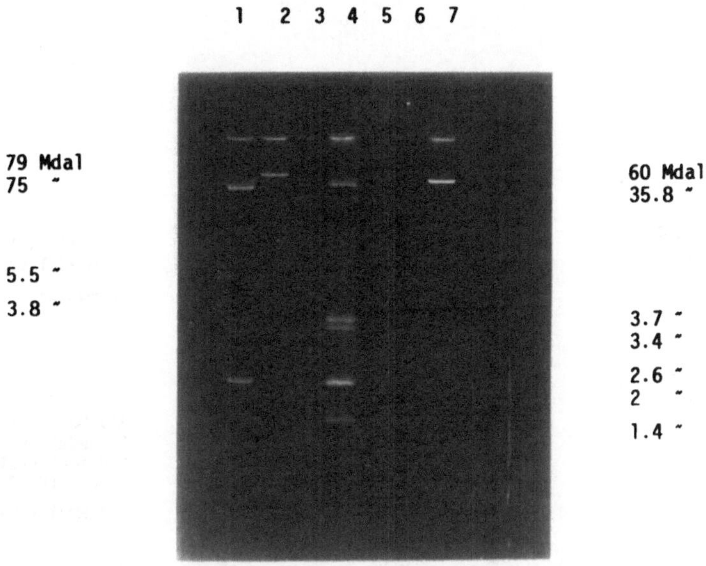

Figure 3. Plasmid patterns of *Salmonella typhi*. Lanes 1 and 4: *E. coli* V 517 control. Lane 7: *E. coli* RP1 control. Lane 2: *S. typhi* multiresistant (plasmids size 75;5.5;3.8 Mdal). Lane 3: *S. typhi* multiresistant (plasmid size 79 Mdal). Lanes 5 and 6: *S. typhi* plasmid free.

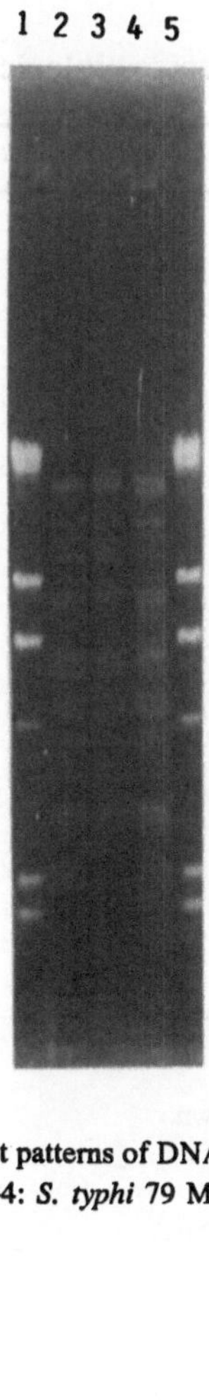

Figure 4. Restriction endonuclease fragment patterns of DNA plasmids from different *S. typhi*. Lanes 1 and 5: λ phage digested with *Hind* III. Lanes 2-4: *S. typhi* 79 Mdal plasmid digested with *Hind* III.

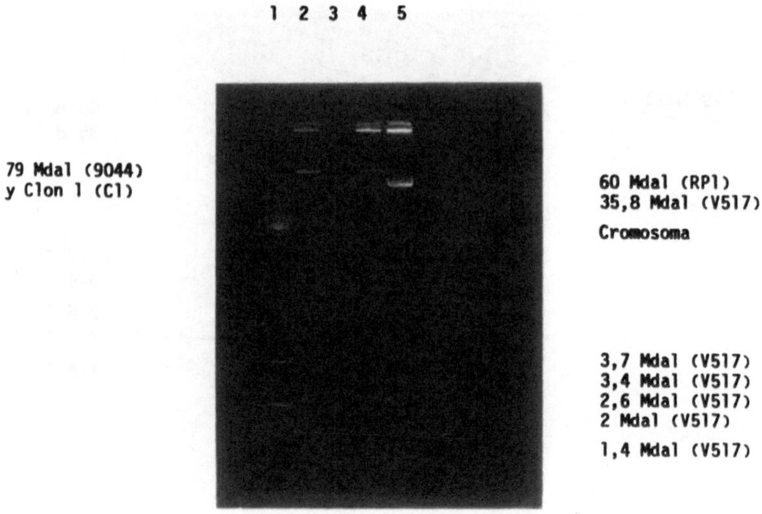

Figure 5. Plasmid patterns of transcojugant. Lane 1: *E. coli* V 517 control. Lane 2: *S. typhi* donor. Lane 3: *E. coli* BM 21 recipient. Lane 4: *E. coli* transconjugant. Lane 5: *E. coli* RP1 control.

Table 4. Outbreaks epidemiological data and markers

Outbreak	Strains	Month/Year	Type[a]	Province	Phagetype
1	1	01/83	C	Murcia	D4
2	1	08/83	U	Madrid	A
3	2	10/83	F	Barcelona	NT-
4	1	04/86	F	Barcelona	C4
5	1	02/89	C	Soria	C1
6	11	08/89	C	Navarra	C1
7	2	09/89	U	Alava	Ela
8	6	10/90	C	Guipuzcoa	C4
9	10	11/89	C	Guipuzcoa	Ela
10	1	07/90	C	Segovia	Ela
11	1	12/90	U	Vizcaya	I+IV
12	30	06/91	C	Madrid	34
13	4	08/91	C	Burgos	D1

Note: NT- = Non typable Vi negative
[a] F = familiar, C = comunitary, U = unknown.

DISCUSSION

Information regarding the incidence of *S. typhi* epidemiological markers in Spain is reasonably good because most provinces have sent strains for study, and is especially good in provinces like Barcelona, Logroño, Madrid and Zaragoza because the strains isolated have been studied annually. There were five isolates reported as being imported, which indicates that most typhoid fever cases in Spain are indigenous.

As in developed countries the age group over fourteen was the one mainly implicated.

Monthly distribution indicates a seasonality of typhoid fever in Spain with two peaks, one at the end of the summer and the beginning of autumm, probably because of the hot weather and the use of contaminated water to irrigate, and another at the end of January and the beginning of February, because of the Christmas Holidays when people probably eat more foodtypes that are susceptible to contamination.

As described in the literature there were more isolates from blood than from faeces and only one of non human origin (pork). It was not possible to determine the origin and possible consecuences of this isolate.

Phagetypes most frequently isolated in Spain are cosmopolitan and frequent in other European countries, except phagetype 34 which is frequent in North Africa and South America. The high incidence of phagetype 34 in Spain was principally due to an outbreak which occurred in Madrid.

Less frequently isolated phagetypes D4 and C4 could be imported from North and Central Africa, respectively, where they are common, although incomplete epidemiological data did not allow us to confirm this assertion.

Phagetype results suggest a different annual incidence of types E1 and C1. Significantly different rates between provinces and some phagetypes were observed. Phagetype D1 was isolated at a high rate in Alicante especially in 1982 and phagetype F1 in Valladolid, which indicates the possibility of considering some phagetypes to be endemic in some provinces.

Colicintyping is not considered useful as an epidemiological marker because only four strains of over 272 tested were positive, these being also of phagetype 40.

Xylose fermentation can be considered as a good complementary marker for E1, A, Ade, B2, D1N and D6 phagetypes.

The prevalence of resistant strains is low in Spain (1.83%), but there is an increasing trend over the last three years. Nearly half the resistant strains were isolated during this period (42.3%). The first multiresistant isolate was described in 1990 probably imported from Asia.

The high incidence (19.2%) of non-typable strains in the resistant group is remarkable. The highest incidence of resistant strains occurred in Madrid and Alicante.

MIC results suggest the presence of a resistance plasmid, CAT production also suggests the presence of an R plasmid in chloramphenicol resistant strains. All strains tested harboured a large plasmid responsable for multiresistance as demonstrated by conjugation. This is probably similar to that described by Karmaker (Karmaker et al., 1991) in India, which leads to the hypothesis that multiresistant strains could be imported from this area.

Outbreaks, especially large ones, are very unusual in Spain. Only four have involved more than 10 people since 1979, three of which occurred in 1989 in the North, coinciding with a drought and one in Madrid due to inappropriate food handling. None occurred in clearly high risk populations and thus none were candidates for the vaccine.

ACKNOWLEDGEMENTS

I am grateful to A. Echeita, A. Aladueña and P. Espinosa for their collaboration in this study and to I. Outschoorn for her assistance in translating the manuscript.

REFERENCES

Bauer, K.W., Kirby, W.M.M., Sherris, J.C., and Turck, M.T, 1966, Antibiotic susceptibility testing by a standardized single disk method, *Am. J. Clin. Pathol.* 45:493.

Birnboim, H.C., and Doly, J., 1979, A rapid alkaline extraction procedure for screening recombinant plasmid DNA, *Nucl. Acid Res.* 7:1513.

Carrasco, J.L., 1986, El método estadistico en la investigación médica. Madrid (Spain): Editorial Ciencia 3 S.A.

Edelman, R., and Levine, M.R., 1986, Summary of an International Workshop on Typhoid Fever, *Rev. Infect. Dis.* 8:329.

Ewing, W.H., 1986, The genus Salmonella, *in:* "Edwards and Ewing's Identification of Enterobacteriaceae," W.H. Ewing, Elsevier Science Publishing, New York, p. 27.

Guinnée, P.A.M., and Van Leeuwen, W.J., 1978, Phagetyping of *Salmonella, Methods Microbiol.* 11:157.

Karmaker, S., Biswas, D., Shaik, N.M., Chaterjee, S.K., Kataria, V.K., and Kumar, R., 1991, Role of a large plasmid of *Salmonella typhi* encoding multiple drug resistance. *J. Med. Microbiol.* 34:149.

Lindberg, A.A., and Le Minor, L., 1984, Serology of *Salmonella, Methods Microbiol.* 15:1-141.

Morshed, M.G., Khan, N.Z., Khan, W.A., and Akbar, M.S., 1986, Multiple drug resistant *Salmonella typhi* in Bangladesh, *J. Diarrhoeal Dis. Res.* 4 (4):241.

National Committe for Clinical Laboratory Standards, 1988, Methods for dilution antimicrobial susceptibility tests for bacteria that grow aerobically. *in:* "Approved Standard M2-T4," NCCLS, Villanova, PA.

Rowe, B., Ward, L.R., and Threlfall, E.J., 1991, Treatment of multiresistant typhoid fever, *Lancet*, 337:1422.

Sambrook, J., Fritsch, E.F., and Maniatis, T.E., 1982, Enzymes used in molecular cloning, *in:* "Molecular Cloning: A Laboratory Manual, Cold Spring Harbor, New York, Cold Spring Harbor Laboratory I.

Usera, M.A., 1991, Estudio de marcadores epidemiológicos de Salmonella enterica subespecie I serotipo Typhi aisladas en España en el periodo 1979-1990. Tesis Doctoral. Universidad Complutense de Madrid, *Facultad de Ciencias Biológicas*, 21.

Vieu, J.F., 1983, Aspects epidemiologiques actuels de la fievre typhoide en France (1982-1983). *Bull. Ass. Anct. Ele. Inst. Pasteur* 98:31.

Ward, L.R., 1990, The geographical distribution of *Salmonella typhi* and *Salmonella paratyphi* A and B phagetypes during the period 1 January 1986 to 31 December 1989. International Federation of Enteric Phage-Typing report. XV Internatinal Congress of Microbiology. Osaka, Japan.

SALMONELLAE AND SALMONELLOSIS IN TURKEY

Özdem Anğ, Kurtuluş Töreci and Mine Anğ-Küçüker

Department of Microbiology
Istanbul Medical Faculty
Çapa, Istanbul, Turkey

Typhoid fever, caused by *S. typhi*, is a rare disease today in developed countries in where personal hygiene is applied in appreciable levels, clean and adequate water supply is offered, and foods and food-handlers are controlled. In countries where these conditions are not set as it happens in our country, typhoid salmonellosis due to *S. typhi* are still among significant bacterial infections in all parts of the country. For instance, 128 cases of typhoid fever were reported from Istanbul in 1989. This may not seem worrisome for a city with a population more than 5 million. But while evaluating these numbers, it is important to keep in mind that infectious diseases are underreported in Turkey and the numbers in the official statistics are well below the real numbers. For this reason, the numbers given for typhoid and paratyphoid cases should only be taken to compare the incidences for different years.

Table 1. Typhoid and paratyphoid fever case-reports from Istanbul (1973-1989).

Year	Typhoid	Paratyphoid	Year	Typhoid	Paratyphoid
1973	249	17	1982	88	17
1974	142	16	1983	204	40
1975	119	14	1984	86	15
1976	73	7	1985	56	16
1977	215	12	1986	85	19
1978	113	75	1987	56	42
1979	125	20	1988	86	16
1980	146	20	1989	128	33
1981	191	33			

Biology of Salmonella, Edited by F. Cabello *et al.*,
Plenum Press, New York, 1993

Table 1 shows the typhoid and paratyphoid cases formally reported from Istanbul in the 17 year-period between 1973-1989 (Istanbul Sağlik Müdürlüğü Istatistikleri, Istanbul, 1989).

Table 2 on the other hand, summarizes the formally reported cases of typhoid and paratyphoid and the associated deaths in Turkey between 1964-1989 (Tükiye Sağlik Istatistik Yilliği 1964-67, 1971; 1968-72, 1975; 1973-74, 1977; 1975-78, 1980).

Table 2. Typhoid and paratyphoid fever cases and the associated deaths in Turkey between 1964-1989.

	Typhoid		Paratyphoid			Typhoid		Paratyphoid	
Year	Cases	Deaths	Cases	Deaths	Year	Cases	Deaths	Cases	Deaths
1964	3425	137	233	1	1977	908	14	332	0
1965	4019	120	258	6	1978	591	4	374	10
1966	4878	127	298	11	1979	862	13	664	8
1967	3354	105	533	14	1980	1423	9	508	6
1968	2982	84	554	4	1981	2402	27	880	6
1969	2250	57	655	10	1982	1273	20	674	13
1970	3402	56	651	4	1983	1471	6	1035	14
1971	1729	45	587	5	1984	1825	10	975	10
1972	1550	40	614	31	1985	2052	10	738	8
1973	1430	37	839	15	1986	3656	5	865	0
1974	1401	35	477	5	1987	4070	4	1313	6
1975	810	20	470	14	1988	3523	8	1206	12
1976	695	23	471	13	1989	6880	13	1521	2

While evaluating these numbers it should also be taken into consideration that the total population of Turkey was 30,400,000 in 1964; 35,600,000 in 1970; 44,900,000 in 1980, and 56,000,000 in 1989. Apart from these, if we consider the reporting procedure and how it works, or rather how it doesn't work, it will be more realistic to put an additional zero at the end of each number. The tables show that paratyphoid cases are reflected with a lower incidence in the statistics than typhoid cases. It should not also be missed that microbiology laboratory services are insufficient, diagnosis are made usually only on clinical grounds, only Gruber-Widal test is used to diagnose typhoid and the number of patients who had their diagnoses by way of positive cultures is quite low.

The distribution of 1126 cases of typhoid and 231 cases of paratyphoid, according to months, in Istanbul in the last decade (1980-1989) is shown in Table 3.

As it is seen in the table, tyhoid cases in Istanbul are encountered more frequently in summer and autumn months. Ministry of Health statistical methods are the same throughout Turkey. The monthly distribution of paratyphoid cases, on the other hand, is not very meaningful.

It is impossible to give statistical data about *Salmonella* gastroenteritis for our country. Food poisoning columns have remained empty for years in the health statistics; and for recent times when some data are put down into them, it is again impossible to guess how many of them are due to *Salmonella* or *Staphylococcus* or any other agent. Up to 1989 in Turkey, the recorded isolates of *Salmonella* serovars from food poisoning cases were only *S. typhimurium*, *S. typhimurium* var. *copenhagen*, *S. reading*, *S. sandiego*, *S. haifa*, *S. enteritidis*, *S. zanzibar*, *S. muenster* and *S. edinburg* (Aksoycan, 1972). The new serovars isolated from human in our country are mostly from sporadic cases and some are from outbreaks in some small and particular populations.

Table 3. The monthly distribution of typhoid and paratyphoid fever cases in Istanbul between 1980-1989.

Month	Jan	Feb	Mar	Apr	May	Jun	Jul	Aug	Sep	Oct	Nov	Dec	Total
Typhoid	87	55	79	60	55	118	126	130	115	97	103	102	1126
Paratyphoid	16	12	16	33	14	12	13	26	19	18	21	31	231

One of the reasons for the scarcity of *Salmonella* serovars isolated in our country is that there is no National *Salmonella* Center in Turkey. Routine laboratories or some reference laboratories are be equipped to only identify some particular serovars found in the Kauffmann-White scheme, as hundreds of sera are needed to make these identifications. It is unfortunate that there still exists no specific Center to prepare and distribute the antisera to laboratories. This makes it clear that a *Salmonella* Center is needed in our country.

For the *Salmonella-Arizona* serovars isolated in Turkey there is Aksoycans's grouping for different years and some other authors have helped to set these groups in their papers (Aksoycan, 1972; Aksoycan, 1977; Aksoycan, 1980a; 1980b; Aksoycan, 1984; Aksoycan, 1985; Aksoycan, 1988; Buget et al., 1988; Töreci and Öztan, 1984; Yücel et al., 1987). Here, we will try to simplify this list for routine use by relying on Aksoycan's study published in "Journal of Turkish Microbiological Society" adding also the new serovars reported in different studies (Aksoycan, 1980b).

Being grouped in C_3, *S. istanbul* 8:z_{10}:e,n,x among all these serovars had been isolated as the first *Salmonella* carrying this unique antigenic structure in the world and thus is named Istanbul (Özek et al., 1969). In the same way *S. adana* from group U has been a serovar isolated first in our country. Other than these, *S. boecker* which had been shown with an antigenic structure 6, 14:1, v:1, 7 formerly in the Kauffmann-White scheme is now shown as "(1), 6, 14, (25)" because of the isolation of strains having the 6,14:1,6,14 and 1,6,14,25 O antigens in Istanbul (Özek et al., 1967; Özek et al., 1971).

The total number of serovars of *Salmonella-Arizonae* isolated in Turkey reaches 94. According to the formal reports, 42 of them were obtained only from human sources, 20 from both human and non human sources and 32 only from non human sources.

Apart from these, a non-motile strain in group C_1 carrying the antigenic formula 6,7:-:- has been isolated from a human source (Töreci et al., 1988). Since this serovar may be one which has lost its ability to produce flagella, has not taken here as a different serovar. Similarly, a *Salmonella* strain with T_1:b:e,n,x antigenic structure and S-R change has been isolated from a nonhuman source (Özek et al., 1967b), and this neither has been taken as a different serovar.

Serovars of *Salmonellae* isolated in our country belong to a total of 21 serogroups and subgroups, being A,B,C_1,C_2,C_3,D_1,E_1,E_2,E_4,F,G_1,G_2H,I,M,P,Q,U,X,Z,O:61. The serogroups isolated from humans are of 13 serogroups or subgroups since there are no isolations from groups G_1,H,I,M,Q,X,Z,O:61 from humans.

The *Salmonella* serovars isolated in Turkey are given in Table 4, according to the order of listing in the Kauffmann-White scheme.

The presence of antibiotic-resistant *Salmonella* strains is an important problem in both developed and developing countries; so, treatment of systemic infections caused by these strains is becoming increasingly difficult in all the world as well as in Turkey.

In Turkey, resistance of *Salmonellae* to chloramphenicol, ampicillin and co-trimoxazole has increased. We had noted a high incidence of resistance and also multiple resistance in *Salmonellae* isolated in 1964 and 1978 (Table 5). Most of these strains belonged to group B and many of the group B Salmonellae were *S. typhimurium* (Anğ, 1976).

In Istanbul, until 1991, all of 23 *S. typhimurium* strains examined were found to exhibit multiple resistance to at least two antibiotics and of these strains, five were resistant to ampicillin, 21 to chloramphenicol, and 16 to co-trimoxazole (Bozok Johansson, in press).

Isolation rate of *S. enteritidis* from cases of gastroenteritis has been notably increasing recently in, Turkey. Of the 23 *S. enteritidis* strains isolated from sporadic cases of gastroenteritis in Istanbul in 1991, one strain was detected as being resistant to ampicillin, one to gentamicin and one to nitrofurantoin. Three strains were reported as showing multiple resistance; among these, two strains were found to be resistant to mexlocillin and carbenicillin and one to mezlocittin, cephalothin, cefuroxime, ceftriaxone, cefoperazone and chloramphenicol (Anğ-Küçüker et al., in press).

Isolation rate of *S. enteritidis* from cases of gastroenteritis has been notably increasing recently in Turkey, too. Of the 23 *S. enteritidis* strains isolated from sporadic cases of gastroenteritis in Istanbul in 1991, one strain was detected as being resistant to ampicillin, one to gentamicin and one to nitrofurantoin. Three strains were reported as showing multiple resistance; among these, two strains were found to be resistant to mezlocillin and carbenicillin and one to mexlocillin, cephalothin, cefuroxime, ceftriaxone, cefoperazone and chloramphenicol (Anğ-Küçüker et al., in press).

Table 4. *Salmonella* serovars isolated in Turkey.

Serogroup	Source	Serovar	Antigenic structure		
A	H	*S. paratyphi* A	$\underline{1}$,2,12	:a	:(1,5)
	H	*S. nitra*	2,12	:g,m	:-
B	H	*S. arechavaleta*	4,(5),12	:a	:(1,7)
	O	*S. bispebjerg*	$\underline{1}$,4,(5),12	:a	:e,n,x
	H	(*) *S. makoma*	4,(5),12	:b	:-
	H+O	*S. paratyphi* B	$\underline{1}$,4,5,12	:b	:1,2
	O	*S. paratyphi* B var. *java* (bioser)	$\underline{1}$,4,5,12	:b	:1,2
	O	*S. paratyphi* B var. *odense*	$\underline{1}$,4,12	:b	:1,2
	O	*S. abony*	$\underline{1}$,4,(5),12,$\underline{27}$	:b	:e,n,x
	O	*S. abony* var. *haifa*	4,12	:b	:e,n,x
	O	(*) *S. sofia*	$\underline{1}$,4,12,$\underline{27}$	:b	:(e,n,x)
	O	*S. abortusbovis*	$\underline{1}$,4,12,$\underline{27}$	:b	:e,n,x
	H	*S. schleissheim*	4,12,$\underline{27}$	:b	:-
	O	*S. abortusovis*	4,12	:e	:1,6
	H	*S. duisburg*	$\underline{1}$,4,12,$\underline{27}$	:d	:e,n,z_{15}
	H	*S. salinatis*	4,12	:d,e,h	:d,e,n,z_{15}
	H	*S. saintpaul*	$\underline{1}$,4,(5),12	:e,h	:1,2
	H+O	*S. reading*	$\underline{1}$,4,(5),12	:e,h	:1,5
	H+O	*S. sandiego*	4,(5),12	:e,h	:e,n,z_{15}
	H	*S. derby*	$\underline{1}$,4,(5),12	:f,g	:(1,2)
	H	*S. agona*	$\underline{1}$,4,12	:f,g,s	:-
	H	*S. essen*	4,12	:g,m	:-
	H+O	*S. typhimurium*	$\underline{1}$,4,(5),12	:i	:1,2
	H+O	*S. typhimurium* var. *copenhagen*	1,4,12	:i	:1,2
	H	*S. tsevie*	4,12	:i	:e,n,z_{15}
	H+O	*S. heidelberg*	$\underline{1}$,4,(5),12	:r	:1,2
	H	*S. haifa*	$\underline{1}$,4,(5),12	:z_{10}	:1,2
	H	*S. ituri*	$\underline{1}$,4,12	:z_{10}	:1,5
	H	*S. tokoin*	4,12	:z_{10}	:e,n,z_{15}

Table 4. Continued

Serogroup	Source	Serovar	Antigenic structure		
C_1	H	*S. edinburg*	6,7	:b	:1,5
	H	*S. paratyphi* C	6,7,(Vi)	:c	:1,5
	H	*S. amersfoort*	6,7	:d	:e,n,x
	O	*S. braenderup*	6,7,14	:e,h	:e,n,z_{15}
	H+O	*S. montevideo*	6,7,14	:g,m,(p),s	:(1,2,7)
	H	*S. oranienburg*	6,7	:m,t	:-
	H	*S. thompson*	6,7,14	:k	:1,5
	H	*S. concord*	6,7	:l,v	:1,2
	H+O	*S. irumu*	6,7	:l,v	:1,5
	H+O	*S. potsdam*	6,7	:l,v	:e,n,z_{15}
	H	*S. virchow*	6,7	:r	:1,2
	H	*S. infantis*	6,7,14	:r	:1,5
	O	*S. richmond*	6,7	:y	:1,2
	O	*S. mikawasima*	6,7	:y	:e,n,z_{15}
	H	*S. tennessee*	6,7,14	:z_{29}	:(1,2,7)
	H	*Salmonella*	6,7	:-	:-
C_2	H+O	*S. muenchen*	6,8	:d	:1,2
	H	*S. manhattan*	6,8	:d	:1,5
	H+O	*S. newport*	6,8	:e,h	:1,2
	H+O	*S. kottbus*	6,8	:e,h	:1,5
	H+O	*S. lindenburg*	6,8	:i	:1,2
	H	*S. aba*	6,8	:i	:e,n,z_{15}
	H+O	*S. bovismorbificans*	6,8	:r	:1,5
	H	*S. glostrup*	6,8	z_{10}	:e,n,z_{15}
C_3	O	*S. virginia*	8	:d	:(1,2)
	H	*S. kentucky*	8,20	:i	:z_6
	H	*S. haardt*	8	:k	:1,5
	O	*S. istanbul*	8	:z_{10}	:e,n,x
D_1	O	*S. durban*	9,12	:a	:e,n,z_{15}
	H	*S. onarimon*	1,9,12	:b	:1,2
	H+O	*S. typhi*	9,12,(Vi)	:d	:-

Table 4. Continued

Serogroup	Source	Serovar	Antigenic structure		
	H	(*) *S. rhodesiense*	9,12	:d	:e,n,x
	H	*S. eastbourne*	1,9,12	:e,h	:1,5
	O	*S. berta*	1,9,12	:f,g,t	:-
	H+O	*S. enteritidis*	1,9,12	:g,m	:(1,7)
	H+O	*S. dublin*	1,9,12,(Vi)	:g,p	:-
	H	*S. mendoza*	9,12	:l,v	:1,2
	O	*S. javiana*	1,9,12	:l,z_{28}	:1,5
	O	(*) *S. canastel*	9,12	z_{29}	:1,5
	O	*S. gallinarum* (bioser)	1,9,12	:-	:-
	H	*S. pullorum* (bioser)	9,12	:-	:-
E_1	H	*S. muenster*	3,10	:e,h	:1,5
	O	*S. anatum*	3,10	:e,h	:1,6
	H	*S. newlands*	3,10	:e,h	:e,n,x
	H+O	*S. zanzibar*	3,10	:k	:1,5
E_2	H	*S. newington*	3,15	:e,h	:1,6
E_4	H+O	*S. senftenberg*	1,3,19	:g,(s),t	:-
F	H	*S. leeuwarden*	11	:b	:1,5
	H	*S. aberdeen*	11	:i	:1,2
G_1	O	(*) *S. clifton*	13,22	:z_{29}	:1,5
G_2	H	*S. ullevi*	1,13,23	:b	:e,n,x
H	O	*S. charity*	(1),6,14,(25)	:d	:e,n,x
	O	*S. boecker*	(1),6,14,(25)	:l,v	:1,7
I	O	*S. hvittingfoss*	16	:b	:e,n,x
M	O	*S. hermannswerder*	28	:c	:1,5
	O	*S. halle*	28	:c	:1,7
	O	*S. nashua*	28	:l,v	:e,n,z_{15}
P	H	*S. roan*	38	:l,v	:e,n,x
Q	O	*S. hofit*	39	:i	:1,5
U	H+O	*S. ahuza*	43	:k	:1,5
	H	*S. adana*	43	:z_{10}	:1,5

Table 4. Continued

Serogroup	Source	Serovar	Antigenic structure		
X	O	(**) *S. arizonae* (Ar 28:32:28)	47	:c	:e,n,x,z_{15}
z	O	(**) *S. arizonae* (Ar9a,9b:24:25)	50	:r	:z_{53}
61	O	(**) *S. arizonae* (Ar26:33:25)	61	:i	z:$_{53}$
	O	(**) *S. arizonae* (Ar26:29:30)	61	:k	:1,5,(7)
	O	(**) *S. arizonae* (Ar26:24:25)	61	:r	:z_{53}

H, human; O, other living or non-living habitats; *, subgenus II; **, subgenus III; ( ), can be absent; -, coded by a prophage.

Table 5. Percentage of resistant *Salmonella* strains isolated from clinical cases in Istanbul in 1964 and 1978

Antibiotics	Percentage of resistant strains	
	1964	1978
Chloramphenicol	6.6	65.0
Tetracycline	12.6	90.0
Streptomycin	26.6	80.0
Kanamycin	26.6	90.0
Ampicillin	-[x]	100.0
Gentamicin	-	0.0

[x]Not tested.

REFERENCES

Aksoycan, N., 1972, Memleketimizde 1972 baslarina kadar tespit edilen *Salmonella* serotipleri ve *bulunduklari* yerler, *Mikrobiyol Bul.* 6:51.

Aksoycan, N., 1977, Memleketimizde 1976 yilina kadar tespit edilen *Salmonella* serotipleri ve *bulunduklari* yerler, *Mikrobiyol Bul.* 11:121.

Aksoycan, N., 1980, Memleketimizde 1979 yili ortalarina kadar tespit edilen *Salmonella* serotipleri, *Mikrobiyol Bul.* 14:73.
Aksoycan, N., 1980, Memleketimizde 1980 yili ortalarina kadar tespit edilen Salmonella serotipleri, *Türk Mikrobiyol Cem Derg.* 10:15.
Aksoycan, N., 1984, Türkiye'de 1983 yili sonuna kadar tespit edilen *Salmonella* serotipleri, *Mikrobiyol Bul.* 18:53.
Aksoycan, N., 1985, Türkiye'de 1984 yili sonuna kadar tespit edilen *Salmonella* serotipleri, *Mikrobiyol Bul.* 19:168.
Aksoycan, N., 1988, Türkiye'de 1987 yili sonuna kadar saptanan *Salmonella-Arizona* serotipleri, *Infec. Derg.* 2:5.
Aksoycan, N., 1989, Türkiyede besin zehirlenmesi olgularindan izol edilen *Salmonella*'lar, *Infec Derg.* 3:209.
Anğ, O., 1976, Changes in antibiotic resistance pattrns in Istanbul, *in:* "Cento Seminar on Use and Misuse of Antimicrobial Drugs", A Publication of Cento.
Anğ-Küçüker, M., Büget, E., Anğ, Ö., and Dinçer, N., in press, Istanbul'da izole edilen *Salmonella enteritidis* suşlarinin özellikleri, *Infec. Derg.*
Bozok Johansson, C., in press, Multiple resistant *Salmonella typhimurium* strains isolated in Istanbul, *Infec. Derg.*
Buget, E., Sariaslan, B., Badur, S., and Töreci, 1988, Türkiye'de ilk defa izole edilen *Salmonella virshow* suşlari, *Türk Mikrobiyol Cem. Derg.* 18:22.
Istanbul Sağlik Müdürlüğü İstatistikleri, Istanbul, 1989.
Özek, Ö., Çetin, E.T., Anğ, Ö., and Töreci, K., 1967a, Yurdumuzda ilk defa rastladiğimiz ve kamlumb ağalardan izole edilen bazi *Salmonella* serotipleri. VI. *Salmonella hvittingfoss*, *Salmonella mikawasima* ve *Salmonella* T_1:b:e,n,x, *Tip. Fak. Mec. (Istanbul)* 30:254.
Özek, Ö., Çetin, E.T., Anğ, Ö., and Töreci, K., 1967b, A *Salmonella* strain (1,6,14,24:1,v:1,7) which completes the antigenic formula of *Salmonella boecker, Int. J. Syst. Bacteriol.* 17:39.
Özek, Ö., Çetin, E.T., Anğ, Ö., and Töreci, K., 1971, *Salmonella boecker* variants with different somatic entigens, *Zbl Bakt I Abt Orig.* 216:268.
Özek, Ö., Çetin, E.T., Hofmann, S., Anğ, Ö., Töreci, K., and Güvener, Z., 1969, A new *Salmonella* species: *S. istanbul* = 8:z_{10}:e,n,x, *Zbl Bakt I Abt Orig.* 211:419.
Töreci, K., Büget, E., Sariaslan, B., and Badur, S., 1988, Türkiye'de ilk defa izole edilen *Salmonella newington* ve *Salmonella* 6,7:-:- suşlari, *Türk Mikrobiyol Cem Derg.* 18:33.
Töreci, K., and Öztan, S., 1984, Türkiye'de izole edilen *Salmonella*'lar listesine ilaveler ve bir *Salmonella leeuwarden* suşu *Türk Mikrobiyol Cem Derg.* 14:30.
Türkiye Sağlik İstatistik Yilliği 1964-67, 1971, SSYB Yayin No. 413, Ankara.
Türkiye Sağlik İstatistik Yilliği 1968-1972, 1975, SSYB Yayin No. 444, Ankara.
Türkiye Sağlik İstatistik Yilliği 1973-1974, 1977, SSYB Yayin No. 456, Ankara.
Türkiye Sağlik İstatistik Yilliği 1975-1978, 1980, SSYB Yayin No. 476, Ankara.
Yücel, A., Samasti, and Mamal, M., 1987, Yurdumuzda ilk defa izole edilen *Salmonella makoma* ve *Salmonella arechavaleta* kökenleri, *Türk Mikrobiyol Cem Derg.* 17:103.

PROPERTIES AND BIOLOGICAL ROLE OF *SALMONELLA* VIRULENCE PLASMIDS

Reiner Helmuth[1], Giovanna Morelli[2], and Maria A. Montenegro[1]

1 Robert von Ostertag Institute, Federal Health Office,
Diedersdorfer Weg 1, D-1000 Berlin 48 Germany

2 Max Planck Institute for Molecular Genetics,
Ihnestr. 63, D-1000 Berlin 33 Germany

ABSTRACT

Isolates of the Salmonella serovars *S. typhimurium*, *S. enteritidis, S. dublin*, and *S. choleraesuis* contain large plasmids. They are required for the successful colonization of organs lying beyond the intestinal tract after oral infection of mice. Electron microscopic heteroduplex analysis was used to show, that these virulence plasmids share large regions of homology. By using a cloned fragment of the *S. choleraesuis* virulence plasmid pRQ20 as a gene probe which is specific to the *Salmonella* serovars mentioned above, 986 isolates of different origins were analyzed for the presence of these plasmids. It turned out, that the frequency of virulence plasmids is nearly 100% when strains have been isolated from animal organs or human blood. Lower frequencies ranging from 48% to 87% in strains of faecal, food or environmental origin were observed. These results show that *Salmonella* virulence plasmids are like in mice also required for systemic infections in humans and livestock.

INTRODUCTION

Several authors could show that, the possession of plasmids is a prerequisite for the virulence of certain *Salmonella* serotypes for mice (Terakado et al., 1983; Helmuth et al., 1985; Barrow et al., 1987; Barrow and Lovell, 1988; Gulig , 1990;). This is especially the case for the most relevant serotypes *S. typhimurium* and *S. enteritidis*.
The plasmids have been shown to be involved in several phenomena related to virulence. In some strains they seem to contribute to serum resistance (Helmuth et al., 1985; Vandenbosch et al., 1987; Terakado et al., 1990) and the resistance of the bacteria to be killed by macrophages (Hackett et al., 1986), eventhough contradicting results have been reported as well (Vandenbosch, et al.,1989; Gulig et al., 1990). Presently it is generally accepted, that the plasmids are necessary for the efficient colonization of spleen, liver and mesenteric lymph nodes of mice after oral inoculation (Pardon et al., 1986; Gulig, 1990;) leading to reduced lethal doses for mice . However, the exact mechanism by which the virulence plasmids contribute to a systemic infection remains to be elucidated.

Biology of Salmonella, Edited by F. Cabello *et al.*,
Plenum Press, New York, 1993

In recent years genetic and physical analysis of these virulence plasmids has lead to the identification of some of the genes specifically involved in virulence expression. A consensus region of about 8 kb has been defined for fully expression of virulence in mice (Williamson et al., 1988a; Gulig, 1990). Further studies permit to assume that this region is also common to virulence plasmids of other serotypes (Williamson et al., 1988b; Korpela et al., 1989; Gulig, 1990;). Gulig and Chiodo (Gulig and Chiodo, 1990) located within this region a gene that encodes a 28 kD protein, which they confirmed as a major virulence factor in *S. typhimurium*. The cloning of a fragment from within the virulence region that is common to the virulence plasmids of *S. typhimurium, S. enteritidis, S. dublin, S. gallinarum pullorum,* and *S. choleraesuis* is presented in this paper. Furthermore the relatedness of these virulence plasmids by electron microscopic heteroduplex analysis is described.
In spite of the increasing data on the molecular properties of *Salmonella* virulence plasmids, the pathogenic importance of plasmid carrying strains for livestock and humans remains still open. Human experiments could only be performed in volunteers and experiments in livestock need large numbers of microbiologically and immunologically defined animals and are therefore very expensive.

In this work we use an epidemiological approach to investigate the importance of plasmid carriage. The results obtained indicate, that the virulence plasmids are necessary for causing systemic infections in livestock and humans as well.

RESULTS

-Electron microscopic heteroduplex analysis of *Salmonella* virulence plasmids

In order to get a quantitative measure and to localize homology regions precisely within the *S. typhimurium* pRQ28, the *S. enteritidis* pRQ29, the *S. dublin* pRQ30, and *S. choleraesuis* pRQ20 plasmids (Helmuth et al., 1985), electron microscopic heteroduplex analysis were performed.
Fig. 1 shows schematically the percentage of homology of all possible heteroduplices between the virulence plasmids mentioned above . All plasmids shared at least 35% homology. The highest degree of homology (99%) was obtained between the *S. typhimurium* plasmid pRQ28 and the *S. enteritidis* plasmid pRQ29. The lowest degree of homology however was found between the *S. dublin* plasmid pRQ30 and the other plasmids. In this case, only 35% hybridized with the *S enteritidis* and *S. choleraesuis* plasmids.

-Development of a virulence plasmid specific gene probe

The observation that four virulence plasmids shared large regions of homology led us to assume that virulence genes are located in these areas. Therefore, we cloned a 3.6 kb *Hind*III fragment from plasmid pRQ20 that is common to the virulence plasmids found in *S. typhimurium, S. enteritidis, S. dublin,* and *S. choleraesuis*, as shown by restriction enzyme and Southern blot analysis. The involvement of this sequence in mouse virulence was confirmed by introducing the resulting recombinant plasmid pRQ51 into the plasmid free *S. typhimurium* strain 955/81 (Helmuth et al., 1985). A thousandfold increase in the ability of this strain to colonize the liver of orally infected mice was found, compared to the plasmid free strain.
The specificity of the gene probe was tested in colony hybridization experiments. It turned out, that only strains belonging to serovars known to harbor virulence plasmids, e.g., *S. typhimurium, S. enteritidis, S. dublin, S. choleraesuis* and *S. gallinarum pullorum*, hybridized to the cloned fragment. No hybridization was found against *Salmonella* isolates of 29 other serovars. Furthermore, no homology was detected to sequences of other species from the *Enterobacteriacea* family . The probe did also not hybridize with other unrelated gram positive or negative bacterial species . From these results we concluded that the cloned fragment in pRQ51 is highly specific for virulence plasmid carrying *Salmonella* serovars, which are able to cause systemic infections in mice.

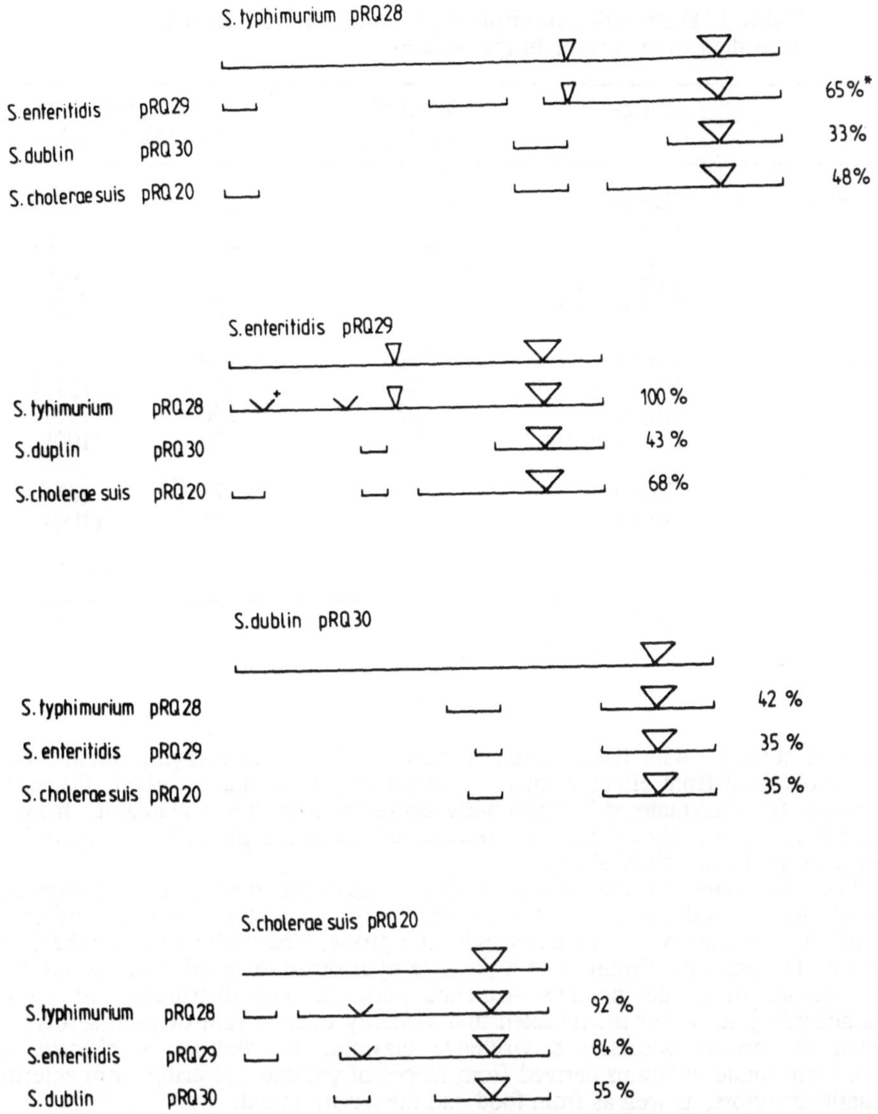

Figure. 1. Schematic representation of homology regions of all possible heteroduplices formed between the four *Salmonella* virulence plasmids pRQ20, pRQ28, pRQ29, and pRQ30 (1). The numbers represent the percentage of homology.

-Distribution of the virulence gene probe among S. *typhi-murium*, S. *enteritidis*, and S. *infantis* isolates from different sources

It is well accepted that *Salmonella* virulence plasmids are necessary in order to provoke a systemic infection after oral inoculation of mice (Helmuth et al., 1985; Pardon et al.; 1986; Gulig, 1990). However, the confirmation for the necessity of virulence plasmids in systemic infections of humans and livestock is still unclear, because especially in the human case experimental confirmation is not easy to fulfill. As an indirect approach, we thought to analyze the occurrence of virulence plasmids in *Salmonella* strains according to their origin of isolation. The hypothesis was that the necessity for the possession of the

Table 1. Distribution of virulence plasmid derived sequences according to the source of the isolates

Serotype	Source	Total No. of strains	No.of strains that hybridized	(%)
S. typhimurium	Faeces	275	187	(68)
	Excretors	33	16	(48)*
	Organs	210	206	(98)§
	Food	55	31	(56)
	Environment	37	22	(59)
S. enteritidis	Faeces	230	199	(87)
	Excretors	44	25	(57)§
	Organs	102	100	(98)§
	Environment	28	28	(100)
S. dublin	Faeces	27	23	(85)
	Organs	21	21	(100)
S. infantis	Organs	180	0	(< 7)

*. $p < 0.05$, as compared to faeces.
§. $p < 0.001$ as compared with the other sources.

plasmid would correlate with the isolation frequency of plasmid carrying strains. Strains known to be obtained from either animal or human organs including blood, faeces of ill patients, faeces from asymptomatic *Salmonella* excretors and, when available, from food and environment were analyzed for the presence of virulence plasmids by hybridization against the gene probe described above.

Isolates belonging to the epidemiologically relevant serotypes *S. typhimurium, S. enteritidis* and *S. dublin* were selected for this study. Table 1 shows that between 76% and 98% of all strains analysed hybridized with the probe. The probe always hybridized to the virulence plasmids , confirming that a positive hybridization reaction correlates always with the presence of an autonomous virulence plasmid. The distribution of virulence sequences according to specimen revealed that virtually every strain of each serotype that has reached the organs possesses a virulence plasmid. In contrast, significant lower frequencies were found in strains derived from faeces of patients suffering from enteritis or asymptomatic excretors, as well as from food and the environment.
Hybridization was also detected in six organ isolates of serogroup Y . In contrast 180 *S. infantis* isolates from the same source did not hybridize.

DISCUSSION

The relatedness of virulence plasmids in *Salmonella* has been documented by several authors (Popoff et al., 1984; Korpela et al., 1989). In this study however, we documented for the first time a quantitative estimate of the homology between the virulence plasmids of the most relevant *Salmonella* serotypes based on electron microscopic heteroduplex analysis.

The highest degree of homology was found between the *S. typhimurium* and *S. enteritidis* plasmids. The sequence of the latter plasmid, pRQ29, was almost completely contained in the *S. typhimurium* plasmid pRQ28, suggesting a common ancestor for both plasmids. They could have evolved by one of the following mechanisms: either the *S. enteritidis* plasmid could have been generated by deletion(s) of the *S. typhimurium* plasmid. Alternatively, the *S. typhimurium* plasmid could have evolved by DNA integration into the smaller *S. enteritidis* plasmid. The lowest degree of homology was detected with the *S. dublin* plasmid pRQ30; less than 50% of its sequences hybridized to each one of the other virulence plasmids.

Part of the homology regions seems to represent common replication and partition functions, because all four plasmids are known to belong to the same incompatibility group (Cerin et al., 1989; Ou et al., 1990). In this respect it is interesting to note that these virulence plasmids are extremely stable; they are not lost by storage of more than 30 years or by frequent subculturing in different environments (Helmuth et al., unpublished results).

In addition to the necessary plasmid replication functions, one can speculate that some, if not all, plasmid encoded virulence genes are located on these highly conserved regions.

Salmonella virulence plasmids have been shown to play a role as virulence factors in mouse infection models, enabling strains to colonize and multiply within the animal organs (Terakado et al., 1983; Helmuth et al., 1985; Pardon et al., 1986; Gulig, 1990). In contrast to the situation in mice, the role of the *Salmonella* virulence plasmids in humans and livestock remains to be elucidated. In the epidemiological study described, we found a strong correlation between the colonization of deeper tissues and plasmid carriage, since virtually all strains isolated from organs possess a virulence plasmid. This shows that also in livestock and humans the presence of the virulence plasmids is a prerequisite for efficient colonization of deeper tissues. The few strains (2%) from organs that failed to hybridize, did not carry virulence plasmids, as shown by plasmid DNA isolation. These few strains all stem from animals and could either have been secondary contaminants of the organs after slaughter or sampling. Alternatively, they could carry different virulence genes not homologous with the probe we used. However, none of these strains was invasive in the mouse model (data not shown).

Plasmid carrying strains were also found in other specimens, like in faeces from excretors, ill persons or animals, food and environmental strains. In contrast to strains originating from organs, they were detected at significant lower frequencies, showing that the virulence plasmid is not a prerequisite for efficient colonization of those habitats. The lowest frequencies of plasmid carriage were found in strains isolated from asymptomatic carriers.

Although virulence plasmids seem to play an important role in systemic infections in humans and livestock, exceptions have been observed. We also analyzed *180 S. infantis* strains from animal organs and all of them did not hybridize with our probe. They might well have different invasion mechanisms which still need further analysis.

From the study presented here we conclude, that a differentiation of *Salmonella* strains according to virulence genes and plasmid content is desirable in order to assess the risks of an infection. This is best achieved by using gene probes, like the one described.

Part of this work has been described in Microbial Pathogenesis 1991;11:391-397

REFERENCES

Barrow, PA., Simpson, JM., Lovell, MA., and Binns, MM., 1987, Contribution of *Salmonella gallinarum* large plasmid toward virulence in fowl typhoid. *Infect. Immun.* 55: 388-392.

Barrow, PA., and Lovell, MA., 1988, The association between a large molecular mass plasmid and virulence in a strain of *Salmonella pullorum*. *J. gen. Microbiol.* 134: 2307-2316.

Cerin, H. and Hackett, J., 1989, Molecular cloning and analysis of the incompatibility and partition functions of the virulence plasmid of *Salmonella typhimurium*. *Microb. Pathogen.* 7: 85-99.

Gulig, PA., 1990, Virulence plasmids of *Salmonella typhimurium* and other *Salmonellae*. *Microb. Pathogen.* 8: 3-11.

Gulig, PA., and Chiodo, VA., 1990 Genetic and DNA sequence analysis of the *Salmonella typhimurium* virulence plasmid gene encoding the 28,000-molecular-weight protein. *Infect. Immun.* 58: 2651-2658.

Hackett, J., Kotlarski, I., Mathan, V., Francki, K., and Rowley, D., 1986, The colonization of Peyer's patches by a strain of *Salmonella typhimurium* cured of the cryptic plasmid. *J. infect. Dis.* 153: 1119-1125.

Helmuth, R., Stephan, R., Bunge, C., Hoog, B., Steinbeck, A., and Bulling, E., 1985, Epidemiology of virulence-associated plasmids and outer membrane protein patterns within seven common *Salmonella* serotypes. *Infect.Immun.* 48:175-182.

Korpela, K., Ranki, M., Sukupolvi, S., Mäkelä, PH., and Rhen, M., 1989, Occurrence of *Salmonella typhimurium* virulence plasmid specific sequences in different serovars of *Salmonella. FEMS Microbiol. Lett.* 58: 49-54.

Ou, JT., Baron, LS., Dai, X., and Life, CA., 1990, The virulence plasmids of *Salmonella* serovars typhimurium, choleraesuis, dublin, and enteritidis, and the cryptic plasmids of *Salmonella* serovars copenhagen and sendai belong to the same incompatibility group, but not those of *Salmonella* serovars durban, gallinarum, give,infantis and pullorum. *Microb. Pathogen.* 8: 101-107.

Pardon, P., Popoff, MY., Coynault, C., Marly, J., and Miras I., 1986, Virulence-associated plasmids of *Salmonella* serotype typhimurium in experimental murine infection. *Ann. Inst. Pasteur/Microbiol.* 137 B: 47-60.

Popoff, MY., Miras, I., Coynault, C., Lasselin, C., and Pardon, P., 1984 Molecular relationships between virulence plasmids of *Salmonella* serotypes typhimurium and dublin and large plasmids of other *Salmonella* serotypes. *Ann. Microbiol.* (Inst. Pasteur) 135 A: 389-398.

Terakado, N., Sekizaki, T., Hashimoto, K., and Naitoh, S.,1983, Correlation between the presence of a fifty-megadalton plasmid in *Salmonella dublin* and virulence for mice. *Infect.Immun.*41: 443-444.

Terakado, N., Ushijima, T., Samejima, T., Ito, H., Hamaoka T., Murayama, S., Kawahara, K., and Danbara, H., 1990, Transposon insertion mutagenesis of a genetic region encoding serum resistance in an 80 kb plasmid of *Salmonella dublin. J. gen. Microbiol.* 136:1833-1838.

Vandenbosch, JL., Rabert, DK., and Jones GW., 1987, Plasmid-associated resistance of *Salmonella typhimurium* to complement activated by the classical pathway. *Infect. Immun.* 55: 2645-2652.

Vandenbosch, JL., Kurlandsky, DR., Urdangaray, R., and Jones GW., 1989, Evidence of coordinate regulation of virulence in *Salmonella typhimurium* involving the rsk element of the 95-kilobase plasmid. *Infect. Immun.* 57: 2566-2568.

Williamson, CM., Pullinger, GD., Lax, AJ.,1988a; Identification of an essentiall virulence region on *Salmonella* plasmids. *Microb. Pathogen.* 5: 469-473.

Williamson, CM., Baird, GD., Manning, EJ., 1988b, A common virulence region on plasmids from eleven serotypes of *Salmonella. J. gen. Microbiol.* 134: 975-982.

CONSTRUCTION OF A PHYSICAL MAP OF THE GENOME OF *SALMONELLA TYPHIMURIUM*

Shu-Lin Liu,[1] K.-K. Wong,[2] Michael McClelland,[2] and Kenneth E. Sanderson[1]

[1] Salmonella Genetic Stock Centre
Department of Biological Sciences
University of Calgary
Calgary, AB, Canada
[2] California Institute of Biological Research
La Jolla, CA, USA

INTRODUCTION

The genus *Salmonella* belongs to the large, medically important eubacterial family Enterobacteriaceae. It is closely related to *Escherichia coli* (Ochman and Wilson, 1987; Riley and Sanderson, 1990), but in *E. coli* most strains are harmless commensals of mammals, whereas the salmonellae are ubiquitous pathogens of both warm-blooded and cold-blooded vertebrates. The salmonellae were classified originally by somatic and flagellar antigens into the many serotypes of the Kauffman-White scheme, then were later shown to be part of a large "geno-species" or reassociation group according to DNA reassociation studies (Crosa et al., 1972). More recently, they have been separated using multi-locus enzyme electrophoresis into a number of electrophoretic types (ETs), which constitute clones (Selander et al., 1991).

Detailed genetic analysis has been done primarily with the strain LT2 of *S. typhimurium;* cellular and molecular biology studies were summarized in Neidhardt et al. (1987). The seventh edition of the linkage map has 680 genes on a circular map (Sanderson and Roth, 1988). The techniques to determine the complete physical map of the genome are now available. This is a three-step process. The first step is the development of a low-resolution restriction map, using digestion of the genome by rare-cutting restriction endonucleases and pulsed-field-gel electrophoresis (PFGE) to separate the fragments, as done in *E. coli* K-12 (Smith et al., 1987) and in other bacteria. The second step, the high-resolution restriction map, is constructed using more-frequently cutting restriction enzymes, usually with 6 bp specificity, as determined in *E. coli* K-12 (Kohara et al., 1987). The third step

Biology of Salmonella, Edited by F. Cabello *et al.*,
Plenum Press, New York, 1993

is the complete nucleotide sequence of the genome. We have accomplished the first step for *S. typhimurium* LT2, developing the restriction maps for *Xba*I and *Bln*I.

The purpose of this report is to describe this work, and to show what further work is needed to carry out the second and third steps of construction of the physical maps. These physical maps will be of great value in furthering molecular genetic analysis of the genus. In addition, the systematic comparison of genomes will help to clarify the genetic differences between the different members of the genus, and to explain the basis for the differences in properties of different serovars.

MATERIALS AND METHODS

The bacterial strains come mostly from the Salmonella Genetic Stock Centre (SGSC). Sources of restriction enzymes are in Table 1. High molecular weight genomic DNA was prepared and digested as in Liu and Sanderson (1992) or as in Wong and McClelland (1992). Separation of fragments was by BioRad CHEF DRII (in Calgary) or by transverse alternating-field electrophoresis using Bechman Instruments equipment (in La Jolla).

CONSTRUCTING A LARGE-SCALE PHYSICAL MAP OF THE SALMONELLA GENOME

The majority of restriction enzymes have recognition sites with 4 bp or 6 bp, and cut bacterial genomes into many fragments. However, a growing number of enzymes are known to cut at rare sites, and produce smaller numbers of fragments. Some of these enzymes are listed in Table 1. We digested genomic DNA of *S. typhimurium* LT2, and separated the fragments by PFGE. *Asc*I, *Not*I, *Pac*I, *Pme*I, *Sfi*I, *Sgr*AI all gave too many fragments for easy interpretation. *Xba*I and *Bln*I gave a small number of evenly distributed fragments that could be separated efficiently by PFGE (Table 1); I-*Ceu*I, an intron-encoded enzyme only recently available, also gave a small number of fragments.

Fig. 1 illustrates the fragments following digestion of *S. typhimurium* LT2 DNA with *Xba*I. The sizes of individual fragments were estimated from the position of the fragments on the gel relative to a standard (a "lambda ladder"), and bands from A (800 kb) to Y (<1 kb) were detected for *Xba*I and A (1500 kb) to K (4 kb) for *Bln*I (=*Avr*II). These summed up to a total genome size of about 4900 kb for both enzymes (Table 2). A list of other restriction enzymes with a small number of sites, and the numbers of fragments obtained from digestion of *S. typhimurium,* where known, is shown in Table 1.

The *Xba*I fragments detected by PFGE were located on the circular physical map of the chromosome by the following methods.

(1) The use of strains with Tn*10* insertions to assign *Xba*I fragments. There is a single *Xba*I site in Tn*10*, thus digestion with *Xba*I causes loss of the fragment in which Tn*10* is inserted, and appearance of two new fragments which should sum to the size of the original fragment. For example, in Fig. 1A, lane 3, the DNA is from a strain with a Tn*10* insertion in band C; two new bands are visible. Lanes 2, 4, and 5 show strains with insertions in bands A, G, and E. The location of several fragments on the linkage map was thus determined (see Fig. 2A).

(2) The use of cloned genes to assign *Xba*I fragments. For example, a probe carrying the *pyr*E gene detected fragment O (Fig. 1C, lane 1); the fragments thus identified are shown in Fig. 2C.

Table 1. Frequencies of restriction sites of restriction enzymes, emphasizing those that cut infrequently.[a]

Enzyme	Sequence	SV40[a]	Lambda[a]	Adeno-2[a]	*S. typhi-murium*[b]	*E. coli*	Source
1. Recognition sequence ≥7 bp							
AscI	GG/CGCGCC	0	2	2	>100	59[c]	Biolabs
BaeI	ACNNNNGTA (CT)C	0	1	4	NT	NT	Biolabs
FseI	GGCCGG/CC	0	0	3	NT	NT	Biolabs
NotI	CG/GGCCGC	0	0	7	>40	22[d]	many
PacI	TTAAT/TAA	0	0	1	>50	48[c]	Biolabs
PmeI	GTT/AAAC	0	2	1	>50	31[c]	Biolabs
RsrII	CG/G(AT)CCG	0	5	2	NT	NT	many
SfiI	GGCCNNNN/NG GCC	1	0	3	>40	31[e]	many
SgrAI	C(AG)/CCGG (TC)G	0	6	6	>50	NT	B-M
Sse8387I	CCTGCA/GG	0	5	3	NT	31[c]	Takara
SwaI	ATTT/AAAT				>40	36[c]	B-M
2. Recognition sequence 6 bp, including CTAG							
BlnI (=AvrII)	C/CTAGG	2	2	2	12	17[d]	many
NheI	G/CTAGC	0	1	4	>50	NT	many
SpeI	A/CTAGT	0	0	3	>30	NT	many
XbaI	T/CTAGA	0	1	5	25	35[f]	many
3. Intron-encoded endonucleases							
I-CeuI	TAACTATAACG GTCCTA/AGGTA GCGA	?	0	0	8-10	8	Biolabs
I-SceI	TAGGGATAA/ CAGGGTAAT	?	0	0	NT	NT	B-M
I-TliI	GGTTCTTTATG CGGACAC/TGA CGGCTTTATG	?	0	0	NT	NT	Biolabs
4. Recognition sequence 6 bp, cut infrequently							
BsiWI	C/GTACG	0	1	4	>100	>100	many
MluI	A/CGCGT	0	7	5	>100	>100	many
NruI	TCG/CGA	0	5	5	>100	>100	many
PvuI	CGAT/CG	0	3	7	>100	>100	many
SalI	G/TCGAC	0	2	3	>100	>100	many
SplI	C/GTACG	0	1	4	>100	>100	many
XhoI	C/TCGAG	0	1	6	>100	>100	many
5. Recognition sequence 6 bp, cut frequently							
EcoRI	G/AATTC	1	5	5	>100	628[g]	many
HindIII	A/AGCTT	6	7	12	>100	545[g]	many

[a] Data on restriction sites from commercial sources, in catalogues. Biolabs is New England Biolabs; B-M is Boeringer-Mannheim; Takara is Takara Biochemicals.
[b] Data from our labs; NT means "not tested".
[c] Number of sites recognized in *E. coli* from sequence; about 35% of the genome sequence is reported (Rudd et al., 1992).
[d] Number of sites from Heath et al. (1992a).
[e] Number of sites from Perkins et al. (1992).
[f] Number of sites from Heath et al. (1992b).
[g] Total number of sites determined by several workers, summarized in Rudd et al. (1992).

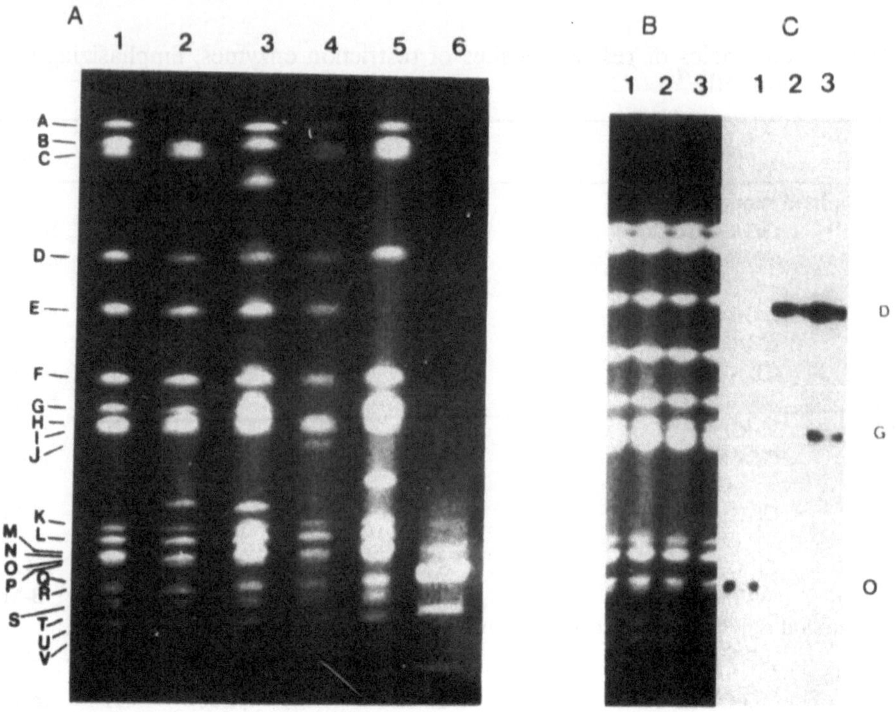

Figure 1. PFGE patterns and Southern blotting analysis of *Xba*I digests of *S. typhimurium* strains. (A) Ethidium bromide-stained gels following *Xba*I digestion. The lanes were digests of DNA from the following strains (fragment with Tn*10* insertion is in parentheses, where appropriate): lane 1, wild-type LT2; lane 2, TT2342 *zbf*::Tn*10* (fragment A); lane 3, TT14835 *srl-1230*::Tn*10* (fragment C); lane 4, TT7235 *ompD159*::Tn*10* (fragment G); lane 5, TT1662 *mel-351*::Tn*10* (fragment E). Lane 6 is DNA of λ monomer plus λ *Hin*dIII digest and some dimers and trimers; the major band is 49 kb, and the smaller bands are 23 and 9.6 kb. Fragments W (6.6 kb) and X (6.4 kb) are not visible on this gel. Complete separation of the smaller bands can be achieved by an extended period of short pulses (5 to 10 s). The molecular weight of each band was determined by comparison with a lane containing λ concatemer. (B) PFGE pattern of *Xba*I digests of DNA from the wild-type LT2 strain. (C) The gel from panel B was blotted to membrane strips, and lanes were probed as follows: lane 1, plasmid pJN13 (*pyrE*); lane 2, DNA from P22 induced from TT1524 *pyrF2690*::Mu*d*Q; lane 3, DNA from P22 induced from TT15246 *pyrF2690*::Mu*d*P. The *Xba*I fragments identified by the probes are indicated on the right.

(3) The use of Mu*d*-P22 prophages as "linking probes." Benson and Goldman (1992) have constructed a set of lysogens with Mu*d*-P22 (Youderian et al., 1988) at various sites on the chromosome; when induced these lysogens replicate and package chromosomal DNA from their insertion site in one direction. This DNA is highly enriched for a chromosomal region of about 100 to 250 kb from the insertion site. This DNA was used to probe the *Xba*I fragments; for example, the Mu*d*-P22 in *pyr*F at 33 min, which replicates counterclockwise, recognized fragments D and G; it is inferred to be inserted in D but to also incorporate G (Fig. 1B, C; lane 3). The same insertion replicating CW detects only D (Fig. 1B, C; lane 2). Data from these probes is summarized in Fig. 2B.

Similar methods were used to assign the *Bln*I fragments to the circular physical map. (1) Tn*10* carries a *Bln*I site, so the location of Tn*10*s in *Bln*I fragments was determined as for *Xba*I. (2) The integration of Mu*d*-P22 introduces 30 kb of DNA that lacks a *Bln*I site, and can easily be detected as a "shift up" in all but the largest *Bln*I fragments. (3) A dot blot array of DNA from induced lysogens with Mu*d*-P22 at known sites around the chromosome (Benson and Goldman, 1992) was probed with *Bln*I fragments from the gels. This generated the circular physical map shown in Wong and McClelland (1992).

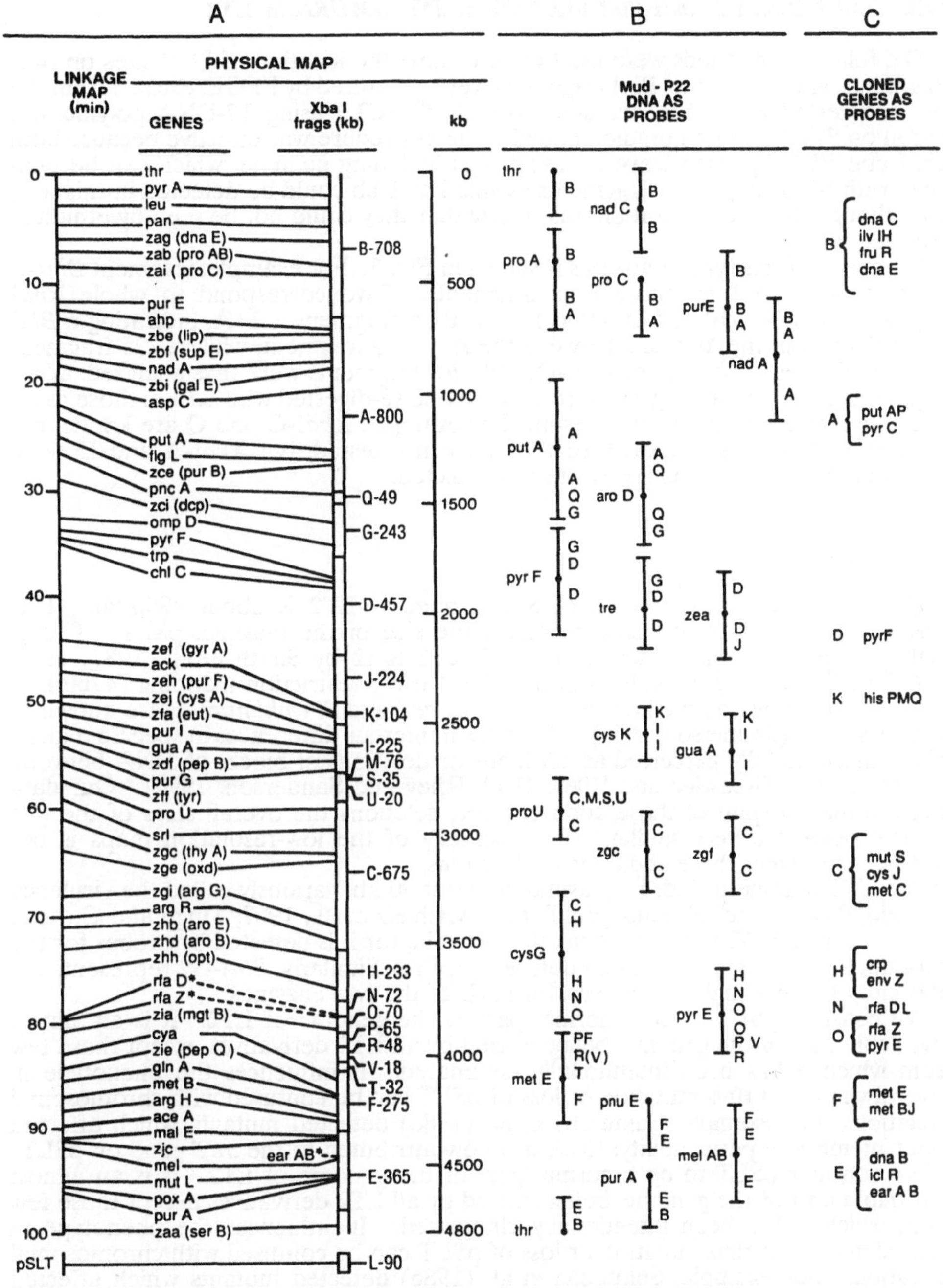

Figure 2. Comparison of the linkage and physical maps of *S. typhimurium,* illustrating the data used to construct the physical map. The linkage map is redrawn from that of Sanderson and Roth (1988). The physical map is from the data in this report, and from Liu and Sanderson (1992). Most of the genes listed in the physical map in part A have been assigned to the *Xba*I fragment and to their locations on that fragment by analysis of Tn*10* strains. The genes marked with asterisks were mapped by probing with cloned genes (*rfaD*, *rfaZ*, and *earAB*), and the detailed data on the genes identified by the use of cloned genes as probes are illustrated in part C. Part B shows a summary of data on the *Xba*I fragment identified by probes produced from induction of Mu*d*-P22 lysogens. The gene into which the Mu*d*-P22 is inserted is to the left of each line, and the letters to the right of the line indicate *Xba*I fragments identified by phage propagated CCW (up) or CW (down). Letters shown in brackets indicate fragments expected but not observed. The length of the line is in some cases arbitrary.

THE CORRELATED *Xba*I-*Bln*I MAP OF *S. TYPHIMURIUM* LT2

The following methods were used to determine the location of *Xba*I sites on *Bln*I fragments, and vice versa. *Bln*I fragments were separated by PFGE, excised from the gel, re-digested with *Xba*I, end-labelled with ^{32}P-dCTP using T7-DNA polymerase, re-run on PFGE, and autoradiographed. This procedure was effective because both *Xba*I and *Bln*I digestion leave 5′ overhangs including guanine, which can be end-filled with labelled dCTP. Fragments as small as 1 kb could be detected in this way, even though their concentration was so low that they could not be seen by ethidium bromide staining.

The result of this type of analysis is shown in Fig. 3. For example, fragment *Bln*I-C digested with *Xba*I produced four fragments. Two correspond to whole *Xba*I fragments, G (243 kb) and Q (49 kb). A 270 kb fragment is 1′A, indicating a *Bln*I site 270 kb from the *Xba*I site between the A and Q fragment, while a 225 fragment is D′, indicating a *Bln*I site in *Xba*I-D. All *Bln*I fragments were thus analyzed. *Xba*I fragments from primary digests were isolated and re-digested with *Bln*I in those cases where useful information could result. For example, *Xba*I-G and Q are known not to have *Bln*I sites (see above), so they were not tested, but *Xba*I-A and D were digested, and each had one *Bln*I site as predicted.

THE GENOME

The size of the chromosome of *S. typhimurium* LT2 is about 4800 kb. This represents the size of the genome minus the size of the plasmid, pSLT. This is slightly larger than that determined for *E. coli* K-12 by Smith et al. (1987) using PFGE (4696 kb) and by Kohara et al. (1987) using restriction mapping (4700 kb). The order of genes on the linkage maps of *E. coli* and *S. typhimurium* are known to be very similar (Sanderson, 1976). There are numerous differences in specific regions of the maps, usually perceived as additions or deletions of blocks of DNA between the two genera (Krawiec and Riley, 1990; Riley and Sanderson, 1990). Our data indicate that in spite of those additions and deletions the overall sizes of the two chromosomes are very similar. The accuracy of the low-resolution maps is not sufficient to detect these additions or deletions.

The LT2 genome includes a plasmid of about 90 kb, variously called the virulence plasmid, the "cryptic" plasmid, pSLT, etc. (Michiels et al., 1987; Tinge and Curtess, 1990). Fragment *Xba*I-L represents this plasmid, for it is detected by probes for the plasmid, and is missing in plasmid-deficient strains. Similarly, *Bln*I-G2 represents the plasmid. The plasmid has one site for each of the two enzymes.

We consider pSLT to be a normal part of the genome of LT2. It is an almost invariable part of the genome, being carried by all LT2 derivatives except those few from which it has been intentionally eliminated. It influences the phenotype in several ways, and thus mutation or loss of pSLT can be confused with chromosomal mutation. For example, Sukupolvi et al. (1986) detected mutants which affected outer membrane permeability; these are now attributed to the *traT* gene on pSLT.

We consider pSLT to be a normal part of the genome of LT2. It is an almost invariable part of the genome, being carried by all LT2 derivatives except those few from which it has been intentionally eliminated. It influences the phenotype in several ways, and thus mutation or loss of pSLT can be confused with chromosomal mutation. For example, Sukupolvi et al. (1986) detected mutants which affected outer membrane permeability; these are now attributed to the *traT* gene on pSLT. In addition, most strains are lysogenic for *Fels-1* and *Fels-2* phages.

POTENTIAL USES OF THE PHYSICAL MAP

Locating Genetic Elements on the Physical Map

Physical map locations can be determined for unmapped Tn*10* insertions, for cloned but unmapped genes, for prophages, and for F-factor insertions. For example, Tn*10* insertions are located by virtue of the fact that Tn*10* has *Xba*I and *Bln*I sites;

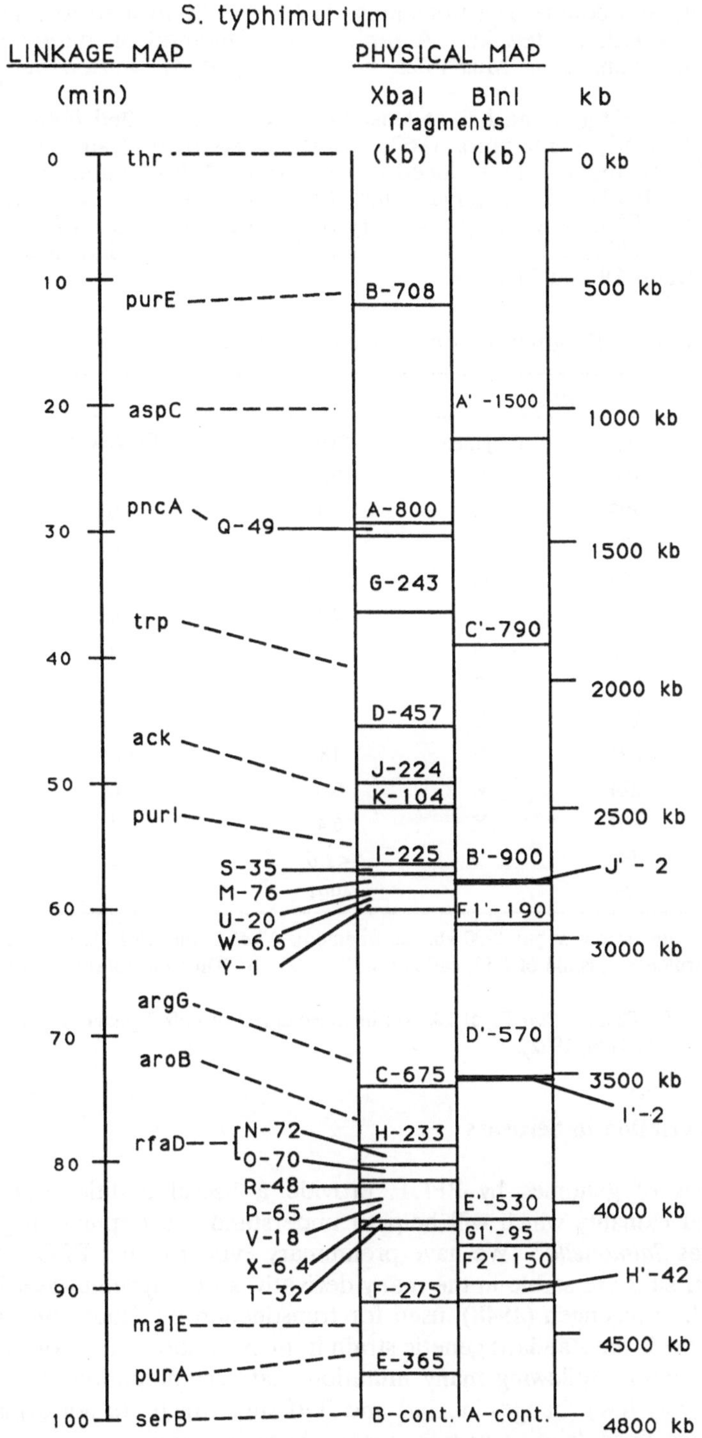

Figure 3. The physical maps of the chromosome of *S. typhimurium* for *Xba*I and *Bln*I, shown in comparison with the linkage map. The linkage map (shown in min) is from Sanderson and Roth (1988). The *Xba*I fragments are as in Liu and Sanderson (1992) and Figure 2. The *Bln*I fragments are in Wong and McClelland (1992). The position of *Bln*I sites on the *Xba*I fragments and vice versa were determined by double digestion and end labelling (see text). The *Bln*I fragments are marked with a (′) to distinguish them from the *Xba*I fragments. The 90-95 kb plasmid pSLT, which is *Xba*I-L and *Bln*I-G2, is not shown in this figure.

following single and double digestions most Tn*10*s can be located to regions of the map, sometimes within a few kb. A rapid genetic method of mapping Tn*10*s to approximate location, using Mu*d*-P22s, was developed by Benson and Goldman (1992).

The wild type LT2 genome has at least two prophages, called *Fels-1* and *Fels-2* (Affolter et al., 1983; Yamamoto, 1967). *Bln*I digests of a strain cured of *Fels-2* showed the F1 fragment to be reduced in size from 190 kb by about 40 kb (Wong and McClelland, 1992), indicating *Fels-2* insertion to be in the F1 fragment (Table 2). Most strains of LT2 in laboratory use carry the *Fels-2* prophage. *Bln*I digestion of strains with *gal*::Tn*10* insertions with or without *Fels-1* prophage indicated that *Fels-1* is located between 18 and 21 mins.

Table 2. Fragments from digestion of *S. typhimurium* DNA

*Xba*I				*Bln*I	
Fragment	Size	Fragment	Size	Fragment	Size
A	800	N	72	A	1500
B	708	O	70	B	900
C	675	P	65	C	790
D	457	Q	49	D	570
E	365	R	48	E	530
F	275	S	35	F1[b]	190
G	243	T	32	F2	150
H	233	U	20	G1	95
I	225	V	18	G2[a]	95
J	224	W	6.6	H	42
K	104	X	6.4	I	2
L[a]	90	Y	*<1.0*	J	*1*
M	76		4901		4864

[a] The total genome size is about 4900 kb, the fragments *Xba*I-L and *Bln*I-G2 represent the cryptic or virulence plasmid of LT2, called pSLT, of 90 kb. Thus the total chromosome is about 4800 kb.

[b] Most strains of LT2 have *Bln*I-F1 of 190 kb; in those cured of *Fels-2* phage it is 150 kb (Wong and McClelland, 1992).

Analysis of Variation in Serovars

Will analysis of genomes by PFGE provide a useful additional measure of variability and clonality which will help to understand the population genetics of groups such as *Salmonella*? We have preliminary evidence that PFGE can do so. *Xba*I and *Bln*I sites are stable in the many derivatives of *S. typhimurium* LT2. LT2 was isolated by Lilleengen (1948), used for transduction by Zinder and Lederberg (1952), and now is the standard genetic strain in many laboratories worldwide. Fifty years after isolation, following many mutation and recombination steps, we have found no (or very few) changes in *Xba*I and *Bln*I sites which are not explainable by known events such as deletion or transposon phage insertion.

However, different serotypes show dramatically changed PFGE fingerprints with both *Xba*I and *Bln*I than with *S. typhimurium*. For example, several strains of *S. typhi* give larger numbers of fragments with both *Xba*I and *Bln*I than with *S. typhimurium*. Strains have been tested from the Salmonella Reference Set A (SARA), composed of 72 strains containing electrophoretic types (ET) from strains in the "typhimurium" complex of Salmonella; these strains were assembled by Beltran et al. (1991) using multi-locus enzyme electrophoresis to define the ETs. The conclusions are as follows. Strains of the same ET have similar, but not always identical PFGE fingerprints, indicating that clones as defined by MLEE can by further subdivided by PFGE. In addition, the PFGE fingerprints obtained from different ETs determined by MLEE become more different as the genetic distance between the ETs increases.

COMPLETION OF THE PHYSICAL MAP OF THE GENOME

The low-resolution restriction map we report here is the first step toward developing a complete physical map. This will require step two, the high-resolution restriction map, and ultimately data on the nucleotide sequence (step 3). We have begun to construct the high-resolution restriction map. This will require cloning the genomic DNA into a suitable vector, and the arrangement of these clones to provide DNA inserts which represent the entire genome as an ordered set of clones. We are using two vectors, because each has unique advantages. λZAP clones small inserts which are amenable to polymerase chain reaction (PCR) amplification, and which can yield rather exact restriction maps. P1 vector (Sternberg, 1990; Pierce et al., 1992) maintains large inserts (up to 100 kb) at low copy number, but amplifies to high number, and will allow complementation analysis. The sorting of these clones into segments representing the chromosome will be greatly facilitated by a set of DNA preparations which represent segments of the genome. We are producing this set by digestion of genomic DNA with *Bln*I and/or *Xba*I and isolation of specific fragments by elution from the gel. As can be seen from Fig. 3, a number of bands of 200-500 kb, representing many segments of the chromosome, can be detected. In addition, we have many strains with Tn*10*s inserted in known regions of the genome which result in a digestion at new sites, yielding appropriate fragments. We have developed a modified element, Tn5pfm, which has *Xba*I and *Bln*I sites, and inserted it in new sites (Wong and McClelland, 1992b). We will use these segments representing parts of the chromosome to clone specific parts, or to probe pooled clones to isolate clones representing that part.

REFERENCES

Affolter, M., Parent-Vaugeois, C., and Anderson, A., 1983, Curing and induction of the *Fels-1* and *Fels-2* prophages in the Ames mutagen tester strains of *Salmonella typhimurium*, *Mutat. Res.* 110:243-262.

Beltran, P. Plock, S.A., Smith, N.H., Whittam, T.S., Old, D.C., and Selander, R.K., 1991, Reference collection of strains of the *Salmonella typhimurium* complex from natural populations. *J. Gen. Microbiol.* 137:601-606.

Benson, N.R., and Goldman, B.S., 1992, Rapid mapping in *Salmonella typhimurium* with Mud-P22 prophages. *J. Bacteriol.* 174:1673-1681.

Crosa, J.H., Brenner, D.J., Ewing, W.H., and Falkow, S., 1972, Molecular relationships among the Salmonelleae. *J. Bacteriol.* 115:307-315.

Heath, J.D., Perkins, J.D., Sharma, B. and Weinstock, G.M., 1992a, *Not*I cleavage map of *Escherichia coli* K-12 strain MG1655, *J. Bacteriol.* 174:558-567.

Heath, J.D., Perkins, J.D., Sharma, B.R., and Weinstock, G.M., 1992b, *Xba*I genomic cleavage map of *Escherichia coli* K-12 strain MG1655 and comparative analysis of other strains, *J. Mol. Biol.* (in press).

Kohara, Y., Akiyama, K., and Isono, K., 1987, The physical map of the whole *E. coli* chromosome: application of a new strategy for rapid analysis and sorting of a large genomic library, *Cell* 50:495-508.

Krawiec, S., and Riley, M., 1990, Organization of the bacterial genome, *Microbiol. Rev.* 54:502-539.
Liu, S.-L., and Sanderson, K.E., 1992, A physical map of the *Salmonella typhimurium* LT2 genome made by using *Xba*I analysis, *J. Bacteriol.* 174:1662-1672.
Michiels, T., Popoff, M.Y., Durviaux, S., Coynault, C., and Cornelis, G., 1987, A new method for the physical and genetic mapping of large plasmids: application to the localization of the virulence determinants of the 90 kb plasmid of *Salmonella typhimurium. Microb. Pathog.* 3:109-116.
Neidhardt, F.C., Ingraham, J.L., Low, K.B., Magasanik, B., Schaechter, M., and Umbarger, H.E., eds., 1987, "*Escherichia coli* and *Salmonella typhimurium*: Cellular and Molecular Biology," American Society for Microbiology, Washington, DC.
Ochman, H., and Wilson, A.C., 1987, Evolutionary history of enteric bacteria, *in:* "*Escherichia coli* and *Salmonella typhimurium*: Cellular and Molecular Biology," F.C. Neidhardt, J.L. Ingraham, K.B. Low, B. Magasanik, M. Schaechter, and H.E. Umbarger, eds., American Society for Microbiology, Washington, DC.
Perkins, J.D., Heath, J.D., Sharma, B.R., and Weinstock, G.M., 1992, *Sfi*I genomic cleavage map of *Escherichia coli* K-12 stain MG1655, *Nucl. Acids Res.* (in press).
Pierce, J.C., Sauer, B., and Sternberg, N., 1992, A positive selection vector for cloning high molecular weight DNA by the bacteriophage P1 system: improved cloning efficiency. *Proc. Natl. Acad. Sci. USA* 89:2056-2060.
Riley, M., and Sanderson, K.E., 1990, Comparative genetics of *Escherichia coli* and *Salmonella typhimurium*, *in:* "The Bacterial Chromosome," K. Drlica and M. Riley, eds., American Society for Microbiology, Washington, DC.
Rudd, K.E., Bouffard, G., and Miller, W., 1992, Computer analysis of *E. coli* restriction map, *in:* "Genetic and Physical Mapping", B. Davies and G. Tilghman, eds., Cold Spring Harbor Press (in press).
Sanderson, K.E., 1976, Genetic relatedness in the family Enterobacteriaceae, *Annu. Rev. Microbiology* 30:327-349.
Sanderson, K.E., and Roth, J.R., 1988, The linkage map of *Salmonella typhimurium*, edition VII, *Microbiol. Rev.* 52:485-532.
Selander, R.K., Beltron, P., and Smith, N.H., 1991, Evolutionary genetics of *Salmonella*, *in:* "Evolution at the Molecular Level," R.K. Selander, A.G. Clark, and T.S. Wittam, eds., Sinauer Assoc., Inc., Sunderland, MA.
Smith, C.L., Econome, J.G., Schutt, A., Keco, S., and Cantor, C.S., 1987, A physical map of the *Escherichia coli* genome, *Science* 236:1448-1453.
Sternberg, H., 1990, Bacteriophage P1 cloning system for the isolation, amplification, and recovery of DNA fragments as large as 100 kilobase pairs, *Proc. Natl. Acad. Sci. USA* 87:103-107.
Sukupolvi, S., O'Connor, D., and Edwards, M.F., 1986, The *Tra*T protein is able to normalize the phenotype of a plasmid-carried mutation of *Salmonella Typhimurium*, *J. Gen. Microbiol.* 132:2079-2085.
Tinge, S.A., and Curtis III, R., Isolation of the replicating and partitioning region of the *Salmonella typhimurium* virulence plasmid and stabilization of the heterologous replicons. *J. Bacteriol.* 172:5266-5277.
Wong, K.K., and McClelland, M., 1992a, Dissection of the *Salmonella typhimurium* genome by use of a Tn*5* derivative carrying rare restriction sites, *J. Bacteriol.* 174:1656-1661.
Wong, K.K., and McClelland, M., 1992b, Tn*5* derivatives carrying a cassette of many rare restriction sites: utility for mapping by pulsed field gel electrophoresis, *J. Bacteriol.* 174 (in press).
Yamamoto, N., 1967, The origin of bacteriophage P221, *Virology* 33:545-547.
Youderian, P., Sugiov, P., Breuer, K.L., Higgins, N.P., and Elliott, T., 1988, Packaging specific segments of the *Salmonella* chromosome with locked-in Mu*d*-P22 prophages, *Genetics* 118:581-592.
Zinder, N., and Lederberg, J., 1952, Genetic exchange in *Salmonella*, *J. Bacteriol.* 64:679-699.

COORDINATION OF GENE EXPRESSION IN PATHOGENIC *SALMONELLA TYPHIMURIUM*

Charles J. Dorman and Niamh Ní Bhriain

Molecular Genetics Laboratory
Department of Biochemistry
University of Dundee
Dundee DD1 4HN
Scotland, U.K.

INTRODUCTION

Salmonella typhimurium is an enteric bacterium capable of adapting successfully to a wide range of environmental conditions both in association with its mammalian host and in the free-living state. In addition it is a sophisticated pathogen capable of causing a typhoid-like disease in mice which has served as a model system for investigations of human typhoid caused by *Salmonella typhi*. Studies carried out over many years in a large number of laboratories have identified many features of the *Salmonella typhimurium* cell that are required for full virulence and these are dealt with in other chapters of this volume. Recently, an awareness has developed that *S. typhimurium* carefully controls the expression of genes whose products contribute either directly or indirectly to virulence. Furthermore, this control of gene expression is coordinated, with expression being governed in response to specific environmental cues. This realisation has led to a search for the regulatory elements responsible for the control of *S. typhimurium* gene expression during the infection process. An emerging pattern from these studies indicates that the regulators tend to be pleiotropic, i.e. they affect the expression of more than one subservient cistron. This is in keeping with the concept of a coordinated regulatory system. This chapter will consider the different levels of virulence gene control employed by *S. typhimurium* and discuss how these fit into an hierarchically-organised, integrated gene control system.

SPECIFIC REGULATORS OF GENE EXPRESSION

Mutagenesis experiments followed by *in vivo* virulence assays have identifed a number of

Biology of Salmonella, Edited by F. Cabello *et al.*,
Plenum Press, New York, 1993

gene regulatory systems as being important contributors to *S. typhimurium* virulence. In 1987, the pleiotropic Cya/CRP system of *S. typhimurium* was found to be required for mouse virulence (Curtiss & Kelly, 1987). This system consists of the cyclic AMP receptor protein (CRP), [also called catabolite activator protein (CAP)] and adenylate cyclase (Cya) and is encoded by the unlinked *crp* and *cya* genes, respectively. Adenylate cyclase is required for the synthesis of cAMP and the site-specific binding of CRP to DNA requires cAMP binding to CRP (Busby, 1986). The interactions of CRP with DNA include an ability to bend the B-DNA duplex and this appears to be essential for its biological function (Lilley, 1991). In general, cAMP-CRP functions as an activator of transcription (although in some cases it can act as a repressor) and when it binds to a canonical sequence found upstream of CRP-dependent promoters it assists RNA polymerase to interact productively with the promoter. The most detailed studies of cAMP-CRP interactions with DNA have been carried out in *Escherichia coli* K-12 where it regulates, inter alia, carbohydrate utilisation operons such as *ara, gal, lac* and *mal* (Magasanik and Neidhardt, 1987). However, it is clear that cAMP-CRP is highly pleiotropic in its effects and may even alter the expression of hundreds of bacterial genes (Busby, 1986). Given its central role in enteric bacterial gene regulation, it is perhaps unsurprising that *S. typhiumurium* mutants deficient in this system are defective in mouse virulence.

The cAMP-CRP system links bacterial gene expression to carbon availability, which is an environmentally-determined parameter. Other systems concerned with the detection and transmission to the genome of information about the extracellular environment have also been found to be required for *S. typhimurium* virulence in the mouse model system. One of these is encoded by the *ompB* locus and consists of the *ompRenvZ* operon. Here, EnvZ is an inner membrane-located histidine protein kinase whose C-terminal domain is phosphorylated in response to changes in growth medium osmolarity as sensed by the periplasmic N-terminal domain. OmpR is a sequence-specific DNA binding protein which participates in the reciprocal regulation of transcription of the *ompC* and *ompF* genes which encode major outer membrane porins (Stock et al., 1989). OmpR is the subject of phosphotransfer from the phosphorylated EnvZ 'sensor' protein. The DNA-binding ability of OmpR is a function of phosphorylation and its differential effects on *ompF* and *ompC* transcription is a function of the distribution pattern of the OmpR binding sites upstream of the porin genes and the affinity of phospho-OmpR for these sites (Slauch and Silhavy, 1991, Slauch et al., 1991; Stock et al., 1989). A transposon insertion mutation which inactivates *ompB* attenuates *S. typhimurium* in a mouse model system (Dorman et al., 1989). Mutations in either *ompC* or *ompF* do not attenuate although mutants deficient in both porins are attenuated (Chatfield et al., 1991). However, the level of attenuation seen in the *ompC ompF* double mutant is not as great as that seen in the *ompB* regulatory mutant, suggesting that other, as yet undiscovered, *ompB*-dependent genes contribute to this virulence 'regulon'. The tripeptide permease operon, *tppB,* also *ompB*-dependent (Gibson et al., 1987) is not one of these genes (Chatfield et al., 1991).

The *phoPphoQ* operon has been shown genetically to be essential for mouse virulence in *S. typhimurium* (Miller et al., 1989). The deduced amino acid sequences of the *phoPphoQ* gene products indicate that these proteins belong to the histidine protein kinase/response regulator family. PhoQ shows homology to the kinase members of the family and its predicted amino acid sequence indicates that it is a transmembrane protein with its phosphorylation site within the cytoplasmic domain. PhoP shares amino acid sequence homology throughout its length with the DNA binding protein, OmpR (Miller et al., 1989). Originally identified as a regulator of *phoN* (the structural gene for acid phosphatase) (Kier et al., 1979), *phoP* has now been shown to regulate expression of *pagC,* a gene required for full virulence in *S. typhimurium.* Mutations in *phoP* cause *S. typhimurium* to become sensitive to defensins (antimicrobial cationic peptides produced by granules within macrophages and neutrophils) (Fields et al., 1989; Groisman and Saier, 1990). Presumably, the *phoPphoQ* system is activated by appropriate environmental signals which herald the onset of defensin attack and PhoP activates the transcription of genes such as *pagC* which are necessary to mount the protective response. Important factors that are known to activate expression of *phoN* (and presumably other PhpoP-dependent genes) include starvation for phosphorous, carbon, nitrogen and a switch to growth at low pH (Kier et al., 1979).

It has been demonstrated biochemically that members of the histidine protein kinase response regulator family can cross-regulate one another. When such 'cross-talk' occurs, a kinase from one pair transfers its phosphate to the response regulator of a second pair, leading to (for example) a response to a change in osmolarity by a system involved normally in responding to nitrogen starvation (reviewed in Stock et al., 1989). This represents one mechanism by which the bacterial cell can integrate its many environmental response systems and suggests that a search for highly pleiotropic effects on gene expression may reveal something of the nature of global gene management in bacteria.

COMMON THEMES IN THE CONTROL OF BACTERIAL GENE EXPRESSION: OPPORTUNITES FOR COORDINATION

The events that occur at the initiation of transcription represent opportunities for the control of expression of individual genes and for the coordination of expression of multiple genes. These events occur in a sequence beginning with the binding of RNA polymerase to the gene promoter to form a closed complex (a reversible step), an irreversible isomerisation to an open complex which is then followed by the elongation of the transcript (Fig. 1). Each of these steps is common to every bacterial promoter and each is potentially a site at which transcription may be regulated. Factors capable of simultaneously affecting these events throughout the genome have the potential to alter gene expression on a global scale. Specific factors such as cAMP-CRP, OmpR and PhoP (discussed above) can alter the expression of multiple genes but are limited in their influence by the need for a specific recognition/binding sequence in the vicinity of the promoters they affect. For this reason, these pleiotropic gene regulators cannot exert truly global effects on transcription. However, negative supercoiling is a feature of the DNA in bacterial cells which has the potential to affect all promoters and thus link them within a global regulon. Its ability to do this in practice will be assessed in the next section.

$$\mathrm{R + P} \underset{K_B}{\rightleftharpoons} \mathrm{RP_C} \underset{k_f}{\longrightarrow} \mathrm{RP_O} \xrightarrow{\mathrm{NTPs}} \mathrm{RNA}$$

Figure 1. Sequence of steps in transcription initiation. R, RNA polymerase; P, promoter; RP_C, closed complex; RP_O, open complex; NTPs, nucleotide triphosphates; K_B, promoter binding constant; k_f, rate constant for the formation of the open complex.

DNA SUPERCOILING AND THE REGULATION OF GENE EXPRESSION

Like other prokaryotes, *Salmonella typhimurium* possesses the type II topoisomerase known as DNA gyrase. This enzyme is composed of two copies of each of two subunits, A and B, encoded by the *gyrA* and *gyrB* genes, respectively and it has the unique ability to introduce negative supercoils into DNA in an ATP-dependent manner. Gyrase can also remove positive supercoils as and when these arise in DNA as a consequence of topological disturbances caused by the passage of transcription complexes or replication forks through the DNA. Negative supercoils are removed form DNA by the type I topoisomerase Topo I, encoded by the *topA* gene.

The counteracting activites of gyrase and DNA topoisomerase I are thought to maintain the level of supercoiling in bacterial DNA at a level suitable for the major processes of DNA (reviewed in Drlica, 1992). These processes (replication, recombination, transcription and transposition) all depend upon strand separation to proceed, and since negative supercoiling involves an underwinding of the DNA duplex, it will tend to promote strand separation and therefore favour these processes. Thus, their efficiencies might be expected to vary if the level of supercoiling varied. However, the homeostatic model predicts that the level of DNA supercoiling tends to be maintained at a fixed value; if too much negative supercoiling is introduced into DNA, the promoter of the *topA* gene is preferentially activated while those of the gyrase genes are de-activated, allowing the relaxing activity (Topo I) to become transiently dominant and to reduce the global level of supercoiling. Conversely, if the DNA becomes too relaxed, the *gyr* promoters are activated while the *topA* promoter is de-activated, permitting extra negative supercoils to be introduced, resetting the supercoiling to the preferred level (Fig. 2). Perturbing the level of DNA supercoiling in *S. typhimurium* either by introducing mutant alleles of the genes encoding the topoisomerases or by inactivating DNA gyrase specifically with antibiotics does alter the expression of many genes (see, for example, the coumermycin experiments of Jovanovich and Lebowitz, 1987). If such wholesale resetting of the transcriptional profile of the cell were to occur in response to environmental change, one might expect a profound change in cellular composition to result.

Recently, there has been an accumulation of data showing that the level of DNA supercoiling in bacterial cells does indeed change in response to variations in the external environment.

Although the level is presumably balanced homeostatically by DNA gyrase and Topo I, the final level of supercoiling appears to be determined in part by environmental factors, such as temperature, anaerobiosis, carbon source availability, growth phase and osmolarity (reviewed in Dorman, 1991; Drlica, 1992; Higgins et al., 1990). All of these environmental stimuli are likely to be experienced by pathogenic bacteria such as *Salmonella typhimurium* during infection and may provide cues for the control of gene expression. Since the effects of DNA supercoiling are likely to be highly pleiotropic, changes in this cellular parameter in response to environmental insult could provide a means of regulating transcription globally.

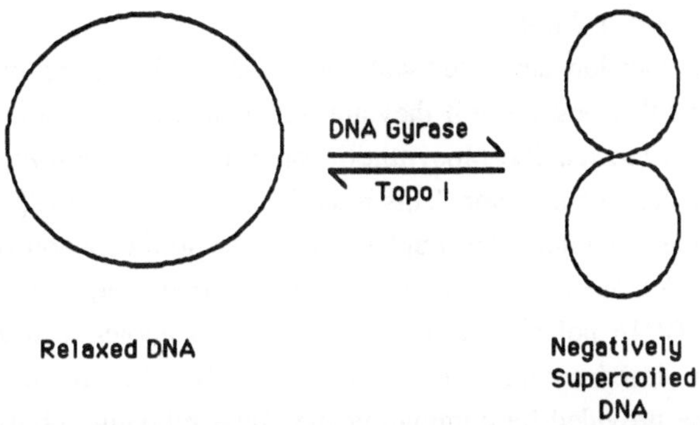

Figure 2. DNA gyrase, DNA topoisomerase I and the control of DNA supercoiling. Relaxed DNA forms a substrate for DNA gyrase, which introduces negative supercoiling. The supercoiled DNA forms a substrate for DNA topoisomerase I (Topo I) which relaxes it. The countervailing activities of these enzymes set the level of negative supercoiling in cellular DNA.

In terms of specific effects on virulence gene expression, it is significant that the ability of *S. typhimurium* to invade mammalian cells is regulated by anaerobiosis and growth phase (Lee and Falkow, 1990) and by osmolarity (Galan and Curtiss, 1990). Furthermore, the *S. typhimurium topA* gene is required for the invasive phenotype and for normal expression of invasion genes (Galan and Curtiss, 1990). As discussed above, the OmpC and OmpF porins are important for the expression of full mouse virulence in *S. typhimurium.* The *ompC* and *ompF* genes which encode these porin proteins possess DNA supercoiling-sensitive promoters (Graeme-Cook et al., 1989). Thus, these latter genes are controlled at more than one level; sensitivity to changes in DNA supercoiling makes them members of a truly global regulon but they are also controlled by the more specific EnvZ/OmpR regulatory system. Presumably, this gives the cell a great deal of flexibility in fine-tuning their expression in response to environmental change.

In general, few genes appear to be regulated exclusively by changes in DNA supercoiling. Most possess additional levels of more specific control, as is the case with the porin genes discussed above or with the *S. typhimurium tonB* gene. This latter gene, which encodes an important component of the iron and vitamin B12 uptake system, is activated by a relaxation of DNA supercoiling levels achieved during growth under aerobic conditions and is negatively regulated by iron acting through the pleiotropic Fur regulatory protein (Dorman et al., 1988; Postle, 1990).

Presumably, DNA supercoiling is too crude a control to be of general use on its own. It affords certain possibilities for transcription initiation to promoters by providing them with an adequate level of free energy of supercoiling for open complex formation. However, the ability of the closed RNA polymerase-promoter complex to take advantage of this is also contingent upon other factors, such as the presence of gene activatiors of the more classical variety (e.g. CRP, OmpR or PhoP).

A further complication associated with DNA supercoiling-sensitive promoters is that they may only display this sensitivity if they are maintained in the correct DNA context. A promoter mutation in the leucine biosynthetic operon of *S. typhimurium* provides an excellent example of this phenomenon. This mutation, *leu500,* is an AT to GC transition in the pribnow box of the promoter. The simplest model for promoter non-function is a failure to form an open complex due to an increased demand for free energy of supercoiling to melt the *leu500* promoter (RNA polymerase binds the mutant promoter with the same efficiency as the wild type, so closed complex formation is not affected). Increased free energy of supercoiling could be provided by eliminating the DNA relaxing acivity from the cell by deleting the *topA* gene. When this is done, the *leu500* promoter is activated. However, when the global level of supercoiling is reduced by the compenstaion of the *topA* mutation by *gyr* lesions, the *leu500* promoter does not cease to form the open complex. Thus, there is no correlation between the global level of DNA supercoiling and transcription from an apparently supercoiling-sensitive promoter (Richardson et al., 1988). Furthermore, when the *leu500* promoter is cloned on a multicopy plasmid, it cannot be activated, even in a strain of *S. typhimurium* deleted for *topA* (Richardson et al., 1988). However, if conditions on the recombinant plasmid are arranged such that the *leu500* promoter is bracketed by two divergently transcribed cistrons so that it is maintained in a domain of perpetual negative supercoiling while in a Δ*topA* mutant strain, the *leu500* promoter will transcribe, in much the same manner as when on the chromosome in a Δ*topA* mutant (Chen et al., 1992). The domain of negative supercoiling is generated by the divergent movement of the coupled transcription/translation complexes of the active genes. These push a region of positive supercoiling ahead of them and generate a region of negative supercoiling behind them, as predicted by the Twin Domain Theory of Liu and Wang (Liu and Wang, 1987). If one of the divergently arranged cistrons is inactivated, the supercoils provided by the other can escape from the *leu500* domain and may even be eliminated by positive supercoils generated by transcription complexes moving elsewhere on the same plasmid molecule (Fig. 3). Under these circumstances, the *leu500* promoter will not transcibe, presumably because the local level of negative supercoiling is too low.

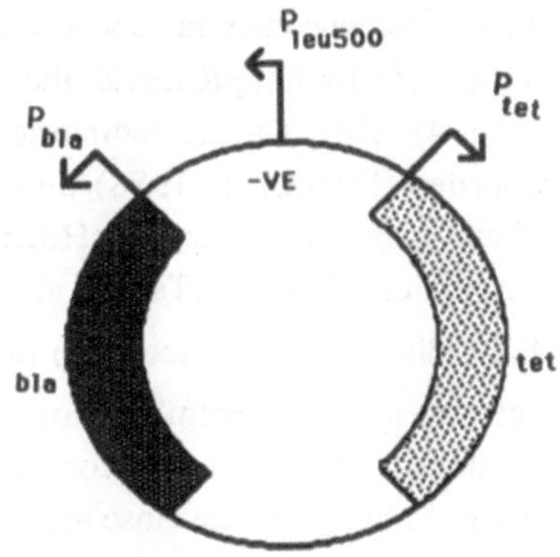

Figure 3. The effect of local DNA supercoiling on the *leu500* promoter. When the *leu500* promoter is bracketed by diverently transcribed genes, it resides within a domain of negatively supercoiled DNA. In the presence of an active *topA* gene, Topo I relaxes this negative supercoiling, depriving the *leu500* promoter of the free energy of supercoiling needed to form an open complex. In a Δ*topA* mutant, the build-up of negative supercoiling in the vicinity of *leu500* is unaffected by Topo I and an open complex forms. If one of the genes bracketing *leu500* is inactivated, the negative supercoils generated by the expression of the other escape from the locality of *leu500* and this promoter (*leu500*) remains silent.

Thus, the context in which the promoter finds itself is all-important to its ability to respond to changes in supercoiling. This example also illustrates the need to differentiate between global levels of DNA supercoiling (which presumably affect the entire genome) and local changes in supercoiling such as those caused by the transcription/translation of adjacent genes. The results obtained from the work with the recombinant plasmid tell us about the situation in which the *leu500* promoter is likely to be on the chromosome with respect to neighbouring genes. Thus, in the future it will be necessary to consider promoters not simply in terms of their primary structure but also in terms of their genetic neighbourhood if we are to understand fully the regulatory influences which they encounter.

GENOME STRUCTURE AND THE COORDINATION OF TRANSCRIPTION

It is now generally accepted that bacteria such as *S. typhimurium* possess a nucleoid consisting of the genetic material and structural proteins analogous to the histones of eukaryotic chromatin as well as topoisomerases with enzymatic and (possibly) structural roles. The ability of topoisomerases to influence gene expression in *S. typhimurium* through the medium of DNA supercoiling has already been discussed. It is becoming clear that the histonelike proteins of prokaryotes can also exert a regulatory role in transcription.

As noted above, the structural genes for the OmpC and OmpF porins are under multifactorial control, making them simultaneously members of several regulons. In addition to regulation by DNA supercoiling and by OmpR/EnvZ, they are also controlled by the integration host factor (IHF) (Fig. 4). This is a heterodimeric site-specific DNA binding protein with histonelike properties (Friedman, 1988) and it is required for normal osmoregulation of both *ompC* and *ompF* transcription (Huang et al., 1990; Tsui et al., 1988). IHF also influences transcription of *ompB* (Tsui et al., 1991). IHF acts as a DNA architectural element in the cell, bending the helix through up to 140° at its binding site and thus assisting in the control of site-specific recombination reactions and transcription (reviewed in Friedman, 1988). The data showing a role for IHF in porin expression come from studies carried out in *E. coli* K-12 but given the close relationship of the *E. coli* and *S. typhimurium* porin systems, the same is likely to obtain in the latter organism.

HNS is another component of the bacterial nucleoid which has been implicated in the control of porin gene expression (Graeme-Cook et al., 1989). This protein is encoded by the *hns* gene (also known as *bglY, drdX, osmZ, pilG, virR* in a variety of enteric bacteria, reviewed in Higgins et al., 1990). Mutations in *hns* are highly pleiotropic and result in expression of *ompF* at high osmolarity and *ompC* at low osmolarity (the reverse of the normal situation) (Graeme-Cook et al., 1989). Furthermore, transposon insertions in the *hns* structural gene attenuate *S. typhimurium* in a mouse model system (see the Chapter by Harrison et al., this volume). Some *hns* mutations alter the level of DNA supercoiling in bacterial cells, although the direction of the change appears to be species-specific; in *S. typhimurium* transposon insertions in *hns* lead to a relaxation of DNA (Higgins et al., 1988; Hulton et al., 1990). Early biochemical work with HNS revealed that it exists in *E. coli* in three isoforms 'a', 'b' and 'c', with form 'a' accumulating in stationary phase cells (Spasskey et al., 1984). The 'a' form strongly compacts DNA in vitro and HNS can act to silence transcription from several promoters (Dorman and Ní Bhriain, 1992; Göransson et al., 1990; Higgins et al., 1990). In fact, an early (and naïve) interpretation of the nature of many *hns* alleles was that, since they relieved transcriptional repression, they were mutations in a repressor gene. It is possible that HNS is a type of global transcriptional repressor and it has been suggested that this protein may bind preferentially to the curved DNA sequences found upstream of many bacterial genes, silencing them in the process (Yamada et al., 1991). How this effect is achieved is not currently understood but it could involve changes in DNA topology (including supercoiling) that lead to promoter shut-down. Structurally, HNS is distinct from the histonelike proteins but has some characteristics in common with the eukaryotic High Mobility Group proteins (HMG). These are nucleoid-associated but appear nonhistonelike in character and have an affinity for single-stranded regions of DNA such as those associated with actively transcribing genes or with extruded cruciforms or with recombination junctions (Bianchi et al., 1991). An association with single-stranded (i.e. 'active') regions of the genome is consistent with the involvement of HNS with pleiotropic effects on transcription control and its role in mutagenesis (Barr et al., 1992; Lejuene and Danchin, 1990).

The observation that the biochemical properties of HNS may vary with the growth phase of the cell is reminiscent of the finding that IHF levels also vary with the bacterial growth cycle. Specifically, IHF accumulates as cells enter stationary phase. In both cases,

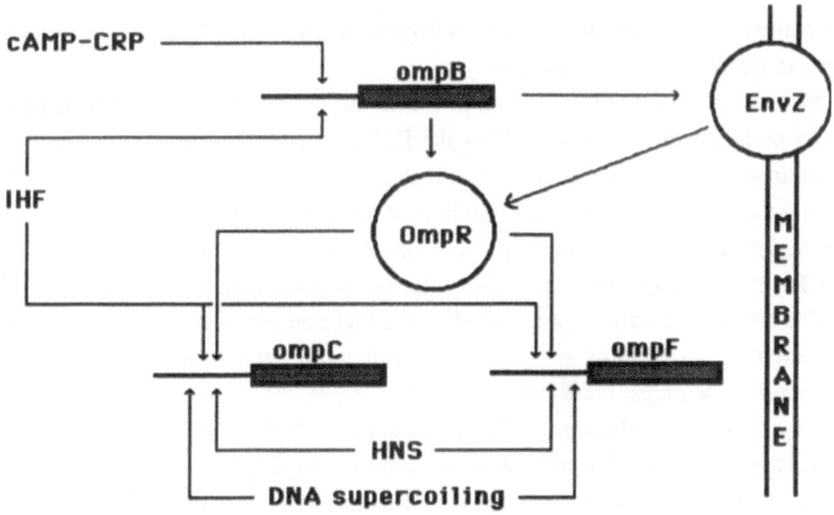

Figure 4. The genes encoding the ompC and OmpF porins are subject to multiple levels of regulation. The *ompB* operon encodes the regulatory proteins EnvZ (a cytoplasmic membrane-located sensor of osmolarity) and OmpR, a cytoplasmic DNA binding protein. Phosphorylation of EnvZ in response to increased osmolarity followed by phosphotransfer to OmpR reciprocally regulates *ompC* and *ompF* transcription (discussed in detail in the text under 'Specific regulators of gene expression'). Transcription of *ompB* is modulated by cAMP-CRP and IHF. IHF also controls expression of *omp*C and *ompF*, as do HNS and DNA supercoiling.

the relationship to the growth state of the cell has the potential to link the molecular events influenced by these nucleoid proteins to the general cellular physiology. This could prove to be immensely valuable in the control of gene expression in response to environmental change, particularly in a pathogenic organism like *S. typhimurium* during the infection processes.

CONCLUSIONS

Salmonella typhimurium possesses a sophisticated gene regulatory programme for adaptation to changes in environmental conditions. Unsurprisingly, much of this programme is also used during pathogenesis. The system is hierarchical from the point of view of the pleiotropic nature of the participating regulators. In this chapter, DNA supercoiling has been discussed as a global regulator of gene expression. Its potential ability to affect any promoter and its sensitivity to the sort of environmental parameters to which *S. typhimurium* needs to adapt make it a particularly attractive candidate for an 'initial response' gene regulator. However, sensitivity to changes in DNA supercoiling is very variable among bacterial promoters and can be influenced by other factors. These include location and orientation with respect to neighbouring genes, as is seen in the case of the *leu500* promoter. Therefore, each promoter must be assessed individually and generalised assumptions about transcriptional responses to DNA supercoiling avoided.

In most instances, the crude control exerted by DNA supercoiling appears to be refined by more specific regulators. Thus the DNA supercoiling-sensitive promoters of the *ompC* and *ompF* genes are controlled specifically by OmpR in response to osmolarity. When considering the make-up of particular regulons, it is important to note that specific regulators may be ranked in terms of pleiotropy. For example, current evidence indicates that the cAMP-CRP protein affects the expression of far more genes than does OmpR.

The situation is further complicated by the fact that genes may belong simultaneously to different regulatory groups. The *ompC* amd *ompF* genes depend upon OmpR for transcription but the *ompB* locus, which includes *ompR,* is regulated in part by cAMP-CRP (Tsui et al., 1991). Thus, the highly pleiotropic cAMP-CRP regulon includes the apparently more restricted *ompB* regulon. Furthermore, *ompB, ompC* and *ompF* transcription are all under IHF control, linking their expression to the other IHF-dependent processes in the cell. So, like other bacteria for which a high degree of molecular detail is available, *S. typhimurium* appears to maintain its environmental response genes (and virulence genes) within multiple regulatory circuits, achieving maximum flexibility in modulating the cell's response to a multifactorial environment. This integrated approach to transcription control offers a partial explanation for the organism's success as both a commensal and pathogenic bacterium.

Acknowledgements

Charles J. Dorman is a Royal Society 1983 University Research Fellow and is also supported by the Agricultural and Food Research Council, the Medical Research Council and the Wellcome Trust.

REFERENCES

Barr, G.C., Ní Bhriain, N. & Dorman, C.J. 1992. Identification of two genetically active regions associated with the *osmZ* locus of *Escherichia coli*: role in regulation of *proU* expression and mutagenic effect at *cya*, the structural gene for adenylate cyclase. J. Bacteriol. **174**:998-1006.

Bianchi, M.E., Falciola, L., Ferrari, S. & Lilley, D.M.J. 1992. The DNA binding site of HMG1 protein is composed of two similar segments (HMG boxes), both of which have counterparts in other eukaryotic regulatory proteins. EMBO J. **11**:1055-1063.

Busby, S.J.W. 1986. Positive regulation in gene expression. In *Regulation of Gene Expression - 25 Years On.* eds, Booth, I.R. & Higgins, C.F., pp.51-77. Society for General Microbiology Symposium 39, Cambridge University Press.

Chatfield, S.N., Dorman, C.J., Hayward, C. & Dougan, G. 1991. Role of *ompR*-dependent genes in *Salmonella typhimurium* virulence: mutants deficient in both OmpC and OmpF are attenuated. Infect. Immun. **59**:449-452.

Chen, D., Bowater, R., Dorman, C.J. & Lilley, D.M.J. 1992. Activity of a plasmid-borne *leu500* promoter depends on the transcription and translation of an adjacent gene. Proc. Natl. Acad. Sci. USA. In Press.

Curtiss, R., III & Kelly, S.M. 1987. *Salmonella typhimurium* deletion mutants lacking adenylate cyclase and cyclic AMP receptor protein are avirulent and immunogenic. Infect. Immun. **55**:3035-3043.

Dorman, C.J. 1991. DNA supercoiling and environmental regulation of gene expression in pathogenic bacteria. Infect. Immun. **59**:745-749.

Dorman, C.J. & Ní Bhriain, N. 1992. Thermal regulation of site-specific recombination and transcription in *Escherichia coli* K-12 *fimA* gene expression: role of *hns*, a gene encoding a bacterial nucleoid protein. Mol. Microbiol. Submitted for Publication.

Dorman, C.J., Chatfield, S.N., Higgins, C.F., Hayward, C. & Dougan, G. 1989. Characterisation of porin and *ompR* mutants of a virulent strain of *Salmonella typhimurium*: *ompR* mutants are attenuated in vivo. Infect. Immun. **57**:2136-2140.

Dorman, C.J., Barr, G.C., Ní Bhriain, N. & Higgins, C.F. 1988. DNA supercoiling and the anaerobic and growth phase regulation of *tonB* gene expression. J. Bacteriol. **170**:2816-2826.

Drlica, K. 1992. Control of bacterial DNA supercoiling. Mol. Microbiol. **6**:425-433.

Fields, P.I., Groisman, E.A. & Heffron, F. 1989. A *Salmonella* locus that controls resistance to microbiocidal proteins from phagocytic cells. Science **243**:1059-1062.

Friedman, D.I. 1988. Integration host factor: a protein for all reasons. Cell **55**:545-554.

Galan, J.E. & Curtiss, R., III. 1990. Expression of *Salmonella typhimurium* genes required for invasion is regulated by changes in DNA supercoiling. Infect. Immun. **58**:1879-1885.

Gibson, M.M., Ellis, E.M., Graeme-Cook, K.A. & Higgins, C.F. 1987. OmpR and EnvZ are pleiotropic regulatory proteins: positive regulation of the tripeptide permease (*tppB*) of *Salmonella typhimurium*. Mol. Gen. Genet. **207**:120-129.

Göransson, M., Sonden, B., Nilsson, P., Dagberg, B., Forsman, K., Emanuelsson, K. & Uhlin, B.E. 1990. Transcriptional silencing and thermoregulation of gene expression in *Escherichia coli*. Nature **344**:682-685.

Groisman, E.A. & Saier, M.H. 1990. *Salmonella* virulence: new clues to intramacrophage survival. Trends Biochem. **15**:30-33.

Graeme-Cook, K.A., May, G., Bremer, E. & Higgins, C.F. 1989. Osmotic regulation of porin expression: a role for DNA supercoiling. Mol. Microbiol. **3**:1287-1294.

Higgins, C.F., Dorman, C.J. & Ní Bhriain, N. 1990. Environmental influences on DNA supercoiling: a novel mechanism for the regulation of gene expression. In *The Bacterial Chromosome*. eds,Drlica, K. & Riley, M. pp. 421-432. American Society for Microbiology, Washington, D.C.

Higgins, C.F., Dorman, C.J., Stirling, D.A., Waddell, L., Booth, I.R., May, G. & Bremer, E. 1988. A physiological role for DNA supercoiling in the osmotic regulation of gene expression in *S. typhimurium* and *E. coli*. Cell **52**:569-584.

Higgins, C.F., Hinton, J.C.D., Hulton, C.S.J., Owen-Hughes, T., Pavitt, G.D. & Seirafi, A. 1990. Protein H1: a role for chromatin in the regulation of bacterial gene expression and virulence? Mol. Microbiol. **4**:2007-2012.

Huang, L., Tsui, P. & Freundlich, M. 1990. Integration host factor is a negative effector of in vivo and in vitro expression of *ompC* in *Escherichia coli*. J. Bacteriol. **172**:5293-5298.

Hulton, C.S.J., Seirafi, A., Hinton, J.C.D., Sidebotham, J.M., Waddell, L., Pavitt, G.D., Owen-Hughes, T., Spasskey, A., Buc, H. & Higgins, C.F. 1990. Histonelike protein H1 (HNS), DNA supercoiling and gene expression in bacteria. Cell **63**:631-642.

Jovanovich, S.B. & Lebowitz, J. 1987. Estimation of the effect of coumermycin A_1 on *Salmonella typhimurium* promoters by using operon fusions. J. Bacteriol. **169**:4431-4435

Kier, L.D., Weppleman, R.M. & Ames, B.N. 1979. Regulation of nonspecific acid phosphatase in *Salmonella* : *phoN* and *phoP* genes. J. Bacteriol. **138**:155-161.

Lee, C.A. & Falkow, S. 1990. The ability of *Salmonella* to enter mammalian cells is affected by bacterial growth state. Proc. Natl. Acad. Sci. USA **87**:4304-4308.

Lejeune, P. & Danchin, A. 1990. Mutations in the *bglY* gene increase the frequency of spontaneous deletions in *Escherichia coli* K-12. Proc. Natl. Acad. Sci. USA **87**:360-363.

Lilley, D.M.J. 1991. When the CAP fits bent DNA. Nature **354**:359-360.

Liu, L.F. and Wang, J.C. 1987. Supercoiling the DNA template during transcription. Proc. Natl. Acad. Sci. USA **84**:7024-7027.

Magasanik, B. & Neidhardt, F.C. 1987. Regulation of carbon and nitrogen utilisation. In Escherichia coli *and* Salmonella typhimurium *Cellular and Molecular Biology.* eds, Neidhardt, F.C., Ingraham, J.L., Brooks, K.B., Magasanik, B., Schaecter, M. & Umbarger, H.E. pp.1318-1325. American Society for Microbiology, Washington D.C.

Miller, S.I., Kukral, A.M. & Mekalanos, J.J. 1989. A two-component regulatory system (*phoP phoQ*) controls *Salmonella typhimurium* virulence. Proc. Natl. Acad. Sci. USA. **86**:5054-5058.

Postle, K. 1990. TonB and the gram negative dilemma. Mol. Microbiol. **4**:2019-2025.

Richardson, S.M.H., Higgins, C.F. & Lilley, D.M.J. 1988. DNA supercoiling and the *leu500* promoter mutation of *Salmonella typhimurium.* EMBO J. **7**:1863-1869.

Slauch, J.M. & Silhavy, T.J. 1991. *cis*-acting *ompF* mutations that result in OmpR-dependent constitutive expression. J. Bacteriol. **173**:4039-4048.

Slauch, J.M., Russo, F.D. & Silhavy, T. 1991. Suppressor mutations in *rpoA* suggest that OmpR controls transcription by direct interaction with the α subunit of RNA polymerase. J. Bacteriol. **173**:7501-7510.

Spasskey, A., Rimsky, S., Garreau, H. & Buc, H. 1984. H1a, an *E. coli* DNA binding protein which accumulates in stationary phase, strongly compacts DNA in vitro. Nucleic Acids Res. **12**:5321-5340.

Stock, J.B., Ninfa, A.J. & Stock, A.M. 1989. Protein phosphorylation and regulation of adaptive responses in bacteria. Microbiol. Rev. **53**:450-490.

Tsui, P., Helu, V. & Freundlich, M. 1988. Altered osmoregulation of *ompF* in integration host factor mutants of *Escherichia coli.* J. Bacteriol. **170**:4950-4953.

Tsui, P., Huang, L. & Freundlich, M. 1991. Integration host factor binds specifically to multiple sites in the *ompB* promoter of *Escherichia coli* and inhibits transcription. J. Bacteriol. **173**:5800-5807.

Yamada, H., Yoshida, T., Tanaka, K.-I., Sasakawa, C. & Mizuno, T. 1991. Molecular analysis of the *Escherichia coli hns* gene encoding a DNA-binding protein which preferentially recognises curved DNA sequences. Mol. Gen. Genet. **230**:332-336.

SOME STRUCTURAL ASPECTS OF THE ANTIGENIC O-POLYSACCHARIDE COMPONENTS OF SALMONELLA SOMATIC LIPOPOLYSACCHARIDES

Malcolm B. Perry

Institute for Biological Sciences, National Research Council of Canada, Ottawa, Ontario, Canada K1A OR6

The lipopolysaccharides (LPS) of Salmonella are not only the most abundant but also the most extensively studied outer membrane constituents of this genus because, in common with the LPSs produced by other gram-negative bacteria, these macromolecules induce a myriad of biological effects many of which are important in the pathogenesis of infections. With few exceptions, LPSs have toxic properties and are often referred to as endotoxins.

The subject of LPSs and Salmonella LPS in particular, has been covered in many papers, books, and excellent reviews (Proctor, 1984; Anderson and Unger, 1983; Meyer et al., 1989; Lüderitz et al., 1982). This work is directed to a discussion of the results of some structural analyses of Salmonella LPS O-polysaccharide (O-PS) components of biological interest.

LPS is an essential element for maintaining the integrity of gram-negative bacteria and acts as a barrier against the entry of small molecules into the organisms. LPSs are also responsible for the induction of pathophysiological effects in infected hosts, including fever, hypotension, leukocytosis, and toxic shock, among other reactions (Raetz, 1990; McCartney and Wardlaw, 1985). They have high immunogenicity with the resulting antibodies being specific for glycosyl epitopes in their polysaccharide components (Lindberg and Le Minor, 1984) and have thus given rise to serological typing systems (Kauffmann, 1966; Le Minor and Popoff, 1988) and proposals of their relationships to pathogenicity.

An understanding of the biological properties of LPSs and the mechanisms of their action have been partially revealed through the elucidation of the molecular structure of intact LPS macromolecules. These LPS studies have disclosed their great compositional and structural diversity and, at the same time, have shown them to be built according to a common architectural form. As structural details of the LPSs of specific bacterial serotypes were determined it became possible to ascribe biological properties to discrete domains in the LPS macromolecules.

The general architecture of LPSs have been established in large measure from the analysis of isolated Salmonella LPSs which are composed of three well-defined regions (Fig. 1) termed lipid A, core oligosaccharide, and O-PS. LPSs

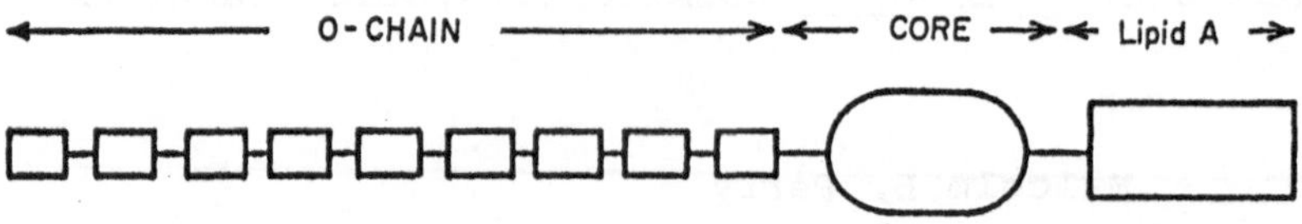

Fig. 1. Diagramatic representation of S-type LPS

composed of these three domains have been described as S-LPS (smooth) whereas, LPS lacking the O-PS domains are referred to as R-LPS (rough) or LOS (lipooligosaccharide) types.

The lipid A region in the genus Salmonella consists of a phosphorylated disaccharide of β- 1,6 linked 2-amino-2-deoxy-D-glucopyranosyl residues to which long-chain fatty acids are attached as ester and amide substituents (Wollenweber and Rietschel, 1990), A generalized picture of a Salmonella lipid A component (Rietschel et al., 1983), which appears to be a conserved in the genus, is depicted in Fig. 2.

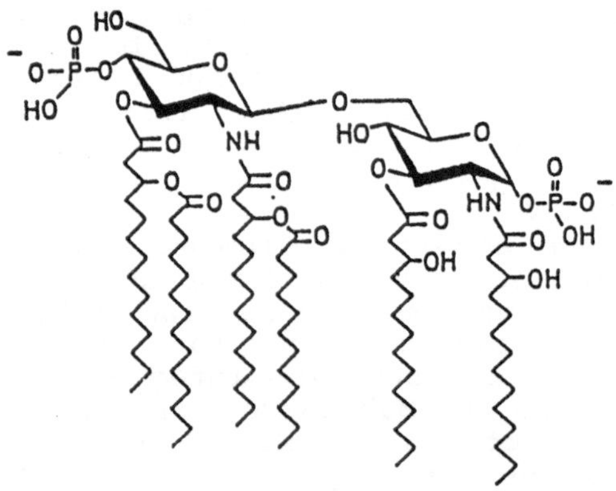

Fig. 2. Basic structure of Salmonella lipid A

The toxic properties of LPS resides in the hydrophobic lipid A moiety of the molecule (Takayama et al., 1984). Recent work on chemically synthesized lipid A (Kuusumoto et al., 1983) as well as the use of positive fast atom bombardment (Qureshi et al., 1988) and plasma desorption mass-spectrometery (Karibian et al., 1991) has provided firmer evidence for the detailed fine structure of lipid A. Syntheses have also pinpointed subunit features associated with particular biological effects.

The core oligosaccharides of LPS differ in composition and structure from one bacterial genus to another however the LPS inner core oligosaccharide portion found in Salmonella LPSs appears to be conserved and is composed of L-glycero-D-manno-heptopyranose (Hep), and 3-deoxy-D-manno-octulosonic acid (Kdo) residues having the structure:

```
←————————— Outer core —————————→  ←——— Inner core ———→
                                                       ¦
α-D-Glcp-(1-2)-α-D-Galp-(1-3)-α-D-Glcp-(1-3)-α-D-Hepp-(1-3)-α-D-Hepp-(1-5)-Kdo---Lipid A
    2                            6              7                          ¦
    1                            1              1                          ¦
α-D-GlcpNAc                  α-D-Galp       α-D-Hepp
```

The outer region of the Salmonella complete core (Ra core) containing D-GlcpNAc, D-Glcp, and D-Galp.residues (Fig. 3) is considered to be common structure (Hellerqvist and Lindberg, 1971) however recent studies indicate that this view may be challenged (Tsang, 1991).

The core oligosaccharides forming the central unit of the LPS are glycosidically linked through the reducing end Kdo residue to the C-6′ hydroxyl group of the 2-amino-2-deoxy-D-glucose residue forming the disaccharide component of lipid A while an O-PS chain, if present, is glycosidically linked through it reducing end to the O-4 position of the Hep residue as shown. The biological properties of the LPS core oligosaccharide region are not understood, although chemical modifications of the core oligosaccharide of the LPS of Bordetella pertussis have been demonstrated to be important in determining biological activity (Caroff et al., 1990), and the adhesion properties of strains of Escherichia coli 0119 having identical O-PS appear to be determined by the nature of their different core structures (Bradley et al., 1991). While several bacterial species such as Neisseria gonorrhoeae, N. meningiditis, Haemophilus influnzeae, and Bordetella pertussis appear to produced R-type LPS others such as E coli, Pseudomonas, Brucella, Actinobacillus pleuropneumoniae and Salmonella produce S-type LPS.

The conserved nature of the LPS core structures within a bacterial species has suggested a rationale for their use as vaccines. however, the results of experimental work have been equivocal (Ziegler, 1988). It is also realized that the structural relationship of LPS to its virulence properties can only be satisfactorily explained through a consideration of the nature of the LPS produced in the appropriate host. It has been demonstrated that cultural conditions can change the nature of LPS produced (Dodds et al., 1987). The demonstration that the core oligosaccharide of the R-LPS of N. gonorrhoeae in the host was highly sialylated, as opposed to the non-sialylated R-LPS produced under laboratory cultural conditions, provides a classic example of how the biological role of LPS may need revision following in vivo growth studies (Fox et al., 1991).

The O-PS is a polymer consisting from 5 to 50 repeating units which, in general, can contain from one to eight monosaccharide residues. The O-PS is glycosidically linked to a core oligosaccharide which in turn, is biosynthesized prior to O-PS attachment by sequential induction of glycoses to lipid A, which anchors the whole LPS assembly in the outer membrane of a gram-negative bacterial cell wall. O-PS homopolymers of a single glycosyl unit such as the 1,2 linked 4-formamido-4-deoxy-D-rhamnopyranosyl of the LPS of Brucella abortus (Caroff et al., 1984) and Yersinia enterocolitica O:9, (Caroff et al., 1984a) the D-galactan of Actinobacillus pleuropneumoniae O:10, (Perry, 1990) and the D-rhamnan Escherichia hermannii (ATCC 33651) (Perry and Richards, 1990) are known, the O-PS of Salmonella LPS, in contrast, are composed of repeating oligosaccharide units. Most structural studies of O-PSs have been made on the chromatographically purified water soluble glycan produced on mild hydrolysis of the acid labile ketosidic linkage between the Kdo residues and lipid A moieties in the native LPSs.

In ascribing the biological properties of O-PS it is important to consider not only their exact fine chemical structures but their conformations and physicochemical properties (Kabat, 1976; Kastoowsky et al., 1992). Many of the earlier conclusions about the structures of Salmonella O-PS have required revision in the light of recent investigations. It is fortunate that the development of new chemical degradation methods, and physical methods such as GLC-MS and ^{1}H and ^{13}C nuclear magnetic resonance, have provided investigators with the tools to elucidate even the most complex structures.

Most of the research conducted on Salmonella LPSs has involved the core oligosaccharide and lipid A regions because of their demonstrated biological properties. Much of the analytical work conducted on O-PS LPS regions was directed towards providing explanatory data for the specificities exhibited by O-PS serological typing antisera and has led to the structural definition of carbohydrate epitopes. Confirmatory evidence for these structural conclusions was frequently obtained from experiments involving inhibition studies (Bundle, 1990) using synthetic reference oligosaccharides, the production of antisera to synthetic oligosaccharide protein conjugates (Bundle, 1990; Lindberg et al., 1983), and more recently to the analysis of the carbohydrate binding sites of the Fab regions of monoclonal antibodies directed to Salmonella O-PS structural features (Rose et al 1990; Cygler et al., 1991). Host antibodies directed to LPS O-PS determinants are osponic and bactericidal and play a role in defense against infections.

The structures of the O-PS of Salmonella serogroups A, B, D, and E, which comprise the largest clinically important groups responsible for gastrointestinal infections, were the first studied. The repeating units of the groups A, B, and D share a common linear -D-Manp-L-Rhap-D-Galp- sequence in which the D-Manp residues are 3-O substituted by different immunodominant 3,6-dideoxyhexose units (D-abequose, D-tyvelose, D-paratose) which define the stereochemistry and serological specificites of the A, B and D_1 groups. The group E_1 -E_4 are composed of linear backbone chains of a trisaccharide sequence -6)-D-Manp-(1-4)-L-Rhap-(1-3)-D-Galp-(1- and variation occurs by changes in the configuration and position of glycosidic linkages as well as branching by α-D-Glcp residues. The correct revised structures of these Salmonella antigens has now been compiled (Bundle, 1990).

The analysis of the Salmonella group LPS O-PS groups C_1, C_3, C_4, and H for example, have been made on selected serotypes expressing assigned antigen factors. Consideration of in these structures can lead to the tentative identification of the epitopes determining the antigenic factors however, confirmations should await the results of inhibition studies using standard oligosaccharides or analysis with monoclonal antibodies having known specificities to carbohydrate epitopes. The following structures were determined in this study of these groups:

Salmonella Group C_1 (O:6,7)

S. livingstone (O:6,7) (DiFabio et al. 1989)

```
-2)-α-D-Manp-(1-2)-α-D-Manp-(1-2)-β-D-Manp-(1-3)-β-D-GlcpNAc-(1-2)-β-D-Manp-(1-
                        3
                        1
                    α-D-Glcp
```

S. ohio (O:6,7) (DiFabio et al. 1989a)

```
-2)-α-D-Manp-(1-2)-α-D-Manp-(1-2)-β-D-Manp-(1-3)-β-D-GlcpNAc-(1-2)-β-D-Manp-(1-
       3
       1
   α-D-Glcp
```

Salmonella Group C_2 (O:8)

S. virginia O:8 (MacLean and Perry 1991)

```
                                                           Ac
                                                           2
-2)-α-D-Manp-(1-2)-α-D-Manp-(1-3)-β-D-Galp-(1-4)-β-L-Rhap-(1-
                                                           3
                                                           1
  [Factor O:8 is α-D-Abep-(1-3)-β-L-Rha-(1-]           α-D-Abe
```

Salmonella Group H (O:6,14)

S. boecker (O:[1], 6,14,[25]) (Brisson and Perry 1988)

O-Chain I

```
-6)-α-D-Manp-(1-2)-α-D-Manp-(1-2)-β-D-Manp-(1-3)-α-D-GlcpNAc-(1-
```

O-Chain II

```
-6)-α-D-Manp-(1-2)-α-D-Manp-(1-2)-β-D-Manp-(1-3)-α-D-GlcpNAc-(1-
                                       3
                                       1
                                   α-D-Glcp
```

S. carrau (O:6,14,[24]) (DiFabio et al. 1988a)

Identical to O-chain II of S. boecker

S. madelia (O:1,6,14,25) (DiFabio et al. 1989b)

O-Chain I

```
-6)-α-D-Manp-(1-2)-α-D-Manp-(1-2)-β-D-Manp-(1-3)-α-D-GlcpNAc-(1-
```

O-Chain II

```
-6)-α-D-Manp-(1-2)-α-D-Manp-(1-2)-β-D-Manp-(1-3)-α-D-GlcpNAc-(1-
                                       3
                                       1
                                   α-D-Glcp
```

O-Chain III

```
-6)-α-D-Manp-(1-2)-α-D-Manp-(1-2)-β-D-Manp-(1-3)-α-D-GlcpNAc-(1-
                    4
                    1
                α-D-Glcp
```

Salmonella Group C_4 (O:6,7,14)

S. eimsbuttel (6,7,14) (DiFabio et al. 1988)

```
-2-α-D-Manp-(1-2)-α-D-Manp-(1-2)-β-D-Manp-(1-3-β-D-GlcpNAc-(1-2)-β-D-Manp-(1-
                                    3
                                    1
                                 α-D-Glcp
```

The above investigations revealed that many of the O-PS were in fact mixtures of structurally unique polysaccharides. Chemical analyses alone usually did not reveal the heterogeneous nature of the O-PS preparations however NMR nuclear Overhauser experiments involving irradiation of anomeric proton resonances and observation of the enhanced resonances of the identified protons in the linked glycoses afforded glycose sequence data, usually giving the first indication of such heterogeneity. In practice, the final proof that two or more distinct PS were present was difficult since chromatographic methods were not normally applicable to separations. A example of a clear cut demonstration that two distinct O-PS were present in the O-PS fraction from S. boecker was proved through chemical separation of the PS I from PS II (see above) could be accomplished since PS I was selectively separated as an insoluble copper complex formed with Fehling's solution (Brisson and Perry, 1988). Other types of separations involving preferential enzymic hydrolysis or the use of specific monoclonal antibody affinity columns were sometimes applicable.

The elucidation of the LPS O-PS structures of Salmonella spp in higher groups in the Kauffmann-White classification scheme is far from complete and for the most part analyses have only been made to determine the basis of serological cross reactions with other bacterial species. An example is the work undertaken to explain the reported serological cross reactions between Brucella abortus. Salmonella group N (O30), and E. coli O157:H7 (Corbel et al., 1984), the causal agent of hamburger disease. This cross-serology arising from LPS O-PSs was surprising since the A and M antigens of Brucella are homopolymers of the rare sugar 4-deoxy-4-formamido-α-D-rhamnopyranose (D-Rhap4NFo) (Caroff et al., 1984). Chemical analyses and the ^1H and ^{13}C NMR spectra of the LPS O-PSs from E. coli O157 (Perry et al., 1985) and S. landau O30 (Bundle et al., 1986) were identical and showed that the O-PS were composed of repeating tetrasaccharide units containing L-Fuc, D-GalNAc, D-Glc, and D-Rhap4NAc and had the linear structure:

-4)-β-D-Glcp-(1-3)-α-D-GalpNAc-(1-2)-α-D-Rhap4NAc-(1-3)-α-L-Fucp-(1-

Subsequent NMR data, drawing on the network of inter-ring nOe's in combination with semi-empirical calculations, allowed a three dimensional model for this antigen to be made which was consistent with the minimum energy model calculated by the HSEA method using GESA programs (Fig. 3) Bundle et al., 1986).

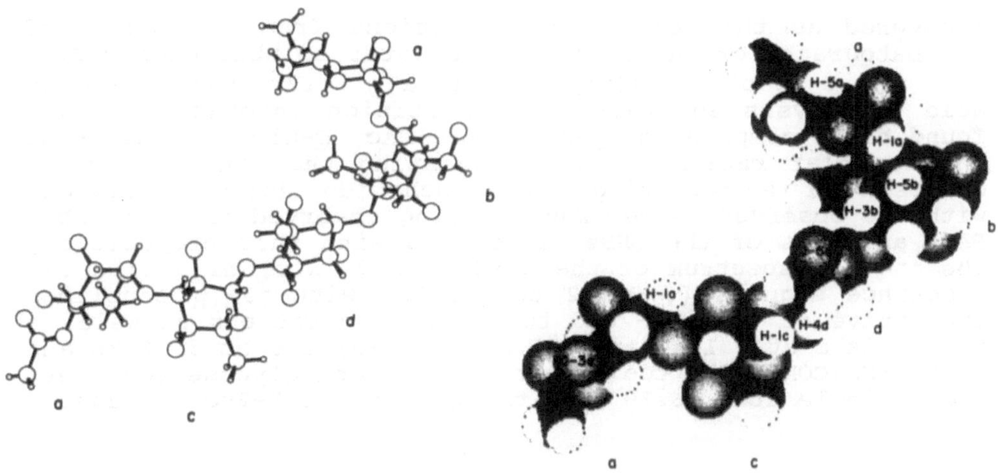

Fig. 3. Three dimensional antigen model

It is concluded that the serological cross reactions between Brucella and the Salmonella O30 antigen resides in a limited epitopic feature in the common N-acylated 1,2 linked α-D-Rhap4NAc residues. The cross-reaction found with polyclonal sera was of course not observed with highly specific monoclonal antibodies to Brucella A and M antigens (Meikle et al., 1989; Bundle et al., 1989).

Studies on S. godesberg and S. urbana (Perry et al., 1986) also belonging to Salmonella group N, revealed that their LPS O-PS were composed of repeating pentasaccharide units having the basic O30 tetrasaccharide backbone to which single β-D-Glcp units are linked to the main chain α-D-GalpNAc residues, thus introducing a second antigenic factor within this group

Early reports of Salmonella LPS having human blood group A and B activities suggested that the responsible serologically common components resided in the O-PS regions. It has been speculated that contact with bacterial antigens from strains of E. coli and Salmonella may be responsible for the production of heterophile blood group specific antibodies. The clinical implication of the protective properties of these antibodies against serologically related strains of gram-negative bacteria has not however been demonstrated.

In order to define the structural basis for the blood group B activity of Salmonella group U (O43) the blood group A activity of Salmonella group R (O40), analyses of the LPS O-PS of S. milwaukee (O43) (Perry and MacLean 1992) and S. riogrande (O40) (Perry and MacLean 1992a) were made.

LPS from S. milwaukee O43 and S. riogrande (O40) were obtained pure by ultracentrifugation of the concentrated dialyzed aqueous phases from phenol-water extractions of washed fermenter grown cells. The LPSs were hydrolyzed under mild conditions (1.5% acetic acid, 100°C, 2 h) and, following removal of precipitated lipid A, the O-PS were recovered as the void volume fractions from Sephadex G-50 chromatography of the concentrated water soluble products.

The O-PS of S. milwaukee had $[\alpha]_D$ +117° (water) and by acid hydrolysis and gas-liquid partition chromatgraphy was found to be composed of D-Gal, D-GalNAc, D-GlcNAc, and L-Fuc in the molar ratios 2:1:1:1 suggesting that the PS may be composed of a repeating pentasaccharide units consistent with the observed ladder rung spacing observed in their SDS-PAGE analysis of the LPSs. In accord with this proposition, the ^{13}C NMR spectrum of the O-PS showed *inter alia* five C-1 resonance signals (δ 93.72 to 103.73) with $^1J_{C,H}$ couplings indicative of three α- and two β- configurations as well as two signals at δ 174. and 175.2 ($CH_3\underline{C}ONH$) and at δ 23.36 and 22.89 ($\underline{C}H_3CONH$) for two 2-acetamido-2-deoxyglycose residues, and a single at δ 16.30 from the C-6 of the L-Fuc residue.

Scheme 1. Stepwise periodate oxidation of Salmonella milwaukee LPS O-chain

[→2)-β-D-Galp-(1→3)-α-D-GalpNAc-(1→3)-β-D-GlcpNAc-(1→4)-α-L-Fucp-(1→]n

↑ Limited $NaIO_4/NaBH_4/H^+$ — Backbone PS

[→2)-β-D-Galp-(1→3)-α-D-GalpNAc-(1→3)-β-D-GlcpNAc-(1→4)-α-L-Fucp-(1→]n
3
|
α-D-Galp — LPS O-chain

↓ $NaIO_4/NaBH_4/H^+$

β-D-Galp-(1→3)-α-D-GalpNAc-(1→3)-β-D-GlcpNAc-(1→2)-4-deoxy-L-threitol — Oligo A

↓ $NaIO_4/NaBH_4/H^+$

α-D-GalpNAc-(1→3)-β-D-GlcpNAc-(1→2)-4-deoxy-L-threitol — Oligo B

↓ $NaIO_4/NaBH_4/H^+$

β-D-GlcpNAc-(1→2)-4-deoxy-L-threitol — Oligo C

Methylation analysis of the O-PS showed it to contain the structural linkage residues of terminal D-Galp, -3)-D-GalpNAc(1-, -3)-D-GlcpNAc-(1-, -4)-L-Fucp-(1-, and a branch point of 2,3)-D-Galp-(1-. The linkage sequence of these residues was determined through the analysis of a basic linear backbone polymers and a series of oligosaccharides (A, B and C) obtained through periodate oxidation and Smith type hydrolyses as outlined in Scheme 1.

Limited periodate oxidation of the O-PS yielded a high molecular weight unbranched backbone PS from which the nonreducing D-Galp endgroups in the native polymer had been removed. Methylation evidence on the new PS showed that these endgroups were 1,3 linked as single units to the 2-O substituted D-Galp residues in the main chain. Full periodate oxidation of the O-PS gave oligosaccharide A whose structure was established through methylation analysis and by two dimensional NMR COSY experiments that allowed assignment of proton signals of all the glycosyl residues. Assignments of ^{13}C NMR signals were subsequently made from a ^{13}C-^{1}H correlation experiment and the $J_{C,H}$ coupling values confirmed the anomeric configurations determined from the ^{1}H NMR spectra. A two-dimensional long range experiment (HMBC) gave three bond coupling across glycosidic linkages that established the sequence shown for A. Sequential periodate degradations gave oligosaccharides B and C whose structural analysis confirmed the characterization of A. Consideration of the above analytical data leads to the unambiguous characterization of the biological repeating unit of S. milwaukee O43 O-PS as:

```
-2)-β-D-Galp-(1-3)-α-D-GalpNAc-(1-3)-β-D-GlcpNAc-(1-4)-α-L-Fucp-(1-
      3
      1
  α-D-Galp
```

This newly established structure for the O43 antigenic O-PS permits the development of a molecular rationale for its blood group B activity. The difference in the blood groups A and B substances resides in the nonreducing terminal trisaccharide structures of the megalosaccharide molecules where the terminal α-D-GalpNAc and the terminal α-D-Galp residues determine the respective A and B specificites as shown:

```
                                             α-L-Fucp
A substance    α-D-GalpNAc-(1-  ┐               1
                                 \              2
    OR                            ──→3)-β-D-Galp-(1──→
                                 /
B substance       α-D-Galp-(1-  ┘
```

The blood group B activity of the O43 polysaccharide would be expressed if the nonreducing end of the O-PS was terminated by the α-D-Galp and α-L-Fucp residues linked 1,3 and 1,2 respectively to the β-D-Galp residues found in the structure of the biological pentasaccharide repeating units as shown:

```
α—D-Galp(1-3)-β-D-Galp-(1──→
                   2
                   1
               α-L-Fucp
```

While it is difficult to established by chemical methods that the O-PS has this proposed terminal end, serological evidence provided supporting evidence for this conclusion.

A selected murine monoclonal antibody to the antigenic determinant of blood group B, prepared by immunization with a protein conjugate of the synthetic group B trisaccharide determinant (Bundle et al., 1982) was reactive against both the S. milwaukee O43 LPS and O-PS in ELISA test systems, and in passive haemagglutination experiments. The activity of the group B antibody was inhibited by O-PS. No positive precipitin reaction between the monoclonal antibody and O-PS was observed in immunodiffusion experiments, consistent with the monovalent antigenic nature of the PS. The O-PS had no serological reaction against monoclonal or polyclonal blood group A antisera.

An analysis of the LPS O-PS of E. coli O86, (Andersson et al., 1989) also reported to have blood group B activity, showed it to be a repeating pentasaccharide unit of almost identical structure to of the Salmonella O43 O-PS except that the β-D-GalpNAc residue in the main chain of the latter O-PS was replaced by a β-D-GlcpNAc residues in the main chain of the E. coli O86 PS. A similar explanation for the O86 blood group B activity also experimentally found it to reside in the same trisaccharide endgroup.

An explanation for the reported blood group A activity of Salmonella group R undertaken through a study of the LPS O-PS of S. riogrande (group R) O40 (Perry and MacLean, 1992a) was less precise than that obtained in the analysis of the blood group B activity of the Salmonella group U O-PS. The O40 O-PS was composed of a repeating pentasaccharide units containing D-GalNAc, D-GlcNAc , D-Man, and D-Glc (2:1:1:1) residues. The absence of L-Fuc and D-Gal residues characteristic of the blood groups A terminal epitope was at first surprising if the blood group A activity is to be interpreted as being due to a common trisaccharide endgroup structure.

A structural analysis of the O40 O-PS using 1D and 2D ^{1}H and ^{13}C NMR, periodate oxidation, and methylation methods similar to those used in the analysis of the O43 O-PS, showed the O40 to be a polymer of repeating branched pentasaccharide units having the structure:

```
-4)-α-D-GalpNAc-(1-3)-β-D-Manp-(1-4)-β-D-Glcp-(1-3)-α-D-Galp-(1-
                          2
                          1
                     β-D-GlcpNAc
```

The serological cross reactions reported between polyclonal Salmonella O40 and blood group A antisera and blood group A erythrocytes and Salmonella O40 cells appears to determined by an epitope involving terminal α-D-GalpNAc endgroups. It is interesting to note that the blood group A activity of the LPS O-PS of E. coli O6 (Jansson et al., 1984) can be similarly related to an epitope in part of its repeating unit as illustrated below:

```
α-D-GalpNAc-(1-3)-β-D-Manp-(1-4)-β-D-Manp-(1→
                      2
                      1
                  β-D-Glcp
```

This suggestion is supported by the fact that methyl 2-acetamido-2-deoxy-α-D-galactopyranoside is an inhibitor of the above blood group A cross reactions using polyclonal antisera. It is probable that the serological cross reactivities reported for blood group A resides in limited similarities of the trisaccharide endgroup as illustrated below:

```
β-D-GlcpNAc-(1-2)-β-D-Manp-(1→   α-L-Fucp-(1-2)-β-D-Galp-(1→
                       3                                 3
                       1                                 1
                 α-D-GalpNAc                       α-D-GalpNAc

 Salmonella O40 O-PS terminus        Blood group A terminus
```

As might be expected, a high specificity blood group A murine monoclonal antibody prepared by immunization and hybridoma selection using a protein conjugate of a synthetic blood group A terminal trisaccharide did not react with LPS or O-PS of Salmonella O40.

Little is known of the role of LPS O-PS in gram-negative infections. Titres of O-PS specific antibody of IgG class against the patients infecting organism correlates with decrease in the frequency of shock and death in bacteraemia (McCabe et al., 1977).

Attempts have been made to relate the role of chemical fine structures, and the lengths of LPS O-PS, to their biological properties and virulence functions. It is considered that the structures of the O-PS of Salmonella determine the rate of compliment activation and C3b deposition on the bacteria (Makela et al., 1988) and studies have implicated LPS as the target site for compliment-mediated killing by serum with the O-PS as determinants of sensitivity or resistance to serum (Porat et al., 1992; Jimenez-Lucho et al., 1990). Monoclonal antibodies to Salmonella LPS O-PS have been reported to protect mice against challenge with virulent S. typhimurium (Colwell et al., 1984).

Information gained through the use of new structural analysis techniques, the application of genetic manipulations and modeling methods, should lead to new evaluations of the importanace of LPS O-PS in bacterial pathogenesis.

REFERENCES

Anderson, L. A., and Unger, F. M., eds. 1983, "Bacterial Lipopolysaccharides," ACS Symposium Series 231, Am. Chem. Soc., Washington, DC.

Andersson M., Carlin, N., Leontein, K., Lindquist, U., and Slettengren, K., 1989, Structural studies of the O-antigenic polysaccharide of Escherichia coli O86 which possesses blood-group B activity, Carbohydr. Res. 185:211.

Bradley, D. E., Anderson, A. N., and Perry, M. B., 1991, Difference between the LPS cores in adherent and non-adherent strains of enteropathogenic Escherichia coli O119, FEMS Microbiol Lett. 80:13.

Brisson, J. R., and Perry, M. B., 1988, The structure of the two lipopolysaccharides produced by Salmonella boecker, Biochem. Cell Biol. 66:1066.

Bundle, D. R., 1990, Synthesis of oligosaccharides related to bacterial O-antigens, in: "Topics in Current Chemistry," Springer-Verlag, Berlin, 154:1.

Bundle, D. R., Gidney, M. A. J., Kassan, N., and Rahman, A. F. R., 1982, Hybridomas specific for carbohydrates: Synthetic human blood group antigens for the production selection, and characterization of monoclonal typing reagents, J. Immunol. 129:678.

Bundle, D. R., Gerken, M., and Perry, M. B. 1986, Two-dimensional nuclear magnetic resonance at 500MHz: the structural elucidation of a Salmonella serogroup N polysaccharide antigens, Can. J. Chem. 64:255.

Bundle, D. R., Cherwonogrodzky, J. W., Gidney, M. A. J., Meikle, P. J., Perry, M. B., and Peters, T., 1989, Definition of Brucella A and M epitopes by monoclonal typing reagents and synthetic oligosaccharides, Infect. Immun. 57:2829.

Caroff, M., Bundle, D. R., Perry, M. B., Cherwonogrodzky, J. C., and Duncan, J. R., 1984, Antigenic S-type lipopolysaccharide of Brucella abortus 1119-3, Infect. Immun. 46:384.

Caroff, M., Bundle, D. R., and Perry, M. B., 1984a,. Structure of the O-chain of the phenol-phase soluble lipopolysaccharide of Yersinia enterocolitica O9, Europ. J. Biochem. 139:195.

Caroff, M., Chaby, R., Karibian, D., Perry, J., Deprun, C., and Szabo, L., 1990, Variations in the carbohydrate regions of Bordetella pertussis lipopolysaccharides: Electrophoretic, serological and structural features, J. Bacteriol. 172:1121.

Colwell, D. E., Michalek, S. M., Briles, D. E., Emilio, J., and McGhee, J. R., 1984, Monoclonal antibodies to Salmonella lipopolysaccharide and O-polysaccharide protect C3H mice against challenge with virulent S. typhimurium, J. Immunol. 132:229.

Corbel, M. J., Stuart, F. A., and Brewer, R. A., 1984, Observation on serological cross reactions between smooth Brucella species and organisms of other genera, Dev. biol. Stand. 56:341.

Cygler, M., Rose, D. R., and Bundle, D. R., 1991, Recognition of a cell-surface oligosaccharide of pathogenic Salmonella by an antibody Fab fragment. Science, 253:442.

DiFabio, J. L., Perry, M. B., and Brisson, J. R., 1988, Structure of the antigenic O-polysaccharide produced by Salmonella eimsbuttel, Biochem. Cell Biol. 66:107.

DiFabio, J. L., Brisson, J. R. and Perry, M. B., 1988a, Structure of the major lipopolysaccharide O-chain produced by Salmonella carrau (O:6,14,24). Carbohydr. Res. 179:233.

DiFabio, J. L., Brisson, J.R., and. Perry, M. B., 1989, Structure of the lipopolysaccharide antigenic O-chain produced by Salmonella livingstone (O:6,7), Biochem. Cell Biol. 67:278.

DiFabio, J. L., Brisson, J. R., and Perry, M. B., 1989a, Structure of the lipopolysaccharide antigenic O-chain produced by Salmonella ohio (O:6.7), Carbohydr. Res. 189:161.

DiFabio, J. L., Brisson, J. R., and Perry, M. B., 1989b, Structural analysis of the three lipopolysaccharide produced by Salmonella madelia (1,6,14,25), Biochem. Cell Biol. 67:78.

Dodds, K. L., Perry, M. B., and McDonald, I. J., 1987 Alterations in lipopolysaccharides produced by chemostat-grown Escherichia coli O157:H7 as a function of growth rate and growth limiting nutrient. Can. J. Microbiol. 33:452.

Fox, A. J., Curry, A., Jones, D. M., Demarco de Hormaecheh, R. Parsons, N. J., Cole, J. A., and Smith, H., 1991, The surface structure seen on gonococci after treatment with CMP-NANA is due to sialylation of surface lipopolysaccharide previously described as a 'capsule.' Microbial Pathogenesis, 11:199.

Hellerqvist, C. G., and Lindberg, A. A., 1971, Structural studies of the commonn core polysaccharide of the cell wall lipopolysaccharide from Salmonella typhimurium 395 MS, Carbohydr. Res. 16:39.

Jansson, P. E., Lindberg, B., Lonngren, J., and Ortega, C., 1984, Structural studies of the Escherichia coli O-antigen 6, Carbohydr. Res. 131:277.

Jimenez-Lucho, V. E., Leive, L. L., and Joiner, K. A., 1990, Role of the O-antigen of lipopolysaccharide in Salmonella in protection against complement action, in; "The Bacteria" vol XI, Academic Press, New York, p.339.

Kabat, E. A., 1976, "Structural Concepts in Immunology and Immunochemistry," Holt Rinehart and Winston, New York.

Karibian, D., Deprun, C., Szabo, L., Le Beyec, Y., and Caroff, M., 1991, ^{252}Cf-Plasma desorption mass spectrometry applied to the analysis of endotoxin Lipid A preparations, Internat. J. Mass Spectrom. Ion Processes, 111:273.

Kastoowsky, M., Gutberlet, T., and Bradaczek H., 1992, Molecular modeling of the three-dimensional structure and conformation flexibility of bacterial lipopolysaccharide, J. Bacteriol. 174:4798.

Kauffmann, F., 1966, "The Bacteriology of the Enterobacteriaceae," Munksgaard, Copenhagen.

Kusumoto, S., Inage, M., Chaki, H., Imoto, M., Shimamoto, T., and Shiba, T., 1983, Chemical synthesis of lipid A for the elucidation of structural-activity relationships, in: "Bacterial Polysaccharides", L Anderson and F. M. Unger eds., ACS Symposium 231, Am. Chem Soc. Washington, p 237.

Le Minor, L., and Popoff, M. Y., 1988, "Antigenic Formulas of the Salmonella serovars," WHO Collaborative Centre for Reference and Research of Salmonella, Institut Pasteur, Paris.

Lindberg, A. A., and Le Minor, L., 1984, Serology of Salmonella, in: "Methods in Microbiology," T. Bergan ed., Academic Press, New York, 15:1.

Lindberg, A. A., Wollin, R., Bruse, G., Ekwall, E., and . Svenson, S. B., 1983, Immunology and Immunochemistry of Synthetic and Semisynthetic Salmonella O-Antigen-Specific Glycoconjugates, in: "Bacterial Lipopolysaccharides," L. Anderson and . M. Unger eds., Am. Chem. Soc. Symposium 231, p 83.

Lüderitz, O., Freudenberg, M. A., Galanos, C., Lehmann, V., Th. Rietschel, E. Th., and Shaw, D. H, 1982, "Current Topics in Membrane and Transport," Academic Press, New York, p 79.

MacLean, L. L., and Perry, M. B., 1991, Characterization of the Salmonella O:8 antigen in the O-chain of the lipopolysaccharide produceed by Salmonella virginia O:8, Biochem. Cell Biol. 69:852.

Makela, P. H., Hovi, M., Saxen, H., Valtonen, M., and Valtonen, V., 1988, Salmonella complement and mouse macrophages, Immunol. Lett. 19:217.

McCabe, W. R., Johns, M. A., Craven, D. E., and Bruins, S. C., 1977, Clinical implications of enterobacterial antigens, in: "Microbiology 1977," D. Schlessinger, ed., Am. Soc. Microbiol. 293.

McCartney, A. C., and Wardlaw, A. C., 1985, Endotoxic activities of lipopolysaccharides, in: "Immunology of the Bacterial Cell Envelope," D. E. S. Stewart-Till & M. Davies, eds. John Wiley & Sons, New York, p 203.

Meikle, P. J., Perry, M. B., Cherwonogrodszky, J. W., and Bundle, D. R., 1989, Fine structure of A and M antigens from Brucella biovars, Infect. Immun. 57:2820.

Meyer, H., Bhat, R., Masoud, H., Radziejewska-Lebrecht, J., Widemann, C., and Krauss, J. H., 1989, Bacterial lipopolysaccharide, Pure Appl. Chem. 61:1271.

Perry, M. B., 1990, Structure of the O-chain of the lipopolysaccharide of Actinobacillus pleuropneumoniae serotype 10. Biochem. Cell Biol. 68:808.

Perry, M. B., MacLean, L. L., 1992, Structural characterization of the O-polysaccharide of the lipopolysaccharide produced by Salmonella milwaukee O:43 (group U) which possesses human blood-group B activity. Biochem. Cell Biol. 70:49.

Perry, M. B., and MacLean, L. L., 1992a, Structure of the polysaccharide O-antigen of Salmonella riogrande :40 (group R) related to blood group A activity, Carbohydr. Res. 232:143.

Perry, M. B., and Richards, J. C., 1990, Identification of the lipopolysaccharide O-chain of Escherichia hermannii (ATCC 33651) as a D-rhamnan, Carbohydr. Res. 205:371.

Perry, M. B., Bundle, D. R., MacLean, L. L., Perry, J. A., and Griffith, D. W., 1986, The structure of the antigenic lipopolysaccharide O-chains produced by Salmonella urbana and Salmonella godesberg, Carbohydr. Res. 156:107.

Perry, M. B., MacLean, L., and Griffith, D. W., 1985, Structure of the O-chain of the phenol-phase soluble lipopolysaccharides of Escherichia coli O157:H7, Biochem. Cell Biol. 64:21.

Porat, R., Mosseri, R., Kaplan, E., Johns, M. A., and Shibolet, S. 1992, Distribution of polysaccharide chains of lipopolysaccharides determine resistance of Escherichia coli to the bactericidal activity of serum, J. Infect. Dis. 165:953.

Proctor, R. A., ed., 1984, "Handbook of Endotoxin," Vols 1-3. Elsevier, Amsterdam, New York and Oxford.

Qureshi, N., Takayama, K., Mascagni, P., Honovich, J., Wong, R., and Cotter, R. J., 1988, Complete structural determination of lipopolysaccharide obtained from deep rough mutant of Escherichia coli, J. Biol. Chem. 263: 11971.

Raetz, C. R. H., 1990, Biochemistry of endotoxin. Annu. Rev. Biochem. 59:129.

Rietschel, E. Th., Sidorezyk, Z., Zahringer, U., Wollenweber, H. W., and Lüderitz, O., 1983, Analysis of the Primary Structure of Lipid A, in: "Bacterial Lipopolysaccharides," L. Anderson and F. M. Unger, eds., Am. Chem. Soc. p 195.

Rose, D. R., Cygler, M., To, R. J., Przybylska, M., Sinnott, B., and Bundle, D. R., 1990, Preliminary crystal structure analysis of an Fab specific for a Salmonella O-polysaccharide antigen, J. Mol. Biol. 215:489.

Takayama, K., Qureshi, N., Raetz, C. R. H., Ribi, E., Peterson, J., Cantrell, J. L., Pearson, F. C., Wiggins, J., and Johnson, A. G., 1984, Influence of fine structure of lipid A on Limulus lysate clotting and toxic activities. Infect. Immun. 45:350.

Tsang, R. S. W., Schlecht, S., Aleksic, S., Chan, K. H., and Chau, P. Y.,.1991, Lack of the α-1-2 linked N-acetyl-D- glucosamine epitope in the other core structures of lipopolysaccharides from certain O serogroups and subspecies of Salmonella enterica, Res. Microbiol. 142:521.

Wollenweber, H. W., and Rietschel, E. T., 1990, Analysis of lipopolysaccharide (lipid A) fatty acids., J. Microbiol. Methods, 11:195.

Ziegler, E. J., 1988, Protective antibody to endotoxin core: The Emperor's new clothes? J. Infect. Dis. 158:286.

FIMBRIAE OF SALMONELLA ENTERITIDIS: MOLECULAR ANALYSIS OF SEF14 AND VACCINE DEVELOPMENT POTENTIAL

Martin J Woodward, Christopher J. Thorns, and Claude Turcotte

Bacteriology R&D Discipline
Central Veterinary Laboratory
Weybridge
Surrey KT15 3NB UK

INTRODUCTION

Salmonella virulence is recognised as being multifactorial. The bacterium possesses phenotypic traits needed for successful colonisation, adhesion, invasion, dissemination, macrophage survival and so on. The host may show varying degrees of susceptibility depending upon the genotype of both the host and pathogen; indeed, there are well recognised host-pathogen relationships with for example the incidence of cattle salmonellosis in the UK being caused almost exclusively by *S. dublin* and *S. typhimurium* DT204C (Anon, 1990). The convenience of the mouse model and the ease of genetic manipulation of salmonellas is leading to a greater understanding of the mechanisms of virulence, although the role of fimbriae expressed by salmonellas is unclear. In *E. coli*, many fimbrial adhesins have been defined (Klemm, 1985) and their role in pathogenesis recognised. In salmonellas, the abiquitous type 1 fimbriae, although recognised as a potential adhesin (Duguid *et al.*, 1966), are not considered essential for virulence (Lockman and Curtiss, 1992). Other classes of fimbriae have been reported to be expressed by salmonellas (Rohde *et al.*, 1975) and fimbrial mediated fibronectin binding has been reported for *S. enteritidis* (Baloda, 1988). Indeed, three distinct fimbriae expressed by an *S. enteritidis* enterotoxin producing isolate from human faeces have been described (Feutrier *et al.*, 1986, 1988; Muller *et al.*, 1989, 1990; Collinson *et al.*, 1991). Concern in this laboratory over the international increase in *S. enteritidis* incidents (Rodrigue *et al.*, 1990) and the zoonotic potential of infected poultry and poultry products (O'Brien, 1988) prompted the development of specific diagnostic and control measures. One of these developments to be described here was the molecular biological analysis of SEF14, a type 3-like fimbriae of *S. enteritidis.* The use of SEF14 to express heterologous antigen as a first step toward vaccine development is described also.

MOLECULAR DEFINITION OF SEF14

Mice were immunized using protocols designed to elicit an antibody response primarily directed against surface antigens of *S. enteritidis* strain 1246/89, isolated from a current field infection of chickens. Of 400 hybridomas produced, one (MAb 69/25) was found in preliminary assays to bind salmonellas of serogroup D (Thorns *et al.*, 1990). Electron microscopic studies confirmed that MAb 69/25 was directed against an epitope on a profuse, fine fimbrial structure. The fimbrial structure was not expressed at incubation temperatures below 22°C whilst maximum yields of the fimbriae were achieved by growth at 37°C on chemically defined minimal media and other nutritionally simple media such as peptone water. A number of standard procedures, such as heat shock and pH2 treatments of cultures followed by size exclusion HPLC, were used to purify the fimbrial monomers which reacted with MAb69/25. A protein product of 14.3 kDa was isolated and used to raise monoclonal antibodies. Competition ELISA between monoclonals of the panel indicate three epitope domains.

MAb 69/25 was used to screen a pUC18 based *S. enteritidis* 1246/89 gene library. One recombinant, designated CT15, was identified as expressing the fimbrial antigen and the plasmid, designated VW400, was extracted and shown to harbour a 2.4 kb *Sau* IIIA generated insert. Subcloning and DNA nucleotide sequencing identified a single ORF, which we designate *sef A,* comprising 528 bp and potentially encoding a protein product of 176 amino acids. The deduced amino acid sequence shared homology with the N-terminal amino acid sequence of a 14 kDa fimbrial antigen originally designated type 1 (Feutrier *et al.,* 1986) but renamed SEF14 (Muller *et al.,* 1988).

Mini-cell and western blot analysis indicated two forms of SEF14 of 14.3 kDa and 16.0 kDa in apparent size in recombinant *E. coli.* The former size is identical to the mature product and indicates peptidase cleavage at the Ala-His-Ala site immediately prior to the known N-terminal sequence. Of the two potential f-met start codons of *sef A*, the first was located in a region with a significant stem-loop structure whilst the second was 9 bp down stream of a consensus ribosomal binding sequence. Translation from the second f-met start codon, giving an ORF of 495 bp, would yield a protein product of 165 amino acids with a leader signal sequence of 21 hydrophobic amino acids. Mature fimbriae were not detected on the surface of the recombinant *E. coli* by IEM or latex agglutination suggesting a lack of accessory genes essential for fimbriae synthesis, export and extension. Analysis of two adjacent ORFS identified on VW400 and designated *sef B* and *sef C* shared homology with genes encoding the *E. coli* CS3 pili synthesis 27 kDa protein and the mannose resistant *Klebsiella pneumoniae* MRKC protein purcusor, respectively. Transfer of pVW400 to CS3+ *E. coli* and MRKC+ K. pneumoniae did not result in surface production of mature SEF14 fimbriae.

SEFA GENE DISTRIBUTION AND EXPRESSION; SPECIFIC DIAGNOSTIC DEVELOPMENT

Over 2000 individual isolates representing 74 salmonella serogroups were tested for expression of SEF14 by latex agglutination and ELISA using MAb 69/25 as the immunoprobe. All *S. enteritidis,* 40% of *S. dublin* and one *S. moscow* isolates reacted but no other isolated did. Here was evidence of restricted expression of SEF14. A *sef A* gene probe used in DNA-DNA colony hybridisation experiment demonstrated homologous sequences encoded by all *S. enteritidis, S. dublin, S. gallinarum, S. pullorum, S. berta, S. rostock* and *S. typhi* isolates all belonging to serogroup D; no other group D isolates (*S. canaster, S. durban, S. ouakam, S. panama, S. wangata*) or any other isolates hybridized. The lack of expression in geneprobe positive immunoprobe negative isolates and variable expression by *S. dublin* is under investigation. PCR was used to amplify the *sef A* genes from a number of isolates, both immunoprobe positive and negative, and the products sequenced. In each experiment, the entire *sef A* ORF was present and no base changes were identified which suggests a genetic or physiologic fault in expression elsewhere. Gene probing of total genomic digests of representative immunoprobe positive and negative isolates showed a lack of restriction length polymorphism indicating conservation of the genetic context of the *sef* gene complex. No plasmids harboured by the strains tested hybridised indicating a chromosomal location for the *sef* gene complex. However, the possible role of the "virulence plasmid" in modulating SEF14 expression should not be overlooked since pleiotropic mutations have been described (Pullinger and Lax, 1992).

The specificity of expression of SEF14 to *S. enteritidis* and *S. dublin* has enabled rapid diagnostic tests to be developed. A simple latex agglutination test (SEFEX) is available for use in the routine serodiagnostic laboratory. Furthermore, SEF14 is highly immunogenic and chick flock tests by ELISA and dipstick technologies are realistic. PCR tests are being developed also.

SEF14 AS A MOLECULAR VECTOR FOR VACCINE DEVELOPMENT

Live attenuated *S. enteritidis* vaccines have been developed in this laboratory (Cooper *et al.*, 1990, 1992) and the potential for their use as vector to generate multivalent vaccines is being pursued. The use of the flagellin of *S. typhimurium* to express protective cholera toxin and streptococcal M protein has been demonstrated (Newton *et al.*, 1989, 1991). SEF14 may be similarly modified to express heterologus epitopes.

Kamamycin marked *sefA* deletion mutants of *S. enteritidis* LA5 (Cooper *et al.*, 1990) were prepared by homogenetisation using pGP704 (gift from G. Dougan) as the suicide vector to introduce modified sef DNA regions. Preliminary data indicated that after oral inoculation of day old chicks (Cooper *et al.*, 1992) no significant difference in colonisation, invasion and clearance was observed between mutant and wild type. To assess whether the SEF14 gene complex could be used to express heterologus antigens, the MPB70 antigen gene of *Mycobacterium bovis* (Harboe and Nagai, 1984) was cloned in-frame into the unique *BamHI* restriction endonuclease site in the *sefA* gene. PCR was used to amplify the full length MPB70 gene and various intragenic regions known to encompass defined B and T-cell epitopes (Radford *et al.*, 1990). The primers were tailed with *BamHI* and *BglII* sites for ease of cloning and rapid orientation of inserts by restriction digests. The various constructs were confirmed by DNA nucleotide sequencing and in each case retained the SEF14 hydrophobic signal sequence, 28 amino acids of the SEF14 N-terminus and fused in frame either with the full length MPB70 gene terminating with a stop codon or with various intragenic fragments of the MPB70 gene and terminating with the SEF14 C-terminus. Each construct was expressed in *E. coli* and *S. enteritidis* and production of MPB70 epitopes was assessed by western blotting using polyclonal and specific monoclonal antibodies. The SEF14 fused with full length MPB70 construct gave significant yield of antigen in both *E. coli* and *S. enteritidis*. None of the MPB70 intragenic subfragments fused in SEF14 were detected suggesting inappropriate presentation, stearic hindrence, proteolysis of the fusion product or reduced expression. Epitope mapping of SEF14 is in progress and rational epitope exchanges are being investigated for heterologous epitope expression. As for the use of the *sef* gene complex to drive expression of heterologous antigens, the *Leptospira hardjo-bovis* flagellin is being examined as a candidate for expression in *S. dublin* (this work is co-sponsored by the Milk Marketing Board of the UK).

FUTURE WORK

A detailed genetic analysis of the *sef* gene complex is in hand specifically to address the correlation between environmental conditions and SEF14 expression and to establish why some group D serotypes, specifically *S. typhi,* do not express SEF14. The biological role of the antigen is unclear; deletion mutants of SEF4 and other fimbrial genes will be made and used in adhesion, invasion and tropism studies.

ACKNOWLEDGEMENTS

The authors wish to acknowledge MAFF and MMB for financial support for this work and to Agriculture Canada for the award of a study fellowship to C Turcotte. Glyn Hewinson and Bill Russell are thanked for advice and technical support with the MPB70 antigen work.

REFERENCES

Anon 1990 Animal salmonellosis. Annual summaries. Ministry of Agriculture, Fisheries and Food. Welsh Office Agriculture Department. Department of Agriculture and Fisheries for Scotland.

Baloda, S.B., 1988, Characterization of fibronectin binding to *Salmonella enteritidis* strain 27655R, *FEMS. Micro. Letts.* 49:483.

Collinson, S.K., Emody, L., Muller, K-H., Trust, T.J. and Kay, W.W., 1991. Purification and characterization of thin, aggregative fimbriae from *Salmonella enteritidis, J. Bact.* 173:4773.

Cooper, G.L., Nicholas R.A.J., Cullen, G.A. and Hormaeche, C.E., 1990, Vaccination of chickens with *Salmonella enteritidis aroA* live oral vaccines, *Micro. Path.* 9:255.

Cooper, G.L., Venables, L.M., Nicholas, R.A.J., Cullen, G.A. and Hormaeche, C.E., 1992, Vaccination of chickens with chicken-derived *Salmonella enteritidis* phage type 4 aroA live oral salmonella vaccines. *Vaccine* 10:247.

Duguid, J.P., Anderson, E.S. and Campbell, I., 1966, Fimbriae and adhesive properties in *Salmonellae, J. Pathol. Bact.* 92:107.

Feutrier, J., Kay, W.W and Trust, T.J., 1986, Purification and characterization of fimbriae from *Salmonella enteritidis, J. Bact.* 168:221.

Feutrier, J., Kay, W.W. and Trust, T.J., 1988, Cloning and expression of a *Salmonella enteritidis* fimbrin gene in *Escherichia coli, J. Bact.* 170:4216.

Harboe, M. and Nagai, S., 1984, MPB70, a unique antigen of *Mycobacterium bovis* BCG, *Am. Rev. Resp. Dis.* 129:444.

Klemm, P., 1984, The *fimA* gene encoding the type 1 fimbrial subunit of *Eschericia coli,* nucleotide sequence and primary protein structure, *Eur. J. Biochem.* 143:395.

Lockman, H.A. and Curtiss III, R., 1922, Virulence of non-type 1-fimbriated and nonfimbriated nonflagellated *Salmonella typhimurium* mutants in murine typhoid fever, *Infect.Immun.* 60:491.

Muller, K-H., Trust, T.J. and Kay, W.W., 1989, Fimbriation genes of *Salmonella enteritidis, J. Bact.* 171:4648.

Muller, K-H., Collinson, S.K., Trust, T.J. and Kay, W.W., 1991, Type 1 fimbriae of *Salmonella enteritidis, J. Bact.* 173:4765.

Newton, S.M.C., Jacob, C.O. and Stocker, B.A.D., 1989, Immune response to cholera toxin epitope inserted in *Salmonella* flagellin, *Science* 244:70.

Newton, S.M.C., Wasley, R.D., Wilson, A., Rosenberg, L.T., Miller, J.F. and Stocker, B.A.D., 1991, Segment IV of a *Salmonella* flagella gene specifies flagellar antigen epitopes, *Mol. Micro.* 5:419.

O'Brien, J.D.P., 1988, *Salmonella enteritidis* infection in broiler chickens, *Vet. Rec.* 122:214.

Pullinger, G.D. and Lax, A.J., 1992, A *Salmonella dublin* virulence locus that affects bacterial growth under nutrient-limited conditions, *Mol. Micro.* 6:1631.

Radford, A.J., Wood, P.R., Billman-Jacobe, H., Greyson, H.M., Mason, T.J. and Tribbick, G., 1990, Epitope mapping of the *Mycobacterium bovis* secretory protein using overlapping peptide analysis, *J. Gen. Micro.* 136:265.

Rodrigues, D.C., Tauxe, R.V. and Rowe, B., 1990, International increase in *Salmonella enteritidis* : a new pandemic?, *Epidem. Inf.* 105:21.

Rohde, R., Aleksic, S., Muller. G., Plavsic, S. and Aleksic, V., 1975, Profuse fimbriae conferring O-inagglutinability to several strains of *S. typhimurium* and *S. enteritidis* isolated from pasta products, Zbl. *Bakt. Hyg. I. Abt. A* 230:38.

Thorns, C.J., Sojka, M.G. and Chasey, D., 1990, Detection of a novel fimbrial structure on the surface of *Salmonella enteritidis* by using monoclonal antibody, *J. Clin. Micro.* 28:2409

THE PATHOGENICITY DETERMINANTS OF *Salmonella typhi*: POTENTIAL ROLE OF FIMBRIAL STRUCTURES

Giuseppe Satta[1], Angela Ingianni[2], Patrizia Muscas[3], Gian Maria Rossolini[3], and Raffaello Pompei[2]

[1]Istituto di Microbiologia - Università Cattolica del S. Cuore, Rome, Italy
[2]Istituto di Medicina Interna - Università di Cagliari, Cagliari, Italy
[3]Dipartimento di Biologia Molecolare - Università di Siena, Siena, Italy

INTRODUCTION

Typhoid fever is a disease endemic in many developing countries and continues to represent a serious public health problem (Edelman and Levine, 1986; Calva et al., 1988). Currently licensed anti-typhoid vaccines are not fully satisfactory in terms of efficacy and duration of protection (Levine, 1988), and there is much interest in developing new and more effective vaccines against the disease. For this purpose, a detailed understanding of the microbial factors contributing to the infection as well as to the disease (the determinants of pathogenicity),which in the case of *S. typhi* remain largely unclear, would be desirable. It has been hypothesized, and in some cases demonstrated, that fimbriae are important for the pathogenicity of Gram-negative bacteria in general and members of the family of *Enterobacteriaceae* in particular (Reid and Sobel, 1987; Finlay and Falkow, 1989). Fimbriae are proteinaceous fibrillar surface appendages that can mediate attachment of the bacterial cell to host tissues (Paranchych and Frost, 1988). Their role in microbial pathogenicity has been attributed to this adhesive function, which could be important for colonization and infection of host tissues (Finlay and Falkow, 1989).

Several types of fimbriae have been described in *Enterobacteriaceae* (Dorman et al., 1980; Clegg and Gerlach, 1987; Paranchych and Frost, 1988). A definite role in pathogenicity, however, has been clearly demonstrated only for some types of fimbriae, while for other types, including type 1 fimbriae (i.e. very common fimbrial structures which are able to recognize eukaryotic cell receptors containing mannose residues) their role in this sense remains unclear (Clegg and Gerlach, 1987; Finlay and Falkow, 1989; Orndorff and Bloch, 1990; Lockman and Curtiss, 1992).

Since type 1 fimbriae are apparently the only adhesin present in *S. typhi* (Duguid et al., 1966; Halula and Stocker, 1987; Satta et al., in preparation), and since most clinical isolates are able to produce type 1 fimbriae (Duguid et al., 1966; Satta et al., in preparation), it would be interesting to understand the possible role of these structures in *S. typhi* pathogenicity.

Biology of Salmonella, Edited by F. Cabello *et al.*,
Plenum Press, New York, 1993

In this work we have evaluated, by means of in vitro experiments, the role of type 1 fimbriae in the interaction of *S. typhi* with some eukaryotic cells.

RESULTS

Type 1 Fimbriae Can Mediate Adhesion of *S. typhi* to Epithelial Cells

In a first set of experiments the adhesion of fimbriated and nonfimbriated *S. typhi* strains to different types of epithelial cells was analyzed.

Fimbriated strains were able to adhere to the human intestinal epithelial cell line INT-407 much more efficiently than the nonfimbriated ones, and a relationship between the degree of fimbriation and adhesiveness was apparent (table 1).

Similar results were obtained with the same *S. typhi* strains using different epithelial cell lines such as HeLa and MDCK (data not shown). These results therefore indicate that type 1 fimbriae can mediate adhesion of *S. typhi* to epithelial cells.

Table 1. Adhesion of some fimbriated and nonfimbriated *S. typhi* strains to the human intestinal epithelial cell line INT-407.

strain	relevant phenotype	No of bacteria/cell[d]
Sty4	Fim+	21
29ty2	Fim+	22
29ty9	Fim+	19
24ty7	Fim+	19
27ty2	Fim++[b]	30
UVP6[a]	Fim++	29
34ty3	Fim-[c]	<1
6S[a]	Fim-	<1

[a] Spontaneous mutant of Sty4.
[b] The degree of fimbriation was semiquantitatively evaluated by haemagglutination assays (Duguid et al., 1966) and confirmed by electron microscopy.
[c] The absence of type 1 fimbriae was assessed both by the haemagglutination assay and by electron microscopy.
[d] Mean value of three experiments; in each experiment 200 cells were subjected to microscopical examination.

Type 1 Fimbriae Are Apparently the Principal Surface Structure Able to Mediate Adhesion of *S. typhi* to Epithelial Cells

The adhesion of fimbriated and nonfimbriated *S. typhi* strains to human intestinal epithelial cells was also evaluated in presence of agents able to specifically inhibit the interaction of type 1 fimbriae with eukaryotic cell receptors, including α-D-mannose (which competes with cellular receptors for interaction with type 1 fimbrial adhesin and is able to inhibit the agglutination of guinea pig red blood cells -GPRBC- by fimbriated strains) or antibodies raised against purified *S. typhi* fimbriae (which were able to agglutinate fimbriated bacteria and to inhibit the agglutination of GPRBC by fimbriated strains).

Adhesion of fimbriated strains to the intestinal epithelial cell line INT-407 was greatly reduced in presence of both inhibitors (table 2).

Table 2. Effect of α-D-mannose and of anti-fimbrial antibodies on adhesion of fimbriated and nonfimbriated *S. typhi* strains to the human intestinal epithelial cell line INT-407.

strain	relevant phenotype	No of bacteria/cell[d]		
			+α-D-mannose[e]	+antibody[f]
Sty4	Fim+	21	1	<1
27ty2	Fim++[b]	30	1	1
UVP6[a]	Fim++	29	1	1
34ty3	Fim-[c]	<1	<1	<1
6S[a]	Fim-	<1	<1	<1

[a, b, c, d] See corresponding footnotes to table 1.

[e] 2% wt/vol.

[f] 1:40 dilution. The agglutinating titre of the serum was 1:36000.

These results therefore suggest that type 1 fimbriae are the principal surface structure of *S. typhi* able to mediate microbial adhesion to epithelial cells.

The Presence of Type 1 Fimbriae Influences the Ability of *S. typhi* to Invade Epithelial Cells

The ability of fimbriated and nonfimbriated *S. typhi* strains to invade the intestinal epithelial cell line INT-407 was also evaluated.

Table 3. Invasion of INT-407 cells by fimbriated and nonfimbriated *S. typhi* strains, and effect of α-D-mannose and anti-fimbrial antibodies on the invasion process.

strain	relevant phenotype	No of bacteria/cell[d]		
			+α-D-mannose[e]	+antibody[f]
Sty4	Fim+	13	<1	<1
27ty2	Fim++[b]	19	2	1
UVP6[a]	Fim++	17	1	<1
34ty3	Fim-[c]	<1	<1	<1
6S[a]	Fim-	<1	<1	<1

[a,b,c] See corresponding footnotes to table 1.
[d] Mean value of three experiments (see under section "experimental procedures" for details on the scoring criterion).
[e,f] See corresponding footnotes to table 2.

Results of these experiments showed that only fimbriated strains were highly invasive in this in vitro experimental system. When interaction of fimbriae with eukaryotic cell receptors was blocked by specific inhibitors (α-D-mannose or anti-fimbrial antibodies), bacterial invasiveness was reduced severalfold (table 3).

The addition of purified type 1 fimbriae of *S. typhi* in invasion experiments was apparently able to increase invasiveness of both fimbriated and nonfimbriated strains (table 4). This effect, however, was evident only when the interaction of bacteria with epithelial cells was facilitated by centrifuging the bacteria on to the cell monolayers. Addition of purified fimbriae did not result in any noticeable effect on invasion when the same experiment was performed without centrifuging the bacteria on to the cell monolayers (data not shown). Its should be noted that adhesion of both fimbriated and non fimbriated bacteria was not affected by the concentrations of purified fimbriae used in these experiments (data not shown).

The above results therefore suggest that type 1 fimbriae are important for the invasion process of intestinal epithelial cells by *S. typhi*.

DISCUSSION

The role played by type 1 fimbriae in *Salmonella* pathogenicity is still largely unclear (Finlay and Falkow, 1988). Virtually all the studies addressing this subject have been focused on *Salmonella typhimurium* , and while earlier reports proposed that type 1 fimbriae played a role in *S. typhimurium* virulence by facilitating intestinal colonization (Darekar and Eyer, 1973; Duguid et al., 1976; Tanaka and Katsube, 1978), it has recently been shown that type 1 fimbriae are not essential virulence factors for *S. typhimurium* and that expression of type 1 fimbriae in this microorganism is neither an advantage nor a disadvantage for bacterial colonization of the mouse intestinal tract (Lockman and Curtiss, 1992).

The role of type 1 fimbriae in *S. typhi* pathogenicity has not been studied before. Even though the murine model of typhoid caused by *S. typhimurium* has often been relied on to study the pathogenesis of *S. typhi* infection, there are differences between the two *Salmonella* serovars (in *S. typhimurium* mannose-resistant adhesins have also been described (Jones and Richardson, 1981; Tavendale et al., 1983; Halula and Stocker, 1987); mice are naturally resistant to *S. typhi* infection (O'Brien, 1982)). These differences along with the fact that type 1 fimbriae are apparently the only adhesin present in *S. typhi* and have been detected in most clinical isolates (Duguid et al., 1966; Halula and Stocker, 1987; Satta et al., in preparation), prompted us to evaluate the possible role played by these structures in *S. typhi* pathogenicity.

Since no suitable in vivo animal models are as yet available to study *S. typhi* pathogenicity determinants, studies in this field must largely rely on in vitro experimental systems.

The results obtained in this work, using an in vitro experimental system to study the interaction of fimbriated and nonfimbriated *S. typhi* strains with human cells of different types, suggest that type 1 fimbriae might play a role in *S. typhi* infection at the point of invasion of the intestinal epithelium. The role of fimbriae in the invasion process is apparently related to

Table 4. Effect of purified fimbriae on invasion of the epithelial cell line INT-407 by fimbriated and nonfimbriated *S. typhi* strains.

strain	relevant phenotype	No of bacteria/cell[d]	
			+fimbriae[e]
Sty4	Fim+	18	23
27ty2	Fim++[b]	27	34
UVP6[a]	Fim++	24	31
34ty3	Fim-[c]	<1	5
6S[a]	Fim-	<1	4

[a,b,c] See corresponding footnotes to table 1.
[d] See corresponding footnote to table 3. It should be noted that in this experiment the interaction of bacteria with epithelial cells was facilitated by centrifugation of the bacteria on to the cell monolayers).
[e] Final concentration: 10 μg/ml of purified fimbriae.

their ability to mediate a close and stable interaction between the two cell surfaces, which in turn could be necessary for the invasion process to occur.

Based on past research and the results presented here, we believe that the role of type 1 fimbriae in *S. typhi* pathogenicity warrants further study. Furthermore, it would be intersting to evaluate whether specific anti-fimbrial immunity can provide protection against *S. typhi* infection in humans.

EXPERIMENTAL PROCEDURES

Bacterial Strains

All *S. typhi* strains used in this study were clinical isolates from patients suffering from typhoid fever. The fimbriated strain Sty4 was selected both for fimbrial purification and as parent strain for selection of spontaneous nonfimbriated and hyperfimbriated mutants. The presence of type 1 fimbriae was evaluated by assaying the ability of bacteria to cause mannose sensitive hemagglutination of guinea pig erythrocytes (Duguid et al., 1966), and by electron microscopy (Korhonen et al., 1980).

Adhesion and Invasion Assays

Adhesion of *S. typhi* strains to eukaryotic cells was assayed as follows. Bacterial cells for the adhesion assay were grown aerobically without agitation at 37°C in Brain Heart Infusion broth (BHI). Stationary phase bacteria were washed, resuspended in Hank's Balanced Salts Solution (HBSS) and susequently applied to eukariotic cell monolayers for 1 hour at 37°C in a 5% CO_2 atmosphere (bacterial cells to eukariotic cells = 200:1). After repeated washing with HBSS, the monolayers were stained with Giemsa stain to score the number of adherent bacteria to each eukaryotic cell by microscopical examination.

Invasion experiments were performed in a manner similar to adhesion experiments, with the following modifications: a) Dulbecco's modified Eagle medium (DMEM) supplemented with 2% fetal calf serum was used instead of HBSS as an incubation medium; b) a time of 2 hours was allowed for interaction of bacteria with epithelial cells, after which the cell monolayers were repeatedly washed with DMEM containing α-D-mannose (2% wt/vol) and gentamicin (100 μg/ml) and incubated for an additional 60 minutes in DMEM containing gentamicin (100 μg/ml); c) the evaluation of the number of bacteria internalized in INT-407 cells was performed by scoring the viable cell count following lysis of monolayers with 1% Triton-X-100 in phosphate buffered saline pH 7.2 and plating appropriate dilutions on a solid medium. The invasion experiments with addition of purified fimbriae were performed both as described and by facilitating the interaction of bacteria with epithelial cells by centrifugation of the bacteria on to the cell monolayers (300 x*g* for the 2 hours allowed for bacteria/cell interaction).

The intestinal epithelial cell line INT-407 (Flow Laboratories, Inc., McLean, Va.) was used for all adhesion and invasion experiments. MDCK and HeLa cells (Flow Laboratories) were also used for some adhesion experiments. The occurrence and degree of fimbriation of bacterial cells to be used for adhesion and invasion experiments were always tested.

Purification of Fimbriae and Preparation of the Polyclonal Antiserum

Type 1 fimbriae were purified essentially according to the protocol described by Korhonen et al. (1985), with the following modifications: a) detachment of fimbriae was obtained by three sequential 20-minute homogenisation steps (4 cycles of 5 minutes separated by 2 minute intervals for cooling); b) deoxycholate treatment was prolonged for 36 hours; c) the ultracentrifugation through a sucrose gradient was performed at 22,000 rpm; d) after the urea treatment the depolimerized flagella were removed by an ultrafiltration step instead of using gel filtration.

The anti-fimbrial polyclonal antiserum was raised in a rabbit by subcutaneous immunization with the purified fimbrial preparation (five 50 µg-doses of antigen administred in incomplete Freund's adjuvant in 15 day-intervals). The 6S nonfimbriated spontaneous mutant of *S. typhi* Sty4 was used for adsorption of the immune serum.

ACKNOWLEDGMENTS

This work was supported in part by grant No. 90.00098.PF70 and No 92.01195.PF70 from the Italian National Research Council (C.N.R.) - Targeted Project "Biotecnologie e Biostrumentazione".

REFERENCES

Calva, E., Puente, J.L., and Calva, J.J., 1988, Research opportunities in typhoid fever: epidemiology and molecular biology, *BioEssays* 9:173.

Clegg, S., and Gerlach, G.F., 1987, Enterobacterial fimbriae, *J. Bacteriol.* 169:934.

Darekar, M.R., and Eyer, H., 1973, The role of fimbriae in the processes of infection. Preliminary report. *Zentralbl. Bakteriol. Mikrobiol. Hyg. Ser. A* 225:130.

Dorman, C.J., Chatfield, S., Higgins, C.F., Hayward, C., Duguid J.P., and Old, D.C., 1976, Adhesive properties of *Enterobacteriaceae*, *in* "Bacterial adherence", E.H Beachey, ed., Chapman and Hall, London.

Duguid, J.P., Anderson, E.S., and Campbell, I., 1966, Fimbriae and adhesive properties in salmonellae, *J. Pathol. Bacteriol.* 92:107.

Duguid, J.P., Darekar, M.R., and Wheater, W.F., 1976, Fimbriae and infectivity in *Salmonella typhimurium*, *J. Med. Microbiol.* 9:459.

Edelman, R., and Levine, M.M., 1986, Summary of an international workshop on typhoid fever, *Rev. Infect. Dis.* 8:329.

Finlay, B.B., and Falkow, S., 1988, Virulence factors associated with *Salmonella* species, *Microbiol. Sci.* 5:324.

Finlay, B.B., and Falkow, S., 1989, Common themes in microbial pathogenicity, *Microbiol. Rev.* 53:210.

Halula, M.C., and Stocker, B.A.D., 1987, Distribution of the mannose-resistant hemagglutinin produced by *Salmonella* species, *Microbial Pathogenesis* 3:455.

Jones, G.W., and Richardson, L., 1981, The attachment to and invasion of HeLa cells by *Salmonella typhimurium*: the contribution of mannose-sensitive and mannose resistant hemagglutinating activities, *J. Gen. Microbiol.* 127:361.

Korhonen, T.K., Nurmiaho, E-L., Ranta, E., and Svanborg Edén, C., 1980, New method for isolation of immunologically pure pili from *Escherichia coli*, *Infec. Immun.* 27:569.

Korhonen, T.K., Rhen, M., Vaisanen-Rhen, V., Pere, A., and Nowicki, B., 1985, Purification and fractionation of enterobacterial fimbriae and fimbriate cells, *in* "Enterobacterial surface antigens: methods for molecular characterization", T.K. Korhonen, E.A. Dawers, and P.H. Makela, eds., Elsevier Science Publishers, Amsterdam.

Levine, M.M., 1988, Typhoid fever vaccines, *in* "Vaccines", S.A. Plotkin and E.A. Mortimer, eds., W.B. Saunders, Philadephia.

Lockman, H.A., and Curtiss III, R., 1992, Virulence of non-type 1-fimbriated and nonfimbriated nonflagellated *Salmonella typhimurium* mutants in murine typhoid fever, *Infect. Immun.* 60:491.

O'Brien, A., 1982, Innate resistance of mice to *Salmonella typhi* infection, *Infec. Immun.* 38:948.
Orndorff, P.E., and Bloch, C.A., 1990, The role of type 1 pili in the pathogenesis of *Escherichia coli* infections: a short review and some new ideas, *Microbial Pathogenesis* 9:75.
Paranchych, W., and Frost, L.S., 1988, The physiology and biochemistry of pili, *Adv. Microb. Physiol.* 29:53.
Reid, G., and Sobel, J.D., 1987, Bacterial Adherence in the pathogenesis of urinary tract infection: a review, *Rev. Infect. Dis.* 9:470.
Tanaka, Y., and Katsube, Y., 1978, Infectivity of *Salmonella typhimurium* for mice in relation to fimbriae, *Jpn. J. Vet. Sci.* 40:671.
Tavendale, A., Jardine, C.K.H., Old, D.C., and Duguid, J.P., 1983, Haemagglutinins and adhesion of *Salmonella typhimurium* to Hep2 and HeLa cells. *J. Med. Microbiol.* 16:371.

THE MALTOSE B REGION IN *SALMONELLA TYPHIMURIUM*, *ESCHERICHIA COLI* AND OTHER *ENTEROBACTERIACEAE*

Elie Dassa[1], Eric Francoz[1], Michael Dahl[1], Erwin Schneider[2], Catherine Werts[1], Alain Charbit[1], Sophie Bachellier[1], William Saurin[1], and Maurice Hofnung[1]

[1] Unité de Programmation Moléculaire & Toxicologie Génétique,
CNRS Ura 1444, Institut Pasteur
75015 Paris (France)
[2] Sonderforschungbereich 171, Fachbereich Biologie
Universität Osnabrück, Osnabrück (Germany)

INTRODUCTION

The envelope of Gram negative bacteria plays a critical role in the primary interactions between bacteria and their environment. Such interactions include the sensing of the presence of substrates (chemotaxis) and their subsequent translocation into the cell (active transport), the relations with other bacteria (mating) and with other prokaryotic organisms such as bacteriophages, and the relations with the host (bacteria-cell interactions, recognition by the host immune system). A large number of these interactions are mediated by envelope proteins. As a model system to study such kinds of interactions, we use the maltose B region which determines the maltose and maltodextrin transport system of *Escherichia coli.* This system is constituted of several proteins belonging to the three layers of the Gram negative bacterial envelope.

The five proteins required for maltose and maltodextrin transport in *E. coli* are encoded by genes located in two divergent operons in the maltose B region at 91 min. on the chromosome (Hofnung, 1974). One operon contains the *malE, malF* and *malG* genes, the other the *malK* and *lamB* genes. This operon also includes a distal gene, *malM*, whose function is not known. Both operons are under direct positive control by the *malT* and *crp* (cyclic AMP receptor protein) gene products (see Saurin *et al.*, 1989; Schwartz, 1987 for review).

Maltose (at micromolar concentrations) and maltodextrins enter the periplasmic space by facilitated diffusion through the specific outer membrane maltoporin LamB, which is also the receptor for phage λ and for a set of other bacteriophages (Charbit and Hofnung, 1985). They are complexed by a periplasmic binding protein MalE (MBP) which is essential for maltose and maltodextrin transport and is also the primary receptor for the chemotaxis towards these substrates. The liganded substrates are then delivered to a hetero-oligomeric complex in the inner membrane constituted by the MalF, MalG and MalK proteins, which achieves their translocation into the cytoplasm by a still unknown mechanism (see Shuman, 1987 for review). MalF and MalG are very hydrophobic integral membrane proteins believed to form a channel. MalK is a hydrophilic peripheral membrane protein located on the cytoplasmic side of the inner membrane. Convincing evidence has been accumulated to support the view that the energy required for transport is provided by the hydrolysis of ATP (Davidson and Nikaido, 1990; Dean *et al.*, 1989). The MalK subunit, which displays the A and B sequence motifs found in ATP-binding proteins and which possesses an ATP binding site (Higgins *et al.*, 1985), functions very likely as an ATPase. The purified MalK protein of *S. typhimurium* present indeen an ATP hydrolyzing activity *in vitro* (Walter *et al.*, 1992). Recent works also suggested that MalF and MalG may transduce a signal from maltose loaded MBP to MalK, so that ATP hydrolysis can occur (Davidson *et al.*, 1992).

Protein-protein interactions seem to be crucial for the function of this complex transport system. For transport, genetic evidence indicated that MalE interacts on one hand with LamB and on the other hand with MalF and MalG. MalK is anchored in the membrane through interaction with MalG. It is very likely that MalF and MalG interact together to form a heterodimer in the membrane. The existence of such interactions has been recently demonstrated biochemically by cross-linking experiments on a proteoliposome-reconstituted transport system (Davidson and Nikaido, 1991). For chemotaxis, MalE interacts with protein Tar, another transmembrane protein, whose gene is not located in the maltose B region, and which transduces a signal to the flagellar apparatus of the bacterium.

The maltose transport system belongs to the family of periplasmic permeases (see Ames, 1988 for review), whose representatives have been found in Gram negative bacteria and, as a paradox in Gram positive bacteria. In these bacteria, which are devoid of periplasmic space, the protein homologous to the periplasmic substrate binding protein is a surface lipoprotein anchored in the membrane (Gilson *et al.*, 1988). Whereas no significant homologies exist between unrelated substrate binding proteins on one hand (Duplay *et al.*, 1984), and only limited sequence similarities have been recorded between the hydrophobic integral inner membrane proteins on the other hand (Dassa and Hofnung, 1985), the membrane peripheral subunits share a highly conserved domain of about 200 amino acid residues which extends far over the ATP-binding motifs (Ames, 1986).

The periplasmic permeases are themselves members of a growing superfamily of transporters, displaying strong similarities in their global organization and in the primary sequence of a protein or a protein domain which is highly conserved and similar to the MalK protein, the "ABC transporters" (Higgins *et al.*, 1990) or "traffic ATPases" (Ames *et al.*, 1990), acting in all the living kingdom, from bacteria to man. The most prominent eukaryotic representatives are the mdr protein (Gros *et al.*, 1986), the CFTR protein (Riordan *et al.*, 1989), and potential peptide transporters involved in the presentation of antigens by the Class I major histocompatibility complex (Bahram *et al.*, 1991 ; Deverson *et al.*, 1990).

Our laboratory is involved in analyzing the functional organization of the periplasmic transport system for maltose, essentially by genetic methods. More recently, we undertook another approach, namely by the comparison at the DNA sequence level of the maltose B regions from different enterobacterial species. This work was initiated to identify regions of functional importance in the proteins of the maltose uptake system and also in the elements that control the expression of those proteins. The rationale was that DNA and protein sequences of greater functional importance should be more highly conserved during evolution. In this paper, we will review the main contributions of this study to the knowledge of the function of the maltose transport system.

RESULTS AND DISCUSSION

We sequenced the complete maltose B region of *Salmonella typhimurium*, the *malE*, *malF*, *malG* and *malK* genes of *Enterobacter aerogenes*, and the *lamB* gene of *Klebsiella pneumoniae* and compared those sequences to those of *E. coli*.

I. The inter operonic region : A new conserved potential regulatory sequence is a previously undetected MalT binding site

When this work was initiated, the DNA sequence of the *E. coli* K12 maltose operon regulatory regions was known to include characteristic hexanucleotides (GGAT/GGA) repeated several times upstream from the transcription initiation sites, the closest of the transcriptional start being at position -35 (Figure 1). Genetic data strongly suggested that these sequences, named MalT boxes, were part of MalT binding sites but direct evidence that MalT binds to these boxes was lacking (see Bedouelle, 1982 and references therein). The comparisons of the regulatory sequences in three enterobacterial species provided additional information on the structure of the putative binding sites.

Four putative MalT-binding sites were strictly conserved but the conserved sequences extended the MalT boxes by a few nucleotides upstream and downstream. The distances between conserved regions were constant but highly variable in sequence. This suggests that selection operates for the maintenance of the exact sequences of the binding sites and of the distances, but not of the nucleotide sequences between them.

A new conserved stretch for which no function had been assigned on the basis of the *E. coli* sequence analysis : GGNGGGGCGTAG was found. We called it the U box (for

Unknown function). A search in Genbank revealed a single occurrence of this U box in the regulatory region of the divergent MalT-regulated operons *pulAB-pulCO* of *K. pneumoniae* (Chapon and Raibaud, 1985). The GGNGGGGCGTAG sequence is strictly conserved in the four species and is followed 6 bp downstream by a MalT box.

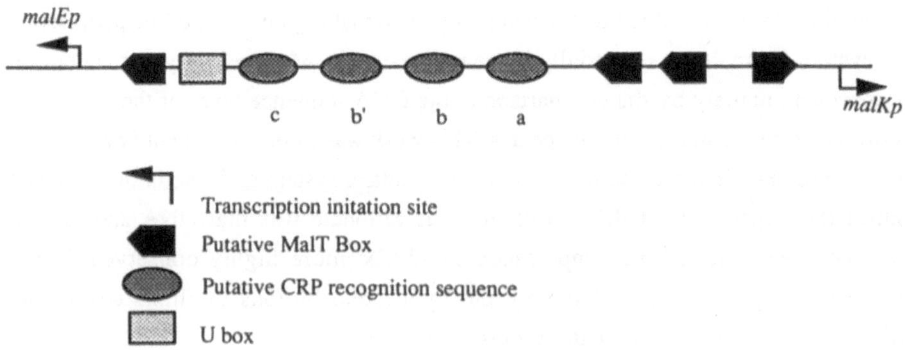

Figure 1. Structure of the maltose B regulatory region showing the distribution of the putative binding sites for regulatory proteins. *malEp* : promoter of the *malEFG operon*; *malKp* : promoter of the *malK, LamB, malM operon.*

Recently, Raibaud and collaborators revised substantially the consensus structure of the MalT-dependent promoters. They found, by DNAse I foot printing experiments, that *E. coli* MalT boxes were larger in size than that assumed previously : GGGGAT/GGAGG, as suggested by the conservation of sequences around the putative MalT boxes. Furthermore, they demonstrated that the U-box was in fact a MalT binding site, and that single base changes in this region abolish the activity of the *malEFG* promoter, and that the activation of a all MalT-dependent promoters needs two MalT binding sites in direct repeat (Vidal-Ingigliardi *et al.*, 1991).

Three out of the four putative binding sites for CRP (CRPb, b' and c) were conserved in the regulatory region. The putative CRPa site was not found in either *S. typhimurium* or *E. aerogenes*. Recently, it has been demonstrated that the CRPa site displays very low affinity towards CRP and is dispensable for the maltose B promoters expression under all growth conditions (Vidal-Ingigliardi and Raibaud, 1991).

The conservation of regulatory elements is consistent with the observation that the control of the maltose B operons appears similar in all three species.

II. Palindromic units

Palindromic units (PU) or REP have been detected for the first time in *Enterobacteriaceae* by comparing the nucleotide sequences of the maltose B region in *E. coli* and the *hisJMQP* operon involved in the transport of histidine in *S. typhimurium* (Higgins *et al.*, 1982). They occur in about 25% of all transcription units in *E. coli* and probably in *S. typhimurium*. Their palindromic nature may allow them to form stem and loop structures. Their function is not known but it has been demonstrated that they are specific binding sites *in vitro* for nucleoid-associated proteins such as DNA polymerase (Gilson *et al.*, 1990) and DNA gyrase (Yang and Ames, 1990). It have also been shown that REP are involved in some recombination events (Shyamala *et al.*, 1990). Our current hypothesis is that they might play a role in nucleoid structure. It appears now that PU belong to a larger repeated DNA element, called BIME for Bacterial Interspersed Mosaic Element, which is a mosaic combination of ten small DNA motifs, including the PU sequence (Gilson *et al.*, 1991).

In the maltose B region, the localization of PU's is conserved between *E. coli.* and *S. typhimurium*, but also in *E. aerogenes*, the latter constituting the first evidence of the existence of PU's in this species. However their number is not conserved. PU consensus sequences for *E. coli* and *S. typhimurium* are similar except for an additional G in *S. typhimurium* (Gilson *et al.*, 1987). This additional base is also observed in *E. aerogenes*, indicating that the PU sequences of *E. aerogenes* may be more closely related to those of *S. typhimurium* than those of *E. coli.*

III. Comparisons at the protein sequence level provide functional informations

1) The high conservation of the *malM* gene suggests a selected physiological role for its product. The *malM* gene encodes a periplasmic protein of 306 amino acids. This protein is abundantly expressed in *E. coli* after induction by maltose. Its physiological role is not known. It is very likely not involved in maltose utilization or chemotaxis (Brass and Manson, 1984 ; Gilson *et al.*, 1986)

The MalM protein is as highly conserved as other maltose B proteins with a known function. Furthermore, by sequencing the DNA 3' to the *lamB* gene of *K. pneumoniae*, an orf displaying strong homologies with *malM* was found (Bachellier, unpublished observations). Therefore, these results strongly suggest that *malM* encodes a protein with a selected function and that *malM* is a general constituent of maltose B regions of *Enterobacteriaceae* (Schneider *et al.*, 1992).

2) Protein MalK. The MalK protein is essential for maltose transport. Furthermore, mutants lacking MalK express constitutively the remaining *mal* genes in the absence of external inducer. Hence, the MalK protein regulates the maltose regulon, most likely by modifying an internal inducer that activates MalT. It is also involved in the $EIII^{Glc}$-mediated inducer exclusion of maltose by glucose.

ATP binding motif A is perfectly conserved between the three proteins, while motif B presents some variable residues The C-terminus region is highly conserved though absent from ATP-binding protein sequences from other periplasmic transport systems (Dahl *et al.*, 1989). It has been recently demonstrated that this region is involved in the regulatory properties of the protein (Schneider and Walter, 1991). Residues involved in inducer exclusion (Kühnau *et al.*, 1991) are perfectly conserved.

3) The MalF and MalG proteins : variable N-terminus versus conserved C-terminus. The two integral inner membrane proteins MalF and MalG are the less characterized components of the maltose transport system since there are poorly expressed, very hydrophobic and hard to purify. In the MalF protein most of the changes are clustered in a single region extending from position 93 to 271 (Dahl *et al.*, 1989; Schneider *et al.*, 1992) suggesting that a strict conservation of the primary sequence of this domain is not essential for function. This region is described as a single periplasmic domain (Froshauer *et al.*, 1988). By contrast, only one change is found within the 140 last amino acids.

Protein MalG also displayed a clustering of changes (20 % of modified residues) in a shorter hydrophilic region (residues 40 to 75) located in the N-terminal third of the protein (Francoz *et al.*, 1990b). This region is facing the periplasmic space, as suggested by the determination of the protein topology thanks to fusions with alkaline phosphatase (Dassa *et al.*, in preparation). A random linker insertion mutagenesis of the *malG* gene indicated that residues 28 to 52 were not strictly required for the function of the protein (Dassa, unpublished results). The C-terminal third is highly conserved. Proteins MalF and MalG display, at a distance of 90 amino acids from the C-terminus, a hydrophilic segment conserved in all integral inner membrane proteins from periplasmic transport systems (Dassa and Hofnung, 1985). Its disruption by linker insertion leads to a complete loss of MalG functions (Dassa, 1990).

4) The MalE protein or MBP : variable residues are located at the protein surface. Missense mutations in *E. coli* that specifically interfere with chemotaxis towards maltose (Kossman *et al.*, 1988) or with maltose transport (Martineau *et al.*, 1990 ; Treptow and Shuman, 1988) have been localized by DNA sequencing. It is noteworthy that all of them affect residues that are absolutely conserved in all three species. This suggests that compensatory double mutations have not occured during evolution. This may also suggest that the maltose transport system is essential for the survival of these species in their natural habitat. A region important for the interactions with membrane proteins MalF and/or MalG, identified by random linker insertion mutagenesis (Duplay *et al.*, 1987), is strictly conserved between the three proteins.

The X-ray structure of MBP has been determined at a resolution of 2.3 Å (Spurlino *et al.*, 1991). As other periplasmic substrate binding proteins, it is constituted by two lobes connected by three flexible polypeptide segments. The substrate binding site is located in a cleft between the two lobes. Non conserved residues are evenly scattered on the surface of the protein, by contrast to conserved functionally important residues which are clustered on the cleft face of the protein.

5) The LamB protein : extensive polymorphism in surface exposed antigenic determinants and variations in bacteriophage susceptibility. Maltoporin consists of a homotrimer of the *lamB* gene product, forming an outer membrane pore specific for the entry of maltose and maltodextrins into the periplasm. In the absence of a three dimensional model, a detailed 2D representation of the protein has been developed, based on the identification of residues involved in phage binding (Charbit *et al.*, 1988), in monoclonal (Desaymard *et al.*, 1986) and antipeptide (Molla *et al.*, 1989) antibody binding, on proteolytic cleavage studies (Ronco *et al.*, 1990; Schenkman *et al.*, 1984), and on foreign epitope insertion (Charbit *et al.*, 1986 ; Charbit *et al.*, 1991).

The sequences of the *lamB* gene of *S. typhimurium* and of *Klebsiella pneumoniae* have been determined in our laboratory (Francoz *et al.*, 1990a; Werts *et al.*, 1992). The LamB protein from *S. typhimurium* can substitute for the *E. coli* LamB protein for maltodextrin binding and transport (Francoz *et al.*, 1990a), and displays the same pore properties in vitro (Schülein and Benz, 1990). The *Klebsiella* protein forms stable trimers and displays a maltoporin function identical to that of *E. coli* (Werts *et al.*, 1992). The first third of LamB is the most conserved between the three proteins with 4-9 % changes. Hence, the conservation of the binding and transport activities can be accounted for by the conservation of the regions known to be directly involved, namely the first third of the protein and a region corresponding to residues 352 to 374 of *E. coli* LamB.

However, none of the known *E. coli* phages using LamB for adsorption were able to infect *S. typhimurium* and *K. pneumoniae*. The last two third of the LamB from *S. typhimurium* contains five variable segments where more than 50% of the residues are changed with respect to LamB from *E. coli*. They alternate with conserved predicted trans-membraneous segments. On the basis of genetic and immunological data, four of these variable regions were predicted to be cell surface exposed loops while one of them was predicted periplasmic (Figure 2). The phage resistance thus can be attributed to the variability of the four cell surface exposed loops, previously identified as essential for phage adsorption (Charbit *et al.*, 1988). Consistent with these results, four monoclonal antibodies directed towards cell surface epitopes located near the C-terminus of the protein did not recognize LamB from *S. typhimurium*. These antibodies have been found highly polymorphic among a number of *Enterobacteriaceae* (Bloch and Desaymard, 1985).

The *Klebsiella* LamB protein displays six variable peptide segments with respect to the *E. coli* protein, five of which are located in the same extra-membraneous loops exposed at the cell surface on the 2D folding model as for LamB from *S. typhimurium*. The additional region of variability encompasses a loop predicted periplasmic and remarkably, a potential transmembrane segment. Such variability suggests that this segment, which would not be essential for maltoporin function, might have a peculiar position, possibly as being

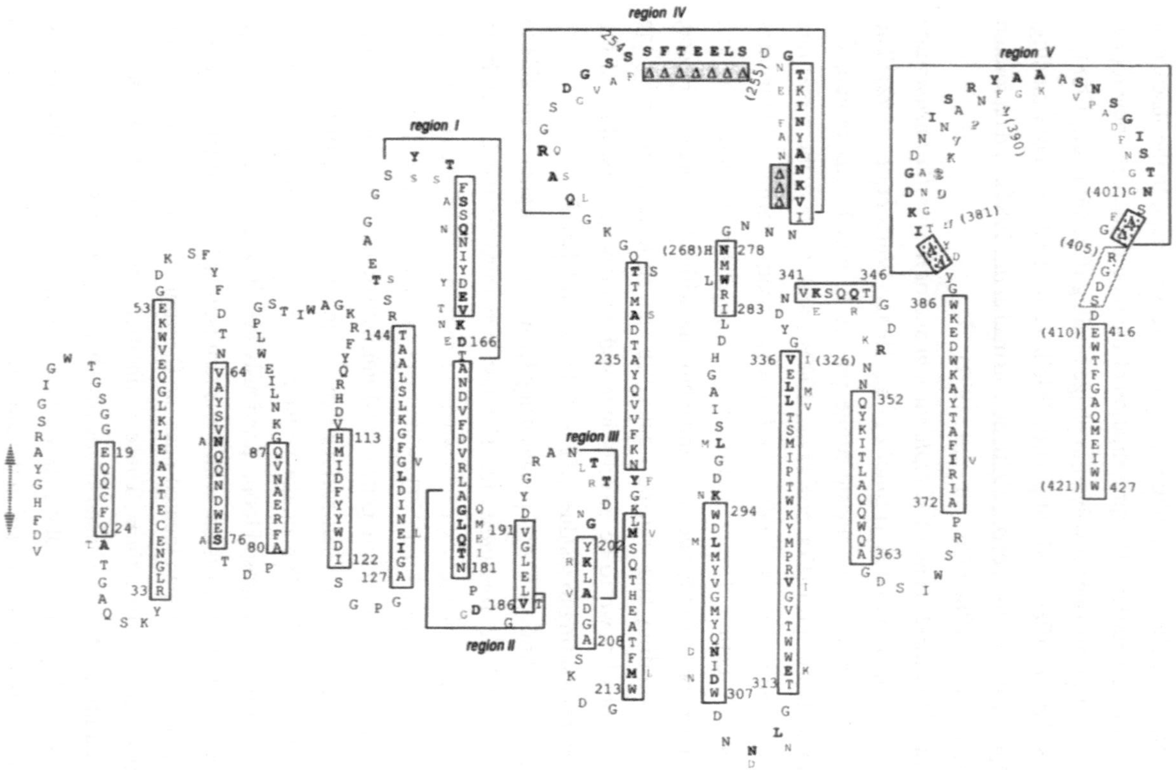

Figure 2. 2D folding model of mature LamB from *Salmonella typhimurium* (S.t.) and sequence variations with respect to *Escherichia coli* (E.c.) and *Shigella sonnei* (S.s.) Regions predicted in beta-structure are included in white rectangles. The width of the apolar part of the membrane is indicated by an arrow.

A T M sequence from S.t. when identical to E.c.

A T M sequence from S.t. when different from E.c.

307 residue number in mature LamB S.t..

A T M sequence from E.c. when different from S.t.

A T M sequence from S.s. when different from E.c.

(326) residue number in mature LamB E.c.or S.s.

[shaded box] residues deleted in S.t.

[white box] residues deleted in E.c.

[dotted box] RGDS sequence

located on the surface of LamB but inside the channel. All these results strongly support the folding model for *E. coli* LamB.

Two hypotheses could be made to explain the origin of local variable regions. The first is that it is the result of selection of mutations leading to resistance to several noxious agents such as bacteriophages or complement. Another cause may be structural, due to the fact that exposed loops of the protein may tolerate important modifications without consequences on protein stability, localization, folding and on maltose transport. This structural and functional tolerance could allow a passive accumulation of mutations in the loops. This explanation is comforted by the fact that the degree of variations of DNA sequences in non-coding segments of the maltose B region can be very high : it reaches 100% in non-conserved segments of the regulatory interval (Dahl *et al.*, 1989). Thus, it seems that the degree of genetic changes between *S. typhimurium*, *K. pneumoniae* and *E. coli* would be enough to explain the differences in the variable regions on the basis of structural and functional tolerance. (Francoz *et al.*, 1990a; Werts *et al.*, in press).

The comparison of LamB protein sequences among four different *Enterobacteriaceae* (*E. coli, S. typhimurium, Shigella sonnei* and *K. pneumoniae*) led to the finding of a RGD conserved motif in the last external loop of LamB. This motif was originally found in fibronectin to be responsible for its ability to aggregate mammalian cells. This observation, in addition to preliminary evidence that LamB is able to bind mammalian cells, suggests a possible functional role for this region of LamB in the adhesion between bacteria and cells of their mammalian hosts (O' Callaghan, in preparation) .

CONCLUSIONS AND PERSPECTIVES

The comparisons of the maltose B DNA sequences among different *Enterobacteriacea* showed that the general genetic organization of this region and the binding sites for the regulatory proteins were conserved. The comparisons of the deduced aminoacid sequences of the proteins showed more than 90% identity. However this study allowed us to identify highly variable regions in some proteins, and particularly in LamB, generally located in predicted surface loops. This is consistent with the fact that "permissive" regions were found in those proteins that can accept important local modifications without major deleterious consequences for their functions and therefore structure (Hofnung *et al.*, 1988). This notion has been exploited in the laboratory to devise "presentation vectors" : permissive sites in LamB or MalE have been used to insert genetically a broad variety of foreign epitopes (Charbit *et al.*, 1988; Martineau *et al.*, 1992). One of the major advantage of these expression vectors is that they can be expressed in other enteric bacteria than *E. coli*. In particular, LamB and MalE hybrids could be expressed in attenuated *aroA* *S. typhimurium* strains, opening the possiblity of developing new generations of live oral vaccines (O'Callaghan *et al.*, 1990).

ACKNOWLEDGEMENTS

This work was supported by grants from the Association pour le Développement de la Recherche sur le Cancer, the Ligue Nationale contre le Cancer, the Fondation pour la Recherche Médicale and World health Organization (Transdisease Vaccinology Programme). We would like to express our gratitude to David O'Callaghan for reviewing the manuscript.

REFERENCES

Ames, G.F.-L., Mimura, C., and Shyamala, V., 1990, Bacterial periplasmic permeases belongs to a family of transport proteins operating from *Escherichia coli* to human traffic ATPases, *FEMS Microbiol. Reviews.* 75:429.

Ames, G.F.L., 1986, Bacterial periplasmic transport systems : structure, mechanism and evolution, *Ann. Rev. Biochem.* 55:397.

Ames, G.F.L., 1988, Structure and mechanism of bacterial periplasmic permeases, *J. Bioenerg. Biomemb.* 20:1.

Bahram, S., Arnold, D., Bresnahan, M., Strominger, J.L. and Spies, T., 1991, Two putative subunits of a peptide pump encoded in the human major histocompatibility complex class-II region, *Proc. Natl. Acad. Sci. USA.* 88:10094.

Bedouelle, H., Schmeissner, U., Hofnung, M., and Rosenberg, M., 1982, Promoters of the *malEFG* and *MalK-lamB* operons in *Escherichia coli* K12, *J. Mol. Biol.* 161:519.

Bloch, M.A., and Desaymard, C., 1985, Antigenic polymorphism of the LamB protein among members of the family oEnterobacteriacae, *J. Bacteriol.* 163:106.

Brass, J.M. and Manson, M.D., 1984, Reconstitution of maltose chemotaxis in *Escherichia coli* by addition of maltose binding protein to calcium-treated cells of maltose regulon mutants., *J. Bacteriol.* 157:881.

Chapon, C. and Raibaud, O., 1985, Stucture of two divergent promoters located in front of the gene encoding pullulanase in Klebsiella pneumoniea and positively regulated by the *MalT* product, *J. Bacteriol.* 164:639.

Charbit, A., and Hofnung, M., 1985, Isolation of different bacteriophages using the LamB protein for adsorption on *E. coli* K12, *J. Virol.* 53:667.

Charbit, A., Boulain, J.C., Ryter, A. and Hofnung, M., 1986, Probing the topology of a bacterial membrane protein by genetic insertion of a foreign epitope. Expression at the cell surface, *EMBO J.* 5:3029.

Charbit, A., Gehring, K., Nikaido, H., Ferenci, T. and Hofnung, M., 1988, Maltose transport and starch binding in phage-resistant point mutants of Maltoporin. Functional and topological implications, *J. Mol. Biol.* 201:487.

Charbit, A., Molla, A., Saurin, W., and Hofnung, M., 1988, Versatility of a vector for expressing foreign polypeptides at the surface of Gram$^-$ bacteria, *Gene* 70:181.

Charbit, A., Ronco, J., Michel, V., Werts, C. and Hofnung, M., 1991, Permissive sites and the topology of an outer-membrane protein with a reporter epitope, *J. Bacteriol.* 173:262.

Dahl, M.K., Francoz, E., Saurin, W., Boos, W., Manson, M.D. and Hofnung, M., 1989, Comparison of sequences from the *malB* regions of *Salmonella typhimurium* and *Enterobacter aerogenes* with *Escherichia coli* K12 : A potential new regulatory site in the interoperonic region, *Mol. Gen. Genet.* 199.

Dassa, E., 1990, Cellular localization of the MalG protein from the maltose transport system in *Escherichia coli, Mol. Gen. Genet.* 222:32.

Dassa, E. and Hofnung, M., 1985, Sequence of *malG* gene in *E. coli* K12 : homologies between integral membrane components from binding protein-dependent transport systems, *EMBO J.* 4:2287.

Davidson, A.L. and Nikaido, H., 1990, Overproduction, Solubilization and reconstitution of the maltose transport system from *Escherichia coli, J. Biol. Chem.* 265:4254.

Davidson, A.L. and Nikaido, H., 1991, Purification and characterization of the membrane-associated components of the maltose transport system from *Escherichia coli, J. Biol. Chem.* 266:8946.

Davidson, A.L., Shuman, H.A. and Nikaido, H., 1992, Mechanism of Maltose Transport in *Escherichia coli* - Transmembrane Signaling by Periplasmic Binding Proteins, *Proc Natl Acad Sci USA.* 89:2360.

Dean, D.A., Davidson, A.L. and Nikaido, H., 1989, Maltose transport in membrane vesicles of *Escherichia coli* is linked to ATP hydrolysis, *Proc. Natl. Acad. Sci. USA.* 86:9134.

Desaymard, C., Débarbouillé, M., Jolit, M. and Schwartz, M., 1986, Mutations affecting antigenic determinants on an outer membrane protein of *Escherichia coli, EMBO J.* 5:1383.

Deverson, E.V., Gow, I.R., Coadwell, W.J., Monaco, J.J., Butcher, G.W. and Howard, J.C., 1990, MHC class II region encoding proteins related to the multidrug resistance familly of transmembrane transporters, *Nature.* 348:738.

Duplay, P., Bedouelle, H., Fowler, A., Zabin, I., Saurin, W. and Hofnung, M., 1984, Sequences of the *malE* gene and of its product, the maltose-binding protein of *Escherichia coli* K12, *J. Biol. Chem.* 259:10606.

Duplay, P., Szmelcman, S., Bedouelle, H. and Hofnung, M., 1987, Silent and functional changes in the periplasmic maltose-binding protein of *Escherichia coli* K12. I. Transport of maltose, *J. Mol. Biol.* 194:663.

Francoz, E., Molla, A., Dassa, E., Saurin, W. and Hofnung, M., 1990a, The Maltoporin of *Salmonella typhimurium* : Sequence and Folding Model, *Res. Microbiol.* 141:1039.

Francoz, E., Schneider, E. and Dassa, E., 1990b, The sequence of the *malG* gene from *Salmonella typhimurium* and its functional implications, *Res Microbiol.* 141:633.

Froshauer, S., Green, G.N., Boyd, D., McGovern, K. and Beckwith, J., 1988, Genetic analysis of the membrane insertion and topology of MalF, a cytoplasmic membrane protein of *E. coli, J. Mol. Biol.* 200:501.

Gilson, E., Alloing, G., Schmidt, T., Claverys, J.P., Dudler, R. and Hofnung, M., 1988, Evidence for high affinity binding protein-dependent transport systems in Gram-positive bacteria and in *Mycoplasma, EMBO J.* 7:3971.

Gilson, E., Clément, J.M., Perrin, D. and Hofnung, M., 1987, Palindromic units : a case of highly repetitive DNA sequences in bacteria, *Trends Genet.* 3:226.

Gilson, E., Perrin, D. and Hofnung, M., 1990, DNA polymerase I and a protein complex bind specifically to *E. coli* palindromic unit highly repetitive DNA : Implications for bacterial chromosome organization, *Nucleic Acids Res.* 18:3941.

Gilson, E., Rousset, J.P., Charbit, A., Perrin, D. and Hofnung, M., 1986, MalM, a new gene of the *malK-lamB* operon in *Escherichia coli* K12. I. *malM* is the last gene of the *malK-lamB* operon and encodes a periplasmic protein, *J. Mol. Biol.* 191:303.

Gilson, E., Saurin, W., Perrin, D., Bachellier, S. and Hofnung, M., 1991, Palindromic units are part of a new bacterial interspersed mosaic element (BIME), *Nucleic Acids Res.* 19:1375.

Gros, P., Croop, J., and Housman, D., 1986, Mammalian multidrug resistance gene : complete cDNA sequence indicates strong homology to bacterial transport proteins, *Cell.* 47:371.

Higgins, C.F., Ames, G.F.L., Barnes, W.M., Clément, J.M., and Hofnung, M., 1982, A novel intercistronic regulatory element of prokaryotic operons, *Nature.* 298:760.

Higgins, C.F., Hiles, I.D., Whalley, K., and Jamieson, D.J., 1985, Nucleotide binding by membrane component of bacterial periplasmic binding protein-dependent transport systems, *EMBO J.* 4:1033.

Higgins, C.F., Hyde, S.C., Mimmack, M.M., Gilaedi, U., Gill, D.R., and Gallagher, M.P., 1990, Binding protein-dependent transport systems, *J. Bioenerg. Biomembr.* 22:571.

Hofnung, M., 1974, Divergent operons and the genetic structure of the maltose B region in *Escherichia coli* K12, *Genetics.* 76:169.

Hofnung, M., Bedouelle, H., Boulain, J.-C., Clément, J.-M., Charbit, A., Duplay, P., Gehring, K., Martineau, P., Saurin, W., and Szmelcman, S., 1988, Genetic approaches to the study and the use of proteins : random point mutations and random linker insertions, *Bull. Inst. Pasteur* 86:95.

Kossman, M., Wolff, C., and Manson, M., 1988, Maltose chemoreceptor of *Escherichia coli* : Interaction of maltose-binding protein and the Tar signal transducer, *J. Bacteriol.* 170:4516.

Kühnau, S., Reyes, M., Sieversten, A., Shuman, H., and Boos, W., 1991, The activities of the *Escherichia coli* MalK protein in maltose transport can be separated by mutations, *J. Bacteriol.* 173:2180.

Martineau, P., Szmelcman, S., Spurlino, J.C., Quiocho, F.A., and Hofnung, M., 1990, Genetic approach to the role of tryptophan residues in the activities and fluorescence of a bacterial periplasmic maltose binding protein, *J. Mol. Biol.* 214:337.

Martineau, P., Guillet, J.G., Leclerc, C., and Hofnung, M., 1992, Expression of heterologous peptides at two permissive sites of the MalE protein : antigenicity and immunogenicity of foreign B and T-cell epitopes, *Gene* 113:35.

Molla, A., Charbit, A., Le Guern, A., Ryter, A., and Hofnung, M., 1989, Antibodies against synthetic peptides and the topology of LamB, an outer membrane protein from *Escherichia coli K12*, *Biochemistry* 28:8234.

O'Callaghan, D., Charbit, A., Martineau, P., Leclerc, C., van der Werf, S., Nauciel, C. and Hofnung, M., 1990, Immunogenicity of foreign peptide epitopes expressed in bacterial envelope proteins, *Res. Microbiol.* 141:745.

Walter, C., Höner zu Bentrup, K. and Schneider, E., 1992, Large scale purification, nucleotide binding properties and ATPase activity of the MalK subunit of *Salmonella typhimurium* maltose trandport complex, *J Biol Chem.* 267:8863.

Ronco, J., Charbit, A. and Hofnung, M., 1990, Creation of targets for proteolytic cleavage in the LamB protein of *E. coli* K12 by genetic insertion of foreign sequence: implication for topological studies, *Biochimie.* 72:183.

Saurin, W., Francoz, E., Martineau, P., Charbit, A., Dassa, E., Duplay, P., Gilson, E., Molla, A., Ronco, G., Szmelcman, S. and Hofnung, M., 1989, Periplasmic binding protein-dependent transport system for maltose and maltodextrins : some recent studies, *FEMS Microbiol. Reviews.* 63:53.

Schenkman, S.A., Tsugita, A., Schwartz, M. and Rosenbusch, J.P., 1984, Topology of phage λ receptor protein, *J. Biol. Chem.* 259: 7570.

Schneider, E., Francoz, E. and Dassa, E., 1992, Completion of the Nucleotide Sequence of the Maltose B Region in *Salmonella typhimurium* - The High Conservation of the *malM* Gene Suggests a Selected Physiological Role for Its Product, *Biochim Biophys Acta.* 1129:223.

Schneider, E. and Walter, C., 1991, A chimeric nucleotide-binding protein, encoded by a *hisP-malK* hybrid gene, is functional in maltose transport in *Salmonella typhimurium, Mol Microbiol.* 5:1375.

Schülein, K. and Benz, R., 1990, LamB (maltoporin) of *Salmonella typhimurium* : isolation, purification and comparison of sugar binding with LamB of *Escherichia coli, Mol. Microbiol.* 4:625.

Schwartz, M., 1987, The maltose regulon, *in*: "*Escherichia coli* and *Salmonella typhimurium* cellular and molecular biology." Neidhardt, F.C. ed. American Society for Microbiology, Washington D.C. vol.2:p. 1482.

Shuman, H.A., 1987, The genetics of active transport in bacteria, *Ann. Rev. Genet.* 21:155.

Shyamala, V., Schneider, E. and Ames, G.F.L., 1990, Tandem chromosomal duplications: Role of REP sequences in the recombination event at the join-point, *EMBO J.* 9:939.

Spurlino, J.C., Lu, G.Y. and Quiocho, F.A., 1991, The 2.3 Å Resolution Structure of the Maltose-Binding or Maltodextrin-Binding Protein, a Primary Receptor of Bacterial Active Transport and Chemotaxis, *J Biol Chem.* 266:5202.

Treptow, N.A. and Shuman, H.A., 1988, Allele-specific *malE* mutations that restore interactions between maltose-binding protein and the inner-membrane components of the maltose transport system, *J. Mol. Biol.* 202:809.

Vidal-Ingigliardi, D. and Raibaud, O., 1991, Three adjacent binding sites for cAMP receptor protein are involved in the activation of the divergent *malEp-malKp* promoters, *Proc. Natl. Acad. Sci. USA.* 88:229.

Vidal-Ingigliardi, D., Richet, E. and Raibaud, O., 1991, Two MalT binding sites in direct repeat. A structural motif involvedin the activation of all the promoters of the maltose regulons in *Escherichia coli* and *Klebsiella pneumoniae, J. Mol. Biol.* 218:323.

Yang, Y. and Ames, G.F.L., 1990, The family of repetitive extragenic palindromic sequences : interactions with DNA gyrase and histonelike protein HU, *in*: "The bacterial chromosome" Drlica, K. and Riley, M. ed. American Society of Microbiology, Washington D.C. 211.

Riordan, J.R., Rommens, J.M., Kerem, B., Alon, N., Rozmahel, R., Grzelczak, Z., Zielinski, J., Lok, S., Plavsic, N., Chou, J.L., Drumm, M.L., Iannuzzi, M.C., Collins, F.S. and Tsui, L.C., 1989, Identification of the cystic fibrosis gene : Cloning and characterization of complementary DNA, *Science.* 245:1066.

Werts, C., Charbit, A., Bachellier, S. and Hofnung, M., 1992, DNA sequence analysis of the *lamB* gene from *Klebsiella pneumoniae* : implications for the topology and the pore functions in maltoporin, *Mol. Gen. Genet.* 233:372.

BIOCHEMICAL AND MOLECULAR CHARACTERIZATION OF SALMONELLA ENTERICA SEROVAR BERTA

John E. Olsen[1], Derek J. Brown[1], Dorte L. Baggesen[1,2] and Magne Bisgaard[1]

[1]Department of Veterinary Microbiology, RVAU
[2]National Veterinary Laboratory
Frederiksberg C., DK-1870, Denmark

INTRODUCTION

Salmonella enterica serovar Berta (*S.* Berta) belongs to *Salmonella* serogroup D_1 and has the antigenic formula O:1,9,12 H:f,g,t:-. It was firt reported in 1938 from pigs in Uruguay (Hormaeche et al. 1938). Since then, little attention has been paid to this serovar outside of America. In 1984, *S.* Berta was introduced to Denmark through imported parent stock chicken (Sørensen et al. 1991). It spread with surprising success through the broiler production system, and quickly became the third most common cause of human salmonellosis in Denmark (Olsen et al. 1992a). An erradication programme was initated, and by 1991, the broiler production system seems to be free from *S.* Berta. Figure 1 shows the annual number of isolations of *S.* Berta from flock of broilers and from human cases of salmonellosis in the period 1984 to 1991.

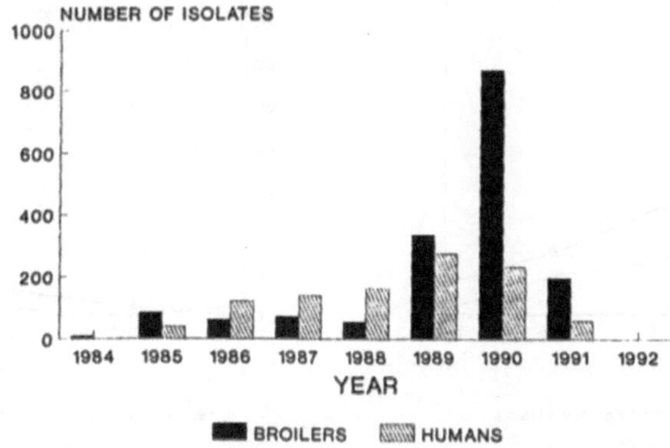

Figure 1. Annual number of isolations of *Salmonella enterica* serovar Berta from infected flocks of broilers and from human cases of salmonellosis in Denmark (data from Olsen et al. 1992a and Gaarslev personal communication).

Biology of Salmonella, Edited by F. Cabello *et al.*,
Plenum Press, New York, 1993

S. berta has very rarely been isolated in Denmark from sources other than broilers and humans (Olsen et al. 1992a) and no reintroduction to the country through infected animals, humans or contaminated foodstuff has been noted (Sørensen et al. 1991). Based on these observation, it has been assumed that broilers are the main, if not only, source for the human infections. Studies, using molecular and biochemical typing methods have been undertaken in order to underline this assumption, and in addition to develope a rational typing strategy in relation to epidemiolgical investigations of *S.* Berta infections. The present paper summarizes the results of these studies.

COMPARISON OF PLASMID PROFILES OF HUMAN AND BROILER ISOLATES OF S. BERTA

Sørensen et al. (1991) described the plasmid content of 610 strains of *S.* Berta isolated from broilers. Olsen et al. (1992a) compared the content of the same strains to the plasmids and the plasmid profiles of 674 strains of *S.* Berta isolated from humans in Denmark over the period 1984 to 1989. A close correlation between the prevalence of predominant plasmids and plasmid profiles in the two population was observed, although some profiles, for unknown reasons, were more common among human isolates than among isolates from broilers. Ninety one percent of the isolates from humans contained plasmid profiles that had previously been observed among isolates from humans.

A time related comparison of prevalences of plasmid profiles was made for several plasmids and plasmid profiles, including the two most common profiles: a 5.7, 2.0 kb profile and a 2.0 kb profile (Figure 2). From 1987 onwards, the 5.7, 2.0 kb profile spread throughout the broiler production system from the primary breeding flock, and the success of this clonal line was clearly reflected in the increasing prevalence among human isolates but with an appeearent time lag. The annual prevalence of the 2.0 kb profile among isolates from broilers and humans was parallel but was always higher

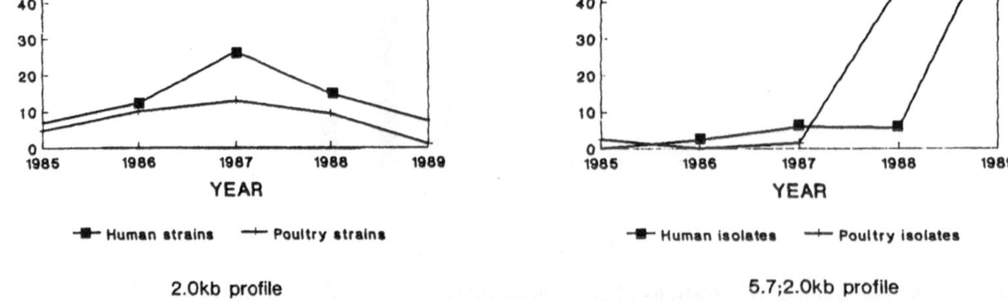

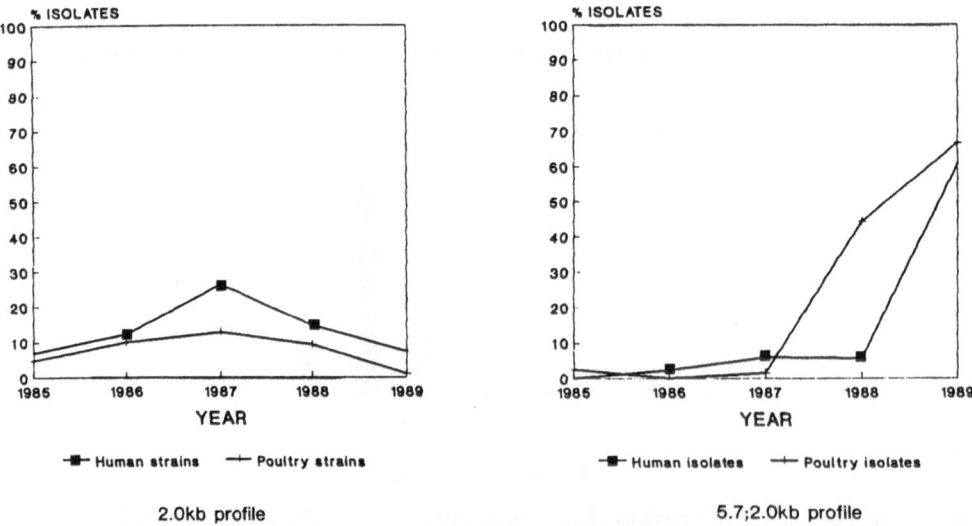

Figure 2. Comparison of yearly prevalence of two common plasmid profiles among isolates of *S.* Berta from poultry and humans in Denmark. Data from Olsen et al. (1992a).

among isolates from humans compared to isolates from broilers. The reason for this observation, whether a matter of selection for virulence, survival capability during processing of broilers or other factors remains unknown.

The studies of plasmid profiles have supported the assumption that broilers in Denmark are the main cause for human cases of salmonellosis caused by *S.* Berta. The study further showed that even in cases where one vehicle appeares to be the sole source for human salmonellosis, an exact correlation between the prevalence of a molecular marker in isolates from the vehicle (the broiler) and isolates from humans cannot be expected.

COMPARISON OF BIOCHEMICAL AND MOLECULAR TYPING METHODS FOR S. BERTA

Biochemical and molecular methods for subspecies characterization have been applied widely in analysis of salmonellae and salmonellosis, expecially in relation to outbreak investigations. For reviews of methods currently applied to salmonellae, the reader is referred to Le Minor (1988) and Threlfall and Frost (1990).

A study, employing typing methods, was undertaken in order establish a rational typing strategy for *S.* Berta (Olsen et al. 1992b). A collection of 213 Danish isolates (148 from broilers, 56 from humans and 9 from other sources) and 25 control strains from foreign countries, including 4 reference strains from the WHO Collaboration centre for research on salmonella (WHO-CCFS) was used in the investigation.

The strains, or a selected subset of the strains, were subjected to the following methods for subspecies characterization: biotyping by 84 criteria, antibiotic resistance testing using 12 antibiotics, whole cell protein profiling on SDS-PAGE, outer membrane protein profiling on SDS-PAGE, multilocus enzyme profiling of 11 enzymes using PAGE, ribotyping using a probe derived from 16S and 23S rRNA of *Legionella pneumophila*, plasmid profiling and plasmid restriction profiling. A detailed description of the methods is given by Olsen and colleagues (1992b).

Ribotyping and multilocus enzyme profiling have been suggested as tools to measure taxonomic/phylogenetic relationship among bacteria (Grimont and Grimont 1986, Selander et al. 1986). As seen from Table 1, strains of *S.* Berta were very uniform when these metods for used for characterization. Five different ribotypes were noted, but more than 90% fell into the most dominant group with both enzymes used; Danish and non-Danish strains possessed the same ribotypes. The four "old" isolates (1936 to 1984) supplied by WHO-CCRS, however, had a different ribotype with the enzyme *Sma*I from the rest of the strains. These four strains also demonstrated a unique phage susceptibility when typed with phages induced from more than 100 different isolates of *S.* Berta (Baggesen, unpublished), and these isolates may represent a different clonal line from the one presently dominating.

Multilocus enzyme profiles were the same in all strains, and the apparent clonal structure of the serovar was underlined by almost identical whole cell protein profiles, identical outer membrane profiles and very little biochemical variation.

Antibiotic resistance was rare among isolates of *S.* Berta. Only 11 strains of 175 tested carried resistance genes, and of these, only 4 were Danish.

Fifty two different plasmid profiles were observed in the study by Olsen and colleageus (1992b), but among 1345 Danish strains analysed, a total of 135 plasmid profiles have been observed (Sørensen et al. 1991, Olsen et al. 1992a). With the exception of a 160 kb R-plasmid encoding ampicillin resistance, no phenotype has been linked to any of the 20 different plasmid sizes demonstrated in Danish strains of *S.* Berta.

Table 1. Molecular and biochemical characterization of *S.* Berta.

Typing method	No. of strains tested	No. of types	% of strains in dominant type	Discrimination[a]
Biotyping	62	2[b]	90	No
Antibiotic resistance	175	4	94	No
Ribotyping	57	5	92/96[c]	No
Plasmid profiling	238	52	8	No[d]
Plasmid restriction profiling (9 plasmids)[e]	20	-	-	Yes
Whole cell protein profiling	58	4	93	No
Outer membrane protein profiling	32	1	100	No
Multilocus enzyme profiling	53	1	100	No

a: Discrimination is indicated as the ability to separate all Danish isolates from foreign isolates, presumably independent from the Danish ones.

b: H_2S positive and H_2S negative strains. Variation was seen in 6 of the 84 criteria tested, but only the H_2S variation was fully reproduceable.

c: Five types were seen with a parallel use of the enzymes EcoR1 and Sma1. The numbers refer to the percent of strains in the most dominant group with each of the enzymes.

d: Three profiles were shared between Danish and non-Danish isolates.

e: The nine plasmids analysed were those contained in the 15 most common plasmid profiles. No plasmid restriction profile was shared between Danish and non-Danish isolates.

To date, plasmid profiling gives the best discrimination between strains. It is, however, not clear how stable the plasmids are in *S.* Berta. Strains isolated in 1985 and stored until 1990 were significantly more often without plasmids than isolates from 1989, which may indicate a curing process during storage. Many of the plasmid profiles of *S.* Berta contain only a limited number of plasmid sizes but in all possible combinations. Such profiles can arise from each other by curing processes and a study has therefore been undertaken in order to measure the stability of plasmids in *S.* Berta during storage. At the present time, after 18 months of storage in stab cultures, loss of one or two plasmids have been observed in three out of five profiles investigated, when the stabculture has been stored at room temperature or at 30°C. Stab cultures kept at low temperature has remained stable (Olsen and Bisgaard, unpublished). It is adviseable always to store stabcultures refrigerated and never to base plasmid analysis of old, stored cultures on single colony results.

A discrimination between Danish strains and unrelated foreign strains was only seen when plasmid profiling was used in combination with restriction enzyme digestion of plasmids, and this method was recommended as the first choice for typing method by Olsen and colleagues (1992b). Plasmid analyses are, however, not very suitable for long term surveillance of bacteria, as only strains analysed on the same gel can be correctly compared. There is therefore a need for a more definitive method for such analysis. Phage typing based on phages induced from *S.* Berta has so far not been very succesful as all strains fall into two types, irrespective of the phages used (Baggesen, unpublished), but work is still ongoing in the field of phages typing of this serovar.

REFERENCES

Hormaeche E., Pelluffo C.A. and Salsaendi R. (1938). Un nuevo tipo del genro *Salmonella*: "S.berta". *Arch. Uruguayos Med. Cirug. & Especial* XII: 377-387.

Grimont F. and Grimont P.A.D. (1986). Ribosomal ribonucleic acid gene restriction patterns as a potential taxonomic tool. *Ann. Microbiol.* 137B:165-175.

Le Minor L. (1988). Typing of *Salmonella* species. Eur. J. Clin. Microbiol. Infect. Dis. 7:214-218.

Olsen J.E., Sørensen M., Brown D.J., Gaarslev K. and Bisgaard M. (1992a). Plasmid profiles as an epidemiolgical marker in *Salmonella enterica* serovar Berta infections. *APMIS* 100:221-228.

Olsen J.E., Brown D.J., Baggesen D.L. and Bisgaard M. (1992b). Biochemical and molecular characterization of *Salmonella enterica* serovar Berta and comparison of methods for typing. *Epidimiol. Infect.* in press.

Selander R.K., Caugant D.A.,Ochman H., Musser J.M., Gilmore M.N. and Whittam T.S. (1986) Methods of multilocus enzyme electrophoresis for bacterial population genetics and systematics. *Appl. Environ. Microbiol.* 51:873-884.

Threlfall E.J. and Frost J.A. (1990). The identification, typing and fingerprinting of *Salmonella*: laboratory aspects and epidemiolgical applications. *J. Appl. Bact.* 88:5-16.

THE *viaB* LOCUS OF *Salmonella typhi*

Michel Y. Popoff, Hervé Waxin, and Suzanne Kolyva

Institut Pasteur
Unité des Entérobactéries
Unité INSERM 199
28, rue du Docteur Roux
75728 Paris Cedex
France

INTRODUCTION

The Vi antigen, discovered by Felix and Pitt (1934), is a capsular polysaccharide found mainly in *Salmonella typhi* and *S. paratyphi* C, as well as in a few strains of *S. dublin* and *Citrobacter freundii* (Felix and Pitt, 1936; Baker et al., 1959; Le Minor and Nicolle, 1964). This surface antigen is a linear homopolymer of *N*-acetyl galactosamine uronic acid (Heyns and Kiessling, 1967), with a high molecular mass (typically, more than 10^6 daltons). Its synthesis is not thermo-regulated. Based on these criteria, the Vi antigen may be classified into the polysaccharides of group I (Jann and Jann, 1990).

Determinants of Vi antigen synthesis occupy two widely separated chromosomal loci, termed *viaA* and *viaB* (Johnson et al., 1965). The *viaA* locus is commonly found in enteric bacteria, such as *Escherichia coli*. In contrast, the *viaB* locus is specific to Vi-expressing strains and maps at 92 min. on the chromosome of *S. typhi* (Johnson and Baron, 1969).

We and others (Cryz et al, 1989; Hashimoto et al., 1991; Kolyva et al., 1992) cloned the *viaB* region from *S. typhi* strain Ty2 and after mutagenesis using Tn*5* transposon, it was demonstrated that the *viaB* locus was present on a 12 kilobases (kb) DNA fragment (Hashimoto et al., 1991; Kolyva et al., 1992).

MATERIAL AND METHODS

Plasmid pVT3 is a recombinant plasmid harbouring the *viaB* locus of *S. typhi* Ty2 (Kolyva et al., 1992). Nucleotide sequence of a DNA fragment of about 10 kb was determined after subcloning of appropriate restriction fragments into M13 vectors (Yanisch-Perron et al., 1985). Single-stranded M13 DNA templates were sequenced by the dideoxy chain-termination method (Sanger et al., 1977) using the modified T7 DNA polymerase, Sequenase version 2.0™ (USB Corp.). All the ends of restriction fragments used overlapped one another and the sequences of both strands were determined.

RESULTS AND DISCUSSION

The sequenced fragment was 9728 bp in length. Examination of this sequence revealed seven open reading frames (ORF). All were transcribed in the same orientation.

ORF1 encoded a putative protein of 179 amino acids with a predicted molecular mass of 21 kiloDaltons (kDa). A typical Shine-Dalgarno sequence was located 5 bases pairs (bp) upstream the ATG start codon. ORF2 encoded a protein of 456 amino acids with a predicted molecular mass of 48 kDa.The third base of the ORF1 stop codon was the first base of the ORF2 start codon. ORF3 encoded a putative protein of 348 amino acids with a predicted molecular mass of 40 kDa. ORFs 2 and 3 were separated by two bp. ORF3 and ORF4 were separated by an inter-genic region of 422 bp. ORF4 encoded a protein of 704 amino acids with a predicted molecular mass of 80 kDa. A Shine-Dalgarno sequence was located 10 bp upstream the ATG start codon of ORF4. ORF5 encoded a protein of 578 amino acids with a predicted molecular mass of 59 kDa.The ATG start codon of ORF5 overlapped the TAA stop codon of ORF4. ORF6 encoded a protein of 355 amino acids with a predicted molecular mass of 39 kDa. The start of ORF6 was separated from the end of ORF5 by 3 bp. ORF7 encoded a 421 amino acids with a predicted molecular mass of 44 kDa. ORF6 and ORF7 were separated by 9 bp.

Homology between the putative proteins encoded by these seven ORFs and protein sequences accessible through GenBank were searched using the Lipmann and Pearson algorithm (1985). None of the proteins encoded by ORF1, 2, 3, 4 and 5 shared any significant homology to protein sequences in the GenBank database. In contrast, the 39 kDa protein (P39) encoded by ORF6 shared homology to the BexD protein, and the 44 kDa protein (P44) encoded by ORF7 shared homology to the BexB protein.

The *bex* locus is a chromosomal region involved in the export of the capsule of *Haemophilus influenzae* (Kroll et al.,1988; Kroll et al., 1990). This locus encodes 4 proteins BexA, B, C and D. It was suggested than these 4 proteins formed an ATP-driven system for the export of the capsule of *H. influenzae* (Kroll et al., 1990).

Comparison of the putative amino acid sequence of BexD and P39 proteins revealed 26% identity in 288 amino acids overlap. Taking into account conservative amino-acids substitution, the homology was extended to 70%. Following the ATG start codon, examination of P39 sequence showed a stretch of 12 non-polar hydrophobic amino acids leading to a glycine in position 16 and a cysteine in position 17. Such sequence appeared typical of a lipoprotein signal sequence with the potential cleavage site for signal peptidase II between the glycine and the cysteine (Hussain et al., 1982). In addition, at amino acid 163, the P39 protein had an ATP-binding consensus sequence (Higgins et al.; 1986). Based on these observations, we suggest that ORF6 encodes an outer-membrane lipoproteine coupling ATP hydrolysis to the transport of the Vi polysaccharide across the outer membrane.

Between the BexB and P44 proteins, there was 24% identity in 217 amino acids overlap. Considering conservative amino acids substitution, the homology was extended to 72%. The hydropathy profil of the P44 protein, generated using the algorithm of Kite and Doolittle (1982), was characteristic of an integral membrane protein with, at least, eight putative membrane-spanning fragments. As proposed for the BexB protein (Kroll et al.; 1990), we suggest that ORF7 encodes an inner-membrane protein, possibly mediating transport of the Vi polysaccharide across the inner membrane.

Very recently, the molecular organization of the *ctr* locus was reported by Frosch et al. (1991). The *ctr* locus was involved in the export of the capsule of *Neisseria meningitidis*. Both the *bex* locus and the *ctr* locus showed a very similar organization and fitted the characteristics of ABC (ATP-binding cassette) transporters (Higgins et al., 1990; Hyde et al.,1990). For group II polysaccharides, these transport systems consisted of three or four membrane-associated proteins: one or two hydrophilic ATP-binding proteins (as are BexA and CtrD proteins); and two hydrophobic proteins located in the inner membrane and mediating the transport of the polysaccharide across the inner membrane (BexB and BexC, CtrB and CtrC correspond to such proteins). In addition, the Ctr system had an outer-membrane protein, CtrA, involved in the export of the polysaccharide across the outer membrane.

The export system of the Vi antigen showed similarity to an ABC transporter since the P44 protein might be associated to the inner-membrane and the P39 protein, to the outer-membrane. These two proteins might act in the energy-dependent export of the Vi polysaccharide acroos the inner- and outer membranes. Moreover, experimental evidence indicated that the putative 80 kDa product might be also involved in the Vi antigen export. After insertion of Tn5 transposons in the sequence encoding the 80 kDa product, the Vi antigen could be detected in the intracellular compartment of bacteria, but not at the cell surface or in the culture supernatant (Kolyva et al.; 1992). Therefore, the 80 kDa protein might be involved, alone or together with the 44 kDa protein, in the transport of the Vi polysaccharide across the inner-membrane.

There is now clear evidence that a conserved mechanism exists for the ATP-dependent export of group II polysaccharides. Our preliminary results indicate that a related system might exist for group I polysaccharides, such as the Vi antigen of *S. typhi*.

REFERENCES

Baker, E.E., Whiteside, R.E., Basch, R., and Derow, M.A., 1959, The Vi antigen of the*Enterobacteriaceae*. II. Immunologic and biologic properties. *J. Immunol.* 83: 680-686.

Cryz,S.J.,Jr, Fürer,E., Baron, L.S., Noon, K.F., Rubin, F.A., and Kopecko,D.J., 1989, Construction and characterization of a Vi-positive variant of the *Salmonella*

typhi live oral vaccine strain Ty21a. *Infect. Immun.* 57: 3863-3868.
Felix, A., and Pitt,R.M., 1934, A new antigen of *B. typhosus*. *Lancet* 227: 186-191.
Felix, A., and Pitt,R.M., 1936, The Vi antigens of various *Salmonella* types. *Brit. J. Exp. Pathol.* 17: 81-86.
Frosch, M., Edwards, U., Bousset K., Krause, B., and Weisgerber, C., 1991, Evidence for a common molecular origin of the capsule gene loci in Gram-negative bacteria expressing group II capsular polysaccharides. *Mol. Microbiol.* 5: 1251-1263.
Hashimoto, Y., Ezaki, T., Li, N., and Yamamoto, H., Molecular cloning of the *viaB* region of *Salmonella typhi*. *FEMS Microbiol. Letters* 90: 53-56.
Heyns, K., and Kiessling, G., 1967, Strukturaufklarung des Vi-antigens aus *Citrobacter freundii* (*E. coli*) 5396/38. *Carbohyd. Res.* 3: 340-352.
Higgins,C.F., Hiles, I.D., Salmond, G.P.C., Gill, D.R., Downie, J.A., Evans, I.J., Holland,I.B., Gray, L., Buckel, S.D., Bell, A.W., and Hermodson, M.A., 1986, A family of related ATP binding subunits coupled to many distinct biological processes in bacteria. *Nature* 323: 448-450.
Higgins, C.F., Hyde, S.C., Mimmack, M.M., Gileadi, U., Gill, D.R., and Gallagher, M.P., 1990, Binding protein dependent transport systems. *J. Bioenerget. Biomembr.* 22: 571-592.
Hussain, M., Ichihara, S., and Mizushima, S., 1982, Mechanism of signal peptide cleavage in the biosynthesis of the major lipoprotein of the *Escherichia coli* outer membrane. *J. Biol. Chem.* 257: 5177-5182.
Hyde, S.C., Emsley, P., Hartshorn, M.J., Mimmack, M.M., Gileadi, U., Pearce, S.R., Gallagher, M.P., Gill, D.R., Hubbard, R.F., and Higgins, C.F., 1990, Structural model of ATP-binding proteins associated with cystic fibrosis multidrug resistance and bacterial transport. *Nature* 346: 362-365.
Johnson, E.M., Krauskopf, B., and Baron, L.S., 1965, Genetic mapping of Vi and somatic antigenic determinants in *Salmonella*. *J. Bacteriol.* 90: 302-308.
Johnson, E.M., and Baron, L.S., 1969, Genetic transfer of the Vi antigen from *Salmonella typhosa* to *Escherichia coli*. *J. Bacteriol.* 99: 355-359.
Jann, B., and Jann, K., 1990, Structure and biosynthesis of the capsular antigens of *Escherichia coli*, in: " Current Topics in Microbiology and Immunology", vol. 150, K. Jann and B. Jann, eds, Springer Verlag, Berlin.
Kolyva, S., Waxin, H., and Popoff, M.Y., 1992, The Vi antigen of *Salmonella typhi*: molecular analysis of the *viaB* locus. *J. Gen. Microbiol.* 138: 297-304.
Kroll, J.S., Hopkins, I, and Moxon, E.R., 1988, Capsule loss in *H. influenzae* type b occurs by recombination-mediated disruption of a gene essential for polysaccharide export. *Cell* 53: 347-356.
Kroll, J.S., Loynds, B., Brophy, L.N., and Moxon,E.R., 1990, The *bex* locus in encapsulated *Haemophilus influenzae*: a chromosomal region involved in capsule polysaccharide export. *Mol. Microbiol.* 4: 1853-1862.
Kyte, J., and Doolittle, R.F., 1982, A simple method for displaying the hydropathic character of a protein. *J. Mol. Biol.* 157: 105-132.
Le Minor, L., and Nicolle, P., 1964, Sur deux souches de *Salmonella dublin* possédant l'antigène Vi. *Ann. Inst. Pasteur* 107: 550-556.
Lipman, D.J., and Pearson, W.R., 1985, Rapid and sensitive protein similarities searches. *Science* 227: 1435-1441.
Sanger, F., Nicklen, S., and Coulson, A.R., 1977, DNA sequencing with chain terminating inhibitors. *Proc. Natl. Acad. Sci. USA* 74: 5463-5467.
Yanisch-Perron, C., Viera, J., and Messing, J., 1985, Improved M13 phage cloning vectors and host strains: nucleotide sequences of M13mp18 and pUC19 vectors. *Gene* 33: 103-119.

THE VIRULENCE PLASMID OF *SALMONELLA* ENCODES A PROTEIN RESEMBLING EUKARYOTIC TROPOMYOSINS

P. Helena Mäkelä, Pertti Koski[2], Petri Riikonen[1], Suvi Taira[1,2], Harry Holthöfer[3] and Mikael Rhen[2]

[1]National Public Health Institute, SF-00300 Helsinki, Finland,

[2]Department of Biochemistry, University of Helsinki, Helsinki and

[3]Department of Bacteriology and Serology, University of Helsinki, Helsinki, Finland

ABSTRACT

The gene *tlp*A located outside the 8 kb virulence determinant region of the virulence plasmid of *Salmonella enterica* serovar Typhimurium has been cloned and sequenced. The sequence predicts a protein of 43 kDa with unique features for a *Salmonella* protein: extensive heptapeptide repeating structure, 80 % alphahelix, and an N-terminal globular portion. Sequence comparisons demonstrate strong homologies to the tropomyosin family of eukaryotic proteins; if so, it would be likely to occur as dimers with a coiled coil structure. Chemical and immunochemical data on the protein expressed in *E. coli* support these predictions.

INTRODUCTION

Although a large part of the nearly 100 kb virulence plasmid of *Salmonella sp.* is shared by plasmids of many serovars (Korpela *et al.*, 1989) only a stretch of 8 kb containing four *spv* genes has been unequivocally demonstrated to be required for the virulence of the bacteria (Williamson *et al.*, 1988; Gulig and Curtiss, 1988; Norel *et al.*,

Biology of Salmonella, Edited by F. Cabello *et al.*,
Plenum Press, New York, 1993

1989; Taira and Rhen, 1989). On the other hand, there are findings that suggest a role in virulence for other genes outside the *spv* region, including the ***tra*** and ***rck*** genes that control resistance to serum complement (Rhen and Sukupolvi, 1988; Van den Bosch *et al.*, 1989); furthermore decrease of virulence in mutants with insertions outside *spv* has been reported (Sizemore *et al.*, 1991). A special difficulty in studying the function of the virulence plasmid genes is that they are largely repressed when the bacteria are grown in the laboratory.

We have recently cloned and sequenced a gene that we call *tlp*A, which is located outside the *spv* region of the plasmid of *S. enteritica* serovar Typhimurium (Koski *et al.*, 1992). This gene appears to code for a protein with extensive homology to an important class of eukaryotic proteins, the tropomyosin superfamily. These proteins are not only essential parts of the contractile apparatus of muscle cells, but also involved in intracellular structures of non–muscle cells, e.g. in stress fibers. Tropomyosins have a very distinctive tertiary structure involving dimer formation with a coiled–coil structure. The resemblance poses intriguing questions both to the function of the TlpA protein and to the origin of the gene.

THE *tlp*A GENE AND PREDICTED STRUCTURE OF THE PROTEIN

Our general approach to identify and express genes of the virulence plasmid has been to subclone DNA fragments of them into suitable vectors in order to release the genes from their regulation in the virulence plasmid. This process allowed us to identify on pKTH3059 a gene encoding a strongly expressed 33 kDa protein (Koski *et al.*, 1992). This gene was located on a *Sma*I–*Xho*I fragment appr. 10 kb transcriptionally upstream of the *spv* genes, and transcribed in the opposite direction of *spv*. The nucleotide sequence of the *Sma–Xho* fragment revealed an open reading frame (ORF) starting 162 basepairs from the *Sma*I site, with a putative ribosome binding site 12 basepairs upstream of the ATG. The ORF extended to the end of the cloned fragment without a transnational stop indicating that only the 5' terminal part of the gene was present in this fragment. A new subclone was then constructed (pTH3061) containing the same *SmaI–XhoI* fragment but extending further towards the 3' end of the ORF. The full ORF (*tlp*A) in this fragment proved to code for 371 amino acids. This is also the size of its predicted protein product, since no apparent secretion signals were present at the N terminus.

Analysis of the *tlp*A gene predicted it to encode a protein with a globular structure for the 61 N–terminal residues, and over 90% α–helix for the rest of the molecule. This region also contained extensive heptapeptide repeats of a structure typical of coiled–coil proteins, the first and fourth residues in the heptapeptide being polar or hydrophobic. In the *tlp*A heptapeptides, this was true for 72% of the first and 59 % of the fourth residues, but only 20–30% of the intervening residues.

Computer comparisons revealed sequence similarity of TlpA with several eukaryotic coiled–coil proteins, notably with the tropomyosin superfamily. The overall identity for the whole protein with the different tropomyosins was 15-20 %, and higher for the alpha-helical part of the protein. A very high Needleman–Wunsch identity score (SD 4.4) was found for a continuous stretch of 90 residues in this region (Koski *et al.*, 1992). On the basis of these homologies it was decided to call the gene *tlp*A for tropomyosin–like protein.

EXPRESSION AND PHYSICOCHEMICAL PROPERTIES OF THE TlpA PROTEIN

The starting point of the analysis of *tlp*A was its strong expression from pKTH3059 in *E. coli*, although as a truncated 33 kDa N–terminal peptide. The expression of the full-length ORF in pKTH3061 was, however, poor so that the expected 371 amino acids long peptide could only be visualized with difficulty by immunoblotting with antiserum to the isolated 33 kDa peptide. We believe that this is another example of the extensive and so far poorly understood regulation of the expression of the genes of the virulence plasmid, and have in fact preliminary evidence of this. However, to express the *tlp*A gene strongly enough to be able to study the protein, we separated it from its own promoter region, by cloning it as a PCR fragment under the *tac* promoter of pKK223–3 (Pharmacia Fine Chemicals, Uppsala, Sweden). A 43 kDa protein was strongly expressed from this plasmid in *E. coli*, and was extensively purified by ammonium sulphate precipitation, heating (for rationale, see below) and gel filtration (Koski *et al.*, 1992).

We were especially interested in testing, on protein level, the suggested homology with tropomyosins. These proteins are known to form dimers in which a long stretch of alpha-helical structure of the two peptides forms a coiled–coil fibrillar structure further stabilized by SS–bridges between cysteine residues. It is possible to demonstrate the fibrillar structure *e.g.* by electron microscopy. Furthermore, a remarkable property of these proteins is that they remain in solution when heated to 100°C. We proceeded to test the TlpA protein for these parameters.

A five minutes' boiling was included in the purification of TlpA, which was one of the few proteins remaining in solution. Gel filtration and polyacrylamide gel electrophoresis (PAGE) under non–denaturing conditions (*i.e.* without SDS) of TlpA suggested a higher molecular mass than 43 kDa, consistent with multimer formation. When the purified protein was allowed to stand in buffer and then run in SDS–PAGE, it had an apparent molecular mass of 90 kDa, but reduced to 43 kDa if mercaptoethanol was added to the sample buffer (Koski *et al.*, 1992). Electron microscopy of the purified protein showed elongated structures as expected of an extended coiled–coil protein (M Rhen *et al.*, data to be published).

IMMUNOCHEMICAL PROPERTIES OF TlpA

Would the homology between TlpA and tropomyosins be reflected in their antigenic properties? We studied this by immunizing rabbits with purified TlpA, and isolating the TlpA–specific immunoglobulins by absorption to a TlpA–sepharose column. These antibodies were found to crossreact in enzyme immunoassays with various tropomyosins and myosins. In immunofluorescence assays a strong staining was seen in several tissues of rats and mice: smooth muscles of various organs and podocytes of rat kidney glomeruli. In cultured cell lines, *e.g.* HeLa cells, intracellular fibrillar elements were also stained.

THE FUNCTION OF TlpA

We do not know at this time the function of TlpA. No TlpA is detected in *Salmonella* carrying the virulence plasmid grown under normal laboratory conditions, indicating a repression *in vitro* of this gene, like of most other genes of the virulence plasmid, and suggesting that their function would be exerted only *in vivo*. Since the phase of infection when the virulence plasmid has been shown to function is during intracellular growth in the first several days of infection, the first guess would be that also TlpA would be expressed and functional at this time. We are currently looking into this with a reporter gene inserted into *tlp*A.

However, mutations of *tlp*A have not been detected in experiments designed to find mutants with reduced ability to grow in the mouse in the first week of infection. Indeed, when a *tlp*A–knockout strain of Typhimurium was compared to the *tlp*A$^+$ parent, no difference in growth or lethality to the mouse was observed in the standard 4 day assay after intravenous injection. We now explore the possibility that TlpA would affect the subsequent phase of infection, when immunity of the host develops and starts curbing the growth of the bacteria.

It is tempting to speculate that the TlpA protein would interfere with the functions of the cytoskeleton of the host cells by mimicking essential components of the tropomyosin family. Indeed, actin polymerization has been shown to be involved in the uptake of several bacteria into host cells, and *Listeria* and *Shigella* species can manipulate and utilize intracellular actin for their locomotion (Tilney and Portnoy, 1989; Clerc and Sansonetti, 1987). However, *Salmonella* has not been shown to penetrate from the phagocytic vesicle into the cytoplasm, and hence its interaction with the cytoskeleton seems unlikely. Another problem is the location of TlpA within *Salmonella*. The protein does not contain secretory signals and thus it could be excreted only by a nonstandard secretory pathway - but this may not be too unlikely, since another virulence-associated protein, hemolysin of *E. coli*, is secreted *via* its own export system (Blight and Holland, 1990).

Finally, TlpA could be an example of bacterial myosins, the presence of which has been suggested by recent studies (Casaregola *et al.*, 1990; Niki *et al.*, 1991). Its function might then be in the segregation of the virulence plasmid, which indeed appears to need a well controlled segregation system during cell division accounting for its extreme stability in bacterial populations in spite of the low copy number. A low level of expression might be sufficient for this function, consistent with our inability to demonstrate TlpA *in vitro*.

ACKNOWLEDGEMENTS

The research was supported by Sigrid Jusélius Foundation, Oskar Öflund's Foundation, the Swedish Cultural Foundation and the Academy of Finland.

REFERENCES

Blight, M.A., and Holland, I.B., 1990, Structure and function of haemolysin B, P-glycoprotein and other members of a novel family of membrane translocators, *Mol. Microbiol.* 4:873.

Casaregola, S., Norris, V., Goldberg, M., and Holland, I.B., 1990, Identification of a 180kD protein in *Escherichia coli* related to a yeast heavy-chain myosin, *Mol. Microbiol.* 4:505.

Clerc, P., and Sansonetti, P.J., 1987, Entry of *Shigella flexneri* into HeLa cells: evidence for directed phagocytosis is involving actin polymerization and myosin accumulation, *Infect. Immun.* 55:2681.

Gulig, P.A. and Curtiss, R. III, 1988, Cloning and transposon insertion mutagenesis of virulence genes of the 100 kilobase plasmid of *Salmonella typhimurium*, *Infect. Immun.* 56:3262.

Korpela, K., Ranki, M., Sukupolvi, S., Mäkelä, P.H. and Rhen, M., 1989, Occurrence of *Salmonella typhimurium* virulence plasmid-specific sequences in different serovars of *Salmonella*, *FEMS Microbiol. Lett.* 58:49.

Koski, P., Saarilahti H., Sukupolvi, S., Taira, S., Riikonen, P., Österlund, K., Hurme, R., and Rhen, M., 1992, A new α-helical coiled coil protein encoded by the *Salmonella typhimurium* virulence plasmid, *J. Biol. Chem.* 267:12258.

Niki, H., Jaffe, A., Imamura, R., Ogura, T., and Hiraga, S., 1991, The new gene *muk*B codes for a 177 kd protein with coiled-coil domains involved in chromosome partitioning of *E. coli*, *EMBO J.* 10:183.

Norel, F., Coynault, C., Miras, I., Hermant, D., and Popoff, M.Y., 1989, Cloning and expression of plasmid DNA sequences involved in *Salmonella typhimurium* virulence, *Mol. Microbiol.* 3:733.

Rhen, M., and Sukupolvi, S., 1988, The role of the *traT* gene of the *Salmonella typhimurium* virulence plasmid for serum resistance and growth within liver macrophages, *Microb. Pathog.* 5:275.

Sizemore, D.R., Fink, P.S., Ou, J.T., Baron, L., Kopecko, D.J., and Warren, R.L., 1991, *Tn5* mutagenesis of the *Salmonella typhimurium* 100 kb plasmid: definition of new virulence regions, *Mcrob. Pathog.* 10:493.

Taira, S., and Rhen, M., 1989, Molecular organization of genes constituting the virulence determinant on the *Salmonella typhimurium* 96 kilobase pair plasmid, *FEBS Lett.* 5:274.

Tilney, L.G., and Portnoy, D.A. 1989, Filaments and the growth, movement, and spread of the intracellular bacterial parasite, *Listeria monocytogenes*, *J. Cell. Biol.* 109:1597.

Vandenbosch, J.L., Rabert, D.K., Kurlandsky, D.R., and Jones, G.W., 1989. Sequence analysis of *rsk*, a portion of the 95-kilobase plasmid on *Salmonella typhimurium* associated with resistance to the bactericidal activity of serum, *Infect. Immun.* 57:850.

Williamson, C.M., Pullinger, G.D., and Alistair, J.L., 1988, Identification of an essential virulence region on *Salmonella* plasmids, *Microbial. Pathog.* 5:469.

IS *Salmonella* AN INTRACELLULAR PATHOGEN ?

Hsiu-Sheng Hsu

Department of Microbiology and Immunology
Medical College of Virginia
Virginia Commonwealth University
Richmond, VA 23298-0678, USA

INTRODUCTION

My assigned topic of presentation in this Workshop is to address the issue of whether *Salmonella* is an intracellular pathogen. The simple answer to that question is yes and no, depending on the host cells into which the pathogen has entered. There are many of us here in this Workshop who are studying the genetic factors governing the invasive or virulent properties of *Salmonella* in eukaryotic cells. The common experimental model is the growth of *Salmonella typhimurium* in cell cultures of epithelial cells or other mammalian cell lines (Finlay and Falkow, 1989). There seems to be reliable evidence that virulent *S. typhimurium* do replicate within these cultured cells. In contrast, I propose to examine this issue using the experimental model of mouse typhoid (Hsu, 1989). Based on direct visual observations in the pathogenic process of the disease, the following narrative is intended to show that virulent *S. typhimurium* do survive and proliferate at least within parenchymal cells, but are effectively killed by inflammatory phagocytes *in vivo*.

PROLIFERATION OF *S. typhimurium* IN ORGANS OF RETICULOENDOTHELIAL SYSTEM

By definition, mouse typhoid is a systemic disease, which can be initiated by the intraperitoneal (i.p.) inoculation of a minimal lethal dose of virulent *S. typhimurium* into susceptible mice (Xu and Hsu, 1992). When 1 x 10^2 and 2 x 10^3 bacteria of *S. typhimurium* SR-11 are introduced i.p. into the highly susceptible C57BL/6J and BALB/c mice, bacteremia can be detected by peripheral blood samples within 6 h and 1 h, respectively. The bacterial growth *in vivo* can be represented by the rate of their multiplication in the liver, which we can follow by viable bacterial counts of the ground-up organ removed at daily intervals from infected mice.

Figure 1 shows that the pathogen has established itself in the liver of the C57BL/6J mice within 2 days post infection and proceeds to multiply rapidly at a

Biology of Salmonella, Edited by F. Cabello *et al.*,
Plenum Press, New York, 1993

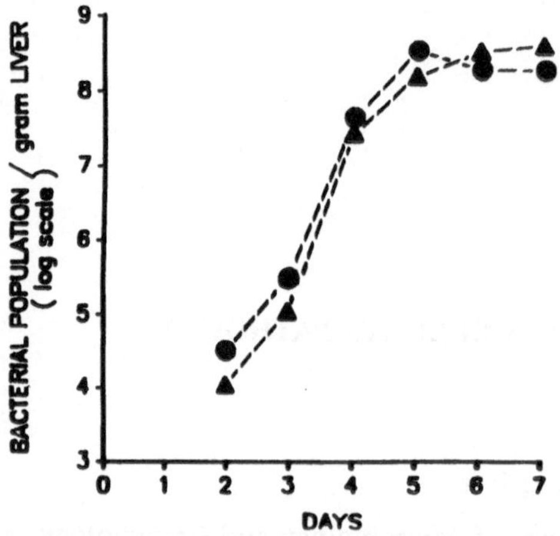

Figure 1. Proliferation of *S. typhimurium* SR-11 in the liver of natively susceptible C57BL/6J mice inoculated i.p. with either 1×10^2 (▲) or 2×10^3 (●) bacteria. Based on data published in the *Journal of Medical Microbiology* (Xu and Hsu, 1992) with permission of the publisher.

logarithmic rate between day 2 and day 5. When the bacterial population reaches 10^8 organisms/g tissues, the animals begin to die between 5 and 7 days post infection. The fatal outcome of this disease is no doubt directly related to the rapid and massive accumulation of the bacterial population in the host. Part of our investigation was an attempt to determine the location of bacterial proliferation in the host tissues.

EARLY TISSUE RESPONSE IN MOUSE TYPHOID

Tissue specimens taken from the liver and spleen of mice infected with a minimal lethal dose of *S. typhimurium* reveal consistently an early influx of polymorphs at the site of infection (Nakoneczna and Hsu, 1980), as illustrated in Figure 2. The formation of multiple, minute microabscesses in the organs within the first 5 days of the infection is marked by the conspicuous absence of macrophages and corresponds with the phase of most rapid bacterial growth as shown in Figure 1. The subsequent enlargement and coalescence of abscesses result in a destructive tissue necrosis.

The transition of tissue lesions into granulomas usually occurs after 5 days post infection and begins with the peripheral infiltration of mononuclear cells (Nakoneczna and Hsu, 1980). Toward the terminal stage of the disease, extensive granulomas with central necrosis become the prominent feature. Occlusions of some blood vessels with massive accumulation of leukocytes, notably mononuclear cells, are indicative of the elicitation of delayed hypersensitivity by the mounting bacterial antigens in the tissues and probably account for the infarcts of the surrounding tissues. Hence, by the time a preponderance of macrophages appears at the scene of infection, the battle against the pathogen has essentially been lost.

In our preparations of histopathological slides, tissue specimens are routinely embedded in paraffin blocks and stained with hematoxylin and eosin. These preparations are normally cut in 5-7 *u*m thickness, such that salmonellae in the tissues are not visible. We could not determine the locations of the organisms in the tissues, intracellular versus extracellular, let alone the predominant site of their multiplication.

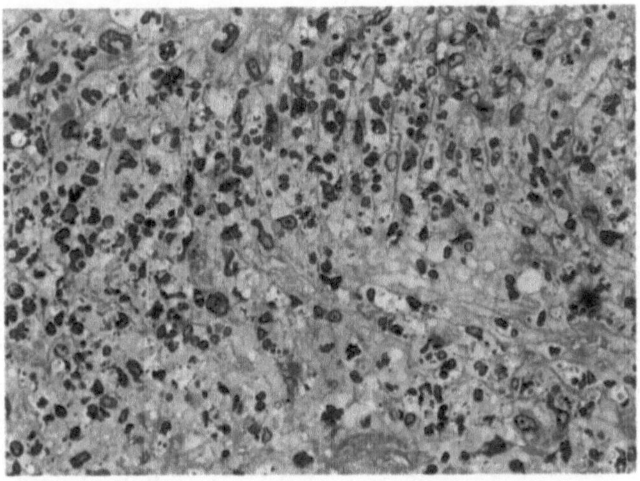

Figure 2. Infiltration of polymorphs forming acute abscess in liver of C57BL/6J mouse 5 days after i.p. inoculation of 2 x 10^3 *S. typhimurium* SR-11. Tissue sections prepared with JB-4 Plus embedding method and stained with Wright's stain by H.R. Xu and C.W. Moncure.

Recently, we have adopted the tissue embedding method using JB-4 Plus plastic resin (Polysciences, Inc.), so that the tissue specimens can be cut in 1-2 *u*m thickness for staining with Wright's stain. Bacteria are clearly visualized in these thin tissue sections.

Using the JB-4 Plus tissue embedding method, tissue sections obtained from C57BL/6J mice infected with the minimal lethal dose of *S. typhimurium* reveal the conspicuous dispersions of bacteria in the extracellular locations among the infiltrating polymorphs and their occasional inclusions within host cytoplasm during the first 5-6 days post infection. The extracellular location of these organisms is certified by the absence of tissue cytoplasm around them and by their proximity to erythrocytes in the lesions.

These tissue sections further reiterate the striking absence of macrophages during the early phase of bacterial infection, during which active bacterial propagation occurs. Colonies of bacteria can also be found extracellularly in the sinusoids and in lesions, as well as within hepatocytes. The thin sectioning also exposes the canaliculi among the hepatocytes and the presence of bacteria in ductular structures of the liver. The latter implies the transportation of salmonellae from hepatocytes through the bile ducts to the gallbladder. Thus, it would be reasonable to assume that the pathogen may be harbored within hepatocytes and excreted to the gallbladder in the carrier state.

The collective experimental evidence from histopathology of the disease lends further credence to the proposed analogy of pathogenesis between mouse typhoid and human typhoid fever. However, it also fails to identify the macrophages as playing a crucial role in the early phase of mouse typhoid, let alone being the site of bacterial multiplication.

FATE OF INTRACELLULAR SALMONELLAE IN MOUSE TYPHOID

By light microscopy, it is impossible to determine definitively the locations of bacterial multiplication in host tissues, even though we can now make the organisms visible in thin tissue sections using the JB-4 Plus embedding method. On the other hand, one can discern the viability and replication of bacteria based on their morphological integrity and cellular division using electron microscopy. By the i.p. inoculation of 10^7 *S. typhimurium* SR-11 into inbred susceptible mice, such as the C57BL/6J mice, we can rapidly initiate an intense inflammatory response in the peritoneal cavity and a diffused dissemination of bacteria into the tissues of RES. The purpose of this protocol is solely to amass a large population of phagocytes expeditiously in the peritoneal cavity for an optimal ingestion of bacteria by host cells and for the scattered distribution of bacteria in host tissues, in order to facilitate observations by EM.

A rapid influx of inflammatory cells, composed predominantly of polymorphs, is readily apparent as early as 1.5 h in the peritoneal washings (Guo et al., 1986a). Occasional macrophages are also observed in the early stage of the infection and their proportion does not increase until 24 h or so later. Pseudopods and cytoplasmic invaginations are seen among the phagocytes in the process of engulfing bacteria. By comparison with the typical morphology of the extracellular bacteria, those found within polymorphs and macrophages appear almost invariably in different stages of disintegration, even as early as 1.5 h post infection. They may be harbored within phagolysosomes or much less likely in direct contact with host cytoplasm. The degenerative changes in salmonellae occur in two possible sequences: more commonly, the bacteria appear with an enlarging central electron-lucent area in their cytoplasm and peripheral condensation of cytoplasmic granules, followed by the disruption of bacterial envelope and the disintegration of cellular structure; or, alternatively at times, they begin to deteriorate with an external compression of bacterial envelope and diffused condensation of cytoplasmic granules.

In the peritoneal washings harvested from 1.5 h through 48 h, neutrophils, eosinophils and monocytes are seen in the process of destroying ingested bacteria within their cytoplasm. It is not unusual to find apparently dividing bacteria undergoing degenerative changes within these phagocytes, which would suggest that they are most likely phagocytized at the time of their replication. At times, a phagocyte continues to engulf additional organisms while it is in the process of destroying others in its cytoplasm. As consistent with our knowledge of histopathology, macrophages may have phagocytized a polymorph which is in turn digesting its own share of ingested bacteria.

The peritoneal cavity therefore provides a desirable *in vivo* setting, superior to cell culture designs, for the study of interactions between salmonellae and inflammatory phagocytes. If the invading pathogens are effectively eliminated by the phagocytes as demonstrated here, it would be reasonable to expect their propagation in the extracellular location in the peritoneal washings. Indeed, morphologically intact and dividing bacteria are found adjacent to phagocytes, some of which may contain disintegrating bacteria. Furthermore, when immune serum is added to salmonellae before the i.p. inoculation, its opsonic function can be ascertained by finding an apparently greater ratio of infected phagocytes in the peritoneal washings and each one of them containing a greater number of intracellular bacteria (Guo et al., 1986b).

A more pertinent site to observe intracellular killing of salmonellae by phagocytes in fact is in the tissues of liver (Lin et al., 1987) and spleen (Wang et al., 1988) during the early stage of the infection. With the large dose of bacterial inoculation in this experimental protocol, extensive tissue involvements in the organs ensue and bacteria can usually be found in tissues within 2-3 days post infection by EM. Figures 3 and 4 illustrate the typical intracellular destruction of salmonellae within a neutrophil and a macrophage, respectively, in the liver of an infected C57BL/6J mouse. They appear to be indistinguishable from the inflammatory phagocytes seen in the peritoneal cavity.

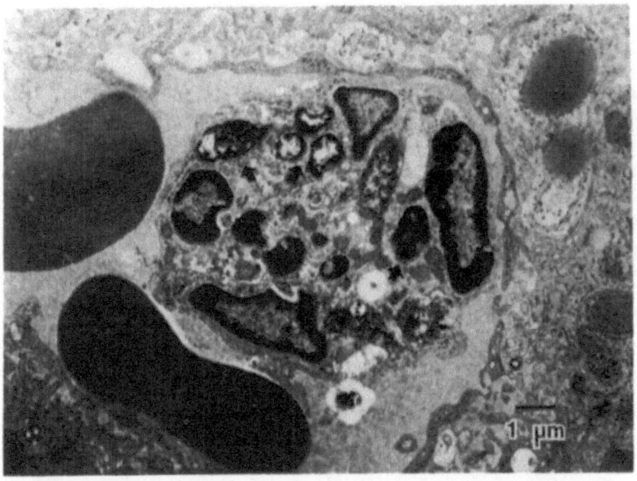

Figure 3. Salmonellae undergoing degenerative changes (arrows) within neutrophil seen in sinusoid of liver in infected C57BL/6J mouse. EM preparation by F.R. Lin, X.M. Wang and V.R. Mumaw.

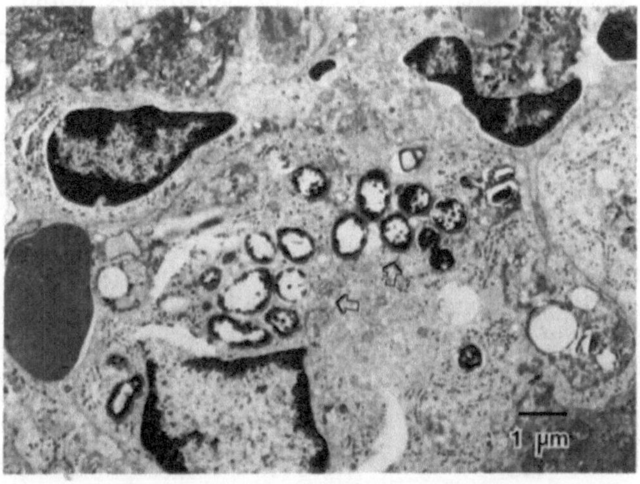

Figure 4. Salmonellae undergoing degenerative changes (arrows) within macrophage seen in lesion of liver in infected C57BL/6J mouse. EM preparation by F.R. Lin, X.M. Wang and V.R. Mumaw.

These defensive host cells can be found vigorously engaged in their antibacterial activities in sinusoids as well as in tissue lesions.

Morphologically intact salmonellae are occasionally seen among other disintegrating bacteria within phagocytes, which would suggest that these are more recently ingested organisms prior to their destruction. In spite of a relatively large infective dose used in this experimental design, inflammatory phagocytes display no evidence of impairment in their antibacterial functions even up to the terminal stage of the disease. In contrast to the intracellular killing by the polymorphs and macrophages, extensive and progressive bacterial multiplication conspicuously takes place in the extracellular space of sinusoids and in the cellular debris of tissue lesions.

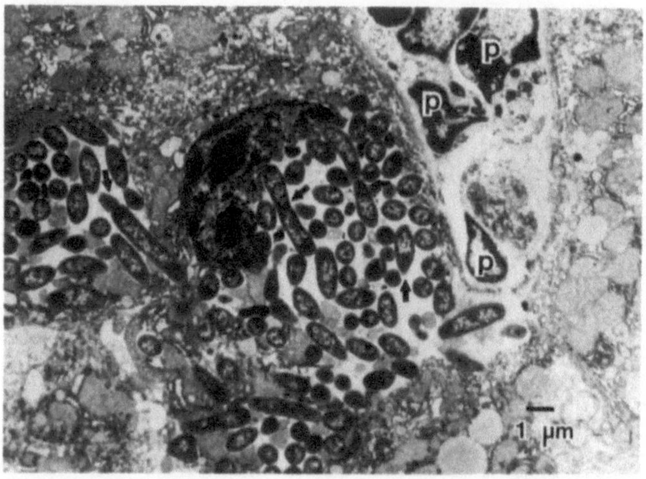

Figure 5. Intact and dividing (arrows) salmonellae within hepatocytes being protected from infiltrating polymorphs (P) seen in infected C57BL/6J mouse. EM preparation by F.R. Lin, X.M. Wang and V.R. Mumaw.

One can also follow the bacterial invasion of hepatocytes, within which massive accumulation of intact and dividing bacteria are visible, as illustrated in Figure 5, leading to the inevitable damage and rupture of the host cells. The intracellular sojourn within hepatocytes also protects the pathogen from hostile actions by the surrounding inflammatory phagocytes.

Thus, based on our observations with EM, the rapid proliferation of bacteria in the mouse at the early stage of infection, as seen in Figure 1, can readily be attributed to their generalized replication in the extracellular locations of sinusoids and among cellular debris in tissue lesions, as well as within hepatocytes and perhaps other parenchymal cells. Even though some bacterial multiplication might occur within a minute portion of the phagocytes (which we have yet to find), its contribution to the overall accumulation of bacterial population *in vivo* will be negligible in view of the overwhelming bacterial propagation elsewhere.

CONFIRMATION OF DIGESTION OF SALMONELLAE WITHIN PHAGOCYTES

We have further confirmed the destruction of salmonellae within inflammatory polymorphs and macrophages using the immunogold labeling technique (Hsu, 1989; Lin et al., 1989). This method is designed to tag the surface antigens of salmonellae with electron-opaque gold particles. Not only that we can verify the intracellular bacteria to be salmonellae, we can in fact visually trace their sequential disintegration within the cytoplasm of a phagocyte beginning from an intact bacterial cell through the disruption of its bacterial envelope to its residual antigenic fragments with no recognizable bacterial structure.

CLASSIFICATION OF SALMONELLAE AS INTRACELLULAR BACTERIA

Our extensive and thorough investigations in the pathogenesis of mouse typhoid as presented above have established that virulent *S. typhimurium* is unlikely to survive, let alone multiply, within phagocytes, i.e., polymorphs and macrophages. However, there is little doubt that hepatocytes are clearly a favorable site of its intracellular replication, which appears to provide a safe haven shielding the pathogen from attack by the infiltrating inflammatory phagocytes. The prominent locations for the rapid and invasive bacterial proliferation in the pathogenic process can be seen primarily in the extracellular space of sinusoids and among cellular debris of tissue lesions.

On the basis of our studies in mouse typhoid and along with data from cell culture studies of others, we are compelled to conclude that, as an exemplary species of *Salmonella*, *S. typhimurium* is an intracellular bacteria only within parenchymal cells, such as hepatocytes and epithelial cells, and perhaps within fibroblasts and cultured macrophage cell lines. In spite of all that is said and written over the years, I am not aware of any unequivocal experimental evidence to document that salmonellae can survive and multiply within host-derived phagocytes, including neutrophils, eosinophils and macrophages.

SALMONELLAE AS FACULTATIVE INTRACELLULAR PATHOGEN

The unfortunate dichotomy in the literature now is that *Salmonella* is commonly referred to as a "facultative intracellular parasite". This classification of pathogenic bacteria was first introduced by Suter (1956). According to his publication, "the organisms classified in this group can, during certain stages of the host-parasite relationship, survive or proliferate within phagocytes or in extracellular spaces" (see pg. 109 of reference Suter, 1956). Although only *S. typhi* was included among this group of organisms in the publication, other species of *Salmonella*, such as *S. typhimurium* and *S. enteritidis*, were added to a subsequent publication (Suter and Ramseier, 1964) and by other authors. Macrophages are generally regarded as the effector host cells involved in the interaction with facultative intracellular bacteria.

On the basis of our observations in the mouse typhoid model, we may offer the following points to demonstrate that, at least with *S. typhimurium*, it does not possess the necessary qualifications to be classified as a facultative intracellular pathogen according to the above stated definition: (a) the early infiltration of polymorphs with a conspicuous absence of macrophages at the site of infection; (b) the predominance of polymorphs coinciding with the most active phase of bacterial proliferation, reminiscent of the pathogenesis with extracellular bacteria; (c) a consistently rapid destruction of salmonellae within inflammatory polymorphs and macrophages at the site of infection; (d) bacterial invasion and replication within hepatocytes resulting in cellular degeneration; and (e) progressive and invasive bacterial proliferation in the extracellular locations of sinusoids and tissue lesions.

Since there is no other substantive evidence to support that any of the species of *Salmonella* can in fact "survive or proliferate within phagocytes", it is clearly inappropriate to continue referring to them as facultative intracellular pathogens as originally classified. Some investigators might argue that this issue is merely a matter of semantics. Regardless of our personal sentiments, I would simply plead for a common understanding in the use of this term "facultative intracellular parasites". Surely, it makes no difference to many of us whether we should adhere to the restricted definition of the term (and thus remove *Salmonella* from this classification) or re-define the term to be more inclusive. It would be immensely helpful if we all agree to use the same language in the literature.

Parenthetically, if I may be permitted to add, no one would be more gratified than myself if someone could provide me with or direct me to a source of definitive experimental evidence showing the survival and/or replication of salmonellae within host-derived polymorphs and macrophages.

PERSPECTIVES OF PATHOGENESIS, IMMUNITY AND VACCINATION IN MOUSE TYPHOID

Regardless of our personal convictions on the present classification of *Salmonella* and for the sake of convenience, we might for a moment accept that *S. typhimurium* does not survive or proliferate within phagocytes. This would imply that its pathogenic property is similar to that of the obligate extracellular bacteria, as in the classical example of *Streptococcus pneumoniae*, in which acquired immunity to the disease is antibody-mediated. Our studies of mouse typhoid can in fact offer some compelling experimental evidence to support such a proposition. Antiserum can passively protect a low dose of infectious challenge in mice and reduce the size of skin lesions in guinea pigs (Hsu, 1989). Both heat-killed salmonellae (Nakoneczna and Hsu, 1983) and their sonicated fragments (Hsu et al., 1984), which induce only a humoral immunity to the bacterial antigens, can provide a highly effective protection against a subsequent challenge. The induction of protective antibody can be attributed primarily to the delipidated O-antigenic polysaccharide of salmonellae (Ding et al., 1990). However, we do recognize that humoral immunity is merely the primary component of acquired immunity in mouse typhoid. Rather, a previous infection or a live attenuated vaccine is likely to provide a full protection against subsequent challenges, since it induces both the antibody-mediated immunity as well as the cell-mediated immunity in the form of delayed hypersensitivity (Hsu, 1989).

The effective protection offered by antibody in mouse typhoid is adequately confirmed by histopathological evidence. Early tissue lesions in susceptible mice, protected with nonviable vaccines and subsequently challenged with virulent *S. typhimurium*, are characterized by an early infiltration of polymorphs, which is indistinguishable from that seen in the infected but unimmunized mice. The significant difference is that the abscess formation in the vaccinated mice is discrete and self-limiting, followed by the granuloma formation and healing. This would imply that the early mobilization of polymorphs at the site of infection is effective in eliminating the pathogen in the presence of opsonizing antibodies, which is analogous to the protective process of humoral immunity in pneumococcal infection. It also confirms the absence of delayed hypersensitivity in the early phase of the infectious process.

In essence, therefore, our observations in the protective action of nonviable vaccines are instructive in supporting the assertion that *S. typhimurium* is unlikely to be a facultative intracellular pathogen as traditionally defined.

Acknowledgments

Our laboratory research in murine salmonellosis was primarily supported by U.S. Public Health Service grants AI 06765 and AI 19434 from the National Institute of Allergy and Infectious Diseases. Faculty colleagues participating in this project include Irene Nakoneczna, Charles W. Moncure and Virgil R. Mumaw.

REFERENCES

Ding, H.F., Nakoneczna, I., and Hsu, H.S., 1990, Protective immunity induced in mice by detoxified salmonella lipopolysaccharide, *J. Med. Microbiol.* 31:95.

Finlay, B.B., and Falkow, S., 1989, *Salmonella* as an intracellular parasite, *Mol. Microbiol.* 3:1833.

Guo, Y.N., Hsu, H.S., Mumaw, V.R., and Nakoneczna, I., 1986a, Electronmicroscopy studies on the bactericidal action of inflammatory leukocytes in murine salmonellosis, *J. Med. Microbiol.* 21:151.

Guo, Y.N., Hsu, H.S., Mumaw, V.R., and Nakoneczna, I., 1986b, Electronmicroscopy studies on the opsonic role of antiserum and the subsequent destruction of salmonellae within murine inflammatory leukocytes, *J. Med. Microbiol.* 22:343.

Hsu, H.S., 1989, Pathogenesis and immunity in murine salmonellosis, *Microbiol. Rev.* 53:390.

Hsu, H.S., Nakoneczna, I., and Guo, Y.N., 1984, Histopathological evidence for protective immunity induced by sonicated *Salmonella* vaccine, *Can. J. Microbiol.* 31:54.

Lin, F.R., Wang, X.M., Hsu, H.S., Mumaw, V.R., and Nakoneczna, I., 1987, Electron microscopic studies on the location of bacterial proliferation in the liver in murine salmonellosis, *Br. J. Exp. Pathol.* 68:539.

Lin, F.R., Hsu, H.S., Mumaw, V.R., and Moncure, C.W., 1989, Confirmation of destruction of salmonellae within murine peritoneal exudate cells by immunocytochemical technique, *Immunology*, 67:394.

Nakoneczna, I. and Hsu, H.S., 1980, The comparative histopathology of primary and secondary lesions in murine salmonellosis, *Br. J. Exp. Pathol.* 61:76.

Nakoneczna, I., and Hsu, H.S., 1983, Histopathological study of protective immunity against murine salmonellosis induced by killed vaccine, *Infect. Immun.* 39:423.

Suter, E., 1956, Interaction between phagocytes and pathogenic microorganisms, *Bacteriol. Rev.* 20:94.

Suter, E., and H. Ramseier, 1964, Cellular reactions in infection, *Adv. Immunol.*, 4:117.

Wang, X.M., Lin, F.R., Hsu, H.S., Mumaw, V.R., and Nakoneczna, I., 1988, Electronmicroscopic studies on the location of salmonella proliferation in murine spleen, *J. Med. Microbiol.* 25:41.

Xu, H.R., and Hsu, H.S., 1992, Dissemination and proliferation of *Salmonella typhimurium* in genetically resistant and susceptible mice, *J. Med. Microbiol.* 36:377.

SELECTION FOR BACTERIAL VIRULENCE GENES THAT ARE SPECIFICALLY INDUCED IN ANIMAL TISSUES

Michael J. Mahan, James M. Slauch, and John J. Mekalanos

Dept. of Microbiology and Molecular Genetics, Harvard Medical School
200 Longwood Ave, Boston, MA 02115

INTRODUCTION

In order to understand the molecular mechanisms by which a bacterial pathogen causes disease, one must identify both the gene products that are required for the various stages of the infection process and the regulatory factors that govern their expression. Clues to this understanding have come from coordinate regulation studies indicating that the expression of most known virulence factors is regulated by environmental signals in the laboratory that presumably reflect similar signals present in host tissues (Mekalanos, 1992). For example, coordinately regulated virulence genes have been isolated that respond to iron levels, calcium levels, carbon sources, temperature, pH, and osmolarity. However, this approach is limited by the knowledge of and ability to reproduce probable host environmental signals in the laboratory. With these limitations in mind, we have developed a genetic system, termed IVET (*in vivo* expression technology), that does not rely on the reproduction of these environmental signals but, rather, is entirely dependent on the induction of genes in host tissues.

RESULTS AND DISCUSSION

The IVET Selection Strategy

Our goal was to select bacterial genes that are specifically induced when a pathogen infects a host. A subset of these genes will encode virulence factors required for the infection process *in vivo*, but which are not required for growth *in vitro*, and are therefore poorly expressed outside of host tissues. The IVET strategy begins with a bacterial strain carrying a mutation in a biosynthetic gene that highly attenuates growth *in vivo*. Growth in the host of the mutant strain is then complemented by operon fusions to the same biosynthetic gene. While several biosynthetic genes can in theory be used in this selection scheme (*e.g.*, *aroA*, *asd*, *thyA*), we have initially focused on the use of the *purA* mutation in *Salmonella typhimurium*. Strains carrying mutations in the *purA* gene are highly attenuated in their ability to cause mouse typhoid and to persist in host tissues (McFarland and Stocker, 1987). For example, when a BALB/c mouse was infected intraperitoneally (i.p.) with a 100:1 mixture of a *purA* mutant and an isogenic wild type strain, we recovered $>10^7$ wild type cells

Biology of Salmonella, Edited by F. Cabello *et al.*,
Plenum Press, New York, 1993

and <20 mutant cells from the spleen 5 days after infection. We describe here a selection strategy based on this PurA requirement, allowing one to positively select for genes that are specifically induced in the mouse.

In devising the IVET selection system, we wanted to fulfill 3 criteria. First, we wanted to build *purA* fusions in single copy on the chromosome in order to avoid any complications such as stability, copy number anomalies, or supercoiling effects that are inherent to plasmids. Second, we wanted to build these fusions in the bacterial chromosome without disrupting any genes. If a gene or operon encodes products required for the infection process, loss of these products would be lethal. Lastly, we wanted a means to conveniently monitor the transcriptional activity of any given gene both *in vivo* and *in vitro*.

Following these 3 criteria, we have constructed a synthetic operon composed of a promoterless *purA* gene and a promoterless *lac* operon that was cloned into the broad host range suicide vector, pGP704 (Miller and Mekalanos, 1988), resulting in plasmid pIVET1 (Figure 1). Replication of pIVET1 is dependent on the Pi replication protein, which can be delivered *in trans*. Random DNA fragments can be cloned 5′ to the *purA* gene, resulting in a pool of recombinant plasmids containing *purA-lac* transcriptional fusions. Upon introduction of this pool of recombinant plasmids into a recipient strain lacking the Pi replication protein, selection for ampicillin resistance demands the integration of the recombinant plasmids into the chromosome by homologous recombination, using the Salmonella DNA as the source of homology (Figure 1). The cloned Salmonella DNA is the only region of homology between the plasmid and the chromosome. The *purA* gene was cloned from *E. coli* and lacks sufficient homology to recombine; moreover, wild type *S. typhimurium* is naturally *lacZYA*⁻. The product of this integration event generates a duplication of Salmonella material in which the native chromosomal promoter drives the *purA-lac* fusion, whereas the cloned promoter drives the expression of the wild type copy of the putative virulence gene, "X". Upon introduction of this pool of fusion strains into a mouse, both copies can be selected for *in vivo*. Chromosomal promoters that are transcriptionally active enough to provide sufficient levels of the *purA* gene product are selected to overcome the parental purine deficiency by demanding growth of the fusion strain in the mouse. Moreover, in those cases where the gene encodes an essential virulence factor, the gene product is selected because it is necessary for infection. The presence of the *lac* operon allows one to monitor the transcriptional activity of the entire synthetic operon. Therefore, this selection scheme results in single copy diploid fusions, whose expression levels can be monitored *in vivo* or *in vitro* by measuring β-galactosidase activity, fulfilling our initial 3 criteria in the design of the IVET selection.

Fusion strains that can be recovered from animal tissues such as the spleen apparently contain *purA-lac* fusions to genes that are expressed at relatively high levels *in vivo*. Note that all genes that have transcriptionally active promoters will answer this selection. However, our goal was to identify genes that are specifically induced in the host, but which are not required for growth *in vitro*, and are therefore poorly expressed outside of host tissues. Thus, among the fusion strains that are recovered from the mouse, we screened for strains that express the fusion at low levels *in vitro* by monitoring the Lac phenotype on MacConkey Lactose indicator medium.

Construction of *purA-lac* Transcriptional Fusions

Random *Sau*3A I restriction fragments of *S. typhimurium* DNA were cloned into the broad host range suicide vector, pIVET1, 5′ to the promoterless *purA* gene, resulting in a "pool" of recombinant plasmids. Plasmid DNA containing the *purA-lac* transcriptional fusions was electroporated into *E. coli* strain SM10 *λpir*, which contains both the Pi replication protein and RP4 conjugative functions necessary for replication and mobilization (Simon et al., 1983). The resultant pool of strains was mated into a streptomycin resistant *purA* deletion strain of *S. typhimurium*, MT168 (DEL2901[*purA874*::IS10]), that does not

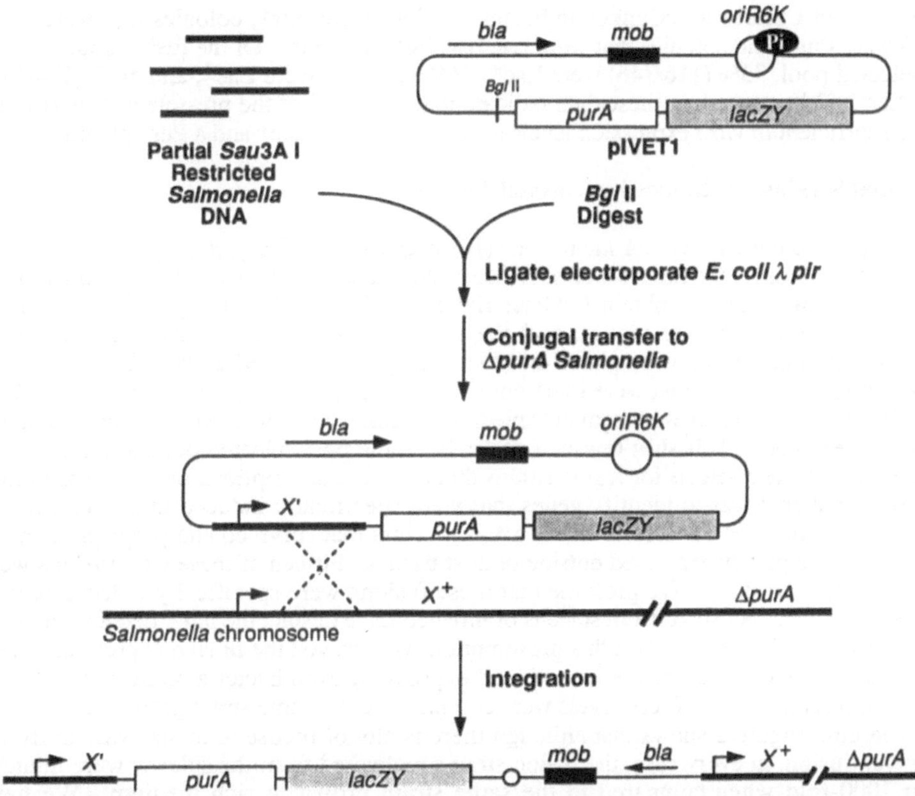

Figure 1. The IVET approach: selection for genes that are specifically induced in the host. Plasmid pIVET is a derivative of pGP704, whose replication is dependent on the Pi protein, which can be delivered *in trans*. This Ap^r plasmid can be mobilized by the RP4 conjugative functions that act at the *mob* site. pIVET1 contains a promoterless *purA* gene and a promoterless *lac* operon. 5′ to the *purA* gene, there are unique sites for cloning random fragments of bacterial DNA. Total *S. typhimurium* DNA was isolated, digested with *Sau3*A I, and cloned into the *Bgl* II site of pIVET1. This pool of recombinant plasmids containing the cloned *purA-lac* transcriptional fusions was electroporated into *E. coli* strain SM10 *λpir*, which contains both the Pi replication functions and the RP4 conjugative functions necessary for replication and mobilization of the recombinant plasmids. The pool of recombinant clones containing the *purA-lac* transcriptional fusions was mated into a *purA* deletion strain of *S. typhimurium*, MT168, which lacks the Pi replication protein necessary for plasmid replication. Selection for Ap^r results in the integration of the plasmid into the bacterial chromosome. This integration events generates a duplication of Salmonella material in which the native chromosomal promoter drives the *purA* gene and the cloned promoter drives the expression of the putative virulence gene, X.

contain the Pi replication protein. Selection for ampicillin resistance demands integration of the recombinant plasmids into the bacterial chromosome, resulting in a pool of chromosomal fusion strains (Figure 1). We plated the pre-selected pool on MacConkey Lactose medium to determine the percentage of integrated fusion strains that were transcriptionally active *in vitro* before being subjected to the *in vivo* PurA selection. The level of *lacZ* expression necessary to give a Lac^+ phenotype appears to correspond with the level of *purA* expression required to supplement the parental chromosomal *purA* deletion. That is, colonies that were

Lac$^+$ (red) on Lactose MacConkey indicator medium were Pur$^+$, colonies that were Lac$^\pm$ (pink) were Pur$^\pm$, and colonies that were Lac$^-$ (white) were Pur$^-$. Of the fusion strains in the pre-selected pool, 33% (116/346) were Lac$^+$, 16% (56/346) were Lac$^\pm$, and 50% (174/346) were Lac$^-$. Taken together, these data indicate that one third of the pre-selected innoculum showed sufficient *in vitro* expression levels to result in both a Lac$^+$ and a Pur$^+$ phenotype.

ivi-Fusion Strains are Induced in Animal Tissues

The pool of integrated *purA-lac* fusions (10^6 organisms) was injected i.p. into a BALB/c mouse. After 3 days, the mouse was sacrificed and bacterial cells were recovered from the spleen, where we routinely obtain 10^8 bacterial cells. These cells were grown overnight in LB, injected into a second mouse, and the process was repeated. The bacterial cells recovered from the spleen were plated on MacConkey Lactose media. We observed a shift in the percentage of cells that were Lac$^+$ compared to the initial innoculum: 86% (235/273) of the bacterial cells recovered from the spleen were Lac$^+$, 9% (24/273) were Lac$^\pm$, and 5% (14/273) were Lac$^-$. This shift toward Lac$^+$ cells in the post-selected population indicates that the PurA system selects for fusion strains that contain transcriptionally active promoters. However, our goal was to identify genes that were specifically induced in animal tissues. Therefore, we focused our efforts on the 5% Lac$^-$ class that survived and propagated in the mouse but were poorly expressed outside of host tissues. Fifteen of these Lac$^-$ fusions were chosen for further study. We presume that these fusions were specifically induced *in vivo* because they were recovered from spleens of infected mice despite the *purA* deletion inherent to the fusion strains. To confirm this presumption, we assayed the *in vivo* expression levels of the fusions by direct measurement of *lacZ* expression from bacteria recovered from the spleens of infected mice. These levels were compared to the same strain grown overnight in rich medium. Figure 2 shows that although there is alot of mouse to mouse variability for any given fusion, in every case, the fusion strains recovered from the spleens were induced 40 to 1000-fold when compared to the same strain grown in rich medium. We have designated genes that are regulated in this fashion *ivi* (*in vivo* induced genes). As a control for this experiment, we chose a random Lac$^+$ fusion, MT222, from our initial pre-selected pool that expressed high levels of β-galactosidase activity on laboratory media. Figure 2 shows that this non-selected Lac$^+$ fusion did not show a dramatic *in vivo* induction compared to growth on rich medium; *i.e.*, the β-galactosidase activity was relatively high in both conditions.

Identification of Genes Defined by the *ivi* Fusions

We have used a genetic approach to clone the 15 chromosomal *ivi*-fusions that survived the *in vivo* PurA selection, but were poorly expressed outside of animal tissues. Bacteriophage P22 is a Salmonella-specific transducing phage that is extremely efficient in moving genes from one strain to another. We have exploited this ability of phage P22 to clone the fusions directly from the bacterial chromosome. A P22 lysate grown on a strain containing a chromosomal *ivi* fusion was used to transduce a recipient that contains the Pi replication protein. Phage P22 packages about one minute of the donor chromosome and, upon infection, delivers this piece as a linear fragment to the recipient cell. The linear fragment that contains a chromosomal fusion can recircularize by homologous recombination, using the duplicated Salmonella sequences that flank the integrated plasmid as a source of homology. The circularized DNA can be maintained as a plasmid in the presence of the Pi replication protein, resulting in the cloned fusion of interest.

In order to identify genes that were selected *in vivo*, we have cloned and partially sequenced the Salmonella DNA contained in the 15 cloned *ivi* fusions. Using a primer homologous to the 5′ end of the *purA* gene, we sequenced from 200 to 400 nucleotides of Salmonella DNA upstream of the *purA* gene. Sequence analysis indicates that the 15 fusions

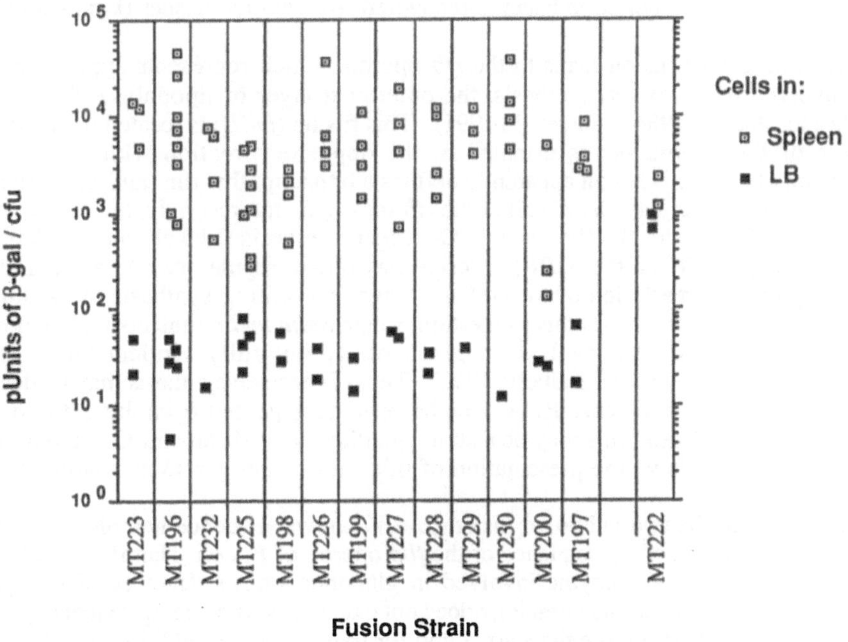

Figure 2. β-galactosidase expression levels from cells recovered from spleens of infected mice versus the same strain grown on rich media. The vertical axis depicts the picounits of β-gal per colony forming unit, where units of β-gal = μmoles o-nitrophenol (ONP) formed/min. The open boxes denote the β-gal activity of cells recovered from the spleen. The closed boxes denote the β-gal activity from cells grown in rich medium. The β-gal activity in each sample was determined by a kinetic assay, using a model SPF-500c spectroflourometer (SLM Aminco). The units of β-galactosidase were obtained by comparing the activity to a standard curve.

represent 5 different genes. Two of five fusions (*ivilV* and *iviV*) are in genes that show no homology to sequences in GenBank (Genbank version 72; Devereaux et al., 1984). This suggests that the selection system may have identified previously uncharacterized genes that are specifically induced during mouse infection. One of the genes in the "unknown" class will be discussed in detail later.

Three of five fusions showed extensive sequence homology to known sequences. One of which is the *carAB* operon (*iviI*), whose genes encode the two subunits of carbamoyl phosphate synthetase, which is involved in arginine and pyrimidine biosynthesis (Nyunoya and Lusty, 1983; Piette et al., 1984; Glandsdorff, 1987). The induction of this operon is consistent with a previous report on the attenuating effect of pyrimidine mutations, suggesting a low availability of pyrimidines in animal tissues (Fields et al., 1986). The second fusion was located in the *pheST himA* operon (*iviII*) that encodes the two subunits of phenylalanyl-tRNA synthetase and one subunit of integration host factor (IHF), which is a DNA binding protein involved in DNA recombination, DNA replication, and gene regulation (Plumbridge and Springer, 1980; Mechulum et al., 1985; Friedman, 1988). While we do not as yet have any evidence whether the other IHF subunit, *himD*, is also induced *in vivo*, it is intriguing to speculate what the effect of enhanced IHF-mediated site specific recombination might have on the infection process. For example, IHF is involved in type 1 pilus phase variation (Dorman and Higgins, 1987) and alterations in type 1 pili expression have been reported in bacteria recovered from animal tissues (Leunk and Moon, 1982).

The third known fusion maps to the *rfb* operon, which represents approximately 20 genes involved in O-antigen synthesis, the outermost layer of lipopolysaccharide (LPS) (Mäkelä and Stocker, 1984; Jiang et al., 1991). This fusion (*iviIII*) is located in the second to last gene of the operon but is oriented in the opposite direction with respect to the transcription of *rfb*. This fusion apparently defines a transcript that runs anti-sense to *rfb* and is complementary to at least the 3′ end of the *rfb* message. Interestingly, the last gene in the *rfb* operon, *rfbP*, encodes the first gene in O-antigen synthesis (Mäkelä and Stocker, 1984; Jiang et al., 1991). Thus, the *iviIII* transcript may play a regulatory role as an anti-sense RNA that prevents translation of the *rfbP* message, preventing synthesis of O-antigen in animal tissues. Consistent with this suggestion, rough *Salmonella* mutants that do not make O-antigen are highly attenuated when given orally but fully virulent when injected intraperitoneally (Nnalue and Lindberg, 1990). Thus, O-antigen synthesis may be decreased by an *in vivo* induced anti-sense RNA once bacteria have passed from the gut into deeper tissues such as the spleen. This may be a strategy either to evade an anti-O antigen immune response or to slow down the presentation of this surface antigen to an immunologically naive host.

Lastly, one of the two unknown fusions, *iviIV*, shows sequence homology to a region immediately 5′ to the coding sequence of the *fbp* operon of *E. coli (*Hamilton et al., 1988). The *fbp* gene encodes an enzyme involved in gluconeogenesis, fructose bisphosphatase Fraenkel, 1987). However, the homology does not extend into the coding sequence; rather, it ends about 20 base pairs before the start codon for translation. In addition, the direction of transcription is in the opposite orientation *vis a` vis* the *fbp* gene of *E. coli.* It is of interest that the *S. typhimurium* and *E. coli fbp* homologs are unusual in that in they do not map in the same position with respect to nearby genes (Sanderson and Roth, 1988; Bachman, 1989). It is possible that we have uncovered a DNA rearrangement between the two bacteria and the fusion is in an as yet unknown gene whose map position in Salmonella corresponds to the position of the *fbp* gene of *E. coli.*

Mutations in Genes Identified by *ivi* Fusions Show a Virulence Defect

The isolation of null mutations in the genes that are induced *in vivo* would determine their overall contribution to Salmonella pathogenesis. The strains containing the *ivi* fusions are phenotypically Lac$^-$ *in vitro*. However, we can still monitor the low level expression of the fusion by using the sensitive chromogenic substrate, 5-bromo-4-chloro-3-indolyl-β-D-galactopyranoside (X-gal). With this screening scheme, we have used transposon mutagenesis and *lac* gene fusion technology to identify Mu*d*-Cm insertion mutations that reduce expression of the operons defined by the fusions. The Mu*d*-Cm insertion element is a defective Mu prophage encoding chloramphenicol resistance (T. Elliot); transposition of this element was according to methods described previously (Hughes and Roth, 1988). Using this system, we have isolated Mu*d*-Cm insertion mutations in all (5/5) of the cloned *ivi* operons. These insertions disrupt the gene at the site of insertion and, in addition, reduce the expression of downstream genes in the same operon (*i.e.*, the insertions are "polar" on downstream genes). Strains containing insertions in such genes decrease the expression of *lacZ*, shifting the colony color from blue (higher levels of expression) to light blue (lower levels of expression).

Since these insertion mutations affect the expression of genes defined by the *ivi*-fusions, we can determine the effect of reduced expression on virulence by crossing the insertion mutations into an otherwise wild type chromosome and subsequently challenging mice with these mutant strains. To date, we have been successful in obtaining strains that contain such polar Mu*d*-Cm insertion mutations in 3 of 5 cases. The oral lethal dose required to kill 50% of infected animals (LD_{50}) for wild type *S. typhimurium* infection of a BALB/c mouse is 5 x 10^4 organisms. BALB/c mice were challenged orally with strains containing Mu*d*-Cm insertion mutations in operons defined by the *ivi*-fusions. Table 1 shows that all 3 classes of insertion-bearing mutants tested have an increase in LD_{50} from 40 to >4000-fold. This indicates that the IVET system selects for attenuated strains of *S. typhimurium*. To determine if this genetic system selects for live attenuated vaccines, we challenged mice that were previously immunized with the attenuated insertion-bearing mutants. One of the highly attenuated strains, strain MT220 (*ivill-7::*Mu*d*-Cm), induced strong immunity to subsequent oral challenge with wild type *S. typhimurium* (LD_{50} was increased 2 x 10^4-fold), indicating that this genetic system will facilitate the construction of live attenuated vaccines.

TABLE 1. Mutations in operons defined by the *purA-lac* fusions confer a virulence defect.

Bacterial strain[a]	Relevant genotype[b]	Oral LD_{50}[c]
14028	wild type	5 x 10^4
MT219	*ivil-6*::Mu*d*-Cm (*carAB*)	>2 x 10^8
MT220	*ivill-7*:: Mu*d*-Cm (*pheST himA*)	>2 x 10^8
MT221	*ivi-IV-9*:: Mu*d*-Cm	2 x 10^6

[a]All bacterial strains used in this study are derivatives of *S. typhimurium* ATCC 14028. Mutant strains are isogenic to wild type except for a Mu*d*-Cm insertion element in an *ivi* operon defined by its respective *purA-lac* fusion. [b]Genes in parenthesis indicate the *ivi* operon whose expression is affected by the Mu*d*-Cm insertion mutation. [c]For each of the mutant strains, groups of 5 mice were perorally infected with a dose of 10^6, 10^7, or 10^8 organisms. The dose required to kill 50% of infected animals (LD_{50}) for each of these strains was compared to that of the wild type.

Additional IVET Selection Systems

Invariably, even subtle changes in a genetic selection results in dramatically different classes of genes. Therefore, another way to identify additional *ivi* genes is to devise additional IVET selection schemes that are based on different biosynthetic genes. For example, *thyA* mutants are deficient in thymine production and are attenuated for virulence. We have constructed a *thyA* fusion vector system, pIVET2, analogous to the *purA* system described above. One advantage of the *thyA* system is the ability to select for *thyA*⁻ mutants, using the antibiotic trimethoprim (Tmp). Thus, the *thyA* system allows the positive selection for genes that are both induced *in vivo* (by selecting $ThyA^+$) and transcriptionally inactive *in vitro* (by selecting $ThyA^-$). Moreover, the level of *thyA* gene expression required to give prototrophy is apparently lower than for *purA* (i.e., even pink colonies on MacConkey agar containing *thyA* fusions are Thy^+). This effect together with trimethoprim selection, allows for the selection of operon fusions *in vivo* that display a completely different dynamic range of derepression than is possible with the PurA selection. Moreover, the *thyA* selection system will identify a different class of genes that are induced at lower levels both in and out of the mouse since low expression is *sufficient* to satisfy the thymine requirement *in vivo*, while low expression is *necessary in vitro* because of the trimethoprim selection.

The IVET approach has been extended to include selections involving antibiotic resistance markers. To this end, we have built a *cat-lac* fusion vector, pIVET8, which allows the construction of *cat-lac* transcriptional fusions. The advantage of this system is that *ivi* genes can be selected *in vivo* by treatment of animals or tissue culture infection systems with chloramphenicol. However, the general use of this approach is limited by the pharmacodynamics of the drug of choice, in this case chloramphenicol (*i.e.*, its tissue distribution, intracellular penetration, toxicity, etc.). With these caveats in mind, pIVET8 should be a useful tool to probe for *ivi* genes in systems where complementation of a biosynthetic gene is not yet feasible, *e.g.*, where either mutations in biosynthetic genes are not readily available or in tissue culture systems that do not involve defined culture medium.

SUMMARY AND APPLICATIONS

The identification of most coordinately regulated virulence genes has been achieved by defining environmental signals in the laboratory that presumably reflect similar signals occurring in host tissues. The major limitation to this strategy, however, is the ability to anticipate the required environmental signals. To overcome this limitation, we have devised a genetic approach, termed IVET (*in vivo* expression technology), that does not rely on the reproduction of these host environmental signals, but rather, is solely dependent on the induction of genes in the host. We have shown that genes identified using the IVET selection are not only induced in animal tissues, but also, mutations in all genes tested confer a clear virulence defect.

The IVET selection system has several applications in the area of vaccine and anti-microbial drug development, some of which are described below. *Live vaccines.* This system selects for mutations in genes involved in virulence and therefore facilitates the construction of live vaccines. Furthermore, the gene products of *ivi* genes may lead to the discovery of new antigens useful as vaccine components. *Delivery of Antigens.* The identification of *ivi* promoters that are specifically induced in animal tissues provides a means of establishing *in vivo* regulated expression of heterologous antigens. This would ensure delivery of the antigen only in the appropriate host tissue. *Drug Development.* The IVET approach will define biosynthetic, catabolic, and regulatory genes that are required for infection and this information should provide new targets for anti-microbial drug development.

ACKNOWLEDGMENTS

We are grateful to Don Wilbur and other members of the animal research facility at the School of Public Health, Harvard Medical School, for their invaluable expertise in breeding mice for these experiments.

This work was supported by NIH research grant AI18045 (J.J.M.), National Research Service Award AI08245 (M.J.M.) and Damon Runyon-Walter Winchell Cancer Research Fund Fellowship DRG-1016 (J.M.S.)

REFERENCES

Bachmann, B.J., 1990, Linkage map of *Escherichia coli* K-12, edition VIII. Microbiol. Rev. 54:130-197.

Devereux, J., Haeberli, P., and Smithies, O., 1984, A comprehensive set of sequence analysis programs for the VAX. Nucleic Acids Res. 12:387-395.

Dorman, C.J. and Higgins, C.F., 1987, Fimbrial phase variation in *Escherichia coli:* dependence on integration host factor and homologies with other site-specific recombinases. J. Bacteriol. 169:3840-3843.

Fields P.A., Swanson, R.V., Hairdaris, C.G., and Heffron, F., 1986, Mutants of *Salmonella typhimurium* that cannot survive within the macrophage are avirulent. Proc. Natl. Acad. Sci. U. S. A. 83: 5189-5193.

Fraenkel, D.G., 1987, Glycolysis, pentose phophate pathway, and Entner-Doudoroff pathway. In *Escherichia* and *Salmonella typhimurium*: cellular and molecular biology, F.C. Neidhart, ed. (Washington, DC: American Society for Microbiology), pp. 142-150.

Friedman, D.I., 1988, Integration host factor: a protein for all reasons. Cell 55:545-544.

Glandsdorff, N., 1987, Biosynthesis of arginine and polyamines. In *Escherichia coli* and *Salmonella typhimurium*: cellular and molecular biology, F.C. Neidhart, ed. (Washington, DC: American Society for Microbiology), pp. 321-344.

Hamilton, W.D., Harrison, D.A., and Dyer, T., 1988, Sequence of the *Escherichia coli* fructose-1,6-bisphosphatase gene. Nucleic Acids Res. 16:8707-8707.

Hughes, K.T. and Roth, J.R., 1988, Transitory *cis* complementation is a method for providing transposition functions to defective transposons. Genetics 119:9-12.

Jiang, X.M., Neal, B., Santiago, F., Lee, S.J., Romana, L.L., and Reeves, P.R, 1991, Structure and sequence of the *rfb* (O antigen) gene cluster of *Salmonella typhimurium* (strain LT2). Mol. Microbiol. 5:695-713.

Leunk, R.D. and Moon, R.J., 1982, Association of type 1 pili with the ability of livers to clear *Salmonella typhimurium*. Infect. Immun. 36:1168-1174.

Mäkelä, P.H., and Stocker, B.A.D., 1984, Genetics of lipopolysaccharide. In: Handbook of endotoxin: volume 1, chemistry of endotoxin, E.Th. Rietschel, ed. (New York, NY: Elsevier), pp. 59-137.

Mekalanos J.J., 1992, Environmental signals controlling expression of virulence determinants in bacteria. J. Bacteriol. 174:1-7.

McFarland, W.C. and Stocker, B.A., 1987, Effect of different purine auxotrophic mutations on mouse-virulence of a Vi-positive strain of *Salmonella dublin* and of two strains of *Salmonella typhimurium*. J. Microbial. Patho. 3(2):129-141.

Mechulam, Y., Fayat, G., and Blanquet, S., 1985, Sequence of the *Escherichia coli pheST* operon and identification of the *himA* gene. J. Bacteriol. 163:787-791.

Miller, V.L. and Mekalanos, J.J., 1988, A novel suicide vector and its use in construction of insertion mutations: Osmoregulation of outer membrane proteins and virulence determinants in *Vibrio cholerae* requires *toxR*. J. Bacteriol. 170:2575-2583.

Nnalue, N.A. and Lindberg, A.A., 1990, *Salmonella choleraesuis* strains deficient in O antigen remain fully virulent for mice by parenteral innoculation but are avirulent by oral administration. Infect. Immun. 58:2493-2501.

Nyunoya, H. and Lusty, C.J., 1983, The *carB* gene of *Escherichia coli:* a duplicated gene coding for the large subunit of carbamoyl-phosphate synthetase. Proc. Natl. Acad. Sci. U.S.A. 80: 4629-4633.

Piette, J., Nyunoya, H., Lusty, C.J., Cunin, R., Weyens, G., Crabeel, M., Charlier, D., Glandsdorff, N. and Pierard, A., 1984, DNA sequence of the *carA* gene and the control region of *carAB*: tandem promoters, respectively controlled by arginine and the pyrimidines, regulate the synthesis of carbamoyl phosphate synthetase in *Escherichia coli* K-12. Proc. Natl. Acad. Sci. U.S.A. 81: 4134-4138.

Plumbridge, J.A., and Springer, M., 1980, Genes for the two subunits of phenylalanyl-tRNA synthetase of *Escherichia coli* are transcribed from the same promoter. J. Mol. Biol. 144:595-600.

Sanderson, K.E. and Roth, J.R., 1988, Linkage map of *Salmonella typhimurium*, edition VII. Microbiol. Rev. 52:485-532.

Simon R., Priefer, U. and Puhler, A., 1983, A broad host range mobilization system for *in vivo* genetic engineering: transposon mutagenesis in gram negative bacteria. Biotechnology 1:784-791.

MECHANISM OF PATHOGENICITY IN *Salmonella abortusovis*

Salvatore Rubino[1], Patrizia Rizzu[1], Giuseppe Erre[1], Mauro M. Colombo[2], Guido Leori[3], Paul A. Gulig[4], Thomas J. Doyle[4], Marina Pisano[1], Sergio Uzzau[1], Piero Cappuccinelli[1].

[1]Ist. di Microbiologia e Virologia, Universita' di Sassari, Italy
[2]Dip. di Biologia Cellulare e dello Sviluppo, Universita' La Sapienza, Roma, Italy
[3]Ist. Zooprofilattico Sperimentale della Sardegna, Sassari, Italy
[4]Dept. of Immunology and Medical Microbiology, University of Florida, Gainsville, USA

INTRODUCTION

Salmonella abortusovis is a serotype of *S. enterica* and causes a contagious infection in sheep leading to bacteraemia and colonization of the placenta followed by abortion. After the infection, in the first pregnancy, a natural immunity is generally acquired by sheep. Frequent mortality in lambs has also been reported. The infection is widely spread to flocks in Europe, and due to the high costly damages to a sheep-based farming, and since antibiotic treatment is ineffective, prevention by vaccination should be encouraged. However most of the dead or living vaccines commercialized at present, give only partial protection, and their effectivenes is often related to specific geographic areas (Pardon et al, 1990). More efforts should be made for the development of a genetically engineered vaccine that would be a better candidate for reliable protection. Since no major phenotypic characters are suitable, the epidemiology based on the molecular features of the pathogenic strains could help to understand the routes of dissemination (Helmuth et al. 1985, Colombo et al. 1992).

In the last years the molecular basis of pathogenicity has been investigated in order to clarify the mechanisms of the infection. A mouse model (BALB/c) has been widely utilized by many authors with reliable results since *S. abortusovis* oral inoculation is lethal for mice. The bacteria can be isolated from sheep vaginal discharges after abortion, from tissues of the placenta, and from the invaded organs of the foetus such as liver, spleen, brain, and

Biology of Salmonella, Edited by F. Cabello *et al.*.
Plenum Press, New York, 1993

stomach. The biochemical profile is very similar to the one belonging to *S. typhi*, and serological identification must be done since the use of biochemical kits is inadequate. Strains resistant to streptomycin are often isolated. *S. abortusovis* shows no fimbriae and does not haemoagglutinate human, chicken, rabbit, guinea pig or sheep red blood cells, but shows adhesion to epithelial cells *in vitro* (Colombo et al. 1992). No chromosomal polymorphism has been demonstrated by restriction analyses, and a ubiquitous plasmid of variable high molecular weight (50-70 kb) is present (Popoff et al. 1984) as in 18 different *Salmonella* species and serovars tested (Poppe et al. 1989). The plasmid is very stable, refractory to curing, and has a common conserved region (related to pathogenicity) and a polymorphic region based on restriction analysis.

It looks as if the pathogenic determinants of *S. abortusovis*, adhesion and invasion of eukaryotic cells, are organized in multiple loci on the chromosome (*inv*) and probably on the plasmid (*spv*) as in other *Salmonella* species or serovars.

ADHESION TO EPITHELIAL CELLS

Stable adherence to epithelial cells by *Salmonella spp.* has been described. This mechanism preludes the cell invasion. Non adhesive mutants can be avirulent. The *S. typhimurium* type 1 fimbriae seem to play a role in the attachment to cells (Ernst et al. 1990).

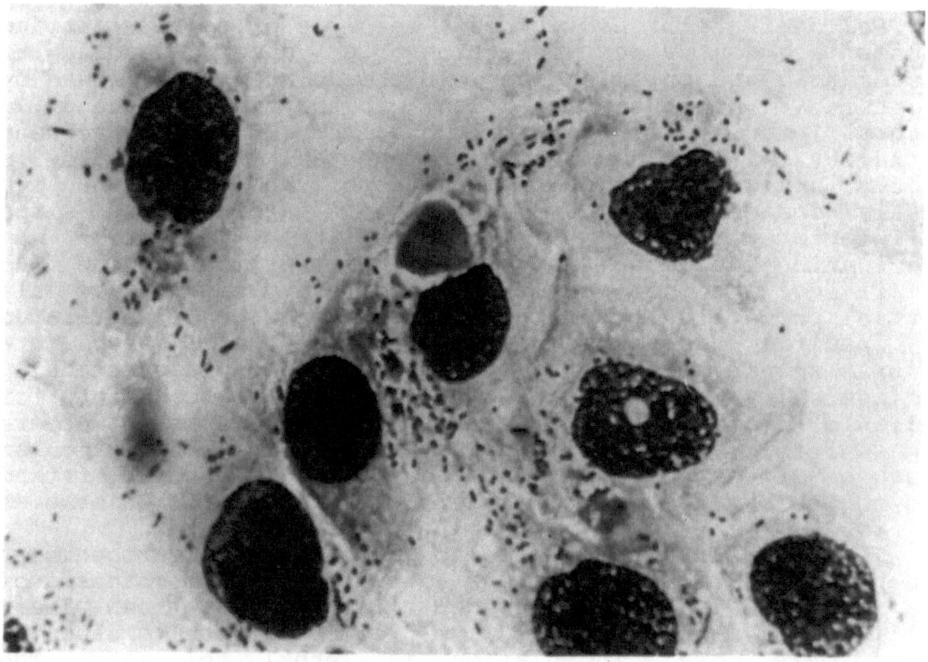

Figure 1. Giemsa stain of HeLa cells infected by *S. abortusovis*. Adhesion can be observed: up to 50 bacteria for cell.

We have demonstrated that strains *of S. abortusovis* from different geographic area are able to adhere to HeLa cells (fig. 1) and to lamb kidney epithelial cells *in vitro* (fig. 2).

In vivo epithelial cell adhesion could allow the colonization of the intestinal wall, the first step for the invasion of the placental chorionic epithelium. Electron microscopy of strains in culture or isolated directly from stomachs of aborted foetuses does not show any filamentous appendages such as fimbriae.

Colonies of non adhesive Tn*Pho*A mutants (see below) are smooth like the wild type and show no perturbation in lipopolisaccharide profile. Adhesion therefore requires no fimbriae and does not involve LPS structures. Some mutants in different chromosomal regions are non adhesive, and we hypothesize the existence of multiple factors acting as adhesins (submitted for publication).

No strict correlation between adhesion and virulence in mouse can be ascertained, since some adhesiveless strains in culture tissue cells are still able to kill mice.

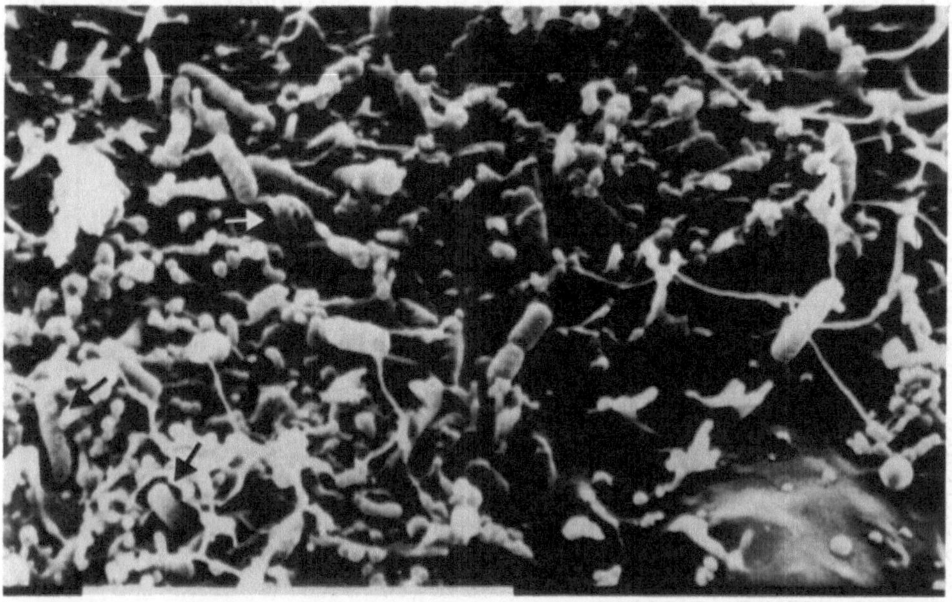

Figure 2. Scanning electron microscopy of lamb kidney epithelial cells infected by *S. abortusovis*. Bacteria adhere to the cell microfibers. Arrows show bacteria invading the cell. The white bar corresponds to 10 microns.

INVASION OF EUKARYOTIC CELLS

A typical pathogenic function of *Salmonella* genus is the ability of invade mammalian cells. After invasion, replication can occur inside the endocytic vacuoles. Generally this mechanism has been revealed in epithelial cells.

A large variety of genetic loci that affect the ability of *Salmonella* to enter eukaryotic cells *in vitro* has been reported. Then a high complexity of the mechanism or different pathways of aggression to mammalian cell can be assumed. Interestingly specific genes whose products allow *S. typhimurium* to penetrate tissue culture cells were cloned and characterized (Galàn and Curtiss III 1989). This is true also for other enterobacteria and analogies among such genes from different species are under focus.

S. abortusovis as well is able to invade eukaryotic cells (Fig. 2).

The role of invasion in a mouse model seems to be quite significant, at least more than adhesion. In fact a strongly adhesive, but invasiveless Tn*pho*A mutant strain (see below) is unable to kill mice.

Experiments were also done with infected HeLa cells.

After treatment with gentamicin to kill extracellular bacteria and triton to lyse the cells, it was still possible to detect the presence of bacteria coming from the inner Hela cell cytoplasm and able to duplicate.

In order to study the genetic organization of such phenotypic feature, we hybridized total *S. abortusovis* DNA with a probe from *S. typhimurium* containing the *inv*A, -B and -C genes. Such a probe already shown to hybridize with 37 different species or serovars of *Salmonella*, and *inv*A mutants of *S. typhi* are deficient for entry into mammalian cells (Galàn and Curtiss III 1991). Homology with *S. abortusovis* DNA was revealed and it can be added to other *Salmonella* species showing this region.

CHARACTERIZATION OF THE RESIDENT LARGE PLASMIDS

Some authors (Popoff et al. 1984, Helmuth et al. 1985, Poppe et al. 1989, Gulig 1990 and Colombo et al. 1992) described high molecular stable plasmids in *Salmonella* and specifically in *S. abortusovis* with a region homologous to the *S. typhimurium spv* virulence region.

The 90 kb virulence plasmid of *S. typhimurium* is responsible for invasion of salmonellae beyond the Peyer's patches of the small intestine to the mesenteric lymph nodes and spleens of mice after oral inoculation (Gulig and Curtiss III 1987). A 7.8 kb region contains six genes in a head-to-tail orientation named *spv*R. -A, -B, -C, -D, and *orf*E (Gulig et al. 1992). The *spv*R gene is a positive regulator of *spv*B and an essential virulence gene activated in vivo during infection.

A stable ubiquitarious plasmid of high molecular mass (50-75 kb) was found in all the *S. abortusovis* strains examined (Popoff et al. 1984 and Colombo et al. 1992).

We submitted to restriction analyses plasmids from epidemic strains belonging to different geographic areas collected over a period of 30 years. The *Hind*III restriction pattern is characterized by eight main fragments distributed in two portions: one with three fragments, stable and common to all strains, and a second polymorphic pattern with five fragments. The stable portion includes a 3.7 kb *Hind*III fragment that hybridizes with the *S. typhimurium* probe containing the *spv*C gene sequence (Gulig and Curtiss III 1988). Such a sequence was carried by plasmid pYA426. *spv*C gene encodes a 28kDa protein that is a major virulence factor for *S. typhimurium* pathogenesis (Gulig et al. 1990). It is possible that a homologous gene to *spv*C could exist in

S. abortusovis, and may have an important role in *S. abortusovis* virulence. This possibility is under investigation.

Since no other phenotypic markers feasible for the purpose have yet been identified, the polymorphic portion of the plasmid could be usefull to correlate different strains in epidemiological studies of different outbreaks (Colombo et al. 1992). In support of this idea, the strains isolated in the same epidemic location show identity in the variable portion too. On the other hand, strains collected in different epidemics or of different clonal origin, show different degrees of polymorphism in the variable portion.

TROPISM FOR MAMMALIAN ORGANS

S. abortusovis's routes of infection in sheep are not well ascertained. It has been demonstrated experimentally that it is not sexually transmitted by infected rams (Sanchis et al. 1986). The most probable way of infection is the oral ingestion of contaminated pasture by vaginal discharge and foetal membranes after abortion. After this step it can be supposed that *S. abortusovis*, resistant to gastric conditions in the four stomachs, adheres and invades intestinal epithelial M cells. These cells are the most probabile target of colonization, at least in *Salmonella* infection. Then a limph nodes route could represent the next step of systemic invasion for reaching the placenta (Pardon et al. 1990) and survival inside the macrophages should be hypotized. Since experimental oral infections of sheep were not reproducible, natural unknown risk factors, leading to abortion, have been hypothesized.

A mouse model for studying the *S. abortusovis* pathogenesys, and the specific tropism for mammalian organs and tissues, has been employed (Pardon and Marly 1979) mostly for vaccine testing. We have also experimented with this approach with encouraging results. We submitted BALB/c mice to LD_{50} ($3X10^6$ bacteria, according to Reed-Muench procedure) in an experimental oral infection with a wild-type strain. The surviving mice showed immunological response by microagglutination test against *S. abortusovis*. In an other experiment, after full lethal dose inoculation, mice were killed at different stages of the infection that eventally leads to death. After 24-48 hours it can be revealed a low number of viable bacteria colonizing only Peyer patches and intestinal cells. After 72-96 hours spleen and liver colonization can be easily recovered that leads to death after a couple of days.

Then it is not possible to compare the intestinal organization of sheep with that of mice, but it can be suggested that *S. abortusovis* is able to easily enter the intestinal barries and to colonize the target organs: liver and spleen in mice, and placenta in sheep.

S. abortusovis MUTANTS UNABLE TO ADHERE AND WITH REDUCED VIRULENCE IN MICE

To better understand the role of genes involved in adhesion, we studied *S. abortusovis* mutants unable to adhere to epithelial cells. We employed the random Tn*pho*A transpositional mutagenesis, and, among others, 24 mutants yielding alkaline phosphatase activity, lost the ability to adhere to epithelial HeLa and LK cells (manuscript submitted for publication).

Tn*pho*A contains a portion of alkaline phosphatase gene whose enzymatic activity is detectable when insertions are in genes coding for periplasm, outer membrane or extracellular proteins.

This approach has been largely and successfully practiced by many authors for studying the superficial determinants of virulence in gram negative bacteria (Manoil et al. 1990).

According to the size of the restriction fragments containing the integration site of the transposon, four classes of adhesiveless mutants were recovered, all in the chromosome.

We studied the phenotypic features of these strains, and among others we analized outer membrane proteins modifications and virulence in mice.

Outer membrane proteins (OMP) were screened for alkaline phosphatase activity in non-denaturing gels of OMP preparations.

Interestingly seven mutants showed such activity, and all had the site of Tn*pho*A integration in the same 13 *Eco*RV restriction fragment.

The virulence of the strains was tested in BALB/c mice using LD_{50} ($3x10^6$ bacteria) via oral infection.

Eighteen mutants were attenuated for virulence. Three mutants revealed an attenuated virulence and two mutants were avirulent.

Two out of the three attenuated strains and the avirulent strains showed a transpositional mutation in the OMP.

Microagglutination tests against *S. abortusovis* on sera from the mice that survived infection revealed an immunological response.

To follow the biological consequences of this response, these mice were then orally challenged with 10 times the LD_{50} ($3x10^7$ bacteria) of the wild type virulent strain. Protection ranging from 60% to 86% to the further challenges was achieved (manuscript submitted for publication).

From these results it can be suggested that adhesion is mediated by several genetic loci, at least four. These loci corrispond to the four classes of Tn*pho*A integration identified by restriction maps and showing mutation in the adhesion performances.

Among these factors, OMPs are relevant and they seem to work in combination.

Anyway, as mentioned above, since not all the adhesiveless mutants in tissue colture cells are avirulent for mice, adhesion cannot be considered a direct evidence of virulence.

This approach not only shows to be helpful to study the pathogenic mechanisms of *S. abortusovis* but is also a good strategy to develop an engineered strain with high protective features to be tested as live vaccine in sheep.

Acknowledgements: This work was supported by grants from: Consiglio Nazionale delle Ricerche (CNR) 89.05352MZ77,

CNR 90.01373.4, Ministero Ricerca Scientifica 60%; Regione Autonoma Sardegna, progetto biotecnologie; Fondazione Istituto Pasteur-Cenci Bolognetti.

REFERENCES

Caldwell, A.L., and Gulig, P.A., 1991, The *Salmonella* regulator of plasmid-encoded virulence gene, *J. Bact.* 173:7176.
Colombo, M.M., Leori, G., Rubino, S., Barbato, A., and Cappuccinelli, P., 1992, Phenotypic features and molecular characterization of plasmids in *Salmonella abortusovis*, *J. Gen. Microbiol.* 138:725.
Ernest. R.K., Dombroski, D.M., and Merrick, J.M., 1990, Anaerobiosis, Type 1 fimbriae, and grouth phase are factors that affect invasion of Hep-2 cells by *S. typhimurium*, *Inf. Imm.* 58:2014.
Galàn, J.E., and Curtiss III, R., 1989, Cloning and characterization of genes whose products allow *S. typhimurium* to penetrate tissue culture cells, *Proc. Natl. Acad. Sci. USA* . 86:6383.
Galàn, J.E., and Curtiss III, R., 1991, Distribution of the invA, -B, -C and -D genes of *Salmonella* serovars: invA mutants of *Salmonella typhi* are deficient for entry into mammalian cells, *Inf. Imm.* 59:2901.
Gulig, P.A., and Curtiss III, R., 1987, Plasmid-associated virulence of *Salmonella typhimurium*, *Inf. Imm.* 55:2891.
Gulig, P.A., and Curtiss III, R., 1988, Cloning and transposon insertion mutagenesis of virulence genes of the 100-kilobase plasmid of *Salmonella typhimurium*, *Inf. Imm.* 56:3262.
Gulig, P.A., 1990, Virulence plasmids of *Salmonella typhimurium* and other salmonellae, *Micr. Pathogen.* 8:3.
Gulig, P.A., Caldwell, A.L., and Chiodo, V.A., 1992, Identification, genetic analysis and DNA sequence of a 7.8-kb virulence region of the *Salmonella typhimurium* virulence plasmid, *Mol. Microbiol.* 6:1395.
Helmuth, R., Stephan, R., Bunge, C., Hoog, B., Steinbeck, A., and Bulling, E., 1985, Epidemiology of virulence-associated plasmids and outer membrane protein patterns within seven common *Salmonella* serotypes, *Inf. Imm.* 48:175.
Lantier, F., Pardon, P., and Marly, J., 1983, Immunogenecity of a low-virulence vaccinal strain against *Salmonella abortus-ovis* infection in mice, *Inf. Imm.* 40:601.
Manoil, C., Mekalanos, J.J., and Beckwith, J., 1990, Alkalyne phosphatase fusions: sensors of subcellulars location, *J. Bacteriol.* 172:515.
Pardon, P., and Marly, J., 1979, Experimental *Salmonella abortusovis* infection of normal or primo-infected CD1 mice, *Ann. Microbiol.* (*Inst. Pasteur*). 130:21.
Pardon, P., Sanchis, R., Marly, J., Guilloteau, L., Buzoni-Gatel, D., Oswald, I., Pepin, M., Kaeffer, B., Berthon, P., and Popoff, M.Y., 1990, Experimental ovine salmonellosis (*Salmonella abortusovis*) : pathogenesis and vaccination, *Res. Microbial.* 141:945.
Popoff, M.Y., Miras, I., Coynault, C., Lasselin, C., and Pardon, P., 1984, Molecular relationships between virulence plasmids of *Salmonella* serotypes typhimurium and dublin and large plasmids of other *Salmonella* serotypes, *Ann. Microbiol. (Inst. Pasteur)* . 135:389.
Poppe, C., Curtiss III, R., Gulig, P.A., and Gyles, C.L., 1989, Hybridization with a DNA probe derived from the virulence region of the 60 Mdal plasmid of *Salmonella typhimurium*, *Can. J. Res.* 53:378.
Sanchis, R., and Pardon, P., 1986, Infection experimentale du belier par *Salmonella abortus ovis*, *Ann. Rech. Vet.* 17:387.

ANTIBIOTIC PROTEINS OF HUMAN NEUTROPHILS: INTERACTION WITH THE SURFACE OF SALMONELLA

John K. Spitznagel and H. Anne Pereira

Department of Microbiology and Immunology
Emory University
Atlanta, GA 30322

INTRODUCTION

Our interest in the biology of Salmonella has centered on their responses to antibiotic proteins carried in the azurophil granules of human neutrophil phagocytes (PMN). Recently we have been working intensively with CAP37 a 37kDa antibiotic glycoprotein that is, as most of them are, cationic and hydrophobic. Several of these cationic, antibiotic proteins have in fact been isolated and characterized following our discovery of their presence in the azurophil granules of guinea pig neutrophils in 1963, in rabbit neutrophils in 1966, and in human neutrophils in 1971. Their molecular sizes range from around 3 kDa to 60 kDa. They are hydrophobic. Two are serine proteases, but the others lack known enzymic activity. Interestingly CAP37 has >47% homology with neutrophil elastase and highly conserved structure placing it in the serine protease family; yet it lacks any serine esterase activity due to a mutation at position 41 replacing histidine with serine (Pereira et. al. 1990). Each has its characteristic antimicrobial spectrum. A concise review of this material was published in November of 1991 (Spitznagel 1991).

We have sequenced CAP37 and cloned its cDNA (Morgan et al 1991) so we know its primary structure. Several of the proteins have been cloned and sequenced by others (Spitznagel 1991). Evidence has been presented that they can damage the permeability of cytoplasmic membranes of gram negative bacteria. The question is how these proteins get through the outer membrane so the cytoplasmic membrane can be attacked. Interaction with the outer membrane requires further investigation. The outer membrane is such a formidable barrier to antimicrobial substances (Nikaido and Vaara 1985) that it is hard to imagine how proteins as large as these can traverse it to reach the cytoplasmic membrane. One of the important aspects of the antibiotic proteins is that although they differ considerably in composition and structure they all are cationic and hydrophobic, Their interactions with bacteria tend to reflect these properties. In the first part of this paper I shall discuss some of these features with respect to the action of CAP37. In the second part I shall describe the antimicrobial actions of a substituent peptide of CAP37 that we have recently synthesized. This 25 amino acid peptide has, mole for mole at least part of the antibiotic activity of the CAP37 protein although it is only one tenth the size of the entire molecule. It is interesting because it may give clues to the molecular basis for antibiotic action of the proteins and promises to be helpful in designing clinically useful antibiotics.

PART I

Conditions for action of CAP37 against *Salmonella typhimurium*

Several conditions maximize the capacities of the antibacterial proteins to kill Salmonella; low pH, low salt concentrations, and low free magnesium ion concentrations (Table 1).

Table 1. Conditions favorable to activity of antibiotic proteins.

pH 5.5-5.7
Salt Concentration <75mM
[Mg+ +] <0.1mM
Aerobic or Anaerobic
Rough Organisms
Growth Conditions
Absence of Free LPS

The protein kills equally well under aerobic or anaerobic conditions. Strain differences in the target bacteria are important; rough strains are much more readily killed than smooth strains. However, smooth strains are not completely resistant just less sensitive. Strains expressing the polymyxin B resistant phenotype, that carry the *pmrA* and *gapA* mutations (see Table 2), are also more resistant to CAP37 than isogenic polymyxin B sensitive parent strains.

Killing has been repeatedly shown to be most active against actively growing cells. Finally, it is important to know that negatively charged complex carbohydrates like endotoxin, especially lipid A, bind and block the biological activities of cationic antibiotic proteins. This implies that endotoxins have a role in binding these proteins in the outer membrane. Their affinity for endotoxin is so strong that can be troublesome when endotoxins contaminate systems *in vitro* and interfere with assays. In the pathogenesis of gram negative sepsis endotoxins released from bacteria *in vivo* could neutralize the antibiotic proteins or be neutralized by them depending on the amounts of each in the system. These interactions may substantially determine the outcome of sepsis.

Model for action of antibiotic proteins

The capacities of antibiotic proteins to bind avidly to gram negative bacteria has been known ever since Hirsch published his work on phagocytin, which we showed to be a crude mixture of antibiotic proteins. Modrzakowski and Spitznagel (1979) showed that antibiotic protein binds to gram negative bacteria through endotoxin. Several investigators have shown that smooth organisms are less sensitive than rough ones. Farley et al (1988) showed that specific binding of cationic proteins is less with smooth Salmonella than with rough ones. The failure of protein to bind was proportional to the resistance of the strain to the protein's antibiotic action. Moreover, the antibiotic proteins, unlike polymyxins depended more on hydrophobic interactions for binding than on charge.

Killing and binding studies suggest that the hydrophilic O and core antigens protect smooth strains against the cationic hydrophobic proteins, but Lipid A, KDO, or both may provide binding sites. Shafer et al (1984) suggested that the binding to endotoxin is due to the negatively charged phosphoryl groups on lipid A. Shafer based his conclusions on the demonstration that the pmrA mutation rendered Salmonella resistant not just to polymyxin B but also to the cationic hydrophobic antibiotic proteins including CAP37, CAP57, and the Defensins.

In order to understand Shafer's conclusions the structures of wild type lipid A, and the lipid A of polymyxin B resistant strains bearing the pmrA mutation described by Makela et al (1978) must be considered. Rietschel's structure for Salmonella lipid A is shown in Figure 1 (Rietschel et al 1984). The important features of lipid A structure are: its hydrophilic diglucosamine groups; the phosphoryl groups on the 1 and 4' positions of the diglucosamines; and several hydrophobic, long chain acyl substitutions on the glucosamines. The important point is that the phosphoryl groups in pmrA, polymyxin B resistant, strains have been shown to be 100% substituted with aminopentose residues (Vaara et al 1981). These 4- aminopentose substitutions were thought to block negative charges and so to interfere with binding of polymyxin B to the outer membrane. The increased resistance to the antibiotic proteins seen in pmrA carrying strains has likewise been attributed by Shafer et al (1984) to the loss of negative charges from the substituted lipid A phosphoryl groups leading to loss of strong ionic binding sites for these basic proteins.

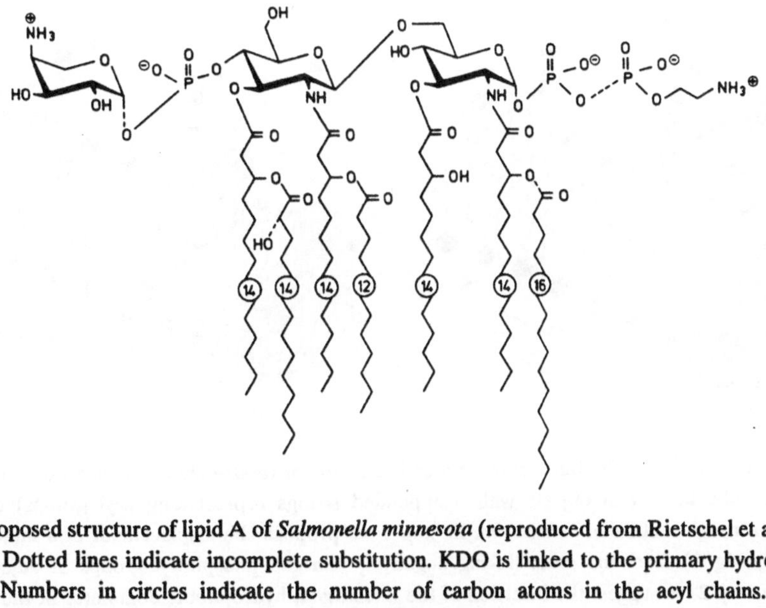

Figure 1. Proposed structure of lipid A of *Salmonella minnesota* (reproduced from Rietschel et al (1984) with permission). Dotted lines indicate incomplete substitution. KDO is linked to the primary hydroxyl group in position 6'. Numbers in circles indicate the number of carbon atoms in the acyl chains. Note the 4 aminoarabinose substitution on 4' phosphoryl of the glucosamine.

Based on this structure and the factors reviewed in Table 1 a model has been developed to explain the interaction of the cationic proteins with gram negative bacteria. In figure 2 is a cartoon of the putative interaction of an antibiotic protein with Salmonella. Based on the work of Schindler and Osborne (1979) it is assumed that the endotoxin or LPS molecules in the outer leaflet of the outer membrane are stabilized by divalent cations that cross link with ionic bonds the negatively charged phosphoryl groups on the lipid A. Schindler suggested that phosphoryl or carboxyl groups of 2-keto-3-deoxyoctonate (KDO) might also serve this purpose. Moreover, Parker et al (1992) have recently concluded that deep rough phenotypes in gram negative rods are likely due to loss of heptose linked phosphoryl and phosphatidyl-ethanolamine groups rather than changes in Lipid A structure. It is therefore with some reservations that we postulate that the phosphoryl groups of the lipd A are necessarily the only or even the critical ones for binding the basic proteins. Nevertheless we feel Mg^{++} is the critical divalent cation involved since increasing its concentration protects the Salmonella against the action of CAP37 (Ca++ has much less effect) and decreasing it makes them much more sensitive. Various investigators also have noted these effects of magnesium.

With respect to the locations of the critical negatively charged groups, we have recently found that the pmrA mutation appears to give additional protection to Salmonella against the basic antibiotic proteins in smooth backgrounds (smooth and Rb1) (see table 2), but in a deep rough background (Re) it fails to do this. This suggests that 4-aminopentose substitutions on lipid A may not be sufficient to confer resistance in the absence of heptose chains.

In summary: Part I relates that the basic antibiotic proteins, such as CAP37, act at least in part by binding to the negatively charged phosphoryl groups of the lipid A or phosphoryl substituents on the heptoses of the LPS. A mutation that increases the resistance of Salmonella to polymyxin B appears to do so by blocking these groups with aminopentose sugars.

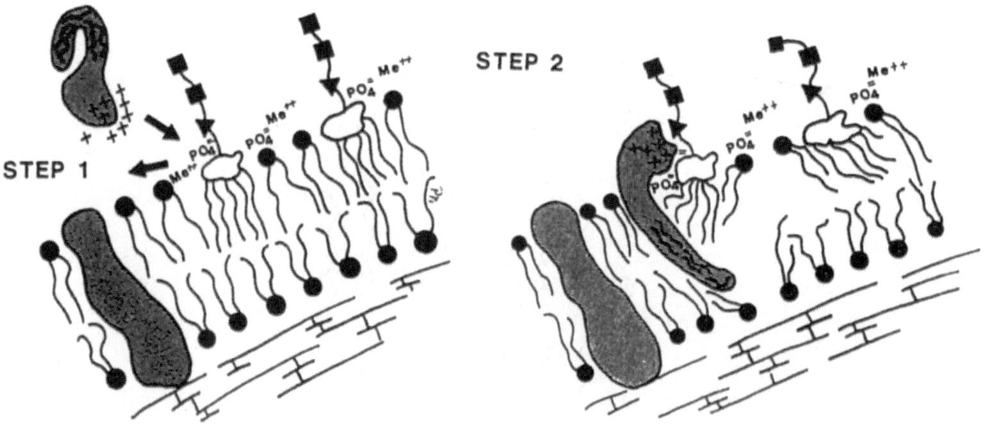

Figure 2. Cartoon depicting the interaction of a cationic protein (gourd shaped object upper left) with the lipid A (pillow-like horizontal objects with 5 appended strings representing acyl groups) of the outer membrane in a rough mutant of *Salmonella.* In step 1 we propose the displacement of a metallic divalent cation ($Me^{++}=Mg^{++}$) from the cross linking negatively charged phosphoryl groups of adjacent lipid A molecules. In step 2 we suggest the insertion of the uncharged, hydrophobic domain(s) of the protein into the phospholipid (closed circles with two appendages) domain of the outer membrane. The membrane is distorted and destabilized by this intrusion.

Table 2. LPS chemotype and pmrA (polymyxin B resistance) interact to influence resistance to antibiotic proteins.

Strain	Geno type	LPS[a]	ED50ug/ ml CGE[b]	Strain	Geno-type	LPS	ED50ug/ ml CGE
LT-2	*gapA*+[c]	O (smooth)	40	LT-2	*gapA*[d]	O (smooth)	>60
SH7585	*galE* *prmA*+[e]	Rc	10	SH7580	*galE* *pmrA*[f]	Rc	30
SH9178	*rfaJ* *pmrA*+	Rb2	5	SH7426	*rfaJ* *pmrA*	Rb2	24
SH5014	*rfaJ* *pmrA*+	Rb2	6	SH5357	*rfaJ* *pmrA*	Rb2	10
SH7518	*rfaE* *pmrA*+	Re	0.2	SH7519	*rfaE* *pmrA*	Re	0.2

a. LPS=lipopolysaccharide chemotype of strain specified.

b. The concentration of crude neutrophil granule extract killing 50% of bacteria in 1 hour at 37°C.

c. gapA+ wild type polymyxin B resistance.

d. gapA mutant allele of gapA+ with polymyxin B resistant phenotype. Spitznagel 1988.

e. pmrA+ wild type polymyxin B resistance.

f. pmrA mutant allele of pmrA+ with polymyxin B resistant phenotype. Makela 1978.

PART II

Peptide 20-44, clues to a structural basis for CAP37 antibacterial action

In order to better understand the antimicrobial action of antibiotic proteins we identified a peptide fragment that accounts for a large part of the antimicrobial action of CAP37. This we did by synthesizing overlapping peptides based on the amino acid sequences we established for purified mature CAP37 protein. The CAP37 was purified to homogeneity from crude granule proteins of human PMN with sequential ion exchange, molecular sieving, HPLC reverse phase, and hydrophobic chromatography. Both the sequencing of CAP37 (Pohl et al 1990) and the synthesis of the peptide subunits (Pereira et al to be submitted) were done with standard techniques. A survey of the antimicrobial capacities of 11 of these synthetic peptides revealed that one near the amino terminus of the mature protein, amino acid positions 20-44, was more active than the others(See Fig. 3). Next most active was the N-terminal peptide, amino acid positions 1-44.

The amino acid sequence of the peptide is shown in figure 4. The presence of two arginine residues suggests the fragment has at least two positive charges in solution at or below physiological pH. There is a substantial preponderance of hydrophobic amino acids (60%). In agreement with these inferences hydropathy and charge plots for the mature CAP37 show this peptide provides the most hydrophobic region of CAP37 and bears two positive charges at pH 7.0. Probably the histidines and asparagine contribute positive charges at pH 5.5 and 5.7, the pH's at which the antibacterial assays were run.

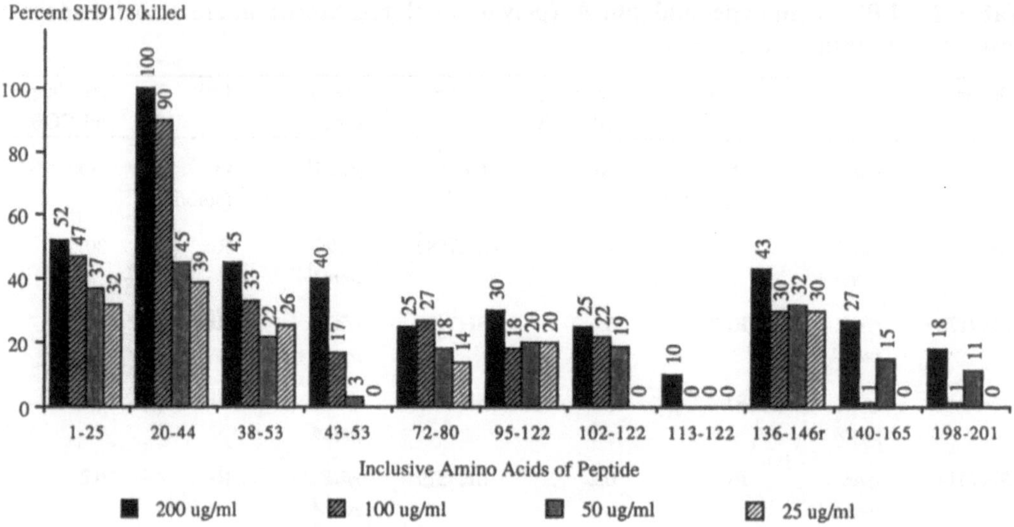

Figure 3. Antibiotic activity of 11 synthetic peptides prepared based on sequences of CAP37.

Asn Gln Gly Arg His Phe Cys Gly Gly Ala

Leu Ile His Ala Arg Phe Val Met Thr Ala

Ala Ser Cys Phe Gln

Figure 4. Amino acid sequence for antibiotic peptide 20-44. The angled lines represent possible disulfide bonding.

The antibiotic capacity of the peptide is seen in figure 5 where four strains of *S. typhimurium* with a range of LPS chemotypes are seen to be sensitive to different concentrations of the peptide. Just as it was very sensitive to CAP37 the deep rough Re LPS strain, SH7518, was very sensitive even to as little as 12.5 uM peptide. SH9178 (Rb1 LPS) is substantially more resistant than the Re strain. Its isogenic, pmrA mutant (SH7426) was still more resistant with only 45% killing even at 75 uM peptide.

It was interesting that the rough SH9178 (Rb_2 LPS) was no more sensitive to the peptide than was LT2 with wild type LPS. These results parallel those seen with the mature CAP37 protein. Most rough strains, however, tend to be more sensitive to both CAP37 and the peptide than is LT2. It is also curious that SH9178 is more sensitive to CAP57 than is LT2 (data not shown), suggesting that CAP37 and CAP57 differ in their actions to some extent.

We have found that the activity of the peptide, like the activity of the protein, is greatest at pH5.5-5.7, and at salt concentrations of 75 mM or less (not shown). In

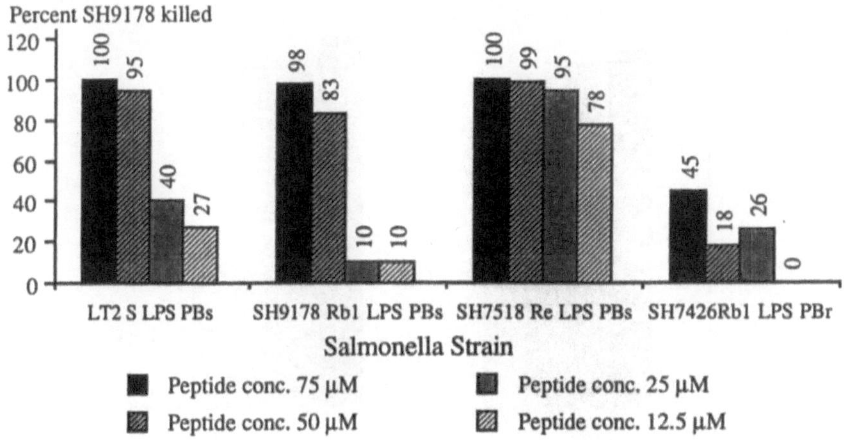

Figure 5. Antibiotic action of peptide 20-44 on *S. typhimurium* of different LPS chemotypes and a pmrA strain SH7426.

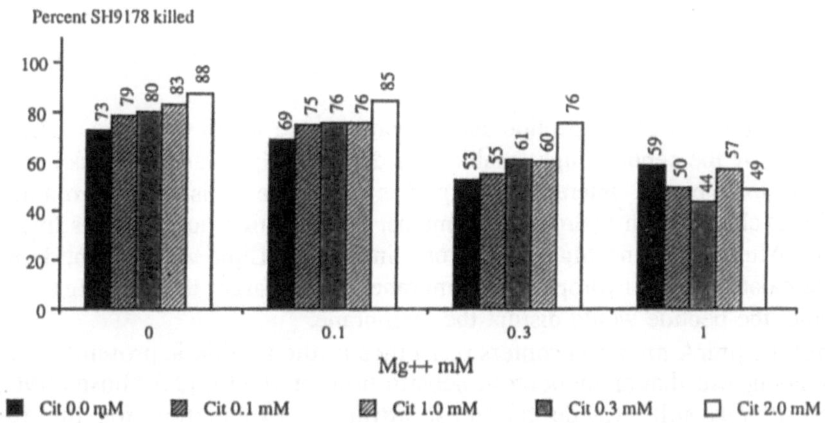

Figure 6. Inhibition of antibiotic activity of peptide 20-44 due to Mg+ + and enhancement due to citrate, possibly due to chelation of Mg+ + by citrate. Mg+ + inhibition is overcome by citrate.

addition Mg+ + inhibits the antimicrobial action of 50uM peptide. Citrate to some extent enhances the action of the peptide. Mg+ + counters the citrate effect (see figure 6). Finally we have found (see figure 7) that LPS has a slight inhibitory effect on the activity of 75uM peptide. Lipid A, however, has a distinctly neutralizing effect on the killing capacity of the peptide although the effects seem less than the comparable effects of endotoxin on the mature CAP37.

DISCUSSION

The results clearly show that Pep20-44 of CAP37 is an active antibiotic; its binding activity is similar to the mature antibiotic proteins of human neutrophil azurophil granules. In general the similarities of optimum pH, salt concentration, effects of Mg+ + and citrate, molar activity, and binding of LPS suggest that the mechanisms

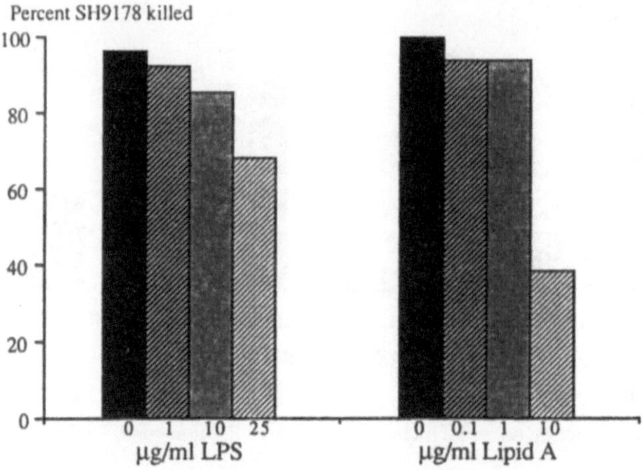

Figure 7. LPS and lipid A bind to antibiotic peptide 20-44 and inhibit its activity against *S. typhimurium* SH9178.

of antimicrobial action of the peptide and the protein are much the same. The bivalent charge carried by the peptide might enable it to displace Mg++ and to cross link Lipid A moieties with ionic bonds. Interestingly Mg++ is very large. It is also hydrophilic so that it would be excluded from hydrophilic domains. The peptide however has hydrophobic sequences. After displacing Mg++ and cross linking the Lipid A, the peptide might be able to intercalate into the hydrophobic membrane. Being larger than the Mg++ and more hydrophobic, the peptide would disrupt the membrane.

That the pmrA mutation confers resistance to the antibiotic proteins as well as to polymyxin B suggests that aminopentose substitutions on the Lipid A phosphoryl or KDO phosphoryl groups still provide the most attractive explanation for the resistance conferred by *pmr*A. This interpretation is consistent with the idea that the positive charges of polymyxin B and CAP37 interact in the pmrA+ strains with the many unsubstituted, negatively charged groups. It was noteworthy, however, that the *pmr*A mutation evidently failed to protect strains with the Re chemotype (results not shown). That suggested that negatively charged groups important for expression of sensitivity are associated not just with the Lipid A but with the substituent KDO(s). These interactions require further study.

We noted that peptide 20-44 has two cysteines. Unpublished experiments showed that when these were blocked to prevent disulfide bond formation the peptide lost antibiotic activity. The importance of the six cysteines found in the defensins is evidence that the formation of internal disulfide bonds is important to their antibiotic activity (Hill et al 1991). Our results show that a single disulfide bond may suffice in a suitable peptide. We have other peptides that also carry two cysteines but they have different amino acids and lack antimicrobial action. Disulfide bond formation is necessary but not sufficient.

SUMMARY

The antibiotic protein of human PMN, CAP37, has several characteristics of other antibiotic proteins. In this paper we report that a substantial portion, perhaps all, of the antibiotic action of CAP37 as well as its capacity to bind endotoxin resides in a peptide defined by the amino acid asparagine in the 20th position to and including the amino acid glutamine in the 44th position. It is noteworthy that the peptide has at least two positive charges. It has two cysteines; therefore it can form an internal disulfide bond, and become cyclic. We found that if these cysteines are blocked so they cannot form disulfide bonds antibiotic activity is lost. The net positive charges and hydrophobicity of the peptide confirm the importance of those characteristics for activity in the antibiotic peptides and proteins.

Acknowledgements

This work was supported by research grants AI 17662 and AI26589 (JKS), and AI28018 (H.A.P.) from the Public Health Service of the United States, and a research grant award from the American Lung Association (HAP).

REFERENCES

Farley, M.M., W.M. Shafer, and J.K.Spitznagel, 1988, Lipopolysaccharide structure determines ionic and hydrophobic binding of a cationic antimicrobial neutrophil granule protein, Infect. Immun., 56:1589.

Hill, C.P., J.Yee, M.E.Selsted, D.Eisenberg, 1991, Crystal structure of defensin HNP-3, an amphophilic dimer: Mechanisms of membrane permeabilization, Science, 251:1481.

Makela, P.H., M.Sarvas, S.Calcagno, K.Lounatma, 1978, Isolation and genetic characterization of polymyxin-resistant mutants of *Salmonella,* FEMS Microbiol. Lett., 3:323.

Modrzakowski, M.C., and J.K.Spitznagel, 1979, Bactericidal activity of fractionated granule contents from human polymorphonuclear leukocytes: Antagonism of granule cationic proteins by lipopolysaccharide, Infect. Immun., 25:597.

Morgan, J.G., T. Sukiennicki, H.A.Pereira, J.K.Spitznagel, M.F.Guerra, and J.W.Larrick, 1991, Cloning of the cDNA for the serine protease homolog CAP37/Azurocidin, a microbicidal and chemotactic protein from human granulocytes, J. Immunol.,147:3210.

Nikaido, H., and M.Vaara, 1985. Molecular basis of bacterial outer membrane permeability, Microbiol. Rev., 49:1.

Parker, C.T., A.W.Kloser, C.A.Schnaitman, M.A.Stein, S.Gottesman, and B.W.Gibson, 1992, Role of the *rfaG* and *rfaP* genes in determining the lipopolysaccharide core structure and cell surface protperties of *Eschericia coli* K-12, J.Bact., 174:2225.

Pereira, H.A., W.M. Shafer, J. Pohl, L.E. Martin, and J.K. Spitznagel, 1990, CAP37, a human neutrophil-derived chemotactic factor with monocyte specific activity, J. Clin. Invest., 85:1468.

Pohl, J., H.A.Pereira, N.M.Martin, and J.K.Spitznagel, 1990, Amino acid sequence of CAP37, a human neutrophil granule-derived antibacterial and monocyte-specific chemotactic glycoprotein structurally similar to neutrophil elastase, FEBS Lett., 272:200.

Rietschel, E.T., H.W.Wollenweber, H.Brade, U.Zahringer, B.Lindner, U.Seydel, H.Bradaczek, G.Barnickel, H.Labischinski, and P.Giesbrecht, 1984, Structure and conformation of the lipid A component of lipopolysaccharides, in Handbook of Endotoxin Vol1, Chemistry of Endotoxin. Eds R.A.Proctor, E.T.Reitschel, Elsevier, Amsterdam, p187 ff.

Schindler, M. and M.F.Osborn, 1979, Interaction of divalent cations and polymyxin B with lipopolysaccharide, Biochemistry, 18:4425.

Shafer, W.M., S.G.Casey, and J.K.Spitznagel, 1984, Lipid A and resistance of *Salmonella typhimurium* to antimicrobial granule proteins of human meutrophil granulocytes, Infect. Immun., 43:834.

Spitznagel, J.K., 1988, Nonoxidative antimicrobial reactions of leukocytes, in Contemporary Topics in Immunology, Regulation of leukocyte function, R. Snyderman (Ed.), Plenum Press N.Y., p283ff.

Spitznagel, J.K., 1991, Antibiotic proteins of human neutrophils, J.Clin. Invest., 86:1381.

Vaara, M., T.Vaara, M.Jensen, I.Helander, M,Numinen, E.T.Rietschel, and P.H.Makela, 1981, Characterization of the lipopolysaccharide from the polymyxin-resistant mutants of *Salmonella typhimurium*, FEBS Lett., 129:145.

EARLY EVENTS IN THE PATHOGENESIS OF ENTERIC FEVER IN MICE

David E. Briles,[1,2,3] Nancy E. Dunlap,[4,1] Edward Swords[1],
and William H. Benjamin, Jr.[5,1]

Departments of [1]Microbiology, [2]Pediatrics, [3]Comparative
Medicine, [4]Medicine, and [5]Pathology, The University of
Alabama at Birmingham, Birmingham, Alabama 35294

INTRODUCTION

Salmonella typhimurium is the best studied of the salmonella which cause mouse typhoid (enteric fever). The portal of entry of salmonella that cause enteric fevers in the mouse and in other species is normally oral. However, the organisms quickly enter the lymph and blood, and most of the organisms that survive blood clearance become lodged in the spleen and liver where they grow and cause disease (Carter and Collins, 1974; Carter and Collins, 1974). Although salmonella which cause enteric fevers have been described as facultative intracellular parasites that resides within macrophages, there is controversy over the major location of *S. typhimurium* multiplication *in vivo* (Benjamin, et al, 1990; Carroll et al, 1979; Dunlap et al, 1991; Guo et al, 1986; Schurr et al, 1989).

A mouse model for the study of blood clearance and the growth of the surviving salmonella within the reticuloendothelial organs involves injection of salmonella directly into the blood. The i.v. route avoids complications that might occur from variable defenses at the mucosal level or abnormal depots of salmonella that might be created by intraperitoneal or subcutaneous routes of infection. The progression of murine salmonellosis following i.v. infection can be divided into different phases, each with its own relationship to host defenses. The first phase is blood clearance with localization of the surviving salmonella in the spleen and liver. This is followed by salmonella growth in the spleen and liver and finally (if the mice live long enough) specific immunity which halts the accumulation of salmonella in these tissues and eventually clears the infection (Briles et al, 1981; Hormaeche, 1979; Lissner et al, 1983; Swanson and O'Brien, 1983). However, the detailed events associated with the different components of the infection have yet to be fully worked out.

Biology of Salmonella, Edited by F. Cabello *et al.*,
Plenum Press, New York, 1993

CELLULAR LOCATION OF SALMONELLA IN MICE

It has long been thought that the bulk of salmonella growth in enteric fever infections in the mouse is within the macrophages. Much of the data for this supposition is indirect. A critical finding was that while immunity to both live and killed salmonella elicited antibody to salmonella, only immunity to live salmonella elicited cell-mediated immunity and protection (Collins and Mackaness, 1968). The association between the elicitation of protection and cell-mediated immunity is reminiscent of similar associations observed with *Listeria monocytogenes* and *Mycobacterium* infection in the mouse, where it seems clear that the bacterial growth is in macrophages and that protection is by transfer of immune T cells (Lane and Unanue, 1972; Orm and Collins, 1983). The conclusion that infecting salmonella were intracellular also fits well with the observations that they can grow in macrophages *in vitro* (Fields et al, 1986; Lissner et al, 1983), and the fact that passive antibody could be shown to be protective only under special circumstances (Colwell et al, 1984; Eisenstein et al, 1984; Muotiala et al, 1989; Rowley and Turner, 1966; Saxen, 1984).

Direct evidence for a cellular location of salmonella during the early stages of enteric fever in mice is lacking. Hsu et al examined mice late in infection, and reported that most of the salmonella in the liver and spleen are either extracellular or in non-professional phagocytes (Nakoneczna and Hsu, 1980), and pointed out the absence of any direct evidence showing an intracellular location for salmonella *in vivo* at any stage of the infection (Lin et al, 1987; Nakoneczna and Hsu, 1980). Although the detection of extra-cellular salmonella in these studies may have been because the mice examined were already septic, the conclusions drawn underscore the lack of direct evidence for the cellular location of salmonella early in infection.

Intracellular Location of Salmonella

One of the first findings that strongly suggested an intracellular location for the bulk of viable intracellular salmonella was the demonstration that mutations in pathways used to acquire iron from transferrin or lactoferrin altered salmonella growth in serum *in vitro*, but failed to affect the ability of *S. typhimurium* to cause typhoid in mice (Benjamin et al, 1985; Stocker and Mäkelä, 1986). These results strongly suggested that growing salmonella in mice were not exposed to serum, and was consistent with the hypothesis that they were growing intracellularly.

To obtain confirmation of the apparent intracellular location of salmonella by a different method we treated salmonella infected mice with gentamicin (Dunlap et al, 1991). This antibiotic quickly kills salmonella outside of cells, but is inefficient at killing intracellular salmonella. We injected this antibiotic before and after i.v. infection with salmonella. In all cases, the *in vivo* exposure to gentamicin was only for 30 minutes. When given prior to infection, gentamicin reduced the numbers of salmonella in the blood by 100 fold but reduced the numbers of salmonella in the liver and spleen by only 10 fold. However, when gentamicin was injected 2 hours after salmonella infection, we observed no reduction in the numbers of salmonella in the liver and spleen. These results suggest that the salmonella in the liver and spleen are intracellular. When the spleens of infected mice were teased and the cells exposed to gentamicin *in vitro,* 41% of the of the salmonella remained resistant to gentamicin, suggesting that they remained intracellular. Sonication of these cells rendered the salmonella sensitive to gentamicin (Dunlap et al, 1991).

Twenty-Four Hours Post Infection, the Majority of Intracellular Salmonella are Within Polymorphonuclear Cells (PMNs). In an attempt to identify the cellular location of salmonella in the spleen 24 hours post infection, we fractionated splenic cells by Ficoll-Hypaque separation and enriched for salmonella containing cells in the Ficoll band and pellet by fluorescence-activated cell sorting for cell surface markers present on macrophages and PMNs. The majority of salmonella containing cells were found in the Mac-1+/ J-11d+ cell fractions of the band and pellet. By direct microscopic visualization of the enriched cell populations it was observed that virtually all of the salmonella were in PMNs. PMNs containing salmonella in the Ficoll band generally contained a single bacteria, those from the pellet generally contained several salmonella (Dunlap, Benjamin, and Briles et al submitted).

A Safe Site for *S. typhimurium* in Mice

By using *S. typhimurium* carrying a temperature sensitive plasmid that shows minimal replication *in vivo*, we have been able to separately study the growth and killing of salmonella for the first few days post i.v. infection. During the first few hours post infection almost all of the injected salmonella are cleared from the blood. Most of them are killed, but about 5-20% survive in the liver and spleen. After the first two hours the salmonella in the liver and spleen were able to grow with little subsequent killing (Benjamin et al, 1990). Thus, it appears that the intracellular location containing most of the salmonella is a "safe-site" situation where they are protected from killing.

The Survival of Salmonella Within the Liver and Spleen is Not the Result of Salmonella Acquiring the Ability to Resist Killing, or Because the Host has Become Unable to Kill. The safe-site might be a privileged location that makes the salmonella relatively inaccessible to killing. Alternatively it could be the result of a suppression of non-specific immunity in the mouse or the activation of salmonella virulence genes shortly after infection that protect the surviving salmonella from subsequent killing. To test the hypothesis that the salmonella adapt to the host by the acquisition of enhanced virulence mechanisms, mice were infected with salmonella freshly recovered (processed at 4 ^{o}C) from other mice after 24 hours of infection (Dunlap et al, 1991). We observed the same degree of killing (approximately 90%) of the *in vivo* grown salmonella and medium-grown salmonella. To examine the possibility that host defenses are overwhelmed or down regulated shortly after infection, mice were infected with wild type salmonella and then reinfected 24 hours later with the same bacteria carrying kanamycin resistance. Control mice (not previously infected) were infected with the kanamycin resistant salmonella. Although slightly faster clearance from the blood was observed by the previously infected animals, the numbers of kanamycin resistant salmonella surviving in the spleen and liver was essentially the same in the infected (Log CFU 4.67 $\pm$ 0.04) versus the uninfected (Log CFU 4.37 $\pm$ 0.11) mice (Dunlap et al, 1990).

Collectively, these results suggest that the "safe site" for salmonella is an intracellular location rather than a change in resistance of the mouse or virulence of the salmonella. This hypothesis is consistent with other studies indicating that salmonella resist both oxygen-dependent and oxygen-independent killing by phagocytic cells and are compatible with the hypothesis that some of the salmonella simply evade the normal killing mechanisms. Salmonella have several enzyme activities such as superoxide dismutase and catalase which may help protect them against the oxidants (Fridovich, 1975). Recently, an outer membrane protein has been identified which protects salmonella from oxidative killing by PMNs (Stinavage et al, 1990). Other microbial structures have been identified which protect salmonella from non-oxidative killing (Fields et al, 1989; Kim et al, 1988; Miller et al, 1990). Electron micrographs (EM) of splenic phagocytes in mice with septic salmonella infections show many bacteria within the cytoplasm, thus providing another mechanism of escaping the destructive mechanisms of the lysome (Linn et al, 1989).

There Appears to be a Critical Number of Salmonella Within the RES of the Mouse Which is Necessary to Elicit an Inflammatory Response

An interesting facet to salmonella infections is that as few as 1-10 bacteria injected i.v. into a susceptible mouse proceeds to kill the animal whereas when 10^6 bacteria are injected i.v., over 90% of the inoculum is killed within 4 hours. Based on these results it seemed likely that mice kill a higher percentage of a large inoculum than a small one. In order to examine the kinetics of salmonella infections in susceptible mice, varying inoculums of virulent salmonella (range: log 0 to 5) were injected i.v. and mice were sacrificed 4 days later. Spleens were removed and the numbers of salmonella present within them were enumerated (Dunlap et al, unpublished). When from 10^1 to 10^2 CFU were injected there was a proportional increase in CFU at 4 days. Inocula of 10^2 to 10^3 CFU all yielded about 10^5 CFU in the spleen and liver at 4 days indicating a plateau had been reached. Inocula above 10^3 again showed proportional increases in CFU at 4 days (Dunlap et al, unpublished). These data are reminiscent of earlier findings by others showing longer time to death when mice were infected with intermediate numbers of salmonella than with either low or high numbers of salmonella (Krishnapillia, 1971).

One interpretation of these data is that at some critical load of salmonella the mouse is able to make an inflammatory response that blunts the course of the infection. One way this could occur is that when infected with 10^3 or more salmonella the inflammatory response increases the percentage of salmonella killed during blood clearance. This could also occur if, as the inoculum reaches a particular number of CFU, the inflammatory response limits the subsequent growth rate. This would have the effect of allowing the numbers of CFU in mice infected with low numbers of CFU to catch up with those in mice infected with larger numbers of CFU.

Evidence for a rapid inflammatory response to salmonella infection is seen in changes in splenic granulocyte population. By 2 hours after injection of 10^6 salmonella the average specific gravity of PMNs (as revealed on Percoll density gradients) had decreased. By 6 hours, PMNs had increased in total number and a peak of dense PMNs was again seen, suggesting that there was an influx of new PMNs into the spleen. At 24 hours, the total number of PMNs remained elevated. In contrast to PMNs, monocytes/macrophages did not appear to increase in number during the first 24 hours after infection, although there was a tendency for the monocytes/macrophages present to be of lower density than those in uninfected spleens (Dunlap et al, 1991).

More support for the hypothesis that a critical number of salmonellae are required to initiate the inflammatory response is demonstrated by our experiments utilizing Percoll density gradients to define the cell types and densities of splenic cell populations of mice inoculated with 10^2 CFU of virulent salmonella and sacrificed on days 3 and 4 post infection. Three days post infection with this small inoculum, there was no change from baseline in the numbers or densities of PMNs or monocytes within the infected spleens. There were approximately 10^3 CFU of salmonella within the spleen at this time point. Four days post infection, when the numbers of bacteria had reached approximately 10^4 - 10^5 CFU, the density of PMNs decreased. No influx of less dense PMNs into the spleen, however, was apparent at this time (Dunlap and Briles, unpublished). We thus hypothesize that there is a minimal number of salmonella which must be present within the spleen to provide the inflammatory signal that elicits an influx of PMNs and their subsequent activation. The cellular location of salmonella within the liver and spleen after injection with very low numbers of salmonella and prior to the PMN influx is unknown.

Ity LOCUS OF THE MOUSE

During the early phase of murine salmonellosis, the *Ity*^{r} (*immunity to typhimurium*^{resistant}) and *Ity*^{s} (*immunity to typhimurium*^{sensitive}) alleles of the *Ity* locus on chromosome 1 of the mouse play a major role in the resistance to systemic *S. typhimurium* infection (Benjamin et al, 1990; Plant and Glynn, 1979; Swanson and O'Brien, 1983). This locus (or closely linked loci) also controls resistance to infections with certain strains of *Mycobacteria* and *Leishmania* (Plant et al, 1982; Skamene, 1983; Skamene, 1982), two organisms known to grow within host macrophage phagolysosomes (Bradley et al, 1979; Hart et al, 1972). The *Ity* gene appears to be responsible for much of the difference in salmonella resistance among inbred strains (Plant and Glynn, 1974; Plant and Glynn, 1976; Robson and Vas, 1972). About half of the inbred strains of mice are *Ity*^{r} and half are *Ity*^{s} (O'Brien and Rosenstreich, 1983).

The effects of the *Ity* locus on the host's ability to control the salmonella infection can be detected as early as 24 hours after injection and can alter the LD_{50} of salmonella from 10^1 to 10^3 or more (Briles et al, 1981; Swanson and O'Brien, 1983). When *Ity*^{s} mice are injected i.v. with 10-100 CFU of virulent *S. typhimurium*, the number of salmonella in the liver and spleen increases exponentially to levels of 10^8-10^9 CFU prior to death at 5-8 days post infection (Hormaeche et al, In Press; Robson and Vas, 1972). In *Ity*^{r} mice, the numbers of salmonella in the liver and spleen also increase exponentially, but at a slower rate than in *Ity*^{s} mice. After infection of *Ity*^{r} mice with 10^1 to 10^3 salmonella the total number of CFU in the liver and spleen seldom exceed 10^4 or 10^5 CFU by days 7-14 when the numbers of salmonella are usually held is check by acquired immunity (Hormaeche, 1979, O'Brien et al, 1984). Subsequently the numbers of CFU in *Ity*^{r} mice decline over the next few weeks. The only other murine locus with a comparable effect on survival of salmonella infected mice is the *nu* locus, which affects induced immunity, but does not play a significant role until as late as two weeks post infection (O'Brien and Metcalf, 1982).

Several observations have led to the idea that the effector cell of *Ity* expression and the cell in which growing salmonella reside is the resident macrophage (Hormaeche, 1979; Plant and Glynn, 1976; Plant and Glynn, 1979; Robson and Vas, 1972). Indirect findings that support this theory include the fact that *S. typhimurium* is a facultative intracellular pathogen that can survive within phagocytes (Lowrie et al, 1979), and the observation that the numbers of viable salmonella increase more rapidly in the spleens and livers of *Ity*^{s} mice than in the same organs of *Ity*^{r} animals. More direct *in vivo* data which implicate a central role for macrophages (or PMNs) in *Ity*-controlled events include the fact that cells involved in conferring early resistance are bone-marrow derived (Hormaeche, 1979) and silica sensitive (O'Brien et al, 1979). *Ity*^{r} nude mice (CD-1 *nu/nu)* are able to restrict early bacterial growth as efficiently as do euthymic mice (CD-1 *nu*/+), indicating that mature T cells are not required for the expression of the *Ity* phenotype (O'Brien and Metcalf, 1982).

Ity Regulation of Growth Versus Killing *in vivo*

Swanson and O'Brien (Swanson and O'Brien, 1983) compared the blood clearance and RES uptake during the first 24 hours of *S. typhimurium* infection in *Ity*^{r} and *Ity*^{s} mice. Their data indicated that the blood clearance and early RES uptake was similarly completed in less than 2 hours in *Ity*^{s} as well as *Ity*^{r} mice. Furthermore, there was a similar decrease in the numbers of splenic bacteria between

2 and 5 hours post infection in *Ity*^{r} and *Ity*^{s} strains. At 12 hours, there were net increases in splenic bacterial content in both strains (as compared to inoculating dose). By 24 hours, the net increase was significantly higher in the *Ity*^{s} strain. Their data indicated that while the *Ity* phenotype influences the pathogenesis of murine typhoid within 24 hours after i.v. inoculation, it does not influence the initial blood clearance, RES uptake or RES killing within the first few hours after infection. Other studies by these investigators demonstrated that *in vitro* cultured macrophages from *Ity*^{r} mice were able to kill intracellular salmonella more efficiently than macrophages from *Ity*^{s} mice (Lissner et al, 1983). These authors thus postulated that differences in macrophage killing, rather than salmonella growth rate were responsible for the disparity in the net growth that they had observed *in vivo* in *Ity*^{r} and *Ity*^{s} mice (Lissner et al, 1983; Swanson and O'Brien, 1983).

Data gathered by other investigators indicate that the primary *in vivo* effect of *Ity* is on salmonella growth. The first studies, by Hormaeche in 1980, comparing inbred mouse strains suggested that the *Ity* locus controls the rate of salmonella growth rather than the rate of killing (Hormaeche, 1980). Since those studies used relatively avirulent salmonella and failed to show linkage of the difference to the *Ity* locus, the results were not conclusive (Lissner et al, 1983).

Our own recent studies have also compared the killing and growth of salmonella in mouse strains congenic for the *Ity*^{r} locus and recombinant inbred strains segregating for the *Ity*^{r} and *Ity*^{s} alleles (Benjamin et al, 1990). We used a *S. typhimurium aroA* mutant, which does not grow *in vivo*, and demonstrated that growth of the infecting salmonella is necessary for manifestation of the *Ity* phenotype. To study the effects of the *Ity* locus on salmonella growth and killing, we used fully virulent salmonella carrying the temperature sensitive plasmid, pHSG422. This plasmid does not replicate at body temperature, and thus, is diluted out during salmonella growth *in vivo*. It is therefore possible, from the total number of salmonella and the number of plasmid-containing salmonella recovered from the mice, to separately estimate the amount of salmonella killing and the amount of salmonella growth. Using such plasmid-containing salmonella, we found that there is approximately three-fold more killing of salmonella in *Ity*^{r} mice compared to *Ity*^{s} mice during the 4 hours post-infection. However, by 48 hours post-infection, there is about 18-fold more growth of salmonella in *Ity*^{s} mice than in *Ity*^{r} mice. These results show that the major *in vivo* effect of the *Ity* locus is on the regulation of salmonella growth (Benjamin et al, 1990).

THE *mviA* LOCUS OF SALMONELLA REGULATES THE ABILITY OF AT LEAST CERTAIN SALMONELLA TO EXPLOIT THE SALMONELLA SUSCEPTIBILITY OF MICE WITH THE *Ity*$^{s/s}$ GENOTYPE

The virulent *S. typhimurium* strain WB600 carries the *mviA* allele of the gene mouse virulence A (Benjamin et al, 1991; Benjamin et al, 1986). We have shown that the virulent phenotype of WB600 is dependent on a non-functional *mviA* gene. As compared to the functional allele, *mviA*$^+$, *mviA* increases virulence in *Ity*^{s} mice but not in *Ity*^{r} mice (Benjamin et al, 1991). A specific *Bgl*II site, *mviA4185*, located at about 35 min on the salmonella chromosome (between *osmZ* and *galU)* was found to be within *mviA* (Benjamin et al, 1991; Benjamin et al, 1986). It was clear that virulence resulted from an a lack of *mviA*$^+$ function because chromosomal insertion of an antibiotic cassette in the *mviA4185* site of *mviA*$^+$, as well as the homologus *mviA4093* site of *mviA* DNA resulted in virulence. When *mviA* and *mviA*$^+$ were both expressed in the same strain, with one carried in the chromosome and the other on a plasmid, avirulence was always dominant (Benjamin et al, 1991). We have recently shown that the mviA+ gene product is a protein of about 37 kDa (Swords and Benjamin, unpublished). Surprisingly, we also observed that independent wild mouse-virulent isolates of *S. typhimurium* all carried the avirulent *mviA*+ allele (Benjamin et al, 1991). We assume that in wild strains the expression of *mviA*+ is suppressed of by a regulatory mechanism, and that this mechanism must be defective in the laboratory strain WB600. Studies are presently underway to identify this suspected regulatory mechanism as well as the mechanism of action of the *mviA* gene product. Hopefully studies to elucidate the mechanism of action of *mviA* will provide insight into the mechanism of action of *Ity*.

SUMMARY

In mouse enteric fever (typhoid) infection with *S. typhimurium* the bacteria were found to grow intracellularly as has been suspected for many years. Our evidence indicates that once salmonella reach the intracellular location they are relatively protected from killing, at least during the first few days of infection. Contrary to expectation the viable salmonella, at least during the early phase of infection are in PMNs rather than macrophages. Whether or not viable salmonella are predominantly in macrophages at later stages of infection, is not yet know. The major *in vivo* effect of the *Ity* locus has been found to be on salmonella growth rather than killing. Since the effect of the *Ity* locus on salmonella growth rate is readily seen during the first few days of infection, these findings indicate that the *Ity* locus is able to mediate its effect on salmonella pathogenesis in PMNs *in vivo*. Finally a salmonella locus, *mviA*, has been identified that can affect the virulence of salmonella in Ity^s but not Ity^r mice. Because the virulence of salmonella in mice depends on the interaction of alleles at the *mviA* and *Ity* loci, we are hopeful that a resolution of the mechanism of action of *mviA* may provide insight into the mechanism of action of *Ity*.

REFERENCES

Benjamin, W. H., Jr., Hall, P. and Briles, D. E., *Salmonella typhimurium* locus *mviA* regulates virulence in Ity^s but not Ity^r mice: functional *mviA* results in avirulence; mutant (nonfunctional) *mviA* results in virulence., *J. Exp. Med.* 174:1073 (1991).

Benjamin, W. H., Jr., Hall, P., Roberts, S. J. and Briles, D. E., The primary effect of the *Ity* locus is on the growth rate of *Salmonella typhimurium* that are relatively protected from killing, *J. Immunol.* 144:3143 (1990).

Benjamin, W. H., Jr., Turnbough, C. L., Jr., Goguen, J. D., Posey, B. S. and Briles, D. E., Genetic mapping of novel virulence determents of *Salmonella typhimurium* to the region between *trpD* and *supD*., *Microbial Pathogenesis.* 1:115 (1986).

Benjamin, W. H., Jr., Turnbough, C. L., Jr., Posey, B. S. and Briles., D. E., The ability of *Salmonella typhimurium* to produce the siderophore enterobactin is not a virulence factor in mouse typhoid., *Infect. Immun.* 50:392 (1985).

Bradley, D. J., Taylor, B. A., Blackwell, J., Evans, E. P. and Freeman, J., Regulation of Leishmania populations within the host. III Mapping of the locus controlling susceptibility to visceral leishmaniasis in the mouse., *Clin. Exp. Immunol.* 37:7 (1979).

Briles, D. E., Benjamin, W. H., Jr., Williams, C. A. and Davie, J. M., A genetic locus responsible for salmonella susceptibility in BSVS mice is not responsible for the limited T-dependent immune responsiveness of BSVS mice., *J. Immunol.* 127:906 (1981).

Briles, D. E., Lehmeyer, J. and Forman, C., Phagocytosis and killing of *Salmonella typhimurium* by peritoneal exudate cells., *Infect. Immun.* 33:380 (1981).

Carrol, M. E. W., Jackett, P. S., Aber, V. R. and Lowrie, D. B., Phagolysosome formation, cyclic adenosine 3':5'-monophosphate and the fate of *Salmonella typhimurium* within mouse peritoneal macrophages., *J. Gen. Micro.* 110:421 (1979).

Carter, P. B. and Collins, F. M., The route of enteric infection in normal mice, *J. Exp. Med.* 139:1189 (1974).

Carter, P. B. and Collins, J. M., Growth of typhoid and paratyphoid bacilli in intravenously infected mice., *Infect. Immun.* 10:816 (1974).

Collins, F. M. and Mackaness, G. B., Delayed hypersensitivity and arthus reactivity in relation to host resistance in salmonella-infected mice., *J. Immunol.* 101:830 (1968).

Colwell, D. E., Michalek, S. M., Briles, D. E., Jirillo, E. and McGhee, J. R., Monoclonal antibodies to salmonella lipopolysaccharide: Anti-O-polysaccharide antibodies protect C3H mice against challenge with virulent *Salmonella typhimurium*., *J. Immunol.* 133:950 (1984).

Dunlap, N. E., Benjamin, W. H., Jr., McCall, R. D., Tilden, A. B. and Briles, D. E., A "safe site" for *Salmonella typhimurium* is within splenic cells during the early phase of infection in mice., *Microbial Pathogenesis.* 10:297 (1991).

Eisenstein, T., Killar, L. and Sultzer, B., Immunity to Infection with *Salmonella typhimurium:* mouse-strain differences in vaccine- and serum-mediated protection, *J. Infect. Dis.* 150:425 (1984).

Fields, P. I., Groisman, E. A. and Heffron, F., A Salmonella locus that controls resistance to microbicidal proteins from phagocytic cells, *Science.* 243:1059 (1989).

Fields, P. I., Swanson, R. V., Haidaris, C. G. and Heffron, F., Mutants of *Salmonella typhimurium* that cannot survive within the macrophage are avirulent., *Proc. Natl. Acad. Sci. (USA).* 83:5189 (1986).

Fridovich, I., The biology of oxygen radicals., *Science.* 201:875 (1975).

Guo, Y.-N., Hsu, H. S., Mumaw, V. R. and Nakoneczna, I., Electronmicroscopy studies on the bacterial action of inflammatory leukocytes in murine salmonellosis, *J. Med. Microbiol.* 21:151 (1986).

Hart, P. D., Armstrong, J. A., Brown, C. A. and Draper, P., Ultrastructural study of the behavior of macrophages toward parasitic mycobacteria, *Infect. Immun.* 5:803 (1972).

Hormaeche, C. E., The natural resistance of radiation chimeras to *S. typhimurium* C5., *Immunology.* 37:329 (1979).

Hormaeche, C. E., Natural resistance to *Salmonella typhimurium* in different inbred mouse strains., *Immunology.* 37:311 (1979).

Hormaeche, C. E., The in vivo division and death rates of *S. typhimurium* in the spleens of naturally resistant and susceptible mice measured by the superinfecting phage technique of Meynell., *Immunology.* 41:973 (1980).

Hormaeche, C. E., Brock, J. and Pettifor, R. A. 1980. Natural resistance to mouse typhoid: Possible role of the macrophage. in Genetic control of natural resistance to infection and malignancy, edited by Skamene et al. New York: Academic Press, 121.

Kim, K., Kim, I. H., Rhee, S. G. and Stadtman, E. R., The isolation and purification of a specific "protector" protein which inhibits enzyme activation by a thiol/Fe(III)/O_2 mixed-function oxydation system., *J. Biol. Chem.* 263:4704 (1988).

Krishnapillia, V., Uridinediphosphogalactose-4-epimerase deficiency in *Salmonella typhimurium* and its correction by plasmid-borne galactose genes of *Escherichia coli* K-12: Effects on mouse virulence, phagocytosis, and serum sensitivity., *Infect. Immun.* 4:177 (1971).

Lane, F. C. and Unanue, E. R., Requirment of thymus (T) lymphocytes for resistance to listeriosis., *J. Exp. Med.* 135:1104 (1972).

Lin, F. R., Hsu, H. S., Mumaw, V. R. and Moncure, C. W., Confirmation of destruction of salmonellae within murine peritoneal exudate cells by immunocytochemical technique., *Immunology.* 67:394 (1989).

Lin, F. R., X. M. Wang, H. S. Hsu, V. R. Mumaw and Nakoneczna, I., Electron microscopic studies on the location of bacterial proliferation in the liver in murine salmonelosis, *Br. J. Exp. Pathol.* 68:539 (1987).

Lissner, C. R., Swanson, R. N. and O'Brien, A. D., Genetic control of the innate resistance of mice to *Salmonella typhimurium*: expression of the *Ity* gene in peritoneal and splenic macrophages isolated *in vitro.*, *J. Immunol.* 131:3006 (1983).

Lowrie, D. B., Abner, V. R. and Carrol, M. E. W., Division and death rates of *Salmonella typhimurium* inside macrophages: Use of penicillin as a probe, *J. Gen. Microbiol.* 110:409 (1979).

Miller, S. I., Rulkkinen, W. S., Selsted, M. E. and Mekalanos, J. J., A characterization of defensin resistance phenotypes associated with mutations in the *phoP* virulence regulon of *Salmonella typhimurium.*, *Infect. Immun.* 58:3706 (1990).

Muotiala, A., Hovi, M. and Mäkelä, P. H., Protective immunity in mouse salmonellosis: comparison of smooth and rough live and killed vaccines., *Microbial Pathogenesis.* 6:51 (1989).

Nakoneczna, I. and Hsu, H. S., The comparative histopathology of primary and secondary lesions in murine salmonellosis., *Br. J. Exp. Pathol.* 61:76 (1980).

O'Brien, A. D. and Metcalf, E. S., Control of early *Salmonella typhimurium* growth in innately salmonella-resistant mice does not require functional T-lymphocytes., *J. Immunol.* 129:1349 (1982).

O'Brien, A. D. and Rosenstreich., D. L., Genetic control of the susceptibility of C3HeB/FeJ mice to *Salmonella typhimurium* is regulated by a locus distinct from known Salmonella response genes., *J. Immunol.* 131:2613 (1983).

O'Brien, A. D., Scher, I. and Formal., S. B., Effects of silica on the innate resistance of inbred mice to *Salmonella typhimurium* infection., *Infect. Immun.* 25:513 (1979).

O'Brien, A. D., Taylor, B. A. and Rosenstreich, D. L., Genetic control of natural resistance to *Salmonella typhimurium* in mice during the late phase of infection, *J. Immunol.* 133:3313 (1984).

Orm, I. M. and Collins, F. M., Protection against *M. tuberculosis* infection by adoptive immunotherapy, *J. Exp. Med.* 158:74 (1983).

Plant, J. and Glynn, A. A., Natural resistance to salmonella infection, delayed hypersensitivity and Ir genes in different strains of mice., *Nature.* 248:345 (1974).

Plant, J. and Glynn, A. A., Genetics of resistance to infection with *Salmonella typhimurium.*, *J. Infect. Dis.* 133:72 (1976).

Plant, J. and Glynn, A. A., Locating salmonella resistance gene on mouse chromosome 1., *Clin. Exp. Immunol.* 37:1 (1979).

Plant, J. E., Blackwell, J. M., O'Brien, A. D., Bradley, D. J. and Glynn, A. A., Are the *Lsh* and *Ity* disease resistance genes at one locus on mouse chromosome 1?, *Nature.* 297:510 (1982).

Robson, H. G. and Vas, S. I., Resistance of inbred mice to *Salmonella typhimurium.*, *J. Infect. Dis.* 126:378 (1972).

Rowley, D. and Turner, J. K., Number of molecules of antibody required to promote phagocytosis of one bacterium., *Nature.* 1:496 (1966).

Saxen, H., Mechanism of the protective action of anti-Salmonella IgM in experimental mouse salmonellosis., *J. Gen. Microbiol.* 130:2277 (1984).

Schurr, E., Skamene, E., Forget, A. and Gros, P., Linkage analysis of the *Bcg* gene on mouse chromosome 1: Identification of a tightly linked marker, *J. Immunol.* 142:4507 (1989).
Skamene, E., Genetic regulation of host resistance to bacterial infection., *Rev. Infect. Dis.* 5:S823 (1983).
Skamene, E., Gros, P., Forget, A., Kongshavn, P. A. L., Charles, C. S. and Taylor, B. A., Genetic regulation of resistance to intracellular pathogens., *Nature*. 297:506 (1982).
Stinavage, P. S., Martin, L. E. and Spitznagel, J. K., A 59 kilodalton outer membrane protein of *Salmonella typhimurium* protects against oxidative intraleukocytic killing due to human neutrophils., *Mole. Microbiol.* 4:283 (1990).
Stocker, B. A. D. and Mäkelä, P. H., Genetic determination of bacterial virulence, with a special reference to *Salmonella.*, *Curr. Topic Microbiol. Immunol.* 124:149 (1986).
Swanson, R. N. and O'Brien, A. D., Genetic control of the innate resistance of mice to *Salmonella typhimurium*: *Ity* gene is expressed in vivo by 24 hours after infection., *J. Immunol.* 131:3014 (1983).

ENTRY OF *SALMONELLA* INTO EPITHELIAL CELLS

Catherine A. Lee[1] and Stanley Falkow[2]

[1]Department of Microbiology and Molecular Genetics
Harvard Medical School
200 Longwood Avenue
Boston, MA 02115
[2]Department of Microbiology and Immunology
Stanford University of School of Medicine
Stanford, CA 94305

INTRODUCTION

In experimental animal infection, ingested *Salmonella* are seen to enter intestinal epithelial cells (Takeuchi, 1967; Worton et al., 1989). The bacteria are able to enter both absorptive epithelial cells as well as specialized microfold (M) cells that lay over gut-associated lymphoid tissue (GALT) (Heffernan et al., 1987). Bacterial contact results in localized disruption of the epithelial microvilli, accumulation of filamentous actin at the site of entry and enclosure of *Salmonella* within membrane-bound vacuoles. Presumably, invasion of the intestinal epithelium allows *Salmonella* access to systemic sites of infection and also stimulates mucosal immunity. It is not known if bacterial entry into absorptive cells, M cells or both is important for *Salmonella* pathogenesis. So in order to understand the role of invasion in pathogenesis as well as understand the mechanism of invasion, *Salmonella* genes have been identified that are involved in the invasion process (Elsinghorst et al., 1989; Finlay et al., 1988; Galán and Curtiss, 1989; Lee et al., 1992). The identification of invasion genes is made possible due to the ability to study the invasion process *in vitro*. *Salmonella* can enter both polarized and non-polarized cultured epithelial cells (Finlay and Falkow, 1990; Finlay et al., 1988). In fact, entry of *Salmonella* into polarized Caco-2 cells morphologically resembles the entry of bacteria into absorptive epithelial cells *in vivo* (C. L. Francis and S. Falkow, unpublished results). Unfortunately, at this time, there is no *in vitro* model for entry of bacteria into the specialized M cells.

Biology of Salmonella, Edited by F. Cabello *et al.*,
Plenum Press, New York, 1993

RESULTS

The ability of *Salmonella* to enter epithelial cells can be quantitated by treatment of infected monolayers with the antibiotic gentamicin. Gentamicin cannot cross the mammalian cell membrane and thus, while treatment with gentamicin will kill extracellular bacteria, any bacteria that have entered the mammalian cells are protected from exposure to the antibiotic and survive the treatment (Vaudaux and Waldvogel, 1979). In this way, the number of intracellular bacteria can be quantitated by measuring the colony-forming units (cfu) which survive treatment with gentamicin.

Modulation of *Salmonella* Entry into Epithelial Cells

In vitro studies revealed that the of ability of a variety of *Salmonella* species to enter cultured epithelial cells is modulated by bacterial growth conditions. Galán and Curtiss discovered that invasiveness is regulated by osmolarity such that growth in media of high osmolarity induce invasion genes (Galán and Curtiss, 1990). We and others discovered that invasiveness is also regulated by oxygen such that *Salmonella* grown under limiting oxygen conditions are more invasive than those grown aerobically (Ernst et al., 1990; Lee and Falkow, 1990; Schiemann and Shope, 1991). In addition, bacterial growth state affects invasion such that bacteria from stationary phase cultures are severely defective in the ability to enter epithelial cells (Table 1) (Ernst, et al., 1990; Lee and Falkow, 1990). Thus, *Salmonella* growing in conditions of high osmolarity and low oxygen availability are highly invasive.

Table 1. Effect of bacterial growth conditions on the ability of *S. typhimurium* SL1344 to enter HEp-2 cells

Bacterial Growth Condition	Bacterial Invasiveness[1]
stationary phase	0.004
aerobic	2.7
low oxygen	100

[1] The ability of bacteria to enter HEp-2 cells during one hour was determined by their survival after treatment with gentamicin. Bacterial invasiveness is normalized such that the invasiveness of bacteria grown under inducing, low oxygen conditions equals 100.

It has been postulated that the relevant site during pathogenesis at which the bacteria would grow under such inducing conditions might be the intestinal lumen. If this is the case, the *Salmonella* invasion genes that are regulated by osmolarity, oxygen and growth state would be those involved in bacterial entry into cells of the intestinal epithelium, possibly absorptive cells, M cells or both. We used a genetic approach to try to identify the *Salmonella typhimurium* oxygen-regulated invasion genes. We reasoned that bacterial mutants that constitutively express the oxygen-regulated invasion genes might enter epithelial cells even when grown aerobically. We have termed this increased invasion phenotype, hyperinvasive (Lee, et al., 1992).

Identification of Hyperinvasive *S. typhimurium* Mutants

The identification of hyperinvasive *S. typhimurium* mutants was simple due to the properties of gentamicin, described above. Large pools of independent mutants were generated and grown under the repressing, aerobic growth condition. Those mutants in the pool that could enter epithelial cells even when grown aerobically were selected for by incubation with HEp-2 cells and subsequent treatment with gentamicin. The desired hyperinvasive mutants should enter the HEp-2 cells and consequently survived the gentamicin treatment even when grown aerobically. We used transposon insertion to generate bacterial mutants. Transposons can insert into genes and disrupt their function. However, we also used transposon Tn5B50 which was engineered with the constitutive *neo* promoter at one end (Simon et al., 1989). In this way, while insertion of Tn5B50 into genes will disrupt them, insertion of Tn5B50 upstream of a gene and in the proper orientation will result in constitutive expression of the gene from the exogenous *neo* promoter.

We identified three different classes of *S. typhimurium* mutants that could enter HEp-2 cells 13- to 74-fold better than wild-type when grown aerobically (Table 2) (Lee, et al., 1992). Interestingly, one class of mutants was only obtained from mutagenesis with Tn5B50. Presumably, this class of mutants can enter HEp-2 cells due to expression of a gene(s) from the exogenous *neo* promoter on Tn5B50. We have termed this locus *hil* for *h*yper-*i*nvasion *l*ocus. In contrast, the other two classes of mutants could be obtained from mutagenesis with Tn*5* and Tn*10*. These mutants are able to enter HEp-2 cells even when grown aerobically due to disruption of genes.

One class of hyperinvasive mutations disrupt the *che* genes and result in a non-chemotactic phenotype. Further study by Jones et al of the effect of *che, fla* and *mot* mutations on bacterial invasion suggests that the disposition of

Table 2. Effect of transposon mutations on ability of *S. typhimurium* SL1344 to enter HEp-2 cells

	Bacterial Invasiveness[1]	
Transposon mutation	Aerobic growth condition	Low oxygen condition
wild-type	1	43
::Tn*10*-181 (*che*)	13	54
::Tn*10*-177 (*rho*)	16	62
::Tn*5*B50-380 (*hil*)	18	96
::Tn*5*B50-378 (*hil*)	74	373

[1] Bacterial invasiveness is normalized such that the invasiveness of aerobically grown wild-type bacteria equals 1.

flagella around the bacterial cell surface impedes bacterial invasion (Jones et al., 1992). Thus, the *che* mutants which are non-chemotactic due to their inability to tumble do not dispose their flagella randomly and therefore are able to enter HEp-2 cells more efficiently than wild-type bacteria. The second class of mutations affect the expression of *rho* which encodes an essential transcription termination factor (Neidhardt et al., 1987). For example, one Tn*5* insertion mutation lay between the -10 and -35 positions of the *S. typhimurium rho* promoter. Mutations in *rho* have pleiotropic effects on gene expression due to changes in transcription termination and the state of DNA supercoiling (Fassler et al., 1986; Korn and Yanofsky, 1976; Storts and Markovitz, 1991). Presumably, the increased invasiveness of *rho* mutants is due to change in expression of invasion genes. Further study is required to identify the specific changes in gene expression that result in the hyperinvasiveness of *rho* mutants.

The *hil* mutations were only obtained after mutagenesis with the Tn*5*B50 transposon. The phenotypic and molecular analysis of the *hil*::Tn*5*B50 mutants suggests that there is a gene(s) downstream of Tn*5*B50-378 that acts positively on expression of *S. typhimurium* invasiveness (Figure 1). If this is the case, deletion or disruption of the *hil* locus might result in loss of invasiveness. Analysis of the effect of a *hil* deletion mutation on bacterial entry into HEp-2 cells showed that this region is, in fact, essential for expression of *S. typhimurium* invasiveness. Even when grown under low oxygen conditions, the *hil* deletion mutant was 1000-fold less able to enter HEp-2 cells than the comparably grown parental strain (Figure 1).

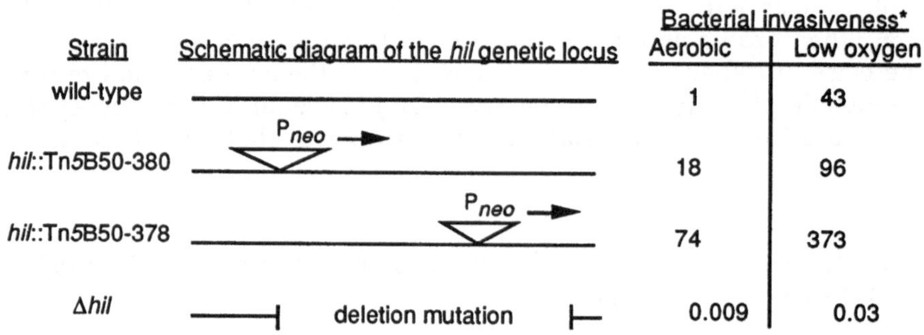

Strain
Schematic diagram of the *hil* genetic locus
Bacterial invasiveness*
Aerobic
Low oxygen
wild-type
1
43
Pneo
hil::Tn5B50-380
18
96
Pneo
hil::Tn5B50-378
74
373
Δ*hil*
deletion mutation
0.009
0.03

*The ability of bacteria to enter HEp-2 cells is normalized such that the invasiveness of aerobically grown wild-type equals one.

Figure 1. Effect of *hil* mutations on *S. typhimurium* SL1344 invasiveness

The fact that both increase and decrease in *hil* expression result in a corresponding change in bacterial invasiveness suggests that the alteration in *hil* mutants is specific for invasion. Furthermore, the finding that deletion of *hil* reduces the invasiveness of bacteria grown under the normally, inducing low oxygen conditions suggests that the *hil* mutants are affected in the oxygen-regulated pathway.

As described above, *hil* is required for *S. typhimurium* entry into non-polarized cultured epithelial cells. The *hil* locus is also involved in entry into polarized epithelial cells. Cultured Caco-2 cells can form tight junctions and a highly organized brush border structure *in vitro,* thus, resembling the absorptive intestinal epithelium (Chantret et al., 1988). A hyperinvasive *hil* mutant, grown under the normally repressing, aerobic conditions, enter polarized Caco-2 cells in a manner morphologically indistinguishable from that of invasive wild-type bacteria grown under the inducing conditions (data not shown; C. L. Francis and S. Falkow, unpublished results). In contrast, wild-type *S. typhimurium* grown aerobically rarely even associate with epithelial cells. This observation indicates that the hyperinvasive *hil* mutant constitutively expresses all of the bacterial factors required for entry into the intestinal epithelium.

Examination of actin in HEp-2 cells incubated with the *hil* deletion mutant revealed that, in addition to its inability to enter epithelial cells, the *hil* deletion mutant does not induce accumulation of actin microfilaments at the site of bacterial contact (data not shown; C. L. Francis and S. Falkow, unpublished results). Actin polymerization appears to be an early event during invasion and likely occurs prior to enclosure of bacteria within vacuoles. For example, electron micrographs show extensive alteration of the cytoskeleton prior to entry (C. Francis, unpublished results). In addition, a pathogenic *E. coli* species can clearly induce actin polymerization within epithelial cells as a result of extracellular bacterial binding at the epithelial cell surface (Knutton et al., 1989). Thus, factors encoded by *hil* appear to be involved in early as well as late cellular events during invasion. Thus, as hoped, the selection for hyperinvasive *S. typhimurium* mutants has identified *hil*, a gene(s) that is central to bacterial invasion.

Location of the *hil* Locus on the Chromosome of *S. typhimurium*

The *hil* locus maps near minute 59 of the *S. typhimurium* chromosome by P22-mediated transduction (Figure 2) (Lee, et al., 1992). Previous studies have identified two gene clusters required for bacterial invasion in this region of the chromosome; the *inv* locus described by Elsinghorst et al (Elsinghorst, et al., 1989) (E. Elsinghorst, unpublished results) encompasses *recA* and *srl* whereas that described by Galán and Curtiss (Galán and Curtiss, 1989) (J. Galán, unpublished results) lies near *mutS*. These results suggest that the *hil* mutations lie between the two *inv* loci.

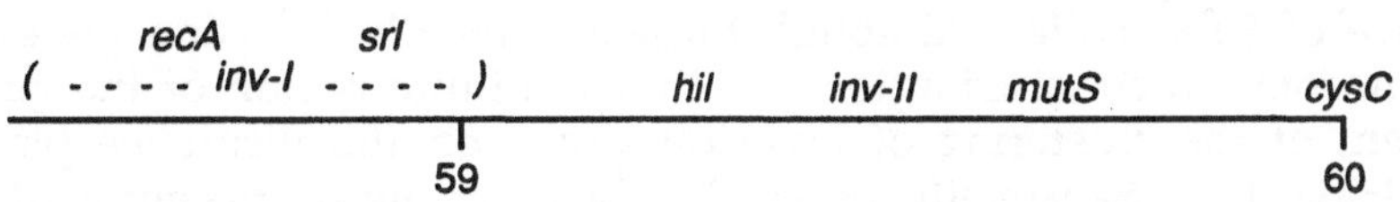

The *inv-I* locus consists of four genes interspersed with *recA* and *srl*. The *inv-II* locus consists of at least seven genes.

Figure 2. Genetic map of genes in the 59 - 60 minute region of the *S. typhimurium* chromosome

One explanation of the invasion phenotypes of the *hil* mutations is that the Tn5B50 and deletion mutations in *hil* act in *cis* by directly increasing or eliminating the expression of one of these previously identified loci. However, it is unlikely that the *hil* mutations directly affect the Elsinghorst *inv* locus since transduction data suggest that 20 - 30 kb intervene between the two loci; the Elsinghorst *inv* genes are >90% linked to *srl* (E. Elsinghorst, unpublished results) whereas the *hil* locus is <0.2% linked to *srl* (Lee, et al., 1992). In addition, the *hil* locus does not appear to encompass the Galán *inv* locus since a cosmid containing these genes does not complement the *hil* deletion mutation (Galán and Curtiss, 1989). Although pYA2217 does not completely complement the Galán *inv* mutation, the *hil* mutation is not complemented to any degree by pYA2217 (Figure 3). The regulation and expression of the Galán *inv* locus may be altered when encoded on a plasmid.

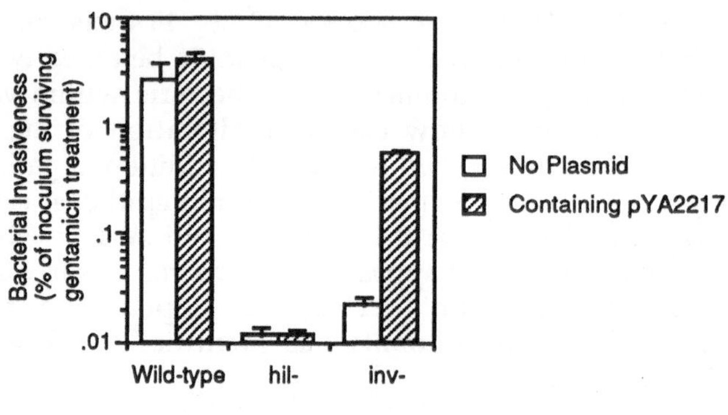

The *hil*⁻ mutation is the deletion mutation described above. The *inv-II*⁻ mutation is the the *invA*::Tn*phoA* mutation described by Galán and Curtiss (1989).

Figure 3. Complementation of *S. typhimurium* invasion mutants by cosmid pYA2217

In summary, the *hil* mutations must affect novel invasion genes which lie between the two previously identified *inv* loci (Figure 2). The grouping of so many invasion genes in one region of the chromosome suggests that they may function together. For example, the genes may encode the proteins required for the expression, assembly and structure of an invasion complex on the surface of *Salmonella*. Although these *Salmonella* invasion genes are on the chromosome, their distribution within a 2 minute region of the genome is reminiscent of the clustering of invasion genes on the virulence plasmid of *Shigella* (Hale, 1991; Sansonetti, 1991). The large virulence plasmids of *Shigella* encode the *Shigella* invasion genes as well as genes required for their expression and assembly. These *Shigella* genes are dispersed over approximately 100 kb of the virulence plasmid.

Although the tissue tropism of *Shigella* and *Salmonella* infection differs, *Shigella* enter epithelial cells in the colon whereas *Salmonella* infect cells of the ileal epithelium, certain aspects of *Shigella* invasion are comparable to those of *Salmonella* invasion. For example, both organisms enter cultured epithelial cells *in vitro* by a mechanism which triggers accumulation of filamentous actin within the host cell (Sansonetti, 1991). At this time, it is not known if any invasion functions are shared between *Shigella* and *Salmonella,* however, a recent study has revealed that the *S. typhimurium invA* gene of the *inv-II* locus is homologous to one of the *Shigella* invasion genes (J. Galán, unpublished results).

It is possible that the region between *proU* and *cysC* contains additional genes involved in *Salmonella* invasion. The three invasion loci in the 59-60 minute region of the *S. typhimurium* chromosome were each identified by different genetic approaches. As described above, *hil* was identified by isolation of hyperinvasive mutations (Lee, et al., 1992). The *inv-I* locus was originally identified in *S. typhi.* Expression of the *S. typhi inv-I* locus in a laboratory strain of *Escherichia coli* allowed the recombinant *E. coli* to enter cultured epithelial cells. The *inv-I* locus was subsequently found to be required for *S. typhimurium* invasiveness (Elsinghorst, et al., 1989). The *inv-II* locus was discovered by analysis of the genetic defect in a non-invasive laboratory strain of *S. typhimurium* (Galán and Curtiss, 1989). Thus, a comprehensive analysis of the genes in this chromosomal region has not been done.

METHODS

Transposon Mutagenesis of *Salmonella typhimurium*

S. typhimurium strains were mutagenized by random transposon insertion (Lee, et al., 1992). *S. typhimurium* strain SL1344 (Hoiseth and Stocker, 1981) was mutagenized with Tn*10* using a phage lysate from DB5204 (Davis et al., 1980). Tails were added to the defective P22 phage particles by incubation with extracts from *E. coli* 294/pPB13 (Sauer et al., 1982). Two SL1344::Tn*10* pools were generated, each containing 1,000 to 5,000 independent transposon mutants.

S. typhimurium strain EE251, a spontaneous streptomycin-resistant derivative of SL4012, (Collins and Stocker, 1976) was mutagenized with Tn*5* or with Tn*5*B50 using a delivery system based on plasmid pRTP1 (Stibitz et al., 1986). Initially, each transposon was moved by transposition from lambda derivatives to pRTP1 in the non-suppressing *E. coli* host strain MC4100. pRTP1::Tn*5* and pRTP1::Tn*5*B50 were selected such that the transposon insertion did not disrupt the *bla* or *rpsL* genes on the plasmid. The mutagenesis system is based on the fact that strain EE251 contains a streptomycin-resistant allele of *rpsL* which is phenotypically recessive to the streptomycin-sensitive *rpsL* allele on the pRTP1 derivatives. In this way, transposition of Tn*5* or Tn*5*B50 onto the genome of strain EE251 and subsequent loss of the plasmid can be selected for by growth of such EE251::Tn mutants in the presence of streptomycin and kanamycin (for Tn*5*; or tetracycline for Tn*5*B50).

For example, to generate one pool of EE251::Tn*5* mutants, an exponentially growing culture, derived from a single colony of EE251/pRTP1::Tn*5*, was diluted and 5,000 to 10,000 cfu were plated onto L agar containing kanamycin.

The plate was incubated overnight at 30°C to allow growth of the colonies. In order to select for the EE251::Tn*5* mutants within each, the colonies were replica plated onto L agar containing streptomycin and kanamycin. The replica plate was incubated overnight at 37°C and the strepomycin-resistant, kanamycin-resistant colonies were scraped from the agar and pooled. The replica plating procedure also serves to reduce the effect of unequal representation of transposon mutations that occurred at different times during growth at 30°C since each independent mutant can only grow to the size of a colony on the selection medium. Using this procedure, almost every original kanamycin-resistant colony gave rise to an independent streptomycin-resistant, kanamycin-resistant colony on the replica plate. The surprisingly efficient selection of these transposon mutants that have lost the dominant strepomycin-sensitive pRTP1 derivative is likely due to the random segregation properties of the colE1 plasmid (Ayala and Gomez, 1989). Approximately 0.2% of the cfu in each pool remained ampicillin-resistant and likely were strepomycin-resistant due to mutation of the *rpsL* gene on the pRTP1 derivative. Ten to 12 independent pools of EE251::Tn*5* and EE251::Tn*5*B50 mutants were generated by this procedure.

Selection for Hyperinvasive *S. typhimurium* Mutants

Our selection procedure (Lee, et al., 1992) is based on the fact that extracellular bacteria are killed by gentamicin whereas intracellular bacteria are protected from exposure to the antibiotic. To enrich for mutants that can enter HEp-2 cells even when grown aerobically, two different aerobic cultures were grown from each independent pool of *S. typhimurium* transposon mutants. One-tenth of a ml of each culture was then inoculated into the medium overlying HEp-2 monolayers. Bacteria were allowed to enter the HEp-2 cells during a one hour incubation at 37°C in 5% CO_2. The medium was then changed to RPMI 1640 containing 5% fetal bovine serum and 100 μg/ml gentamicin, so that the extracellular bacteria were preferentially killed during an additional two hour incubation. To release any intracellular bacteria, the monolayers were rinsed twice with phosphate-buffered saline (PBS) and incubated with 50 ml 1% Triton X-100 for 10 minutes at room temperature. The viable bacteria were recovered as a saturated bacterial culture by adding 1 ml of LB broth to each well and agitating the entire dish overnight. The next day, the saturated culture was used to prepare an aerobic culture and the enrichment procedure was repeated. After four sequential enrichment cycles, intracellular bacteria were released from the HEp-2 cells and directly plated onto LB agar. Enumeration of the cfu released from each monolayer indicated which wells contained mutant strains that were more invasive than the wild-type *S. typhimurium*. Single colonies from such wells were purified and analyzed.

Mutant Analysis

S. typhimurium transposon mutants obtained from the selection procedure were 1) assayed for their ability to enter HEp-2 cells (see below), 2) reconstructed into both SL1344 and EE251 strain backgrounds, crossing out the transposon mutation by P22-mediated transduction and 3) re-tested after reconstruction for invasiveness. From this analysis, independent transposon mutations which were found to result in at least a 10-fold increase in aerobic invasiveness were saved and analyzed further.

Invasion Assay

Bacterial invasiveness was assayed by inoculating ~10^7 cfu into the medium overlying a HEp-2 monolayer. Bacteria were allowed to enter the HEp-2 cells during a one hour incubation. After a two hour treatment with gentamicin, the cell monolayer was rinsed with PBS and disrupted by incubation with 0.2 ml 1% Triton X-100. Each sample was vigorously mixed with 0.8 ml LB broth using a pasteur pipet and the viable bacteria were quantitated by plating for cfu on LB agar. Differences in bacterial invasiveness were verified by Giemsa staining of infected monolayers and direct microscopic observation of bacterial association with cells.

Bacterial and Tissue Culture Growth Conditions

Bacterial strains were grown in LB broth (Luria-Bertani; 1% Bacto-tryptone (Difco), 0.5% Bacto-yeast extract (Difco), 0.8% sodium chloride) or on LB agar (GIBCO). Media were supplemented with antibiotics when necessary; 100 μg/ml ampicillin, 25 μg/ml kanamycin, 100 μg/ml streptomycin and/or 10 μg/ml tetracycline. Strains were grown at 37°C. Different growth conditions were used to assess the effect of bacterial growth state on *S. typhimurium* invasiveness. Stationary phase cultures were prepared by inoculating 2 ml of LB broth in 16 x 150 mm borosilicate tubes with bacteria from a colony. After rolling the tubes on a rotator drum for approximately 20 hours, the bacteria were in stationary phase for at least 12 hours. Aerobic cultures were prepared by first growing bacteria to stationary phase. Approximately 10^6 cfu from such stationary phase cultures were then inoculated into 1 ml of LB broth in tubes. Aerobic cultures with a final density of ~10^8 cfu/ml were obtained after placing the tubes on a rolling rotator drum for 2 to 3 hours. Oxygen-limited cultures were prepared by inoculation of 5 ml of LB broth in 16 x150 mm tubes with ~10^4 cfu/ml. The tubes were incubated without agitation overnight until the cultures reached a density of 5×10^8 to 10^9 cfu/ml.

HEp-2 cells (ATCC CCL23), a line established from a human epidermoid carcinoma, were grown without antibiotics in RPMI 1640 medium (Whittaker, Walkerville, MD) supplemented with 5% fetal bovine serum (GIBCO). Monolayers for bacterial invasion were prepared by seeding ~10^5 cells into each well of a 24-multiwell tissue culture plate and incubating overnight at 37°C in 5% CO_2.

Acknowledgments

We thank Jorge Galán and Eric Elsinghorst for providing information about the *inv-II* and *inv-I* invasion loci prior to publication. We also thank Jorge Galán for generously providing the cosmid pYA2217 for the experiment represented by Figure 3. Carol L. Francis analyzed the effect of *hil* mutations on entry into polarized Caco-2 cells and actin polymerization in HEp-2 cells. Bradley D. Jones identified and characterized the *che* mutants. Much of this work was previously published in the Proceedings of the National Academy of Sciences in June 1990 by Lee and Falkow and in March 1992 by Lee, Jones and Falkow.

REFERENCES

Ayala, S. J., and Gomez, E. M., 1989, Stability of ColE1-like and pBR322-like plasmids in *Escherichia coli*, *Mol. Microbiol.* 3:1745.

Chantret, I., Barbat, A., Dussaulx, E., Brattain, M. G., and Zweibaum, A., 1988, Epithelial polarity, villin expression, and enterocytic differentation of cultured human colon carcinoma cells: a survey of twenty cell lines, *Cancer Res.* 48:1936.

Collins, A. L., and Stocker, B. A., 1976, *Salmonella typhimurium* mutants generally defective in chemotaxis, *J. Bacteriol.* 128:754.

Davis, R. W., Botstein, D., and Roth, J. R., 1980, "Advanced Bacterial Genetics: A Manual for Genetic Engineering," Cold Spring Harbor Laboratory, Cold Spring Harbor, New York.

Elsinghorst, E. A., Baron, L. S., and Kopecko, D. J., 1989, Penetration of human intestinal epithelial cells by *Salmonella*: molecular cloning and expression of *Salmonella typhi* invasion determinants in *Escherichia coli*, *Proc. Natl. Acad. Sci. USA* 86:5173.

Ernst, R. K., Dombroski, D. M., and Merrick, J. M., 1990, Anaerobiosis, type 1 fimbriae, and growth phase are factors that affect invasion of HEp-2 cells by *Salmonella typhimurium*, *Infect. Immun.* 58:2014.

Fassler, J. S., Arnold, G. F., and Tessman, I., 1986, Reduced superhelicity of plasmid DNA produced by the *rho-15* mutation in *Escherichia coli*, *Mol. Gen. Genet.* 204:424.

Finlay, B. B., and Falkow, S., 1990, *Salmonella* interactions with polarized human intestinal Caco-2 epithelial cells, *J. Infect. Dis.* 162:1096.

Finlay, B. B., Gumbiner, B., and Falkow, S., 1988, Penetration of *Salmonella* through a polarized Madin-Darby canine kidney epithelial cell monolayer, *J. Cell. Biol.* 107:221.

Finlay, B. B., Starnbach, M. N., Francis, C. L., Stocker, B. A., Chatfield, S., Dougan, G., and Falkow, S., 1988, Identification and characterization of Tn*phoA* mutants of *Salmonella* that are unable to pass through a polarized MDCK epithelial cell monolayer, *Mol. Microbiol.* 2:757.

Galán, J. E., and Curtiss, R. 3., 1989, Cloning and molecular characterization of genes whose products allow *Salmonella typhimurium* to penetrate tissue culture cells, *Proc. Natl. Acad. Sci. USA* 86:6383.

Galán, J. E., and Curtiss, R. 3., 1990, Expression of *Salmonella typhimurium* genes required for invasion is regulated by changes in DNA supercoiling, *Infect. Immun.* 58:1879.

Hale, T. L., 1991, Genetic basis of virulence in *Shigella* species, *Microbiol. Rev.* 55:206.

Heffernan, E. J., Fierer, J., Chikami, G., and Guiney, D., 1987, Natural history of oral *Salmonella dublin* infection in BALB/c mice: effect of an 80-kilobase-pair plasmid on virulence, *J. Infect. Dis.* 155:1254.

Hoiseth, S. K., and Stocker, B. A., 1981, Aromatic-dependent *Salmonella typhimurium* are non-virulent and effective as live vaccines, *Nature* 291:238.

Jones, B. D., Lee, C. A., and Falkow, S., 1992, Invasion of *Salmonella typhimurium* is affected by the direction of flagellar rotation, *Infect. Immun.* in press.

Knutton, S., Baldwin, T., Williams, P. H., and McNeish, A. S., 1989, Actin accumulation at sites of bacterial adhesion to tissue culture cells: basis of a new diagnostic test for enteropathogenic and enterohemorrhagic *Escherichia coli*, *Infect. Immun.* 57:1290.

Korn, L. J., and Yanofsky, C., 1976, Polarity suppressors defective in transcription termination at the attenuator of the tryptophan operon of *Escherichia coli* have altered *rho* factor, *J. Mol. Biol.* 106:231.

Lee, C. A., and Falkow, S., 1990, The ability of *Salmonella* to enter mammalian cells is affected by bacterial growth state, *Proc. Natl. Acad. Sci. USA* 87:4304.

Lee, C. A., Jones, B. D., and Falkow, S., 1992, Identification of a *Salmonella typhimurium* invasion locus by selection for hyperinvasive mutants, *Proc. Natl. Acad. Sci. USA* 89:1847.

Neidhardt, F. C., Ingraham, J. L., Low, K. B., Magasanik, B., Schaechter, M., and Umbarger, H. E., 1987, "*Escherichia coli* and *Salmonella typhimurium*: Cellular and Molecular Biology," American Society for Microbiology, Washington, D. C.
Sansonetti, P. J., 1991, Genetic and molecular basis of epithelial cell invasion by *Shigella* species, *Rev. Inf. Dis.* 13(Suppl 4):S285.
Sauer, R. T., Krovatin, W., Poteete, A. R., and Berget, P. B., 1982, Phage P22 tail protein: gene and protein sequence, *Biochemistry* 21:5811.
Schiemann, D. A., and Shope, S. R., 1991, Anaerobic growth of *Salmonella typhimurium* results in increased uptake by Henle 407 epithelial and mouse peritoneal cells *in vitro* and repression of a major outer membrane protein, *Infect. Immun.* 59:437.
Simon, R., Quandt, J., and Klipp, W., 1989, New derivatives of transposon Tn*5* suitable for mobilization of replicons, generation of operon fusions and induction of genes in gram-negative bacteria, *Gene* 80:161.
Stibitz, S., Black, W., and Falkow, S., 1986, The construction of a cloning vector designed for gene replacement in *Bordetella pertussis*, *Gene* 50:133.
Storts, D. R., and Markovitz, A., 1991, A novel *rho* promoter::Tn*10* mutation suppresses and *ftsQ1*(Ts) missense mutation in an essential *Escherichia coli* cell division gene by a mechanism not involving polarity suppression, *J. Bacteriol.* 173:655.
Takeuchi, A., 1967, Electron microscope studies of experimental *Salmonella* infection. I. Penetration into the intestinal epithelium by *Salmonella typhimurium*, *Am. J. Pathol.* 50:109.
Vaudaux, P., and Waldvogel, F. A., 1979, Gentamicin antibacterial activity in the presence of human polymorphonuclear leukocytes, *Antimicrob. Agents Chemother.* 16:743.
Worton, K. J., Candy, D. C., Wallis, T. S., Clarke, G. J., Osborne, M. P., Haddon, S. J., and Stephen, J., 1989, Studies on early association of *Salmonella typhimurium* with intestinal mucosa *in vivo* and *in vitro*: relationship to virulence, *J. Med. Microbiol.* 29:283.

PLASMID GENES INVOLVED IN VIRULENCE IN *SALMONELLA*

Alistair J. Lax, Gillian D. Pullinger, Jayne M. Spink, Fakhar Qureshi, Michael W. Wood and Philip W. Jones

Agricultural and Food Research Council
Institute for Animal Health
Compton
Berkshire
RG16 0NN, United Kingdom

INTRODUCTION

Salmonella virulence plasmids were first recognised in 1970 (Dowman and Meynell, 1970), but it was 10 years before a role in virulence was identified (Jones *et al.*, 1982). A series of papers followed which indicated that several serotypes contained such plasmids and that curing reduced virulence in mice or chickens, which could be restored by reintroduction of the plasmid (Table 1: Chikami *et al.*, 1985; Gulig and Curtiss, 1987; Barrow *et al.*, 1987; Barrow and Lovell; 1988, Kawahara *et al.* 1988; Hovi *et al.*, 1988). Originally referred to as "cryptic" or serotype-specific, they are now generally and more accurately referred to as virulence plasmids. Although virulence plasmids are generally serotype-specific, there are exceptions. The closely related *S. rostock* and *S. dublin* carry an identical plasmid (Platt, D.J., personal communication), and different virulence plasmids are found within both *S. enteritidis* and *S. pullorum* (Williamson *et al.*, 1988a). The virulence plasmids range in size from about 54kb to 98kb, and therefore represent up to 2% of the genetic information of the *Salmonella* genome. So far only 12 serotypes have been found to contain virulence plasmids (Table 1), and within a serotype there are phage-type differences in plasmid carriage (Brown *et al.*, 1986; Woodward *et al.*, 1989). The relative importance of these plasmids for virulence in mice varies in different serotypes (Williamson *et al.* 1988a). Discussion of the role of the plasmid genes in virulence must also take into account that serotypes without virulence plasmids are not avirulent.

A VIRULENCE REGION COMMON TO *SALMONELLA* VIRULENCE PLASMIDS

Early experiments with these plasmids sought to identify regions of interest using transposon insertion with Tn*A* (Baird *et al*, 1985), which preferentially transposes to plasmids (Kretschmer and Cohen, 1977). These transposon mutants were screened in mice

Biology of Salmonella, Edited by F. Cabello *et al.*,
Plenum Press, New York, 1993

for loss of virulence, and several avirulent mutants were identified. The locations of about 250 *S. dublin* Tn*A* mutants, 3 of which were avirulent, were identified (Baird *et al.*, 1985; Lax *et al.*, 1990). Several of these were targeted to the region where the first avirulent mutations had occurred (Lax *et al.*, 1990). The region around these avirulent mutations corresponded to a similar region in the *S. typhimurium* plasmid (Baird *et al.*, 1985), and an 8kb probe from this region reacted in colony blots and Southern hybridisation with serotypes containing plasmids (Williamson *et al* 1988a). An 8kb fragment hybridised with the probe in most situations, whereas stained gels of the fragments showed that the rest of these plasmids differed greatly. Other groups published similar results using probes (Woodward *et al.*, 1989; Korpela *et al.*, 1989; Poppe *et al.*, 1989), or other methods (Platt *et al.*, 1988; Montenegro *et al.*, 1991). Differences among serotypes can be used for serotype specific detection by PCR (Wood, M.W. *et al.*, in preparation).

Table 1. *Salmonella* serotypes which contain virulence plasmids

Role in mouse virulence known	Role in mouse virulence not known
Salmonella typhimurium	*Salmonella abortusovis*
Salmonella dublin	*Salmonella blegdam*
Salmonella enteritidis	*Salmonella paratyphiC*
Salmonella choleraesuis	*Salmonella rostock*
Salmonella gallinarum	*Salmonella newport*
Salmonella pullorum	*Salmonella moscow*

The clustering of mutations which affected mouse virulence and the near identity of the restriction patterns around this region suggested that not all of the plasmid was essential for virulence. Indeed the *S. typhimurium* plasmid was able to restore virulence to a cured *S. dublin* strain (Williamson *et al.*, 1988b) even though these plasmids were considerably different. Large fragments which encompassed the common region were unable to restore virulence when cloned into high copy cloning vectors. The use of a low copy vector led to a near 100% restoration of virulence to either cured *S. dublin* or *S. typhimurium* with either the 8kb fragment which had been found to be common to all the serotypes with virulence plasmids (Williamson *et al.*, 1988b) or with the 14kb *Sal*I B fragment in which all the avirulent mutations had been identified. Other groups made similar findings (Gulig and Curtiss, 1988; Beninger *et al.*, 1988; Norel *et al.*, 1989a; Krause *et al.*, 1991b). The reason why these genes have to be present in low copy has still to be addressed, but the virulence plasmids are believed to occur at a few copies per cell (Tinge and Curtiss, 1990b).

Thus the hunt for plasmid virulence genes had to a large extent been simplified by the identification of an essential virulence region.

GENES ON THE ESSENTIAL VIRULENCE REGION

Various groups sequenced genes from the virulence region from several serotypes (Figure 1). Six open reading frames (ORFs) encoded off one strand were identified. A different nomenclature was used by each group (Figure 1), but a nomenclature *spv* (for *s*almonella *p*lasmid *v*irulence) has now been agreed by all the groups which have contributed sequence (Gulig, P.A., Danbara, H., Guiney, D., Lax, A., Norel, F. and Rhen, M., in preparation). It is clear that, where comparisons can be made, the sequences from different serotypes and from different laboratories are very similar. With the exception of *spvR* none of the genes shows extensive homology to any sequences on DNA or protein databases. Mutational analysis has identified a clear role in virulence for some genes, but for others polar effects may account for loss of virulence. Similarly, expression in minicells or maxicells or N terminal amino acid sequencing has confirmed that the predicted open reading frames are functional only for some of these ORFs. None of the ORFs has a signal sequence. A summary of information published by mid-1992 on these genes is given below.

						Laboratory
vsdA	*vsdB*	*vsdC*	*vsdD*	*vsdE*	*vsdF*	A
spvR	*spvA*	*spvB*	*virA*	*spvB*	*orfE*	B
vagA						C
ORF1	*ORF2*	*mkfB*	*mkfA*			D
mkaC	*mkaB*	*mkaA*	*mkaD*			E
	mba2	*mba3*	*mba4*			F
spvR	*spvA*	*spvB*	*spvC*	*spvD*	*orfE*	Agreed nomenclature

Figure 1. Sequenced ORFs from the 8kb essential virulence region from different laboratories, and agreed new nomenclature. Sequence A is from *S. dublin*, B - E from *S. typhimurium*, and F is from *S. choleraesuis*. The references are: A - Krause *et al.*, 1991; B - Gulig and Chiodo, 1990, Caldwell and Gulig, 1992, Gulig *et al.*, 1992; C - Pullinger *et al.*, 1989; D - Norel *et al.*, 1989b, Norel *et al.*, 1989c, Norel *et al.*, 1989d; E - Taira and Rhen, 1989a, Taira and Rhen, 1989b, Taira and Rhen, 1990; F - Matsui *et al.*, 1990a, Matsui *et al.*, 1990b, Matsui *et al.*, 1991.

orfE

Expression of this ORF, which potentially encodes a 13.5kDa protein, has been followed using Northern analysis (Fang *et al.*, 1991), and it is believed to be constitutively expressed from a promoter immediately upstream of the ORF. Deletion of this gene does not affect virulence (Krause *et al.*, 1991b; Gulig *et al.*, 1992).

spvD

This gene, which encodes a 24.8kDa protein, has been expressed in minicells (Gulig *et al.*, 1992), probably as part of the *spvA* operon (Krause *et al.*, 1992), although there is a possible promoter immediately upstream of the ORF (Krause *et al.*, 1991b). Transposon insertion affects its production and reduces virulence (Gulig *et al.*, 1992).

spvC

This gene, which encodes a 27.6kDa protein, has been expressed in minicells (Taira and Rhen, 1989b; Williamson *et al.*, 1990; Gulig and Chiodo, 1990) and maxicells (Norel *et al.*, 1989a). Over-expression followed by N terminal sequencing shows that the first AUG in the ORF is the start site (Taira *et al.*, 1991a), as is also the case for *spvR*, *spvA* and *spvB*.

Transposon mutations within the coding region affect virulence (Gulig & Chiodo, 1990), although an insertion near the C terminus does not (Pullinger, G.D., unpublished observations). The gene may be expressed from 2 promoters; one is upstream of *spvA* (Krause *et al.*1992); there is possible promoter within the coding frame of *spvB* which shows a good consensus fit with the standard *Escherichia coli* promoter (Gulig and Chiodo, 1990). The former will be discussed below. The latter promoter may be constitutively expressed - a similar arrangement is found in the *his* operon of *S. typhimurium* (Alifano *et al.*, 1992).

spvB

This gene, which encodes a 65.6kDa protein has been expressed in minicells (Taira and Rhen, 1989a; Williamson *et al.*, 1990; Gulig *et al.*, 1992) and maxicells (Norel *et al.*, 1989a). Transposon mutations within the ORF affect virulence (Krause *et al.*, 1991b; Gulig *et al.*, 1992), but produce polar effects (Gulig *et al.*, 1992). An internal deletion mutant of *spvB* did not affect expression of *spvC*, but resulted in loss of mouse virulence (Williamson *et al.*, 1990). The gene appears to be expressed from the *spvA* promoter.

spvA

This gene, which encodes a 28.1kDa protein has been expressed in minicells (Taira and Rhen, 1989b; Williamson *et al.*, 1990; Gulig *et al.*, 1992). Transposon insertion within the gene affects virulence, but results in polar effects (Gulig *et al.*, 1992). Deletion of this gene prevents expression of *spvR*, which suggests that this SpvA activates expression of *spvR* (Williamson *et al.*, 1990). Recent experiments support this view (Spink, J.M. *et al.*, in preparation). There is evidence that a promoter - yet to be identified - lies upstream of this gene.

spvR - a Regulatory Gene

This gene, which encodes a 33.8kDa protein has been expressed in minicells (Taira and Rhen, 1989b; Williamson *et al.*, 1990). Transposon insertions within the gene affect virulence (Pullinger *et al.*, 1989; Taira and Rhen, 1989b; Krause *et al.*, 1991b). The sequence of SpvR shows significant similarity to the LysR family of bacterial activator proteins (Pullinger *et al.*, 1989). Members of this group contain a helix-turn-helix motif, and some have been shown to bind to DNA (Henikoff *et al.*, 1988). Several groups have now identified that SpvR functions as an activator of the *spvA* operon (comprising *spvA*, *spvB*, *spvC* and perhaps *spvD*) and of itself (Caldwell and Gulig, 1991; Fang et al., 1991; Matsui *et al.*, 1991; Taira *et al.*, 1991b). Coactivators have been found for some members of the LysR family (Henikoff *et al.*, 1988); none has been identified for *spvR*.

Regulation of *spvR*. It has been shown that expression of these genes is dependant on the *in vitro* growth conditions of the bacteria (Fang *et al.*, 1991; Gulig *et al.*, 1992; Spink *et al.*, in preparation); nothing has been published yet on *in vivo* regulation. The region upstream of *spvR* is of interest. A mutation just upstream of the predicted promoter increases *spvR* expression and leads to a strain of higher virulence (Caldwell and Gulig, 1991). Various possible motifs which could bind regulatory factors have been identified (Spink, J.M., *et al.*, in preparation). In addition it is likely that there exist sites for binding SpvR and SpvA. The possible regulatory circuits for these genes are illustrated in Figure 2.

IS*630*

Upstream of *spvR* is an insertion-like element related to IS*630* (Krause *et al.*, 1991a). This does not appear to have a role in virulence, but may relate to the original *in vivo* evolution of these plasmids; however only one copy is found on the virulence plasmid of *S. dublin* and *S. typhimurium* (Qureshi, F., unpublished observations).

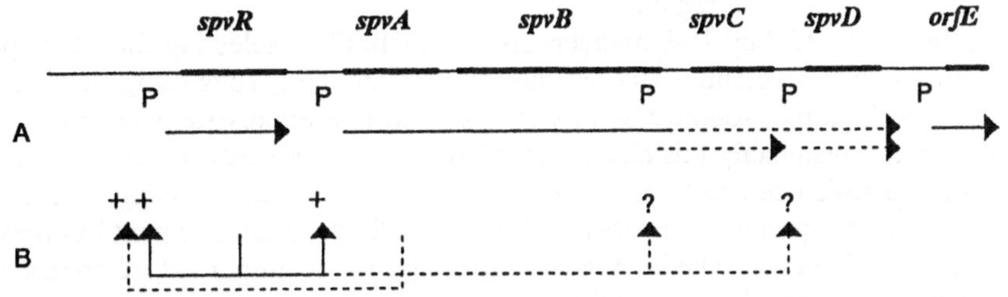

Figure 2. Regulatory circuits within the essential virulence region. A: "P" represents potential promoters, and the lines below the genes indicate the possible transcriptional organisation, with dotted lines signifying unproven or possible transcripts B: Solid lines indicate regulatory pathways which are known and dotted lines indicate pathways which have been postulated or where the evidence is not comprehensive.

VIRULENCE GENES OUTSIDE THE ESSENTIAL VIRULENCE REGION

Various genes which map outside the essential virulence region affect virulence in both *S. dublin* and *S. typhimurium* (Williamson *et al.*, 1988b; Norel *et al.*, 1989a). The *S. typhimurium* genes are believed to be involved in replication and partition (Tinge and Curtiss, 1990b), but little is known about their organisation or distribution among serotypes. We have located two regions outside the essential virulence region in *S. dublin* which affect virulence in mice. Little is yet known about one of these, marked by transposon insertion M173 (Baird *et al.*, 1985), but an extensive investigation has been carried out on the other, G19 (Lax *et al.*, 1990; Pullinger & Lax, 1992).

vagC - Sequence and Analysis

The mutant G19 contains a transposon insertion within the *Sal*I B fragment. This could be complemented by various cloned fragments which encompass this mutation, including a small *Hpa*I fragment which has now been sequenced (Pullinger & Lax, 1992). An ORF which potentially could encode a protein of 16kDa was identified, although resection experiments suggest that the third AUG codon is used, and the resulting gene encodes a 12kDa protein which runs anomalously on SDS PAGE at 16kDa. The region upstream of the start codon shows 7 imperfect 17 base-pair direct repeats, whose function is not known, but which are postulated to be involved in regulation of expression.

Several transposon insertions close to the gene have been located; the insertion in the mutant G19 is within the coding region. The protein has been expressed in minicells. It is expressed by mutants which map close to the ORF, but not by mutant G19. The ORF shows no significant similarity to sequences in databases.

vagD

Several types of experiment suggested that a second gene was required for the phenotype of *vagC* to be expressed: the most informative was the isolation of suppressor mutations. It had been found that G19 was unable to grow on citrate minimal media. However 2 - 3 days after cells were spread on this media, several colonies grew which had regained the ability to grow on citrate and in some cases had regained mouse virulence. These fell into 4 classes: (i) had large deletions; (ii) had 11kb deletions starting at Tn*A* deleting the C terminal of *vagC*; (iii) had a deletion within Tn*A*; (iv) had no detectable alterations. (ii), (iii) and (iv) had regained virulence. The most obviously useful were those where deletion of about 11kb had occurred. This suggested that another gene, designated *vagD* was located within this region.

This gene was located by Tn*5* mutagenesis of mutant G19, selecting for citrate positive revertants. In all such revertants, Tn*5* had inserted within Tn*A*, suggesting that *vagD* was being read from the *tnpR* promoter of Tn*A* in G19 and that expression was interrupted by the Tn*5* insertion. Presumably this also occurred with the suppressor mutation within Tn*A*. An ORF was identified downstream of *vagC*, which overlapped the coding sequence of *vagC* by one base. It is proposed that this ORF is *vagD*. Such an arrangement has previously been identified with genes where translational coupling is important (Oppenheim and Yanofsky, 1980; Schumperli *et al.*, 1982).

Model for *vagC*/*vagD* Action

G19 was also unable to utilise other sugars in minimal medium (glucose, isocitrate, and glutamate), showing that the effect was not via a block on citrate uptake.

Microscopic examination of G19 or wild-type cells which had been on minimal medium for varying lengths of time showed that the mutant cells were highly elongated, implying that they had been inhibited in cell division. Naturally occurring citrate negative strains did not show this phenotype. This phenomenon has been found with genes which couple plasmid replication to cell division (Miki *et al.*, 1984). It is therefore proposed that the *vagC*/*vagD* locus is involved in linking plasmid replication to cell division under certain *in vivo* and *in vitro* conditions. This might be necessary for maintenance of the plasmid under conditions when it is required *in vivo*. The mutation G19 disturbs the normal operation of this locus. The factors which regulate expression of *vagC*/*vagD* are under investigation as is the possible interaction between the 2 proteins.

Comparison of the replication systems found on the various *Salmonella* plasmids reveals that both the *par* and *repC* loci found on the *S. typhimurium* and *S. enteritidis* virulence plasmid hybridise weakly to the *S. dublin* plasmid (Tinge and Curtiss, 1990a), and it is possible that the *vagC*/*vagD* locus, which is only found on the *S. dublin* plasmid, supplies an alternative stability system although there is no evidence that it has replication functions. There is also a multimer resolution system which is present on *S. dublin*, *S. typhimurium* and other serotypes (Krause and Guiney, 1991).

ROLE OF PLASMID VIRULENCE GENES IN PATHOGENESIS

The potential role of these genes in pathogenesis remains controversial whilst it is still unclear which cells are important for growth of salmonellas *in vivo*. Nevertheless, it is clear that the plasmid genes are not required for translocation of the salmonellas across the gut wall in mice (Manning *et al.*, 1986; Pardon *et al.*, 1986), and that they play a role in the growth and accumulation of bacteria in liver and spleen.

However the mouse model of salmonellosis does not mimic all the signs of disease found in other species; in particular gastroenteritis is not a feature of the disease. In order to address this issue, wild-type strains of *S. dublin* and *S. typhimurium*, cured strains derived from these and some of the mutant strains were used to infect calves (Jones *et al.*, 1988b;

Table 2 Virulence of *S. dublin* mutants in calves

	Strains				
	wild-type	cured	mutants		
	2229	173c	173	G19	M242
Number survived[1]	0	4	5	3	2
Mean maximum temperature (°C)[2]	41.4	40.5	40.2	40.8	40.9
Number with anorexia[3]	5	3	0	3	5
Anorexia score[4]	NT[5]	4.8	0.1	3.4	4.3
Number with scour[6]	5	5	5	5	5
Scour score[6]	NT	2.7	1.4	1.8	2.0
Mean excretion (days)[7]	NT	12	7	14	18
Number with bacteraemia[8]	5	0	0	3	0
Number infected at necropsy[9]	5/5	3/4	1/5	1/3	2/2

[1]Five calves (male, Freisian) aged 7 days were each given 10^{10} salmonellas orally of the strain specified (Jones *et al.*, 1988a).
[2]Rectal temperature was measured twice daily and a mean calculated from the highest temperature recorded for each calf.
[3]Calves were fed 2l of milk daily. Anorexia was defined as refusal to consume more than half the milk offered on more than one occasion.
[4]Anorexia was scored as the mean daily volume of milk refused from 2 to 10 days post-infection (Jones *et al.* 1991).
[5]Not tested.
[6]Scour was defined as the production of abnormally liquid faeces. Scour score was calculated according to the method of Jones *et al.* (1991).
[7]Samples of faeces were removed daily and examined for the presence of salmonellas as previously described (Jones *et al.*, 1988a).
[8]Calves were bled daily for 14 days post-infection and salmonellas isolated as previously described (Jones *et al.*, 1988a).
[9]Number of animals from which salmonellas were isolated at necropsy 28 days post-infection/ number of animals which survived until 28 days post-infection. The tissues and fluids have been described previously (Jones *et al.*, 1988a).

Jones, P.W. *et al.*, in preparation). All measures of disease taken indicated that (i) the cured and mutant strains were still virulent and (ii) the cured and mutant strains caused significantly less invasive disease (Table 2). Thus the plasmid virulence genes are involved in the systemic spread of the bacteria, but do not play a part in the enteric stage of salmonellosis.

OTHER FUNCTIONS ENCODED BY THE VIRULENCE PLASMIDS

Various other potential functions have been ascribed to the *Salmonella* virulence plasmids. In particular several groups have identified various loci which appear to be involved in resistance to serum killing (Vandenbosch *et al.*, 1989; Heffernan *et al.*, 1992; Sukupolvi *et al.*, 1992). These functions are not found on plasmids from all serotypes, and their role in disease is unclear. It appears likely that as for example with invasion, *Salmonella* has several different pathogenic mechanisms.

A 45bp region between *spvB* and *spvC* is very similar to an intergenic region on several *Yersinia* plasmids (Krause *et al.*, 1991a), but it has no known function. The *samAB* locus on the *S. typhimurium* plasmid affects resistance to UV radiation (Nohmi *et al.*, 1991), and is unlikely to be involved in pathogenesis. A locus on the *S. typhimurium* virulence plasmid with similarity to *E. coli* membrane proteins involved in the export and assembly of fimbria has been described but its function is not known (Rioux *et al.*, 1990).

It is therefore clear that very little is known about the majority of the coding capacity of these virulence plasmids.

OUTSTANDING QUESTIONS

Within the essential virulence region, there remain many questions to be addressed. It is not clear which genes are involved in virulence, there are few clues as to the *in vivo* role of these genes, and patterns of regulation are just beginning to emerge.

Outside this region several genes involved in replication and plasmid maintenance have been identified as well as several other functions, but the genes on the rest of the plasmid remain of unknown function.

REFERENCES

Alifano, P., Piscitelli, C., Blasi, V., Rivellini, F., Nappo, A.G., Bruni, C.B. and Carlomagno, M.S., 1992, Processing of a polycistronic mRNA requires a 5' *cis* element and active translation, *Mol. Microbiol.* 6:787-798.

Baird, G.D., Manning, E.J. and Jones, P.W., 1985, Evidence for related virulence sequences in plasmids of *Salmonella dublin* and *Salmonella typhimurium*, *J. Gen. Microbiol.* 131:1815-1823.

Barrow, P.A. and Lovell, M.A., 1988, The association between a large molecular mass plasmid and virulence in a strain of *Salmonella pullorum*, *J. Gen Microbiol.* 134:2307-2316.

Barrow, P.A., Simpson, J.M., Lovell, M.A. and Binns, M.M., 1987, Contribution of *Salmonella gallinarum* large plasmid toward virulence in fowl typhoid, *Infect. Immun.* 55:388-392.

Beninger, P.R., Chikami, G., Tanabe, K., Roudier, C., Fierer, J. and Guiney, D.G., 1988, Physical and genetic mapping of the *Salmonella dublin* virulence plasmid pSDL2, *J. Clin. Invest.* 81:1341-1347.

Brown, D.J., Munro, D.S. and Platt, D.J., 1986, Recognition of the cryptic plasmid, pSLT, by restriction fingerprinting and a study of its incidence in Scottish salmonella isolates, *J. Hyg. Camb.* 97:193-197.

Caldwell, A.L. and Gulig, P.A., 1991, The *Salmonella typhimurium* virulence plasmid encodes a positive regulator of a plasmid-encoded virulence gene, *J. Bacteriol.* 173:7176-7185.

Chikami, G.K., Fierer, J. and Guiney, D.G., 1985, Plasmid-mediated virulence in *Salmonella dublin* demonstrated by use of a Tn*5*-*oriT* construct, *Infect. Immun.* 50:420-424.

Dowman, J.E. and Meynell, G.G., 1970, Pleiotropic effects of de-repressed bacterial sex factors on colicinogeny and cell wall structure, *Mol. Gen. Genet.* 109:57-68.

Fang, F.C., Krause, M., Roudier, C., Fierer, J. and Guiney, D., 1991, Growth regulation of a *Salmonella* plasmid gene essential for virulence, *J. Bacteriol.* 173:6783-6789.

Gulig, P.A., Caldwell, A.L. and Chiodo, V.A., 1992, Identification, genetic analysis and DNA sequence of a 7.8-kb virulence region of the *Salmonella typhimurium* virulence plasmid, *Mol. Microbiol.* 6:1395-1441.

Gulig, P.A. and Curtiss III, R., 1987, Plasmid-associated virulence of *Salmonella typhimurium*, *Infect. Immun.* 55:2891-2901.

Gulig, P.A. and Curtiss III, R., 1988, Cloning and transposon insertion mutagenesis of virulence genes of the 100-kilobase plasmid of *Salmonella typhimurium*, *Infect. Immun.* 56:3262-3271.

Gulig, P.A. and Chiodo, V.A., 1990, Genetic and DNA sequence analysis of the *Salmonella typhimurium* virulence plasmid gene encoding the 28,000-molecular-weight protein, *Infect. Immun.* 58:2651-2658.

Heffernan, E.J., Harwood, J., Fierer, J., and Guiney, D., 1992, The *Salmonella typhimurium* virulence plasmid complement resistance gene *rck* is homologous to a family of virulence-associated outer membrane protein genes, including *pagC* and *ail*, *J. Bacteriol.* 174:84-91.

Henikoff, S., Haughn, G.W., Calvo, J.M. and Wallace, J.C., 1988, A large family of bacterial activator proteins, *Proc. Natl. Acad. Sci. USA* 85:6602-6606.

Hovi, M., Sukupolvi, S., Edwards, M.F. and Rhen, M., 1988, Plasmid-associated virulence of *Salmonella enteritidis*, *Microb. Pathogen.* 4:385-391.

Jones, G.W., Rabert, D.K., Svinarich, D.M. and Whitfield, H.J., 1982, Association of adhesive, invasive and virulent phenotypes of *Salmonella typhimurium* with autonomous 60-megadalton plasmids, *Infect. Immun.* 38:476-486.

Jones, P.W., Collins, P. and Aitken, M.M., 1988a, Passive protection of calves against experimental infection with *Salmonella typhimurium, Vet. Rec.* 123:536-541.

Jones, P.W., Collins, P. and Lax, A., 1988b, The role of large plasmids in the pathogenesis of salmonellosis in cattle, *J. Med. Microbiol.* 27:x.

Jones, P.W., Dougan, G., Hayward, C., Mackenzie, N., Collins, P. and Chatfield, S.N., 1991, Oral vaccination of calves against experimental salmonellosis using a double *aro* mutant of *Salmonella typhimurium, Vaccine* 9:29-34.

Kawahara, K., Haraguchi, Y., Tsuchimoto, M., Terakado, N. and Danbara, H., 1988, Evidence of correlation between 50-kilobase plasmid of *Salmonella choleraesuis* and its virulence. *Microb. Pathog.* 4:155-163.

Korpela, K., Ranki, M., Sukupolvi, S., Makela, P.H. and Rhen, M., 1989, Occurrence of *Salmonella typhimurium* virulence plasmid-specific sequences in different serovars of *Salmonella, FEMS Microbiol. Lett.* 58:49-54.

Krause, M. and Guiney, D., 1991, Identification of a multimer resolution system involved in stabilisation of the *Salmonella dublin* virulence plasmid pSDL2, *J. Bacteriol.* 173:5754-5762

Krause, M., Fang, F.C. and Guiney, D.G., 1992, Regulation of plasmid virulence gene expression in *Salmonella dublin* involves an unusual operon structure, *J. Bacteriol.* 174:4482-4489.

Krause, M., Harwood, J., Fierer, J. and Guiney, D., 1991a, Genetic analysis of homology between the virulence plasmids of *Salmonella dublin* and *Yersinia pseudotuberculosis, Infect. Immun.* 59:1860-1863.

Krause, M., Roudier, C., Frier, J., Harwood, J. and Guiney, D., 1991b, Molecular analysis of the virulence locus of the *Salmonella dublin* plasmid pSDL2, *Mol. Microbiol.* 5:307-316.

Kretschmer, P.J. and Cohen, S.N., 1977, Selected translocation of plasmid genes: frequency and regional specificity of translocation of the Tn*3* element, *J. Bacteriol.* 130:888-899.

Lax, A.J., Pullinger, G.D., Baird, G.D. and Williamson, C.M., 1990, The virulence plasmid of *Salmonella dublin*: detailed restriction map and transposon mutant analysis, *J. Gen. Microbiol.* 136:1117-1123.

Manning, E.J., Baird, G.D. and Jones, P.W., 1986, The role of plasmid genes in the pathogenicity of *Salmonella dublin, J. Med. Microbiol.* 21:239-243.

Matsui, H., Kawahara, K., Terakado, N. and Danbara, H., 1990a, Nucleotide sequence of a gene encoding a 29kDa polypeptide in mba region of the virulence plasmid, pKDSC50, of *Salmonella choleraesuis, Nucleic Acids Res.* 18:1055.

Matsui, H., Kawahara, K., Terakado, N. and Danbara, H., 1990b, Nucleotide sequences genes encoding 32kDa and 70kDa polypeptides in mba region of the virulence plasmid, pKDSC50, of *Salmonella choleraesuis, Nucleic Acids Res.* 18:2181-2182.

Matsui, H., Abe, A., Kawahara, K., Terakado, N. and Danbara, H., 1991, Positive regulator for the expression of Mba protein of the virulence plasmid, pKDSC50, of *Salmonella choleraesuis, Microb. Pathog.* 10:459-464.

Miki, T., Yoshioka, K. and Horiuchi, T., 1984, Control of cell division by sex factor F in *Escherichia coli.* I. The 42.84-43.6 F segment couples cell division of the host bacteria with replication of plasmid DNA, *J. Mol. Biol.* 174:605-625.

Montenegro, M.A., Morelli, G. and Helmuth, R., 1991, Heteroduplex analysis of *Salmonella* virulence plasmids and their prevalence in isolates of defined sources, *Microb. Pathog.* 11:391-397.

Nohmi, T., Hakura, A., Nakai, Y., Watanabe, M., Murayama, S.Y. and Sofuni, T., 1991, *Salmonella typhimurium* has two homologous but different *umuDC* operons: cloning of a new *umcDC*-like operon (*samAB*) present in a 60-megadalton cryptic plasmid of *S. typhimurium, J. Bacteriol.* 173:1051-1063.

Norel, F., Coynault, C., Miras, I., Hermant, D. and Popoff, M.Y., 1989a, Cloning and expression of plasmid DNA sequences involved in *Salmonella* serotype typhimurium virulence, *Mol. Microbiol.* 3:733-743.

Norel, F., Pisano, M.-R., Nicoli, J. and Popoff, M.Y., 1989b, Nucleotide sequence of the plasmid-borne virulence gene *mkfA* encoding a 28kDa polypeptide from *Salmonella typhimurium, Res. Microbiol.* 140:263-265.

Norel, F., Pisano, M.-R., Nicoli, J. and Popoff, M.Y., 1989c, Nucleotide sequence of the plasmid-borne virulence gene *mkfB* from *Salmonella typhimurium, Res. Microbiol.* 140:455-457.

Norel, F., Pisano, M.-R., Nicoli, J. and Popoff, M.Y., 1989d, A plasmid-borne virulence region, (2.8kb) from *Salmonella typhimurium* contains two open reading frames, *Res. Microbiol.* 140:627-630.

Oppenheim, D.S. and Yanofsky, C., 1980, Translational coupling during expression of the tryptophan operon of *Escherichia coli, Genetics* 95:785-795.

Pardon, P., Popoff, M.Y., Coynault, C., Marly, J. and Miras, I., 1986, Virulence-associated plasmids of *Salmonella* serotype typhimurium in experimental murine infection, *Ann. Pasteur. Microbiol.* 137:47-60.
Platt, D.J., Taggart, J. and Heraghty, K.A., 1988, Molecular divergence of the serotype-specific plasmid (pSLT) among strains of *Salmonella typhimurium* of human and veterinary origin and comparison of pSLT with the serotype specific plasmids of *S. enteritidis* and *S. dublin*, *J. Med. Microbiol.* 27:277-284.
Poppe, C., Curtiss III, R., Gulig, P.A. and Gyles, C.L., 1989, Hybridisation studies with a DNA probe derived from the virulence region of the 60MDal plasmid of *Salmonella typhimurium*, *Can. J. Vet. Res.* 53:378-384.
Pullinger, G.D., Baird, G.D., Williamson, C.M. and Lax, A.J., 1989, Nucleotide sequence of a plasmid gene involved in the virulence of salmonellas, *Nucleic Acids Res.* 17:7983.
Pullinger, G.D. and Lax, A.J., 1992, A *Salmonella dublin* virulence plasmid locus that affects bacterial growth under nutrient-limited conditions, *Mol. Microbiol.* 6:61-66.
Rioux, C.R., Friedrich, M.J. and Kadner, R.J., 1990, Genes on the 90-kilobase plasmid of *Salmonella typhimurium* confer low-affinity cobalamine transport: relationship to fimbria biosynthesis genes, *J. Bacteriol.* 172:6217-6222.
Schumperli, D., McKenney, K., Sobieski, D.A. and Rosenberg, M., 1982, Translational coupling at an intercistronic boundary of the *Escherichia coli* galactose operon, *Cell* 30:865-871.
Sukupolvi S., Riikonen, R., Taira, S., Saarilahti, H. and Rhen, M., 1992, Plasmid-mediated serum resistance in *Salmonella enterica*, *Microb. Pathog.* 12:219-225.
Taira, S. and Rhen, M., 1989a, Identification and genetic analysis of mkaA-a gene of the *Salmonella typhimurium* virulence plasmid necessary for intracellular growth. *Microb. Pathog.* 7:165-173.
Taira, S. and Rhen, M., 1989b, Molecular organisation of genes constituting the virulence determinant of the *Salmonella typhimurium* 96 kilobase pair plasmid, *FEBS Lett.* 257:274-278.
Taira, S. and Rhen, M., 1990, Nucleotide sequence of *mkaD*, a virulence-associated gene of *Salmonella typhimurium*, containing variable and constant regions, *Gene* 93:147-150.
Taira, S., Baumann, M., Riikonen, P., Sukupolvi, S. and Rhen, M., 1991a, Amino-terminal sequence analysis of four plasmid-encoded virulence-associated proteins of *Salmonella typhimurium*, *FEMS Microbiol. Lett.* 77:319-324.
Taira, S., Riikonen, P., Saarilahti, H., Sukupolvi, S. and Rhen, M., 1991b, The *mkaC* virulence gene of the *Salmonella* serovar typhimurium 96kb plasmid encodes a transcriptional activator, *Mol. Gen. Genet.* 228:381-384.
Tinge, S.A. and Curtiss III, R., 1990a, Conservation of *Salmonella typhimurium* virulence plasmid maintenance regions among *Salmonella* serovars as a basis for plasmid curing, *Infect. Immun.* 58:3084-3092.
Tinge, S.A. and Curtiss III, R., 1990b, Isolation of the replication and partitioning regions of the *Salmonella typhimurium* virulence plasmid and stabilisation of heterologous replicons, *J. Bacteriol.* 172:5266-5277.
Vandenbosch, J.L., Rabert, D.K., Kurlandsky, D.R. and Jones, G.W., 1989, Sequence analysis of *rsk*, a portion of the 95-kilobase plasmid of *Salmonella typhimurium* associated with resistance to the bactericidal activity of serum, *Infect. Immun.* 57:850-857.
Williamson, C.M., Baird, G.D. and Manning, E.J., 1988a, A common virulence region on plasmids from eleven serotypes of *Salmonella*, *J. Gen. Microbiol.* 134:975-982.
Williamson, C.M., Pullinger, G.D. and Lax, A.J., 1988b, Identification of an essential virulence region on *Salmonella* plasmids, *Microb. Pathog.* 5:469-473.
Williamson, C.M., Pullinger, G.D. and Lax, A.J., 1990, Identification of proteins expressed by the essential virulence region of the *Salmonella dublin* plasmid, *Microb. Pathog.* 9:61-66.
Woodward, M.J., McLaren, I. and Wray, C., 1989, Distribution of virulence plasmids within salmonellae, *J. Gen. Microbiol.* 135:503-511.

COLONIZATION AND INVASION OF THE INTESTINAL TRACT BY *SALMONELLA*

Roy Curtiss III[1], David L. MacLeod[1], Hank A. Lockman[1,3], Jorge E. Galan[1,2], Sandra M. Kelly[1], and Gregory G. Mahairas[1]

[1]Department of Biology, Washington University
St. Louis, Missouri 63130
[2]Department of Microbiology
State University of New York at Stony Brook
Stony Brook, New York 11794
[3]Department of Microbiology & Immunology
University of North Carolina
Chapel Hill, North Carolina 27599

INTRODUCTION

There are over 1,800 serotypes of *Salmonella* that are capable of colonizing and infecting a diversity of cold-blooded and warm-blooded animal hosts (LeMinor, 1984). *Salmonella* infection is a major cause of food-borne and water-borne illness (Hook, 1985) and leads to gastroenteritis, bacteremia, or enteric fever. *S. enteritidis* serotype *typhimurium*, *S. choleraesuis* and *S. typhi*, respectively, are most often studied as causing these three disease states. Although *S. enteritidis* and its 1,000 or more serovars are the principal cause of gastroenteritis in humans, most studies on pathogenesis have used the murine animal model in which the disease symptomology is more analogous to enteric fever caused by *S. typhi* in humans. *Salmonella* infection in rabbits constitutes a better system for studying the causes of gastroenteritis as reported by Stephens and colleagues (1985).

Salmonella species, and especially *S. typhimurium*, are readily manipulated by a diversity of genetic methods (Davis, et al., 1980; Miller, 1972). In addition, there are excellent cell culture, organ culture and animal model systems such that it is likely that this pathogen will be most fully studied with regard to the genetic basis for its ability to infect and cause disease.

While other groups have concentrated their attention on means by which *Salmonella* survives in macrophages (Fields et al., 1986; Fields et al., 1989; Buchmeier and Heffron, 1990; Miller and Mekalanos, 1990) and in internal lymphoreticular organs (Gulig, 1990; Dunlap et al., 1991; Nnalue et al., 1992), we have concentrated our efforts on understanding the first events following infection, namely colonization in the intestinal tract which involves adherence to and invasion of enterocytes and the M cells of the gut-associated lymphoid tissue (GALT) (Galan and Curtiss, 1989; Gulig and Curtiss, 1987; see Gulig, 1990; Lockman and Curtiss, 1990, 1992a, 1992b). Although we make extensive use of both avian and murine cell culture and animal models in studying colonization by *S. enteritidis* and *S. typhimurium*, this review will be restricted to our studies with mammalian cells and the murine animal model.

Biology of Salmonella, Edited by F. Cabello *et al.*,
Plenum Press, New York, 1993

ROLE OF CELL SURFACE APPENDAGES IN COLONIZATION OF THE INTESTINAL TRACT

S. typhimurium produces a variety of surface structures that are likely candidates for mediating colonization of the intestine. The best known of these are type-1 fimbriae which are morphologically similar but antigenically different than the type-1 fimbriae of *E. coli* (Duguid and Campbell, 1969; Crichton et al., 1989). *S. typhimurium* type-1 fimbriae mediate a mannose-sensitive agglutination to guinea pig erythrocytes and attachment to a variety of epithelial cells in culture (Jones and Richardson, 1981; Lockman and Curtiss, 1992a). However, their role in colonization of the intestine is not clear. *S. typhimurium* sometimes produces type-2 fimbriae which are morphologically and antigenically identical to type-1 fimbriae, but do not possess the adhesin along their tips and length (Crichton et al., 1989). A mannose-resistant hemagglutinin has also been identified and correlated with the ability of *S. typhimurium* to attach to HeLa cells (Jones and Richardson, 1981; Tavendale et al., 1983; Halula and Stocker, 1987). Mannose-resistant agglutination of mammalian erythrocytes has been associated with the presence of colonization fimbriae in pathogenic *E. coli* (Jones and Rutter, 1974; Duguid et al., 1978) and may be the case with *S. typhimurium*. Another candidate for mediating attachment to the intestine are thin aggregative fimbriae termed curli. These structures have been best characterized in *S. enteritidis* (Collinson et al., 1991) but have been identified in *S. typhimurium* (MacLeod and Curtiss, unpublished observation). *S. typhimurium* mutants lacking curli have not been examined in tissue culture for adherence and invasion properties or in animal models for virulence. Flagella and motility also have been shown to be important to invasion of intestinal epithelial cells (Henle-407) in culture but do not appear to be necessary for invasion of the mouse intestine (Lockman and Curtiss, 1990; Jones et al., 1992).

We sought to extend studies of the role of these surface structures in colonization and invasion of the intestine. Transposon mutagenesis with Tn*10* and Tn*5* has been used to construct nontype-1 fimbriated (Lockman and Curtiss, 1992a; 1992b), nonflagellated and nonmotile flagellated (Lockman and Curtiss, 1990) and nonmannose-resistant hemagglutinating (Halula and Stocker, 1986) mutants in the mouse virulent SR-11 and SL1344 strains of *S. typhimurium*. Strains with single mutations or combinations of several mutations were compared to wild-type *S. typhimurium* for their ability to adhere to and invade into the intestinal epithelial cell line Henle-407 and for virulence in mice.

Table 1 summarizes the data from assays for adherence to and invasion of Henle-407 cells by *S. typhimurium* strains. Nontype-1 fimbriated (Fim^-) and nonmannose-resistant hemagglutinating ($MRHA^-$) mutants exhibit wild-type abilities to attach to and invade into Henle-407 cells. Strains with single mutations eliminating synthesis of flagella (Fla^- Mot^-) or resulting in nonmotile flagella (Fla^+ Mot^-) were somewhat diminished in their ability to attach to Henle-407 cells but had a more significant reduction in ability to invade into these cells. The combination of mutations conferring Fla^+ Mot^- Fim^- or Fla^- Mot^- Fim^- phenotypes resulted in a significant reduction in attachment and invasion, particularily with the Fla^- Mot^- Fim^- phenotype. The addition of a single mutation abolishing the mannose-resistant hemagglutinin ($MRHA^-$) in combination with other mutations did not alter attachment or invasion of Henle-407 cells.

Table 2 summarizes the 50% lethal dose values (LD_{50}) for *S. typhimurium* strains following oral inoculation of eight-week-old female BALB/c mice. As we have previously reported, *S. typhimurium* mutants lacking type-1 fimbriae (Lockman and Curtiss, 1992a; 1992b) or flagella (Lockman and Curtiss, 1990) have LD_{50}s slightly lower than wild-type SR-11, while a nonmotile-flagellated mutant (Lockman and Curtiss, 1990) had an LD_{50} slightly higher than wild type. The combination of mutations that abolish synthesis of type-1 fimbriae and flagella or type-1 fimbriae in association with nonmotile flagella resulted in significantly elevated LD_{50}s. Elimination of the mannose-resistant hemagglutinin alone or in combination with other mutations is without notable effect on LD_{50} values.

It is evident from these studies that flagella and flagellar motility play a significant role in invasion in cell culture but have little effect on LD_{50} values. This effect is most evident when type-1 fimbriae are also eliminated by mutation. A mutation eliminating the mannose-resistant hemagglutinin alone or in combination with other mutations did not alter attachment to or invasion into Henle-407 cells or LD_{50} values. This isn't to say that the mannose-resistant hemagglutinin does not contribute to cell attachment and invasion, but at least not

Table 1. Adherence and invasion of Henle-407 cells by *S. typhimurium* strains.

Strain	Phenotype	Percent adherence[1]	Percent invasion[2]
χ3181	Wild-type SR-11	24 ± 3.3[3]	1.0 ± 0.33
χ3419	Fla+ Mot-	14 ± 2.9	0.10 ± 0.05
χ3422	Fla- Mot-	16 ± 1.3	0.15 ± 0.04
χ4253	Fim-	20 ± 4.8	0.64 ± 0.14
χ4602	Fla+ Mot- Fim-	7.6 ± 0.24	0.08 ± 0.02
χ4308	Fla- Mot- Fim-	8.9 ± 2.0	0.0008 ± 0.0001
χ4441	MRHA-	21 ± 4.7	1.0 ± 0.27
χ4579	Fla+ Mot- MRHA-	10 ± 3.3	0.07 ± 0.015
χ4580	Fla- Mot- MRHA-	17 ± 1.4	0.11 ± 0.026
χ4605	Fim- MRHA-	11 ± 4.4	1.0 ± 0.11
χ4606	Fla+ Mot- Fim- MRHA-	12 ± 2.7	0.10 ± 0.02
χ4607	Fla- Mot- Fim- MRHA-	13 ± 4.3	0.0007 ± 0.0001

[1]Percentage of inoculum adherent to cells after incubation for 1h.
[2]Percentage of inoculum recovered from cells after 2h incubation in gentamicin (100 μg/ml).
[3]Mean percentage ± standard deviation of CFU recovered from quadruplicate wells in triplicate experiments.

Table 2. Virulence of *S. typhimurium* strains in BALB/c mice.

Strain	Phenotype	LD_{50} (CFU)[1]
χ3181	Wild-type SR-11	2.0×10^5
χ3419	Fla+ Mot-	5.8×10^5
χ3422	Fla- Mot-	9.7×10^4
χ4253	Fim-	4.7×10^4
χ4602	Fla+ Mot- Fim-	3.8×10^6
χ4308	Fla- Mot- Fim-	4.5×10^6
χ4441	MRHA-	9.0×10^4
χ4579	Fla+ Mot- MRHA-	2.6×10^5
χ4580	Fla- Mot- MRHA-	7.6×10^4
χ4605	Fim- MRHA-	2.4×10^5
χ4606	Fla+ Mot- Fim- MRHA-	2.2×10^6
χ4607	Fla- Mot- Fim- MRHA-	4.6×10^6

[1]Data were obtained 30 days postchallenge for orally infected mice.
LD_{50} values were determined by the method of Reed and Muench (1938).

appreciably under the conditions employed in this study. It would therefore appear that *Salmonella* is able to use alternate means of attaching to and invading into enterocytes and/or M cells, and some of these might be partially supplied by the mammalian cell interacting with *Salmonella*.

CELL INVASION MECHANISMS

We have previously reported the cloning and characterization of a cluster of genes controlling the ability of *S. typhimurium* to invade cells in culture and cells lining the intestinal tract of mice (Galan and Curtiss, 1989). The initial work identified an operon of three genes designated *invABC* and a closely linked gene *invD* which were involved in the invasion process. Mutants with a polar defect in the *invA* gene were fully capable of attaching to a diversity of cell types in cell culture assays but were impaired about one-hundred fold in ability to invade the cells (Galan and Curtiss, 1989). The recent work by Galan and associates has identified additional *inv* genes playing a key role in *Salmonella*'s ability to invade cells in culture and presumably into cells lining the intestinal tract (Galan et al., 1992; Ginocchio et al., 1992). The genetic map of these genes is depicted in Figure 1. These genes map to minute 59 on the *Salmonella* chromosome. The *invABC* operon is expressed optimally under conditions of high osmolarity and this expression is influenced by the extent of negative DNA superhelix coil density (Galan and Curtiss, 1990). *invA* mutants of *S. typhimurium* have a diminished ability to invade cells lining the intestinal tract and have an elevated oral LD_{50} (Galan and Curtiss, 1989). The *inv* genes of *Salmonella* are unique to *Salmonella* (Galan and Curtiss, 1990) being found in 1,717 of 1,721 *Salmonella* strains evaluated and being absent in 162 non-*Salmonella* bacterial strains, representing 21 different genera (Galan and Curtiss, 1991; Rahn et al., 1992). Generation of *invA* mutants of a diversity of *Salmonella* serotypes leads to a significant reduction in ability to invade Henle-407 cells (Table 3).

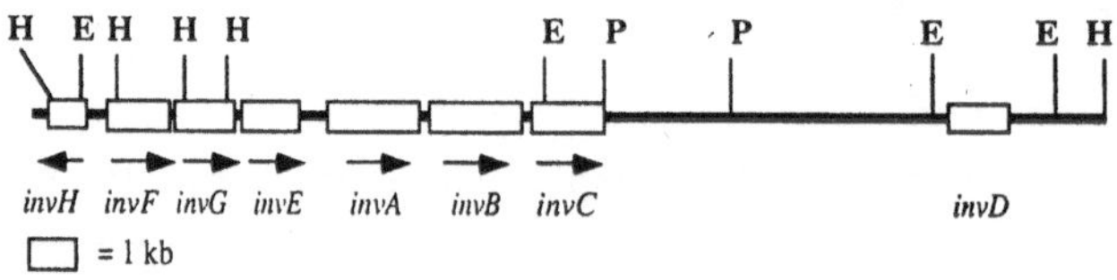

Fig. 1. *Salmonella typhimurium inv* locus

Table 3. Invasion of *Salmonella* strains into Henle-407 cells

Strain	Serovar	Relevant genotype	Percent invasion[1]
SL1344	*S. typhimurium*	wild type	19.2 ± 6.1[2]
SB103	*S. typhimurium*	*invA::kan*	0.02 ± 0.0008
Ty2	*S. typhi*	wild type	7.46 ± 5.2
SB129	*S. typhi*	*invA::kan*	0.07 ± 0.04
ISP2825	*S. typhi*	wild type	25.2 ± 1.9
SB130	*S. typhi*	*invA::kan*	0.03 ± 0.003
7193	*S. enteritidis*	wild type	12.9 ± 2.15
SB131	*S. enteritidis*	*invA::kan*	0.04 ± 0.006
Stock	*S. gallinarum*	wild type	30.6 ± 5.6
SB132	*S. gallinarum*	*invA::kan*	0.1 ± 0.02
Lane	*S. dublin*	wild type	27.6 ± 2.4
SB133	*S. dublin*	*invA::kan*	0.26 ± 0.054
77-85	*S. arizonae*	wild type	0.52 ± 0.01
875-84	*S. arizonae*	wild type	0.07 ± 0.01

[1]Percentage of inoculum recovered from cells after 2h incubation in gentamicin (100 μg/ml).

[2]Values are mean ± SD of triplicate samples. Similar results were observed in several repetitions of this experiment. From Galan and Curtiss (1991).

CELL ATTACHMENT AND INVASION BY *S. CHOLERAESUIS*

Although *S. choleraesuis* contains DNA encoding all of the *inv* functions (Galan and Curtiss, 1991), results on knocking out *inv* functions in regard to cell attachment and invasion have given ambiguous results. In line with this, Finlay and colleagues (1988), who had generated an extensive collection of *S. choleraesuis* mutants which had been screened for cell infectivity, never found any mutants with specific invasion defects associated with wild-type abilities to attach to cells in culture. In other words, all Inv^- mutants were defective in attachment and were therefore blocked in an event that normally precedes invasion in *S. choleraesuis*. Some years ago, we had established that *S. typhimurium* mutants with deletion mutations abolishing the genes for adenylate cyclase and the cyclic AMP receptor protein rendered *S. typhimurium* avirulent and immunogenic (Curtiss and Kelly, 1987). The Δ*cya* Δ*crp S. typhimurium* strains exhibited wild-type abilities to colonize and invade into cells of the intestinal tract (Curtiss and Kelly, 1987). It is revealed by the data in Table 4 that Δ*cya* Δ*crp S. typhimurium* strains sometime display wild-type abilities of attaching to and invading both CHO and MDCK cells but exhibited a prolonged generation time. This latter attribute is due to the fact that Δ*cya* Δ*crp* mutants are defective in the transport and catabolism of carbohydrates, amino acids and peptides (Botsford and Harman, 1992) thus diminishing their growth rate under a diversity of in vitro and in vivo conditions. In other experiments, Δ*cya* Δ*crp S. typhimurium* strains displayed reduced abilities to attach to and invade CHO cells compared to their wild-type parent. These results might have been due to the altered growth of the Δ*cya* Δ*crp* strain relative to the wild-type parent prior to addition of bacteria to the CHO cell monolayer. However in almost every experiment, the reduction in percentage of bacteria invading was correlated with the percentage attaching. We therefore conclude that cell invasion of *S. typhimurium* in vivo and in vitro is not appreciably dependent on the function of the *cya* and *crp* genes. Recent studies on the effects of Δ*cya* Δ*crp* mutations on the virulence and immunogenicity of *S. choleraesuis* (Kelly et al., 1992) revealed that Δ*cya* Δ*crp S. choleraesuis* strains are somewhat diminished in ability to attach to cells in culture and are more diminished (i.e. about ten-fold) in ability to invade cells in culture (Table 5). These mutants also display a ten-fold reduced ability to transcytose MDCK polarized cell layers (Kelly et al., 1992). In line with this, the Δ*cya* Δ*crp S. choleraesuis* strain also displayed a diminished ability to attach to and invade cells lining the intestinal tract, including cells of the GALT (Kelly et al., 1992). These results suggest that some genes influencing the ability of *S. choleraesuis* to attach to and invade cells may be controlled in total or in part by catabolite repression.

Table 4. Infection of CHO and MDCK cells by wild type and Δ*cya* Δ*crp* *S. typhimurium*

Strain	Genotype	Cell line	Percent adherence[1]	Percent invasion[2]	MGT[3]
χ3306	wild type	CHO	14.5 ± 1.0	35.3 ± 9.6	5.8 ± 0.48
χ4064	Δ*cya* Δ*crp*	CHO	18.0 ± 2.7	27.8 ± 22.6	14.7 ± 3.15
χ3306	wild type	MDCK	19.6 ± 5.6	27.9 ± 3.0	5.4 ± 0.15
χ4064	Δ*cya* Δ*crp*	MDCK	25.8 ± 5.5	35.1 ± 2.4	10.2 ± 0.62

[1] Percentage of inoculum adherent to cells after incubation for 1h.

[2] Percentage of inoculum recovered from cells after 2h incubation in gentamicin (100 μg/ml).

[3] Mean generation time.

Values are mean ± SD of triplicate samples. Similar results were observed in several repetitions of this experiment.

Table 5. Infection of CHO cells by wild-type *S. choleraesuis* $\chi3246$ and Δcya Δcrp $\chi3781$

Experiment	Strain	Percent adherence [1]	Percent invasion [2]
1	$\chi3246$	6.6 + 0.6	11.7 + 3.1
	$\chi3781$	5.1 + 0.6	1.1 + 0.4
2	$\chi3246$	29.5 + 5.2	7.2 + 2.5
	$\chi3781$	19.7 + 2.5	0.4 + 0.2
3	$\chi3246$	10.1 + 0.2	3.5 + 0.3
	$\chi3781$	7.2 + 1.4	0.2 + 0.02

[1] Percentage of inoculum adherent to cells after incubation for 1h.

[2] Percentage of inoculum recovered from cells after 2h incubation in gentamicin (100 µg/ml).

Values are mean ± SD of triplicate samples. Similar results were observed in several repetitions of this experiment.

IDENTIFICATION OF *SALMONELLA* UNIQUE GENETIC FUNCTIONS THAT MIGHT SPECIFY VIRULENCE ATTRIBUTES

Based on the foregoing results, it appears that *S. choleraesuis* possesses a unique means not exhibited by other *Salmonella* species of being invasive. One can consider this to be due to differential expression of common genetic information or to the existence of genetic information not present in other *Salmonella.* We opted to test the second possibility and have recently begun to use genomic subtraction (Straus and Ausubel, 1990) to identify DNA sequences unique to *S. choleraesuis* that are absent from a prototrophic wild-type *Escherichia coli* K-12 strain (Mahairas and Curtiss, 1992). We thus prepared amplified radioactive *S. choleraesuis* DNA which was used to probe two *S. choleraesuis* libraries. This permitted us to identify nine different regions of unique DNA ranging from 10-60 kb in size and which map to different regions of the chromosome. Additionally, two blocks of DNA (20 and 60 kb) are present in *S. choleraesuis* but absent in *S. typhi.* Experiments are underway to determine if some of the unique cell adherence and invasion characteristics of *S. choleraesuis* are encoded by these regions. Toward this end, methods have been developed to generate *S. choleraesuis* mutants having deletions in different portions of the unique sequence DNA. These mutants are being evaluated in cell culture, transcytosis and animal infectivity systems.

CONCLUSION

Mutant generation and isolation and gene cloning have been used to identify *Salmonella* cell surface attributes important in the ability of *S. typhimurium* and *S. choleraesuis* strains to attach to and invade cells in culture and cells lining the intestinal tract of mice. It is evident that *Salmonella* strains possess, in many instances, duplicative means to facilitate colonization and invasion. Although not discussed in the results presented above, it is apparent from our studies with avian systems that *S. typhimurium* and *S. enteritidis* may use additional means not needed or at least different from those used for cell attachment and invasion in mice. It will thus be through a rigorous study of several *Salmonella* serotypes and several animal systems that will eventually provide a complete understanding of the mechanisms by which *Salmonella* successfully infects such a diversity of animal species.

ACKNOWLEDGMENTS

The research has been supported by research grants from the U.S. Public Health Service (NIH ROI-AI24533), United States Department of Agriculture (GAM9001702 and 9102452) and Bristol-Myers Squibb. We thank Betty Smith for help in the preparation of this manuscript.

REFERENCES

Botsford, J.L., and Harman, J.G., 1992, Cyclic AMP in prokaryotes, *Microbiol. Rev.* 56:100.

Buchmeier, N.A., and Heffron, F., 1990, Induction of *Salmonella* stress proteins upon infection of macrophages, *Science* 248:730.

Collinson, S.K., Emody, L., Muller, K.-H., Trust, T.J., and Kay, W.W., 1991, Purification and characterization of thin, aggregative fimbriae from *Salmonella enteritidis.*, *J. Bacteriol.* 173:4773.

Crichton, P.B, Yakubu, D.E., Old, D.C., and Clegg, S., 1989, Immunological and genetical relatedness of type-1 and type-2 fimbriae in salmonellas of serotypes Gallinarum, Pullorum, and Typhimurium, *J. Appl. Bacteriol.* 67:283.

Curtiss, R. III, and Kelly, S.M., 1987, *Salmonella typhimurium* deletion mutants lacking adenylate cyclase and cyclic AMP receptor protein are avirulent and immunogenic, *Infect. Immun.* 55:3035.

Davis, R.W., Botstein, D., and Roth, J.R., 1980, Advanced microbial genetics, a manual for genetic engineering, Cold Spring Harbor Laboratory, Cold Spring Harbor, New York.

Duguid, J.P., and Campbell, I., 1969, Antigens of the type-1 fimbriae of salmonellae and other enterobacteria, *J. Med. Microbiol.* 2:535.

Duguid, J.P., Çlegg, S., and Wilson, M.I., 1978, The fimbrial and non-fimbrial haemagglutinins of *Escherichia coli.*, *J. Med. Microbiol.* 12:213.

Dunlap, N.E., Benjamin Jr., W.H., McCall, R.D., Tilden, A.B., and Briles, D.E., 1991, A 'safe-site' for *Salmonella typhimurium* is within splenic cells during the early phase of infection in mice, *Microbial Pathog.* 10:297.

Fields, P.I., Swanson, R.V., Haidaris, C.G., and Heffron, F., 1986, Mutants of *Salmonella typhimurium* that cannot survive within the macrophage are avirulent. *Proc. Natl. Acad. Sci. USA* 83:5189.

Fields, P.I., Groisman, E.A., and Heffron, F., 1989, A *Salmonella* locus that controls resistance to microbicidal proteins from phagocytic cells, *Science* 243:1059.

Finlay, B.B., Starnback, M.N., Francis, C.L., Stocker, B.A.D., Chatfield, S., Dougan, G., and Falkow, S., 1988, Identification and characterization of Tn*phoA* mutants of *Salmonella* that are unable to pass through a polarized MDCK epithelial cell monolayer, *Molec. Microbiol.* 2:757.

Galan, J.E., and Curtiss, R. III, 1989, Cloning and molecular characterization of genes whose products allow *Salmonella typhimurium* to penetrate tissue culture cells, *Proc. Natl. Acad. Sci. USA* 86:6383.

Galan, J.E., and Curtiss, R. III, 1990, Expression of *Salmonella typhimurium* genes required for invasion is regulated by changes in DNA supercoiling, *Infect. Immun.* 58:1879.

Galan, J.E., and Curtiss, R. III, 1991, Distribution of the *invA*, *-B*, *-C*, and *-D* genes of *Salmonella typhimurium* among other *Salmonella* serovars: *invA* mutants of *Salmonella typhi* are deficient for entry into mammalian cells, *Infect. Immun.* 59:2901.

Galan, J.E., Ginocchio, C., and P. Costeas, 1992, Molecular and functional characterization of the *Salmonella* invasion gene *invA*: homology of InvA to members of a new protein family, *J. Bacteriol.* 174:4338.

Ginocchio, C., Pace, J., and Galan, J.E., 1992, Identification and molecular characterization of a *Salmonella typhimurium* gene involved in triggering the internalization of salmonellae into cultured epithelial cells, *Proc. Natl. Acad. Sci. USA* 89:5976.

Gulig, P.A., 1990, Virulence plasmids of *Salmonella typhimurium* and other salmonellae, *Microbial Pathog.* 8:3.

Gulig, P.A., and Curtiss, R. III, 1987, Plasmid-associated virulence of *Salmonella typhimurium, Infect. Immun.*55:2891.

Halula, M., and Stocker, B.A.D., 1986, Mannose-resistant haemagglutination gene(s) of *Salmonella typhimurium.*, *in*: Protein-Carbohydrate Interactions in Biological Systems. Academic Press Inc. (London) Ltd.

Halula, M.C., and Stocker, B.A.D., 1987, Distribution and properties of the mannose-resistant hemagglutinin produced by *Salmonella* species, *Microbial Pathog.* 3:455.

Hook, E.W., 1985, *Salmonella* species (including typhoid fever), *in*: Principles and practice of infectious diseases, G.L. Mandell, R.G. Douglas, and J.E. Bennett, eds., John Wiley and Sons, New York.

Jones, B.D., Lee, C.A., and Falkow, S., 1992, Invasion by *Salmonella typhimurium* is affected by the direction of flagellar rotation, *Infect. Immun.* 60:2475.

Jones, G.W., and Richardson, L.A., 1981, The attachment to, and invasion of HeLa cells by *Salmonella typhimurium*: the contribution of mannose-sensitive and mannose-resistant haemagglutinating activities, *J. Gen. Microbiol.* 127:361.

Jones, G.W., and Rutter, J.M., 1974, The association of K88 antigen with haemagglutinating activity in porcine strains of *Escherichia coli*, *J. Gen. Microbiol.* 84:135.

Kelly, S.M., Bosecker, B.A., and Curtiss R. III, 1992, Characterization and protective properties of attenuated mutants of *Salmonella choleraesuis*, *Infect. Immun.* In press.

LeMinor, L, 1984, *Salmonella* Lignieres 1900, 389^{AL}, *in*: Bergey's Manual of Systemic Bacteriology, N.R. Kreig and J.G. Holt, eds., The Williams and Wilkins Co., Baltimore.

Lockman, H.A., and Curtiss, R., III, 1992a, Isolation and characterization of conditional adherent non-type 1 fimbriated *Salmonella typhimurium* mutants, *Mol. Microbiol.* 6:933.

Lockman, H.A., and Curtiss, R., III, 1992b, Virulence of non-type-1 fimbriated and nonfimbriated nonflagellated *Salmonella typhimurium* mutants in murine typhoid fever, *Infect. Immun.* 60:491.

Lockman, H.A., and Curtiss, R., III, 1990, *Salmonella typhimurium* mutants lacking flagella or motility remain virulent in BALB/c mice, *Infect. Immun.* 58:137.

Mahairas, G.G., and Curtiss, R., III, Identification of chromosomal sequences encoding putative virulence genes in *Salmonella choleraesuis* by subtractive hybridization, Abstr. 92nd Annu. Meet. Am. Soc. Microbiol. American Society for Microbiology, Washington, D.C.

Miller, J.H., 1972, Experiments in molecular genetics, Cold Spring Harbor Laboratory, Cold Spring Harbor, New York.

Miller, S.I., and Mekalanos, J.J., 1990, Constitutive expression of the PhoP regulon attenuates *Salmonella* virulence and survival within macrophages, *J. Bacteriol.* 172:2485.

Nnalue, N.A., Shnyra, A., Hultenby, K., and Lindberg, A.A., 1992, *Salmonella choleraesuis* and *Salmonella typhimurium* associated with liver cells after intravenous inoculation of rats are localized mainly in kupffer cells and multiply intracellularly, *Infect. Immun.* 60:2758.

Rahn, K., DeGrandis, S.A., Clarke, R.C., McEwen, S.A., Galan, J.E., Ginocchio, C., Curtiss, R. III, and Gyles, C.L., 1992, Amplification of an *invA* gene sequence of *Salmonella typhimurium* by polymerase chain reaction as a specific method of detection of *Salmonella*, *Molec. Cell Probes* 6:271.

Reed, L.J., and Muench, H., 1938, A simple method for estimating fifty percent endpoints, *Am. J. Hyg.* 27:493.

Stephens, J., Wallis, T.S., Starkey, W.G., Candy, D.C.A., Osborne, M.P., and Haddon, S.J., 1985, Salmonellosis: in retrospect and prospect, *in*: Microbial toxins and diarrhoeal disease, D. Evered, and J. Whelan, eds., Pitman, London (Ciba Foundation Symposium 112).

Straus, D., and Ausubel, F.M., 1990, Genomic subtraction for cloning DNA corresponding to deletion mutations, *Proc. Natl. Acad. Sci. USA* 87:1889.

Tavendale, A., Jardine, C.K.H., Old, D.C., and Duguid, J.P., 1983, Haemagglutinins and adhesion of *Salmonella typhimurium* to HEp2 and HeLa cells, *Med. Microbiol.* 16:371.

EXPERIMENTAL *SALMONELLA TYPHIMURIUM* - INDUCED GASTROENTERITIS

John Stephen, Iqbal I. Amin, and Gillian R. Douce

Microbial Molecular Genetics and Cell Biology Group
School of Biological Sciences
University of Birmingham
PO Box 363
Birmingham B15 2TT, United Kingdom

INTRODUCTION

The rising incidence of salmonellosis in developed countries has refuelled intensive research into the pathogenesis of disease caused by Salmonella *spp*. These organisms cause two types of disease in man - acute gastroenteritis and systemic typhoid disease - as well as other important infections in domestic animals. It is highly desirable therefore, that we seek to understand the detailed biological and biochemical mechanisms involved in the pathogenesis of disease caused by these organisms. The gains from fundamental studies could be twofold. First, a greater ability to control infection caused by such a wide-spread pathogen whose eradication from the food chain would seem to be an impracticable aspiration. Second, one would hope that deeper insights into the fundamental mechanisms of disease causation would help generate greater confidence in the use of live metabolically crippled *Salmonellae* as *Salmonella* vaccines (O'Callaghan et al., 1988) or as vectors for delivering extraneous immunogens (Charles and Dougan, 1990).

In this contribution a brief overview is given of work designed to elucidate the mechanisms involved in gastroenteritic disease. The progress of events will be followed from the introduction of the pathogen into the lumen of the small intestine through to the induction of a diarrhoeal fluid response. This requires a working grasp of the physiology of fluid transport in the gut, consideration of the pathogenesis of the disease, a discussion concerning potential virulence attributes of *S. typhimurium* and the combined impact of such facts and ideas on the elucidation of the pathophysiology of acute gastroenteritis.

PHYSIOLOGY OF FLUID TRANSPORT

It is desirable to appreciate the normal structure and function of gut before attempting to understand the processes whereby such are perturbed (Fig. 1; see also Stephen and

Biology of Salmonella, Edited by F. Cabello *et al.*,
Plenum Press, New York, 1993

Pietrowski, 1986). The main function of the gut is the active inward transport of ions and nutrient solutes which is followed by the passive movement of water. The driving force is the Na^+/K^+ ATPase situated in the basolateral membrane of enterocytes which maintains a low intracellular [Na^+] thus creating the electrochemical gradient favourable for Na^+ entry and, a high regional [Na^+] in the intercellular spaces; Cl^- follows Na^+. A similar situation exists in crypt cells: Na^+/K^+ ATPase drives secretion. The key difference is the location of the carrier systems responsible for the facilitated entry of the actively transported species. In villus cells the carriers are present in the brush border whereas in crypt cells they are located in the basal membrane: this is responsible for the vectorial aspects of ion/fluid traffic in villus/crypt assemblies. However, from the work of Lundgren's group in Sweden it is clear that several factors in addition to enterocytes are involved in regulating fluid transport in the gut; these include the enteric nervous system and the anatomy of the microcirculation . The latter plays a profoundly important role in the uptake of fluid (Lundgren, 1988). This is illustrated in Figure 1, which shows the existence of zones of graded osmotic potential. Lundgren and

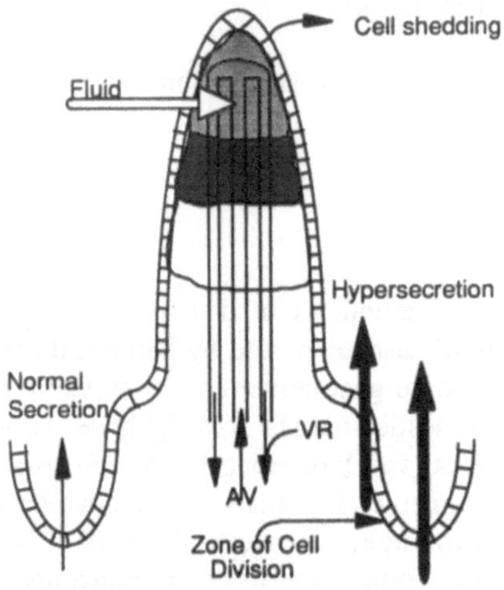

Figure 1. Schematic representation of a villus. Note the central arterial vessel (AV) which arborizes at the tip into a capillary bed drained by a subepithelial venous return (VR). Movement of sodium into VR creates a concentration gradient between VR and AV causing absorption of water from AV and surrounding tissue. This results in a progressive increase in the osmolarity of incoming blood moving into the tip region through to VR. Tip osmolarity is about 3 times higher than normal. This counter current system has been demonstrated in man and can be inferred in mice (see text).

The shaded areas indicate a vertical increase in osmolarity. Left crypt: represents normal physiological secretion. Right crypt: represents hypersecretion. Based on Hallback et al., (1978).

coworkers have measured the osmolality in adult human gut. At the tips of villi they found osmolalities ranging from 700 - 800 mOsm kg^{-1} H_20 which would generate huge osmotic forces (Lundgren, 1988). Thus current perceptions are that enterocytes are responsible for generating this gradient; the blood supply acts as a countercurrent multiplier which amplifies the gradient in a manner analagous to the loops of Henle in the kidney. In our laboratory the hypertonic zone has been demonstrated directly in whole villi of infant mice in terms of the changing morphology of erythrocytes: in the lower regions of villi they show characteristic discoid morphology whereas in the upper region they are crenated indicating a hyperosmotic environment (Osborne et al., 1991a,b). Another consequence of the microcirculatory anatomy is that villus tip regions are relatively hypoxic. In addition, neonatal brushborders contain disaccharidases (principally lactase) which break down non-absorbable disaccharides (e.g. lactose) into constituent absorbable monosaccharides.

Villus tips and crypts are regarded as the anatomical sites of physiological absorption and secretion respectively. Fluid transport is a bidirectional process in the healthy animal with net absorption in health and net secretion in disease. The balance between absorption and secretion is poised at different points throughout the intestinal tract reflecting differences in both structure and function. Proximal small intestine is relatively leaky; in contrast the colon is a powerfully absorptive organ. Finally, crypts are the principal sites of cell regeneration replacing cells which migrate up the epithelial escalator. The epithelium is renewed in approximately 3 to 5 days. At villus tips senescent cells are shed. Most of these points will be referred to in the ensuing discussion.

PATHOGENESIS OF EXPERIMENTAL GASTROENTERITIS

In our laboratory we have made extensive use of the ligated rabbit ileal loop (RIL) model originally used for *Salmonella* research by Giannella and coworkers in the early 1970's which they validated in comparative studies in monkeys. The arguments justifying its use have been fully spelt out elsewhere (Stephen et al., 1985) and will not be repeated here. In addition, we have also used an organ culture system to study initial attachment of *S. typhimurium* to gut epithelium (Worton et al., 1989). Currently we are using an improved version of that organ culture system to study primary invasion of gut epithelium in parallel with a HEp-2 cell culture system recently optimised for this purpose (Douce et al., 1991). The selection of one's model for study is vitally important. While some non-specific symptoms are common (e.g. nausea and abdominal pain) the clinical features of Salmonella-induced diarrhoea and systemic typhoid infections (for which the mouse is the standard model) are quite different. For example, gastroenteritis may follow 8-36 hours after ingestion of contaminated food, whereas typhoid may follow an incubation period of 10-20 days. Diarrhoea, (which is usually watery, but may be severe, and sometimes bloody,) is the predominating feature of gastroenteritis, whereas in the case of adults constipation may occur in the early clinical stages of typhoid; diarrhoea may occur much later. In gastroenteritis, fever may occur whereas in typhoid this may be so severe as to cause delirium. Current perceptions are that gastroenteritis results from the initial interactions of *S. typhimurium* for example (and/or it's products) with the gut mucosa, whereas typhoid fever is produced by *S. typhi* organisms which translocate the mucosa, survive within macrophages, multiply and release endotoxin which triggers the highly complex endotoxin-cascade. Gastroenteritis is usually self-limiting, whereas in untreated typhoid mortality can be as high as 10% (Candy and Stephen, 1989). It follows, therefore, that the choice of organism/host combination for experimental study and the precise definition of the question(s) to be addressed, are of crucial importance particularly when, as is the case with *Salmonella spp*., virulence is multifactorial. In much contemporary research on Salmonella infections, appreciation of these basic points is missing, or has become blurred or ignored. The rest of this article is devoted mainly (but not exclusively) to work on *S. typhimurium* as an agent of gastroenteritis.

Invasiveness

In the context of facultative or obligate intracellular pathogens we consider invasiveness to be the biological integral of several components: attachment to, invasion of, intracellular survival and/or multiplication within and release from eucaryotic cells. Each differentiated cell type in the *in vivo* environment must be considered separately. As outlined below, the importance of this perception cannot be over estimated. Currently, a major international effort is being directed towards understanding the mechanisms and determinants whereby Salmonella attach to, invade, survive/multiply within and translocate eucaryotic cells (Elsinghorst et al.,1989; Ernst et al.,1990; Fields et al.,1986, 1989; Finlay and Falkow, 1988; Finlay et al.,1988a, 1988b, 1989; Gahring et al., 1990; Galan and Curtis, 1989, 1990, 1991; Groisman et al., 1989; Groisman and Saier, 1990; Lee and Falkow, 1990; Lee et al., 1992; Miller S.I. et al.,1989; Miller I. et al.,1989; Small et al., 1987). Invasiveness is regarded by many as a *sine qua non* for virulence in Salmonella infections - self-evidently true for disseminated infections but till very recently of unproven relevance in gastroenteritis. Since invasive non-diarrhoeagenic strains of *S. typhimurium* (for example, SL1027 and LT-7) have been known for many years it was important to determine if there were different routes or degrees of invasion; to date we have tackled the latter question and are about to tackle the former.

In whole animal experiments gut invasion data are difficult to interpret since the distinction between cellular invasion of gut epithelia and tissue spread through the lamina propria and beyond is often hard to make or not taken into account in the first place. For example avirulent *S. typhimurium* strains SL1027 and LT-7 are defined as invasive on the basis of their recovery (18h post loop inoculation) from internal organs by culture in enrichment media (Wallis et al., 1986a). In this laboratory an asymmetric organ culture system was developed with which it was hoped to measure quantitatively epithelial invasiveness of strains of known different virulence. Initially, it could only be shown that virulence (defined as the ability to elicit a fluid response) did <u>not correlate completely</u> with the ability of *S. typhimurium* to associate rapidly with intestinal mucosae (Worton et al., 1989). Virulent strains (TML, W118) associated at a level which was greater than that of avirulent strains (SL1027, Thax-1) but <u>similar</u> to that of avirulent strains (LT-7, M206) within the first 30 min of infection *in vitro*. Unfortunately it was not possible to extend the time period to allow invasive studies to be made since continued absorption of fluid resulted in the destruction of villi. This was due to the continued inward movement of fluid and the low hydraulic conductivity of the submucosa. In the absence of a functioning blood supply, fluid accumulates and disrupts villi (Worton et al., 1989). Most of the problems associated with the asymmetric organ culture system have now been overcome and conditions established which slow down - but do not abolish - fluid uptake by gut tissue, for sufficiently long to enable <u>**quantitative**</u> measurements of *S. typhimurium* to be made in distal ileal tissue freshly removed from rabbits. The effects on gut mucosa *in vitro* of changes in temperature, and composition of bathing media were examined by light and electron microscopy. The conditions sought were those which would 1) slow down but not completely inhibit uptake of fluid, and 2) extend the viability of enterocytes. The criteria used were the absence of gross oedema (as judged by light microscopy) and the state of enterocytes (as judged by electron microscopy). The compositions of the best solutions found to date are as follows. The serosal solution is the WHO rehydration formulation: NaCl, 60 m*M*; $NaHCO_3$, 30 m*M*; KCl, 20 m*M*; and glucose, 111 m*M*. The mucosal solution was based on the same solution with two important changes. <u>First</u>, all the sodium was replaced by choline which is non-absorbed. <u>Second</u>, tissue culture medium (Tcm; consisting of commercial Minimum Essential Medium with Earle's salts without glutamine, to which glutamine and foetal calf serum were added to final concentrations of 2.0 m*M* and 10 % (v/v) respectively) was added to the choline-containing medium to a final concentration of 10% (v/v); this is designated Mucosal medium.

The rationale for these changes was as follows. Previous work with neonatal mouse gut (Starkey et al., 1990a,b) and recent work with rabbit gut (unpublished) showed that Na^+ is by far the most important single factor regulating fluid uptake. Substitution of Na^+ with $choline^+$ in WHO solution abolished fluid uptake even in the presence of glucose, whereas substitution of non-absorbable mannitol for glucose in WHO solution still resulted in fluid uptake in the presence of Na^+. However, villus architecture was preserved only up to but not beyond 3 h when the mucosal side was bathed in Na^+-depleted/choline$^+$-containing medium. In addition, *S. typhimurium* only survived for *c.* 2 h in this medium; thereafter there was a dramatic fall in bacterial viability. Empirically it was found that addition of Tcm to a final concentration of 10% preserved the gut explants, at least for 4 h, during which fluid accumulated very slowly (i.e. the tissue was still working); this medium also supported the growth of organisms. Under these conditions the initial invasiveness of the virulent and avirulent strains were examined as follows. Organisms were grown in Hartley Digest Broth, removed when 3 h into log phase, and resuspended at an appropriate concentration in the Mucosal medium. This suspension was introduced to the mucosal side of the organ culture. After 2 h the supernatant was removed and fresh medium containing gentamicin added for 1 h. Medium was then removed, the tissue homogenised and organisms counted. Throughout all incubation stages the tissue was constantly gassed with 95% O_2/ 5% CO_2. A summary of preliminary comparative data from the new test and the optimised HEp-2 cell-based test is shown in Figure 2a,b. Note that,

1).The anomalous result in HEp-2 cells with avirulent strain SL1027 disappeared.

2).This is the first time to our knowledge that quantitative studies of invasiveness have been carried out *in vitro* on freshly isolated functioning gut.

3).There is a clear correlation between initial mucosal invasion and virulence of Salmonellae in a model which is relevant to human gastroenteritis.

4).The short period (2h) allowed for invasion means that one is almost certain to be measuring primary invasion of epithelial cells only and not the translocation of the lamina propria as well.

This last point is potentially very significant. Osmo-regulated genes are now beginning to be reported as being important in invasion of the gut (Galan and Curtiss, 1990). The products of these genes are believed to play a role in epithelial invasion as judged by comparison of LD_{50}s measured by oral and parenteral routes of administration. It is highly likely that such studies reflect the ability of organisms to translocate the upper regions of the lamina propria as well as invading epithelial cells. It is less likely that they are responding to changes in gut content osmolarity. As shown in Figure 1 the most hypertonic region of the gut is in gut tissue otherwise fluid absorption would be impossible. Similar points could be made concerning the implication of oxygen-regulated genes (Lee and Falkow, 1990; Lee et al., 1992; Ernst et al., 1990) and their role in gut invasion. Admittedly, the gut lumen becomes increasingly anaerobic as one descends the alimentary tract. Hence on *a priori* grounds one might expect that oxygen limitation would induce the expression of invasion genes. However, it is clear that there are non-growth-limiting factors other than oxygen governing the expression of the invasion phenotype: Douce et al., (1991) showed that *S. typhimurium* grown under the same static (and hence microaerophilic conditions) in two different media resulted in organisms with significantly different invasion potentials. Again it could be that it is the ability to translocate the lamina propria that is facilitated by the anaerobic induction of 'spreading' genes since villus apices are relatively hypoxic in health. Moreover, murine villi undergoing response to rotavirus infection become demonstrably transiently anoxic as judged by their ischaemic appearance (Osborne et al., 1991a,b). It would not be surprising if similar conditions were to obtain in non-haemorrhagic invasive bacterial infections. This is currently under investigation.

Significance of Invasion

Having established (albeit with an as yet limited number of strains) that invasiveness correlates with 'gastroenteritic' virulence the question must be addressed as to what role can/must be ascribed to invasion in the causation of fluid secretion induced by an enterotoxigenic pathogen. There are several possibilities illustrated in Figure 3.

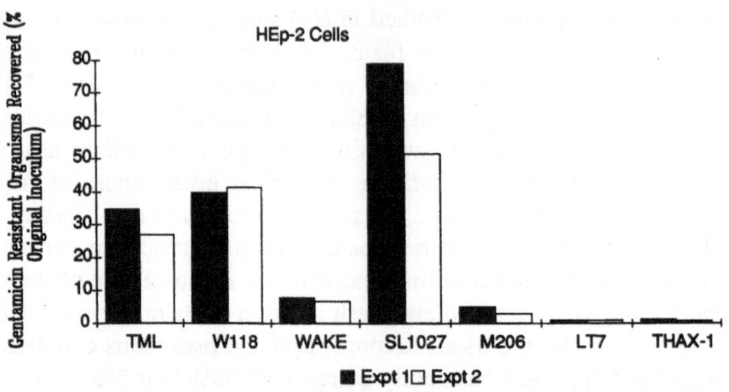

Figure 2a. Data taken from Douce et al., (1991). Strains TML, W118, and WAKE are virulent, and strains SL1027, M206, LT7, and THAX-1 are avirulent. Note (a) virulent strains show differential degrees of invasiveness, and (b) avirulent strain SL1027 is apparently more invasive than all the virulent strains; the latter is a demonstrable artifact of the system attributable to the centrifugation step routinely used in all such test systems. The inoculum size was 10^5 bacteria/monolayer of 10^5 cells.

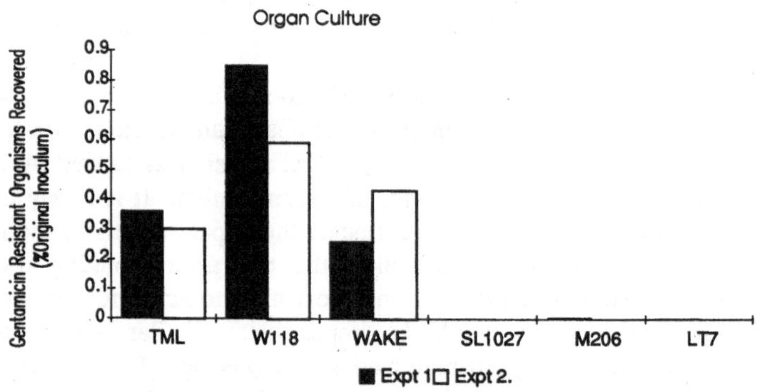

Figure 2b. Unpublished preliminary data from our laboratory. Ileal tissue was held under conditions described above, and challenged with the same strains as in Fig. 2a. The principal differences between the two systems is that avirulent strain SL1027 is now - as one might expect on the basis of histological analysis of gut infected *in vivo* - non-invasive. WAKE is now just as invasive as TML. The inoculum size was 10^8 bacteria/50mm^2. The area of tissue was comparable to that of the tissue culture wells. No centrifugation was used in this system.

The first concerns enterotoxin whose nature and presumed role in the causation of *S. typhimurium*-induced gastroenteritis has never been satisfactorily resolved (Stephen et al., 1985; Wallis et al., 1986a), and may only be resolved when both the multifactorial nature of Salmonella virulence, and, the variability of the host response to Salmonella infection are fully recognised. The toxin bears partial resemblance to cholera toxin (Finklestein et al., 1983; Chopra et al.,1987; Clarke et al.,1988; Prasad et al., 1990; Chopra et al.,1991) but unlike cholera and *E.coli* LT toxins Salmonella enterotoxin is not readily secreted into culture media; it requires to be extracted (Wallis et al.,1986a). Enterotoxin in RILs elicits a clear non-haemorrhagic non-purulent fluid.

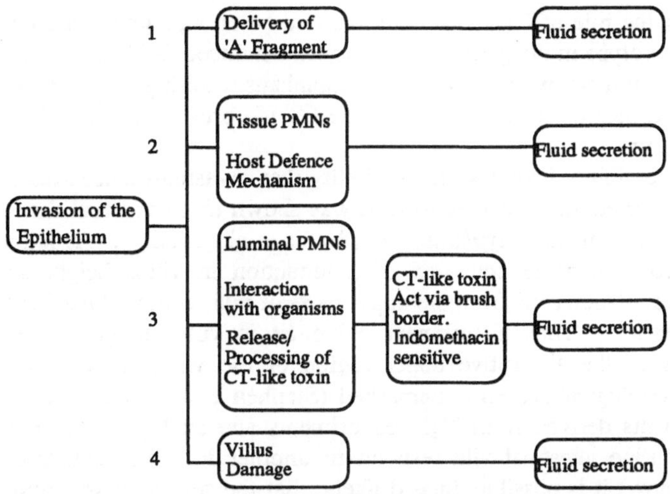

Figure 3. Possible sequelae to mucosal invasion by *S. typhimurium* in a gastroenteritic context.

Both avirulent and virulent strains of *S. typhimurium* can make enterotoxin (Wallis et al., 1986a); thus enterotoxigenicity *per se* can not be the sole criterion of virulence. However, there is a correlation between the expression of a cholera-toxin-related antigen (CTRA) and the ability to elicit a fluid response in RILs (Clarke et al.,1988; Qi et al.,1989; Wallis et al.,1989). This leads to the first possible explanation of the need for invasiveness. It has been suggested that invasion of enterocytes is the means of delivering the equivalent of A fragment to its presumptive intracellular site: the assumption is that toxin release occurs more readily in the intracellular environment.

The second possibility concerns the role of the massive influx of polymorphonuclear cells (PMNs) which is always observed in this infection. It could be that this is simply a host defence mechanism in response to the presence of invading bacteria and any damage they may cause. However, Giannella et al., (1973; 1975) proposed that the difference in the ability of virulent and avirulent strains to cause diarrhoeal disease lay in their capacity to induce an inflammatory response the cellular component of which was mainly PMNs. The evidence implicating PMNs was based on rendering rabbits neutropenic by treatment with nitrogen mustard (N_2M): this abolished *S. typhimurium*-induced fluid secretion without (it was claimed) damaging the gut or inhibiting CT-induced secretion; CT is nearly always used as a positive control in such experiments. Gots et al., (1974) proposed that prostaglandins of presumed PMN origin were the mediators of fluid secretion based on the ability to inhibit secretion by indomethacin. In this laboratory, the magnitude of the leucocyte influx has been quantified in relation to virulence. This was assessed in rabbit ileal loops with ^{111}Indium-labelled leucocytes (Wallis et al., 1989). Virulent strains (TML, W118, and WAKE) induced a dose-dependent leucocyte influx; avirulent strains failed to induce a significant leucocyte influx.

The third possibility involves PMNs in a different way. Fluid secretion was never observed in the absence of a leucocyte influx, but leucocyte influx *per se* did not induce fluid secretion; even in the presence of masses of leucocytes in some experiments, loops were flat. The phenotype of the challenge inoculum influenced fluid secretion since young log-phase organisms induced fluid secretion with a higher frequency than overnight cultures. These findings demonstrated a positive correlation between virulence and the ability to elicit a leucocyte influx, thus supporting the earlier evidence implicating leucocytes (see Wallis et al.,

1989). However the role of PMNs is more likely to be an interactive one involving bacteria of the correct phenotype in the genesis of *Salmonella*-induced fluid secretion. It may well be that luminal PMNs interact with toxin-bearing organisms resulting in the release/processing of toxin which would then be capable of acting like CT *via* brush borders and an indomethacin-sensitive mechanism.

The fourth possibility involves the possibility that invasion-induced damage itself could give rise to fluid secretion. In earlier work it was shown that structural damage occurred to villi challenged with virulent *S. typhimurium* (Wallis et al., 1986b). As judged by light and electron microscopy, damage consisted of a reduction in villus height arising from the shedding of bacteria-laden cells from the upper parts of villi, without breaching the epithelium or causing ulceration. Thus if sufficient reduction in villus height were to occur with concomitant loss of the absorptive upper region of the villus, absorption could become impaired and physiological secretion unmasked (Stephen et al., 1985; Wallis et al., 1986b). Based on arguments derived from N_2M experiments and analogies drawn from a detailed study of how murine intestinal villi respond to another invasive 'epithelial' pathogen (i.e. rotavirus; see below) it is possible, indeed likely, that the regenerative response required to rebuild truncated villi (when that occurred since, there was a difference in the degree of damage and the timing of it's onset observed in Dutch and New Zealand rabbits) could of itself give rise to hypersecretion .

Dissection Of The Models Proposed

If one assumes that CTRA is related to or part of a CT-like toxin and if one could make a CTRA-minus mutant which was invasion-positive and diarrhoeagenic, then one could effectively rule out any involvement of a CT-like toxin-mediated fluid secretion; this would eliminate all the possibilities described above involving this toxin. If such a mutant were non-diarrhoeagenic then one would have to decide between those possibilities which did and did not involve PMNs. Since it has been demonstrated by Au-labelled antibody that CTRA is a periplasmic protein then a CTRA-minus mutant should be capable of being picked up in our Tn*pho*A bank of mutants with anti-CT as already described (Clarke et al.,1988; Qi et al.,1989). However, if, as could be the case, expression of the invasive and CTRA-positive phenotypes is co-ordinately regulated (both invasiveness and CTRA expression increase when organisms are grown in HDB medium) in a manner similar to that reported for CT expression and adhesion in *Vibrio cholerae* (Miller, et al., 1987) then it would be essential to inactivate the structural effector CTRA genes and not the control genes governing the co-ordinated expression of CTRA and invasion. This is being actively pursued.

How might one prove the validity of or eliminate those possibilities implicating PMNs? Superficially this is simple: render animals neutropenic and challenge them with fully virulent CTRA-positive wild type organisms or isogenic CTRA-minus mutants. In practice rendering rabbits neutropenic has proved to be extremely difficult. A number of attempts have already been made to resolve this controversial matter (Wallis et al., 1989, 1990). The earlier experiments involving indomethacin and nitrogen mustard (N_2M) were repeated. Pretreatment of rabbits with indomethacin almost completely abolished Salmonella-induced secretion and significantly reduced secretion induced by CT as was found by Gots et al., (1974). They suggested that the data implicated prostaglandins of leucocytic origin. However, if *S. typhimurium* and CT share a common or partially common mechanism for inducing fluid secretion, that common part cannot involve neutrophils: there is no evidence for a necessary influx of neutrophils in CT-induced secretion. Moreover, in every case examined by Wallis et al., (1986*b*; 1989; 1990), either quantitatively or qualitatively, virulent strains elicited a leucocyte response but not always a fluid response. Therefore, any common indomethacin-sensitive step must be mediated by cells other than neutrophils (possibly enterocytes) as has been claimed for CT (Duebbert and Peterson, 1985).

In agreement with Giannella (1979), Wallis et al., (1990) showed that rabbits may be rendered leucopenic by N_2M-treatment, and that the ability of such animals to mount a secretory response to challenge by virulent strains of *S. typhimurium* was almost completely abolished. However in two important respects the data differed from those of Giannella (1979). First, as judged both by light and scanning electron microscopy, uninfected villi treated with N_2M showed significant morphological damage as compared with untreated villi. Neither uninfected-N_2M- or infected-N_2M-treated loops showed significant fluid secretion despite (i) N_2M-induced damage in both, and (ii) the huge numbers of invading organisms of strains that regularly induced fluid secretion in normal animals. Thus the basic premise that N_2M did not damage or upset the gut in a manner prejudicial to the experimental design - claimed by Giannella to have been fulfilled - was not fulfilled in our work. The second major difference was that N_2M reduced by 45% the expected fluid secretion induced by CT. While the N_2M experiments left the role of leucocytes unresolved they highlighted other pathophysiological possibilities summarised as follows.

N_2M is used therapeutically in the treatment of certain neoplasias by virtue of its alkylating effect on DNA and hence on dividing cells (Salmon, 1980). It is therefore possible that the reduction of villus height and other morphological changes (Wallis et al., 1990) are a direct consequence of N_2M affecting crypt cell mitoses. This would affect normal epithelial cell replacement and hence height of villi, and also crypt functions which, by general consensus (Roggin et al., 1972; Field et al., 1980; Welsh et al., 1982), are involved in normal secretion. Recently, Collins et al., (1988) and Spencer et al., (1990) demonstrated that increased rates of villus base/crypt cell division are involved in the hypersecretory phase of rotavirus-induced diarrhoea in mice. Thus it could be that the inhibition of secretion by N_2M seen in this work is due to it's ability to inhibit cell division in villus base/crypt assemblies. The significance of this is to be appreciated in the context of a new concept of the pathophysiology of infectious diarrhoea (see Spencer et al., 1990). The intestine responds to insult/injury by shedding affected enterocytes which are rapidly replaced. Dividing cells accumulate transiently high levels of NaCl the secretion of which is perceived to be the basis of fluid loss in diarrhoeal disease. These pathophysiological principles have been developed most fully for murine rotavirus diarrhoea and as suggested by Stephen and Osborne (1988) could be extrapolated to invasive nonhaemorrhagic diarrhoeas such as that induced by *S. typhimurium*.

In conclusion, though we have not yet fully elucidated the mechanisms and determinants of gastroenteritic disease caused by this highly sophisticated pathogen, some progress has been made in defining the biological parameters of *S. typhimurium*-induced diarrhoeal disease. We can now more accurately formulate questions concerning the relative importance of initial invasiveness, the role of enterotoxin, the inflammatory response and the homeostatic response of the gut reacting to insult/injury.

REFERENCES

Candy, D. C. A. and Stephen, J., 1989, *Salmonella*, *in*: "Enteric Infecton. Mechanisms, Manifestations and Management", Farthing M. J. G., Keusch G. T., (eds), Chapham and Hall Medical, London, p 289.

Charles, I. and Dougan, G.,1990, Gene expression and the development of live enteric vaccines, *TIBTECH.*, 8: 117.

Chopra, A. K., Houston, C. W., Peterson, J. W., Prasad, R. Mekalanos, J. J., 1987, Cloning and expression of the *Salmonella* enterotoxin gene, *J.Bact.*, 169: 5095.

Chopra, A. K., Peterson, J.W., Houston, C. W., and Prasad, R., 1991, Enterotoxin-associated DNA sequence homology between *Salmonellae* species and *Escherichia coli*, *FEMS (Fed Eur Microbiol Soc) Microbiol Lett*, 77: 133.

Clarke, G. J., Qi, G.-M., Wallis, T. S., Starkey, W. G., Collins, J., Spencer, A. J., Haddon, S. J., Osborne, M. P., Worton, K. J., and Stephen, J., 1988, Expression of an antigen in strains of *Salmonella typhimurium* which reacts with antibodies to cholera toxin, *J. Med.Microbiol.*, 25: 139.

Collins, J., Starkey, W. G., Wallis, T. S., Clarke, G. J., Worton, K. J., Spencer, A. J., Haddon, S. J., Osborne, M. P., Candy, D.C.A., and Stephen, J., (1988), Intestinal enzyme profiles in normal and rotavirus-infected mice, *J. Pediatr. Gastroenterol. Nutr.*,7: 264.
Douce, G. R., Amin, I. I.,and Stephen, J., 1991, Invasion of HEp-2 cells by strains of *Salmonella typhimurium* of different virulence in relation to gastroenteritis, *J.Med.Microbiol.* 35: 349.
Duebbert, I. E., and Peterson, J. W., 1985, Enterotoxin-induced fluid accumulation during experimental salmonellosis and cholera: involvement of prostaglandin synthesis by intestinal cells, *Toxicon* 23: 157.
Elsinghorst, E. A., Baron, L. S., Kopecko, D.J., 1989, Penetration of human intestinal epithelial cells by *Salmonella:* molecular cloning and expression of *Salmonella typhi* invasion determinants in *Escherichia coli, Proc. Natl. Acad. Sci. USA*, 86: 5173.
Ernst, R. K., Dombroski, D. M., and Merrick, J. M., 1990, Anaerobiosis, type 1 fimbriae, and growth phase are factors that affect invasion of HEp-2 cells by *Salmonella typhimurium, Infect. Immun.* 58: 2014.
Field, M., Smith, P. L., Bolton, J. E., 1980, Ion transport across the isolated intestinal mucosae of the winter flounder, *Pseudopleuronectes americanus* II; effects of cyclic AMP, *J. Membrane Biol.*, 55: 157.
Fields, P. I., Groisman, E. A., and, Heffron, F., 1989, A *Salmonella* locus that controls resistance to microbicidal proteins from phagocytic cells, *Science,* 243: 1059.
Fields, P. I, Swanson, R. V., Haidaris, C. G., and Heffron, F., 1986, Mutants of *Salmonella typhimurium* that cannot survive within the macrophage are avirulent, *Proc. Natl. Acad. Sci.,USA*, 83: 5189.
Finklestein, R. A., Marchlewicz, B. A., Mcdonald, R. J., Boesman-Finklestein, M.,1983, Isolation and characterizaton of a cholera-related enterotoxin from *Salmonellae typhimurium* , *FEMS (Fed Eur Microbiol Soc) Microbiol Lett,* 17: 239.
Finlay, B. B., and Falkow, S., 1988, Comparison of the invasion strategies used by *Salmonella choleraesuis, Shigella flexneri* and *Yersinia enterocolitica* to enter cultured animal cells: endosome acidification is not required for bacterial invasion or intracellular replication, *Biochimie,* 70: 1089.
Finlay, B. B., Gumbiner, B., and Falkow, S., 1988*a*, Penetration of *Salmonella* through a polarized Madin-Darby canine kidney epithelial cell monolayer, *J. Cell Biol.*, 107: 221.
Finlay,B. B., Heffron, F., and Falkow, S., 1989, Epithelial cell surfaces induce *Salmonella* proteins required for bacterial adherence and invasion, *Science*, 243: 940.
Finlay, B.B., Starnbach, M. N., Francis, C. L., Stocker, B. A. D., Chatfield, S., Dougan, G., and Falkow, S., 1988*b*, Identification and characterization of Tn*pho*A mutants of *Salmonella* that are unable to pass through a polarized MDCK epithelial cell layer, *Mol. Microbiol.*, 2: 757.
Gahring, L. C., Heffron, F., Finlay, B. B., and Falkow, S., 1990, Invasion and replication of *Salmonella typhimurium* in animal cells, *Infect. Immun.*, 58: 443.
Galan, J. E., and Curtiss III, R., 1989, Cloning and molecular characterization of genes whose products allow *Salmonella typhimurium* to penetrate tissue culture cells, *Proc. Natl. Acad. Sci. USA*, 86: 6383.
Galan, J. E., and Curtiss III, R., 1990, Expression of *Salmonella typhimurium* genes required for invasion is regulated by changes in DNA supercoiling, *Infect. Immun.*, 58: 1879.
Galan, J. E., and Curtiss III, R., 1991, Distribution of the *invA*, *-B*, *-C*, and *-D* genes of *Salmonella typhimurium* amoung *Salmonella* serovars: *invA* mutants of *S.typhi* are deficient for entry into mammalian cells. *Infect Immun* 59:2901.
Giannella, R. A., 1979, The importance of the intestinal inflammatory reaction in *Salmonella*-mediated intestinal secretion, *Infect. Immun.*, 23: 140.
Giannella, R. A.. Formal, S. B., Dammin, G. J., and Collins, H., 1973, Pathogenesis of Salmonellosis. Studies of fluid secretion, mucosal invasion and morphological reaction in the rabbit ileum, *J. Clin. Invest.* 52: 441.
Giannella, R. A., Gots, R. E., Charney, A. N., Greenough, W. B., and Formal, S. B., 1975, Pathogenesis of *Salmonella* mediatedintestinal fluid secretion, *Gastroenterol*, 69: 1238.
Gots, R. E., Formal, S. B., Giannella, R. A., 1974, Indomethacin inhibition of *Salmonella typhimurium, Shigella flexneri,* and cholera-mediated rabbit ileal secretion, *J. Inf. Dis.*, 130: 280.
Groisman, E. A., Chiao, E., Lipps, C. J., and Heffron, F., 1989, *Salmonella typhimurium phoP* virulence gene is a transcriptional regulator, *Proc. Natl Acad. Sci. USA,* 86: 7077.
Groisman, E. A., and Saier, M. H., 1990, *Salmonella* virulence: new clues to intramacrophage survival, *TIBS*, 15-January 1990, 30.
Hallback, D. A., Hulten, L., Jodal, M., Lindhagen, J., and Lundgren, O., 1978, Evidence for the existence of a countercurrent exchanger in the small intestine of man. *Gastroenterol.* 74: 683.
Lee, C. A., and Falkow, S., 1990, The ability of *Salmonella* to enter mammalian cells is affected by bacterial growth state. *Proc. Natl Acad. Sci. USA.* 87:4304.
Lee, C. A., Jones, B. D., and Falkow, S., 1992, Identification of a *Salmonella typhimurium*, invasion locus by selection for hyperinvasive mutants, *Proc. Natl Acad. Sci. USA,* 89: 1847.
Lundgren, O., 1988, Factors controlling absorption and secretion in the small intestine, *in*: "Bacterial infections of respiratory and gastrointestinal mucosae", W Donachie, E Griffiths, J Stephen, eds., special publication of the Society for General Microbiology, vol 24, , UK, IRL Press, Oxford, p 97.

Miller, S. I., Kukral, A. M., and Mekalanos,J. J., 1989, A two component regulatory system (*phoP, phoQ*) controls *Salmonella typhimurium* virulence, *Proc. Natl. Acad. Sci. USA*, 86: 5054.

Miller I., Maskell, D., Hormaeche, C., Pickard, D., and Dougan, G., 1989, The isolation of orally attenuated *Salmonella typhimurium* following Tn*phoA* mutagenesis, *Infect. Immun.*, 57: 2758.

Miller, V.L., Taylor, R.K. and Mekalanos, J.J. (1987). Cholera toxin transcriptional activator ToxR is a transmembrane DNA membrane protein. *Cell* 48: 271-279.

O'Callaghan, D., Maskel, D., Liew, F., Easmon, C.S.F., and Dougan G., 1988, Characterisation of aromatic- and purine-dependent *Salmonella typhimurium* :attenuation, prsistance and ability to induce immunity in BALB/c mice. *Infect. Immun.* 56: 419.

Osborne M P, Haddon S J, Worton K J, Spencer A J, Starkey W G, Thornber D, and Stephen J (1991a). A study of the microcirculation in whole villi of neonatal mice using a peroxidase histochemical staining method. *J. Pediatr. Gastroenterol. Nutr.* 12: 105.

Osborne M P, Haddon S J, Worton K J, Spencer A J, Starkey W G, Thornber D, and Stephen J (1991b). Rotavirus-induced changes in the microcirculation of intestinal villi of neonatal mice in relation to the induction and persistence of diarrhea. *J. Pediatr. Gastroenterol. Nutr.* 12: 111.

Prasad, R., Chopra, A.K., Peterson, J.W., Pericas, R. and Houston, C.W. (1990). Biological and immunological characterization of a cloned cholera toxin-like enterotoxin from *Salmonella typhimurium. Microbial Pathogenesis* 9:315.

Qi, G.-M., Clarke, G. J., Wallis, T. S., and Stephen, J., 1989, The influence of cultural conditions on the expression in *Salmonella typhimurium* of an antigen related to cholera toxin, *J. Med. Microbiol.*, 30: 213.

Roggin, G. M., Banwell, J. G., Yardley J. H., and Hendrix T. R., 1972, Unimpaired response of rabbit jejunum to cholera toxin after selective damage to villus epithelium, *Gastroenterology*, 63: 981.

Salmon, S. E., 1980, Cancer Chemotherapy, *in*: "Review of medical pharmacology," Meyers F.H., et al. eds. 7th edn. Lange Medical Publications, Los Altos, CA., p 477.

Small P. L. C., Isberg, R. R., and Falkow, S., 1987, Comparison of the ability of enteroinvasive *Escherichia coli, Salmonella typhimurium, Yersinia pseudotuberculosis*, and *Yersinia enterocolitica* to enter and replicate within HEp-2 cells, *Infect. Immun.*, 55: 1674.

Spencer, A. J., Osborne, M. P., Haddon, S. J., Collins, J., Starkey, W. G., Candy, D. C. A., and Stephen. J., 1990, X-ray microanalysis of rotavirus-infected mouse intestine: a new concept of diarrhoeal secretion, *J Pediatr Gastroenterol Nutr.* 10: 516

Starkey W G, Candy D C A, Collins J, Spencer A J, Osborne M P and Stephen J (1990a). An *invitro* model to study aspects of the pathophysiology of murine rotavirus-induced diarrhoea. *J.Pediatr.Gastroenterol. Nutr.* 10, 361.

Starkey W G, Candy D C A, Collins J, Spencer A J, Osborne M P and Stephen J (1990b). Transport of water and electrolytes by rotavirus-infected mouse intestine. A time course study*J.Pediatr.Gastroenterol.Nutr.* 11, 254-260.

Stephen, J., Osborne, M. P., 1988, Pathophysiological mechanisms in diarrhoeal disease, *in*: "Bacterial infections of respiratory and gastrointestinal mucosae", W Donachie, E Griffiths, J Stephen, eds., special publication of the Society for General Microbiology, vol 24, , UK, IRL Press, Oxford, p 149.

Stephen, J., and Pietrowski, R. A., 1986, "Bacterial Toxins," 2nd. edn. Van Nostrand Reinhold (UK) Co. Ltd., Wokingham, UK.

Stephen, J., Wallis, T. S., Starkey, W. G., Candy, D.C.A., Osborne, M.P., and Haddon, S. J., 1985, Salmonellosis: in retrospect and prospect, *in*: "Microbial toxins and diarrhoeal disease," eds. Evered, D., Whelan, J., Pitman, London (Ciba Foundation Symposium 112) p 175.

Wallis, T. S., Hawker, R. J. H., Candy, D.C.A., Qi, G.-M., Clarke, G. J., Worton, K. J., Osborne M. P. and Stephen, J., 1989, Quantification of the leucocyte influx into rabbit ileal loops induced by strains of *Salmonella typhimurium* of different virulence, *J.Med. Microbiol.*, 30: 149.

Wallis, T. S., Starkey, W. G., Stephen. J., Haddon, S. J., Osborne, M. P., Candy, D. C. A., 1986*a* Enterotoxin production by *Salmonella typhimurium* strains of different virulence, *J. Med. Microbiol.*, 21: 19.

Wallis, T. S., Starkey, W. G., Stephen, J., Haddon, S. J., Osborne, M. P., Candy, D. C. A., 1986*b*, The nature and role of mucosal damage in relation to *Salmonella typhimurium*-induced fluid secretion in the rabbit ileum, *J. Med. Microbiol.*, 22: 39.

Wallis, T. S., Vaughan, A. T. M., Clarke, G. J., Qi, G.-M., Worton, K. J., Candy, D.C.A., Osborne, M. P., and Stephen, J., 1990, The role of leucocytes in the induction of fluid secretion by *Salmonella typhimurium, J. Med. Microbiol.*, 31: 27.

Welsh, M. J., Smith, P. L., Fromm M., and Frizell, R. A., 1982, Crypts are the site of intestinal fluid and electrolyte secretion, *Science*, 218: 1219.

Worton, K. J., Candy D. C. A., Wallis T.S., Clarke, G. J., Osborne, M. P., Haddon, S. J., and Stephen, J., 1989, Studies on early association of *Salmonella typhimurium* with intestinal mucosa *in vivo* and *in vitro*: relationship to virulence, *J. Med. Microbiol.*,29: 283.

CELLULAR MEDIATORS OF ANTI-MICROBIAL RESISTANCE

Frank M. Collins

Trudeau Institute Inc.
Saranac Lake, NY 12983

INTRODUCTION

Most members of the normal gut microflora which accidentally gain entry to the tissues will be immediately taken up and killed by the resident macrophages before they can cause anything more than a localized inflammatory reaction. However, some of these commensal organisms have the ability to elude these normal cellular defenses allowing them to colonize (at least temporarily) the intestinal sub-mucosa and its associated lymphoid tissues (Collins, 1987). These organisms are often referred to as 'opportunistic pathogens' because they can cause active disease if the normal cellular defenses have been depleted as a result of aging, immunosuppressive chemotherapy or HIV infection (Collins, 1990). Once these normally benign organisms become established within the immunodepleted tissues, they quickly assume life-threatening proportions, involving virtually every organ of the body.

A few members in this group have the ability to actively *invade* the mucosal cells, multiplying within the sub-mucosa until they induce the pathological changes associated with systemic disease (Collins and Campbell, 1982). Such pathogens may produce toxins (aggressins) which promote their intracellular growth and spread within the affected tissues unless they are neutralized by specific anti-toxins. Opsonic antibodies promote the uptake and intracellular inactivation of these pathogens by the macrophages which pour into the developing lesion from the bloodstream. such protective humoral factors can be readily demonstrated in both actively and passively immunized animals following parenteral or oral challenges (Collins and Carter, 1974).

The immune response to infections caused by such facultative intracellular pathogens as *Listeria monocytogenes, Brucella abortus* or *Salmonella typhimurium* is complicated by the fact that these organisms can multiply within the cells whose normal function is to kill them. In fact, some intracellular pathogens such as *M. leprae* and *R. rickettsii* can only grow in such an environment. The precise mechanism(s) by which these intracellular pathogens elude these highly effective cellular defenses is only poorly understood (Lowrie and Andrew, 1988). Some organisms produce so-called 'virulence' factors which enable them to block or inhibit lysosomal

Biology of Salmonella, Edited by F. Cabello *et al.*,
Plenum Press, New York, 1993

fusion with the phagosomal membrane (Buchmeier and Heffron, 1991). Others produce enzymes which break down the phagosomal membrane, releasing the organisms into the cytoplasm where they will no longer be exposed to the bactericidal (Bacteriostatic) action of the lysosomal enzymes (Collins and Campbell, 1982). Some organisms (*M. tuberculosis* for example) may neutralize the bactericidal action of the hydrogen peroxide and superoxide anion which the phagocyte produces as a result of the oxidative burst. Unfortunately, it has not always been possible to show a direct correlation between the presence of catalase and superoxide dismutase and virulence (Lowrie and Andrew, 1988).

Many of these infections are characterized by the development of a chronic carrier state in which small numbers of viable bacilli may persist within the tissues for months or even years (Collins, 1987). Paradoxically, many of these chronically infected carriers are able to prevent second attacks of the disease although they are unable to eliminate the residual 'vaccinating' infection. This selective infection-immunity constitutes one of the most fascinating aspects of the cell-mediated immune response to these pathogens, a phenomenon which will be discussed further in later sections of this review.

INNATE IMMUNITY AND MACROPHAGE ACTIVITY

Innate resistance depends upon a number of apparently unrelated factors, both host and microbial in nature and involves humoral and cellular factors in an integrated type of defense system. The dominant component in this response may vary extensively as the bacterial infection first progresses and later wanes. Thus, to ask which of these components is more important is like trying to compare the wheels of a bicycle.

Normal mice infected with virulent *S. typhimurium* or *S. enteritidis* exhibit a growth cycle with 4 separate phases extending over a period of weeks or even months (Figure 1). *Phase I* (innate immunity) may only last for a few minutes to an hour or so, during which as much as 90% of the challenge inoculum may be inactivated. The protective value of this non-specific killing phase varies depending upon the number of surviving organisms within the sub-mucosa. The extent of this initial kilo will be greatly accelerated by the presence of specific

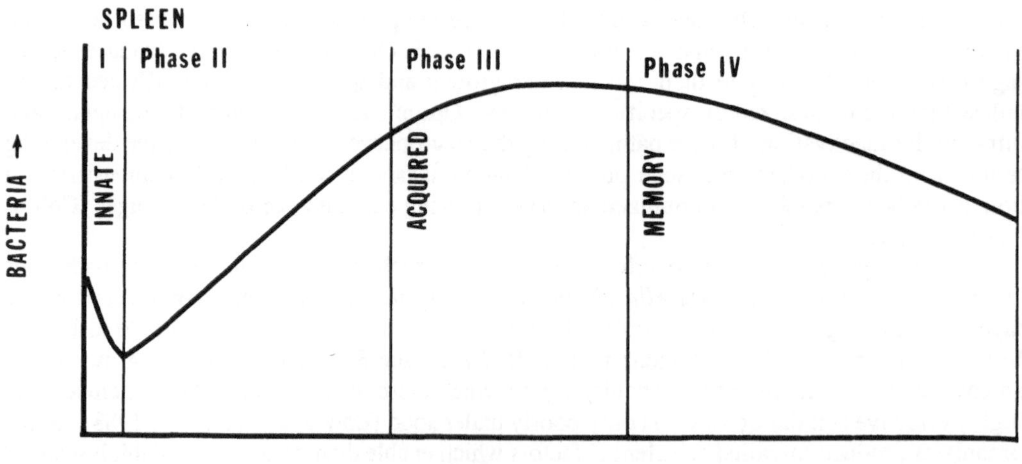

Figure 1. Diagram showing the *Salmonella* growth phases in the spleens of intravenously infected mice.

psonic antibodies and/or complement (Collins, 1971). The extent of this early kill will also lepend upon the genetic background of the host (Hormaeche, *et al.*, 1985). Thus, nacrophages harvested from C3H (*Ity*^{r}) mice are more bactericidal for *S. typhimurium* than imilar cells harvested from C57BL/6 (*Ity*^{s}) mice (Blumenstock and Jann, 1981). Pre-treatment f these mice with dextran sulfate or silica (both widely recognized as macrophage poisons) blates this innate resistance and increases the severity of the resulting infection (Hof, *et al.*, 982). The early bactericidal activity expressed by the *Ity*^{r} mice seems to be independent of unctional T-cells (O'Brien and Metcalf, 1982) and survival may be roughly proportional to he virulence of the test organism (O'Brien, 1986).

Once this initial kill is complete, the surviving bacilli begin to multiply logarithmically *Phase II*) and the infection may reach clinically significant proportions before it can be rought under control by the emerging cell-mediated immune response (*Phase III*). The reason or the sudden change in growth during this transition is largely unclear although it may be the nost important aspect of the disease process. Several explanations for this phenomenon have een advanced but it has proven surprisingly difficult to verify them experimentally. The most lausible explanation postulates that the normal macrophage population consists of a diverse roup of permissive and non-permissive phagocytes, with most of the intracellular growth ccurring in the former group. Some organisms may also infect parenchymal cells within the ffected organ where they can continue to multiply without further exposure to the professional hagocytes entering the lesion from the bloodstream. This type of evasive mechanism was ecently described in *Listeria monocytogenes*-infected mice (Tilney and Portnoy, 1989) and may ccur with a number of other intracellular pathogens. Some of these organisms can adapt to heir *in vivo* environment, undergoing some sort of phenotypic change which enables them to row more readily within the tissues compared to *in vitro*-grown organisms. Such *in vivo*-rown bacteria exhibit a number of subtle changes in their organ preference leading to an pparent increase in their virulence (Collins and Montalbine, 1976).

Logarithmic growth (*Phase II*) will continue until the emerging cell-mediated immunity imits further *in vivo* growth and the infection passes into its stationary *Phase III*. uperinfection of mice at this stage using a genetically-tagged strain of *S. typhimurium* (SM^r) eveals the presence of an emerging cell-mediated immunity which can eliminate a secondary hallenge from both the liver and spleen. We still know very little about the factors which are esponsible for this bacteriostatic response (Kagaya, *et al.*, 1989) or why this bacterial opulation passes into a slow decline during *Phase IV*. In many cases, a few of these rganisms will persist in the tissues to be reactivated years later if the patient is subjected to mmunosuppressive chemotherapy. This 'carrier state' is also characterized by the presence f a population of long-lived memory T-cells which bear unique Pgp-1 (Ly24) membrane narkers (Cerottini and MacDonald, 1989). It is not clear whether this Tm population requires ontinuous stimulation from this residual infection which can also be a source of reactivational lisease in immunosuppressed individuals (Collins, 1990).

Salmonellosis is a water and food-borne disease in which the pathogen enters the tissues ia the intestinal mucosa, infecting the regional Peyer's patch before spreading to the draining nesenteric lymph node (Carter and Collins, 1974). Experimentally, C57BL/6 (*Ity*^{s}) mice can e readily infected with *S. enteritidis* using the intragastric route (Collins and Niederbuhl, 983) or even by rectal inoculation (Carter and Collins, 1974). These Salmonellae are equally dept at crossing the nasopharyngeal mucosae and can heavily infect the lung following an erogenic challenge (Carter, *et al.*, 1975). Protection data obtained using parenterally hallenged mice should also be confirmed in orally challenged animals to ensure that the results re equally relevant to the naturally acquired disease. Such oral challenges will be more lifficult to quantitate reproducibly when a biologically relevant inoculum is used. However, ral infection of C57BL/6 (*Ity*^{s}) mice with 10^6 virulent *S. enteritidis* suspended in 5% icarbonate solution yielded counts of 10^3-10^4 CFU per homogenate of washed ilial mucosa 12 hours after inoculation (Collins and Carter, 1974). This mucosal infection quickly spreads to the GALT organs and from there to the liver and spleen where the main systemic infection develops. This limited early intestinal involvement makes it virtually impossible to visually demonstrate the sub-mucosal infection route, even with specific immunofluorescence staining (Gahring, *et al.*, 1990). On the other hand, if this local infectious load is enhanced by means of oral chemotherapy or persistalsis inhibition, the resulting data will have limited relevance to the naturally acquired disease (Collins and Carter, 1974).

When this same oral challenge is applied to vaccinated mice, the early growth of the *Salmonella* within the GALT organs will be largely unaffected although later, the anamnestic immune response limits further growth *in vivo*. However, the vaccinated host is unable to prevent the establishment of the initial mucosal infection and this seems consistent with earlier data obtained with other intracellular pathogens (Collins, 1987).

Oral virulence may be expressed at several different levels, ranging from an increased attachment to the gut mucosa (colonization), to more efficient invasion of the mucosal cells (infection), to enhanced growth and survival by the organisms after they reach the liver and spleen (disease). Immunity may be expressed at any of these stages although it will generally be more effective at the systemic, rather than the mucosal level. Demonstration of such mucosal immunity is a time-consuming and labor intensive process but may provide data leading to the development of more effective immunotherapeutic measures for use in immunosuppressed patients suffering with the naturally acquired enteric disease.

T-CELLS AND ACQUIRED RESISTANCE

The immunocompetent T-cell constitutes the cornerstone of any antibacterial immune response to infections caused by intracellular pathogens. This role has been established by countless passive transfer studies involving both delayed hypersensitivity (DTH) and cell-mediated immunity (CMI) in naive syngeneic recipients of immune splenic T-cells (Collins, 1979). Such studies can now be carried out by infusing selected T-cell subsets obtained by panning on monoclonal antibodies able to recognize specific T-cell membrane markers (Baldridge, *et al.*, 1990). In general, $L3T4^+$ ($CD4^+$) T-cells seem to be associated with the expression of DTH and CMI whereas the $Lyt\text{-}2^+$ ($CD8^+$) T-cells are responsible for CMI only. A third group of $CD4^+$ T-cells are associated with immune memory and consist of long-lived lymphocytes resistant to pre-treatment with cyclophosphamide. They may also provide cross-protection to infections caused by a number of related species (Orme and Collins, 1986).

Immune T-cells bearing the surface markers $Thy\text{-}1.2^+$, $L3T4^+$, $Lyt\text{-}2^-$ ($CD4^+$) or $Thy\text{-}1.2^+$, $L3T4^-$, $Lyt\text{-}2^+$ ($CD8^+$) make up the dominant immune population which may change as the infection first progresses, then wanes (Kaufmann, 1988; Orme, 1987). Some of these T-cell interactions are represented schematically in Figure 2. As the pathogen multiplies within the tissues it will release a number of specific sensitins (stress proteins) which then stimulate the appropriate $CD4^+$ or $CD8^+$ T-cell sub-populations depending upon the presence of Class I or II MHC membrane determinants on the APC (Bierer, *et al.*, 1986). The dominant T-cell response will depend upon the nature of the stimulus (particulate, soluble, live, dead, adjuvanted), its composition (polysaccharide, protein, glycolipid), its concentration and rate of release from the granuloma. However, the precise factors which are responsible for the development of humoral vs. a DTH responses or CMI vs. unresponsiveness are still largely undefined.

Antigenically stimulated T-cells release a battery of lymphokines (IL-2, IL-4, IL-6, IFNγ, TNFα) which may activate or depress (modulate) monocyte activity within the developing lesion. These immunologically activated macrophages are larger and more metabolically active than normal (resident) phagocytes and can inactivate (or at least limit the growth of) the fully virulent pathogen. However, these activated macrophages may also be responsible for substantial tissue damage (fibrosis, necrosis, consolidation) in the chronically infected tissues (Dannenberg and Tomashefski, 1988). To minimize this effect, the T_{cmi} population will be down-regulated once the acute phase of the infection has been brought under control. The expressor T-cells are replaced by a population of long-lived memory immune T-cells (T_m) which protect the host against a second attack of the disease. At present, we know very little about the bacterial antigens which are responsible for this T_m response.

Cytotoxic ($CD8^+$) T-cells also express anti-bacterial resistance, especially during the chronic phase of the infection (Kaufmann, 1988). The sensitized $CD8^+$ T-cells recognize heavily infected macrophages within the chronic lesion and lyse them, releasing the intracellular bacteria to fresh macrophages within the chronic lesion and lyse them, releasing the intracellular bacteria to fresh macrophages which can presumably inactivate them more efficiently. These $CD8^+$ cells are known to be active against virus and *Listeria*-infected macrophages but their defensive role in *Salmonella*-infected mice has yet to be explored in detail.

Recently, there has been an increasing amount of interest in a group of gamma-delta bearing T-cells which are present in small (though significant) numbers in normal lung, skin and intestine and their associated lymphoid organs where they may serve a sentinel function (Janeway, *et al.*, 1988). These cells can recognize the stress proteins released by a variety of cells (Born, *et al.*, 1990), particularly those produced by *M. tuberculosis* and *S. typhimurium* (Havlir, *et al.*, 1991). Sublethal infection of B6D2 mice with virulent *S. enteritidis* increased the number of gamma-delta T-cells in the draining lymph node from control levels of around 1% to as high as 30% over a 10 day infection period (Flory and Collins, to be published). The

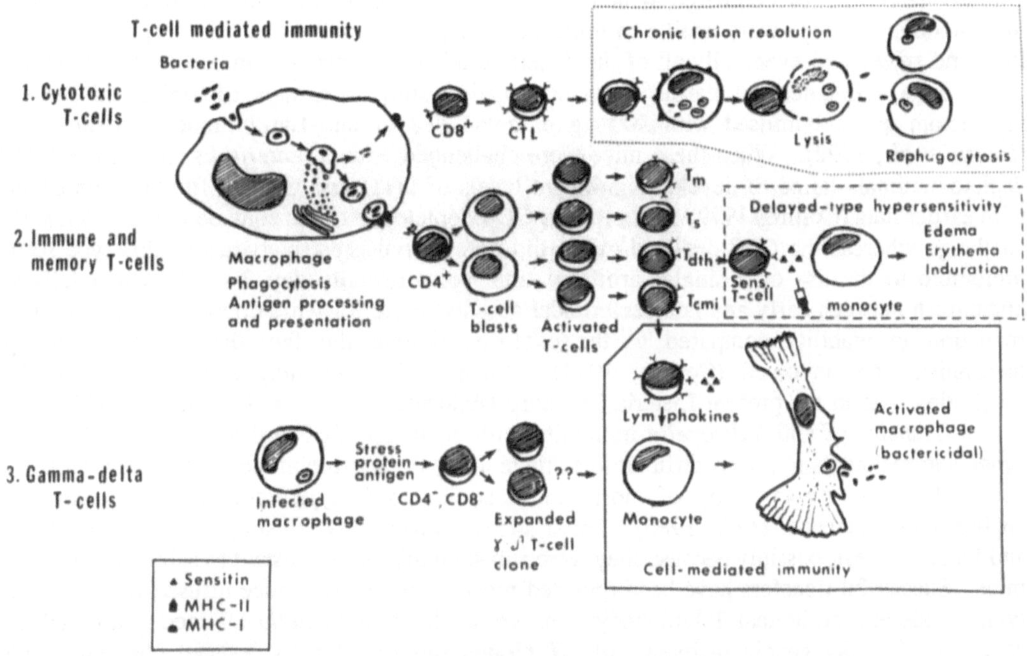

Figure 2. Diagram showing T-cell responses to *Salmonella* infection.

gamma-delta T-cells may be part of an early immune response against a number of important pathogens, although a lot more work will be required before their protective role is fully understood.

IMMUNOSUPPRESSION AND SALMONELLOSIS

Human salmonellosis normally results in little more than a localized intestinal inflammation, although the associated diarrhea may be severe enough to cause dehydration and absorptive problems, especially in the very young and the very old (Turnbull, 1979). The *S. typhimurium* infection seldom progresses beyond the gut-associated lymphoid tissues in immunocompetent adults, although it can result in severe systemic disease in rodents (Collins and Carter, 1974). On the other hand, *S. typhi* produces minimal intestinal inflammation, the infection passing rapidly through the GALT system to reach the liver and spleen. Interestingly, the typhoid bacillus is virtually avirulent for orally infected mice, even when gnotobiotic animals are used (Collins and Niederbuhl, 1983). This inability to grow in normal mouse tissues seems to be due to nutritional rather than bactericidal factors (O'Brien, 1982).

With the emerging AIDS epidemic in the United States, there has been a dramatic increase in the severity of *S. typhimurium*, *S. enteritidis* and even *Shigella flexneri* infections which can result in a disseminated typhoid-like disease in these immunosuppressed individuals once their peripheral $CD4^+$ counts drop below 500 per mm^3 (Sperber and Schenpuer, 1987). While these systemic *Salmonella* infections are still amenable to conventional chemotherapeutic intervention, active disease will often recur when the treatment is discontinued as a result of increasing drug toxicity. There is very little quantitative information regarding the effect of progressive immunodepletion on these *Salmonella* infections (Collins and Niederbuhl, 1983; Izhar, *et al.*, 1990). Experimentally, T-cell depleted *Ity*^{r} mice still showed a substantial resistance to a modest *Salmonella* challenge (Hormaeche, *et al.*, 1983). On the other hand, congenitally athymic (nu/nu) BALB/c mice infected with 10^3 viable *S. enteritidis* showed reduced early growth compared with their immunocompetent controls (Figure 3). These immunodeficient mice were unable to limit the later growth of the challenge organism in the liver and spleen and eventually all of the T-cell deficient animals succumbed to the infection.

In a refinement of earlier immunodepletion studies, groups of B6D2 mice were thymectomized and infused with 500 μg of anti-L3T4 or anti-Lyt-2 monoclonal antibody (Flory, *et al.*, 1992). When these mice were challenged with *S. enteritidis*, only the $CD4^+$-depleted animals failed to develop significant levels of DTH and CMI, a finding compatible with earlier data (Collins, 1979). All of these CD4-depleted mice succumbed to the *Salmonella* challenge whereas the CD8-depleted group did not. Serial growth studies of this type seem preferable to the use of a single, arbitrary time point (usually day 2 or 3) which fails to distinguish between early and late bactericidal activity, either of which may result in a similar reduction in viability compared to the controls, despite the fact that widely differing mechanisms are involved (Collins, 1974). Such differences may explain some of the conflicting data in the present Listeria literature (Baldridge, *et al.*, 1990; Orme, 1989).

Transfer of anti-*Salmonella* immunity with an infusion of purified T-cells into naive syngeneic recipients has been difficult to achieve using the murine model. This may be partly due to the high susceptibility of most mouse strains to *S. typhimurium* or *S. enteritidis* challenge (Hormaeche, *et al.*, 1985). Better success was achieved using inbred rats (Hougen and Jensen, 1990), possibly because they were substantially more resistant to salmonellosis than mice. Successful transfers have been reported recently in BALB/c mice infused with immune lymph node and peritoneal T-lymphocytes harvested from *Salmonella*-infected donors (Paul, *et al.*, 1985). Syngeneic recipients of 10^6 cloned immune T-cells resisted challenge with virulent *S. enteritidis* both in terms of reduced growth within the liver and spleen and by an increase in survival time.

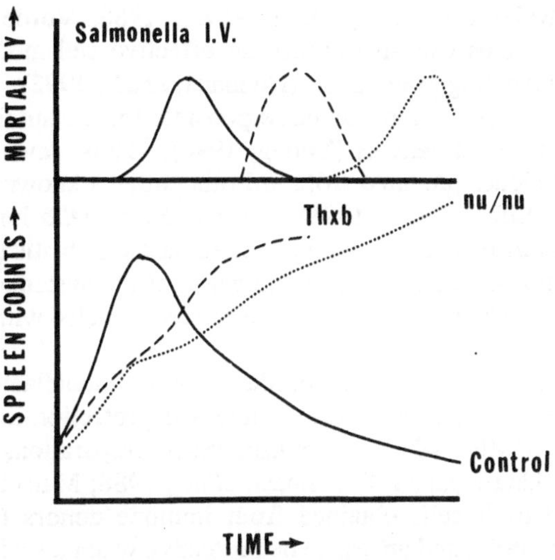

Figure 3. Diagram showing *S. enteritidis* growth in the spleens of athymic (nu/nu), thymectomized (Thxb) and sham thymectomized control mice, together with their corresponding mortally curves (top).

Most of these transfer studies have been carried out using intravenously or intraperitoneally challenged mice. Passive protection studies should also be carried out in orally challenged mice, which could yield important new information about the T-cell defenses operating at the mucosal level (MacDonald and Carter, 1980). Such information will also be essential to the development of new immunotherapeutic strategies in the treatment of *Salmonella*-infected AIDS patients.

SALMONELLA IMMUNOGENS AND IMMUNITY

Live attenuated *Salmonella*are more effective as vaccines than killed suspensions, whether against a parenteral and an oral challenge (Collins and Carter, 1974; George, *et al.*, 1987). This increased effectiveness has generally been ascribed to the ability of the viable organisms to produce larger quantities of protective sensitins within the tissues than will be present in a killed vaccine. The slow, steady release of these immunogens may result in the development of a cell-mediated rather than a humoral response of the type normally associated with dead vaccines (or their sonic extracts). These dead preparations produce cellular immune responses only when presented in a suitable adjuvant (Collins, 1987). It is not clear how much of this difference is quantitative vs. qualitative in nature. Virulent organisms will multiply extensively within the host tissues, leading to the release of large quantities of the protective antigen(s). On the other hand, attenuated organisms will multiply less extensively, producing more limited amounts of the protective antigen. Finally, avirulent organisms will be unable to multiply *in vivo* at all and so will induce little or no protective immunity despite the presence of the necessary protective antigen(s) within the bacterial cell (Gahring, *et al.*, 1990).

There is increasing evidence that these protective antigens are 'stress' proteins produced by the vaccinating organism in response to the inimical growth conditions encountered within the phagosome (Young, *et al.*, 1988). These protective proteins can also be recovered from culture filtrates of *in vitro*-stressed organisms, some of which are recognized by sensitized T-cells harvested from convalescent mice (Collins, *et al.*, 1988; Vordermeier and Kotlarski, 1990). Some of these proteins can also induce an effective cell-mediated immunity when presented in a suitable adjuvanting preparation (Hubbard, *et al.*, 1992). Not all of these stress proteins are beneficial, however. Some may be responsible for the autoimmune complications seen in some chronically infected patients (Young, 1990). Thus, several of these heat-shock proteins react with T-cell clones prepared from arthritis patients known to have been infected earlier with *Salmonella* (Life, *et al.*, 1991). Some *Salmonella*-infected rats develop a progressive type of polyarthritis resembling human rheumatoid arthritis (Volkman and Collins, 1973). These lesions do not arise during the acute phase of the systemic infection but may be induced by stress proteins which interact with sensitized T-cells within the synovial fluid (Young, 1990).

The nature of the protective antigens produced by Salmonellae has been a matter for some controversy. Some investigators reported increased protection in mice vaccinated with outer membrane proteins (OMP), while others used porin preparations from *S. typhimurium* (Udhayakumar and Muthukkaruppan, 1987; Isibasi, *et al.*, 1988; Matsui, *et al.*, 1990). These proteins were recognized by T-cells obtained from immune donors (Young, *et al.*, 1989; Vordermeier and Kotlaski, 1990) and appear to be protective when tested in suitably vaccinated mice (Foulaki *et al.*, 1989).

Development of gene cloning techniques now makes it possible to transfer the genes which are responsible for the production of these 'protective' antigens into suitable vectors using a γgt11 expression library (Warren, *et al.*, 1990). The resulting *E. coli* recombinants are first screened for the presence of specific proteins using T-cell clones prepared from *Salmonella*-immune donors. These proteins, present in culture filtrates (and whole cell sonicates) can be separated by electrophoresis and then transferred onto nitrocellulose sheets where they can be visualized by immunostaining with patient serum or mouse monoclonal antibodies (Warren, *et al.*, 1990; Young, *et al.*, 1986). The separated proteins can be solubilized (Abou-Zied, *et al.*, 1987) and tested for their ability to stimulate T-cell lines or clones using a standard lymphocyte blast assay. Finally, the gene(s) responsible for the synthesis of the most protective sensitins can be transferred to a suitably attenuated 'carrier' vaccine using the type of protocol described recently by Curtiss, *et al.*, (1989). These recombinant DNA vaccines may provide new immunotherapeutic and prophylactic tools needed to combat these important human pathogens in the severely immunodepleted host.

CONCLUSION

Our understanding of the role played by the somatic antigens of *S. typhimurium* during the induction of acquired resistance has undergone a number of important changes over the past 25 years or so. Specific antibodies directed against the polysaccharide antigens of the cell wall have both taxonomic and diagnostic importance, and their opsonic and bactericidal properties may help to protect the vaccinated host. However, most of these antibodies have produced relatively late in the infection, some time after the infection has been brought under control by the cellular defenses. This immune response is mediated by a population of sensitized T-cells which release a battery of lymphokines (IL-2, IL-4, IL-6, IFNγ, TNFα) which modulate the bactericidal activity of the macrophages as they enter the developing lesion from the bloodstream. These activated cells are able to kill (or at least prevent from growing) fully virulent organisms growing within the immune host. Different T-cell subsets are involved in the expression of delayed-type hypersensitivity, cell-mediated immunity and immune memory. At present, we know relatively little about the sensitins (or their epitopes) which ar responsible for triggering these different T-cell responses. However, as these immunogens become better defined, it should be possible to clone the genes responsible for their production, eventually ading to the development of new and more protective recombinant vaccines for human use. hus application of recent advances in the molecular biology of salmonellosis should lead to nproved prophylactic and immunotherapeutic reagents for use against these important enteric athogens, both in immunocompetent and immunosuppressed patients.

cknowledgements – Work described in this review was supported by Grants AI 14065 and I 27156 from the NIH, Bethesda.

EFERENCES

bou-Zied, C., Filey, E., Steele, J., and Rook, G.A.W., 1987, A simple new method for using antigens separated by polyacrylamide electrophoresis to stimulate lymphocytes *in vitro* after converting bands cut om western blots into antigen-bearing particles, J. Immunol. Methods, 98: 5.

aldridge, J.R., Barry, R.A., and Hinrichs, D.J., 1990, Expression of systemic protection and delayed-type hypersensitivity to *L. monocytogenes* is mediated by different T-cell subsets, Infect. Immun., 58: 654.

ierer, B.E., Sleckman, B.P., Ratnofsky, S.E., and Barakoff, S.J., 1989, The biologic roles of CD2, CD4, and CD8 in T-cell activation, Ann. Rev. Immunol., 7: 579.

lumenstock, E., and Jann, K., 1981, Natural resistance of mice to *S. typhimurium*: bactericidal activity and chemiluminescence response of murine peritoneal macrophages, J. Gen. Microbiol., 125: 173.

orn, W., Happ, M.P., Dallas, A., Reardon, C., Kubo, R., Shinnick, T.E., Brennan, P., and O'Brien, R., 1990, Recognition of heat-shock proteins and gamma-delta cell function, Immunol. Today, 11: 40.

uchmeier, N.A., and Heffron, F., 1991, Inhibition of macrophage phagosome-lysosome fusion by *S. typhimurium*, Infect. Immun., 59: 2232.

arter, P.B., and Collins, F.M., 1974, The route of enteric infection in normal mice, J. Exp. Med., 139: 1189.

arter, P.B., Woolcock, J.B., and Collins, F.M., 1975, Upper respiratory tract involvement in orally induced Salmonellosis in mice, J. Infect. Dis., 131: 570.

erottini, J.-C., and MacDonald, H.R., 1989, The cellular basis of T-cell memory, Ann. Rev. Immunol., 7: 77.

ollins, F., 1971, Mechanisms in antimicrobial immunity, J. Reticuloendothel. Soc., 10: 58.

ollins, f.M., 1979, Cellular antimicrobial immunity, Crit. Rev. Microbiol., 7: 27.

ollins, F.M., 1987, Immunological responses to microbial antigens by the infected host, In "Microbial Antigenodiagnosis," K. Wisher, ed., CRC Press, Florida.

ollins, F.M., 1990, Bacterial cofactors in AIDS, In "Cofactors in HIV-1 infection and AIDS," R.R. Watson, ed., CRC Press, Florida.

ollins, F.M., and Campbell, S.G., 1982, Immunity to intracellular bacteria. Vet. Immunol. Immunopathol., 3: 5.

ollins, F.M., and Carter, P.B., 1974, Cellular immunity in enteric disease, Am. J. Clin. Nutrition, 27: 1424.

ollins, F.M., Lamb, J.R., and Young, D.B., 1988, Biological activity of protein antigens isolated from *M. tuberculosis* culture filtrate, Infect. Immun., 56: 1260.

ollins, F.M., and Montalbine, V., 1976, Distribution of *in vivo*-grown *Mycobacteria* in the organs of intravenously infected mice, Am. Rev. Resp. Dis., 113: 281.

ollins, F.M., and Niederbuhl, C.J., 1983, Mechanisms of defense against intestinal pathogens: a review, In "Experimental Bacterial and Parasitic Infections," G. Keusch, T. Waldstrom, eds., Elsevier Press, ork.

urtiss, R., Nakayama, K., and Kelly, S.M., 1989, Recombinant avirulent *Salmonella* vaccine strains with stable maintenance and high level expression of cloned genes *in vivo*, Immunol. Invest., 18: 583.

annenberg, A.M., and Tomashefski, J.F., 1988, Pathogenesis of pulmonary tuberculosis, In "Pulmonary diseases and Disorders," Sec., Ed., McGraw-Hill, New York.

lory, C.M., Hubbard, R.D., and Collins, F.M., 1992, Effects of *in vitro* T-lymphocyte subset depletion on mycobacterial infection in mice, J. Leuk. Biol., 51: 225.

oulaki, K., Gruber, W., and Schlecht, S., 1989, Isolation and immunological characterization of a 55-kilodalton surface protein from *S. typhimurium*, Infect. Immun., 57: 1399.

ahring, L.C., Heffron, F., Finlay, B.B., and Falkow, S., 1990, Invasion and replication of *S. typhimurium* in animal cells, Infect. Immun., 58: 443.

eorge, A., Nair, R., Rath, S., Ghosh, S.N., and Kamat, R.S., 1987, Regulation of cell-mediated immunity in mice immunized with *S. enteritidis*, J. Med. Microbiol., 23: 239.

Havlir, D.V., Ellner, J.J., Chervenak, K.A., and Boom, W.H., 1991, Selective expansion of human gamma-delta T-cells by monocytes infected by live M. tuberculosis, J. Clin. Invest., 87: 729.

Hof, H., Emmerling, P., Hacker, J., and Hughs, C., 1982, The role of macrophages in primary and secondary infection of mice with *S. typhimurium*, Ann. d'Immunol., 133C: 21.

Hormaeche, C.E., Harrington, K.A., and Joysey, H.S., 1985, Natural resistance to Salmonellae in mice: control by genes within the major histocompatibility complex, J. Infect. Dis., 152: 1050.

Hormaeche, C.E., Maskell, D.J., Harrington, K.A., Joysey, H., and Brock, J., 1983, Mechanisms of natural resistance to mouse typhoid, Bull. Europ. Physiopath. Resp., 19: 137.

Hougen, H.P., and Jensen, E.T., 1990, Experimental *S. typhimurium* infections in rats. III. Transfer of immunity with primed lymphocyte subpopulations, Acta. Path. Microbiol. Immunol. Scand., 98: 1015.

Hubbard, R.D., Flory, C.M., and Collins, F.M., 1992, Immunization of mice with mycobacterial culture filtrate proteins, Clin. Exp. Immunol., 87: 94.

Isibasi, A., ortiz, V., Vargas, M., Paniagua, J., Gonzalez, C., Moreno, J., and Kumate, J., 1988, Protection against *S. typhi* infection in mice after immunization with outer membrane proteins isolated from *S. typhi*, 9,12,d,Vi, Infect. Immun., 56: 2953.

Izhar, M., DeSilva, L., Joysey, H.E., and Hormaeche, C.E., 1990, Moderate immunodeficiency does not increase susceptibility to *S. typhimurium* aroA live vaccines in mice, Infect. Immun., 58: 2258.

Janeway, C.A., Jones, B., and Hayday, A., 1988, Specificity and function of T cells bearing gamma/delta receptors, Immunol. Today, 9: 73.

Kaufmann, S.H.E., 1988, $CD8^+$ T lymphocytes in intracellular microbial infections, Immunol. Today, 9: 168.

Kagaya, K., Watanabe, K., and Fukazawa, Y., 1989, Capacity of recombinant gamma interferon to activate macrophages for *Salmonella*-killing activity, Infect. Immun., 57: 609.

Life, P.f., Bassey, E.O.E., and Gaston, J.S.H., 1991, T-cell recognition of bacterial heat-shock proteins in inflammatory arthritis, Immunol. Rev., 121: 114.

Lowrie, B.D., and Andrew, P.W., 1988, Macrophage antimycobacterial mechanisms. Br. Med. Bull., 44: 624.

MacDonald, T.T., and Carter, P.B., 1980, Cell-mediated immunity to intestinal infection, Infect. Immun., 28: 516.

Matsui, K., and Arai, T., 1990, Protective immunities induced by porins from mutant strains of *S. typhimurium*, Microbiol. Immunol., 34: 917.

O'Brien, A.D., 1982, Innate resistance of mice to *S. typhi* infection, Infect. Immun., 38: 948.

O'Brien, A.D., 1986, Influence of host genes on resistance of inbred mice to lethal infection with *S. typhimurium*, Curr. Topics Microbiol. Immunol., 124: 37.

O'Brien, A.D., and Metcalf, E.S., 1982, Control of early *S. typhimurium* growthin innately *Salmonella*-resistant mice does not require functional T lymphocytes, J. Immunol., 129: 1349.

Orme, I.M., 1987, The kinetics of emergence and loss of mediator T lymphocytes acquired in response to infection with *M. tuberculosis*, J. Immunol., 138: 293.

Orme, I.M., 1989, Active and memory immunity to *L. monocytogenes* infection inmice is mediated by phenotypically distinct T-cell populations, Immunol., 68: 93.

Orme, I.M., and Collins, f.M., 1986, Cross-protection against non-tuberculous mycobacterial infections by *M. tuberculosis*-memory immune T lymphocytes, J. Exp. Med., 163: 203.

Paul, C.C., Shalala, K., Warren, R., and Smith, R.A., 1985, Adoptive transfer of murine host protection to salmonellosis with T-cell growth-factor-dependent Salmonella-specific T-cell lines, Infect. Immun., 48: 40.

Sperber, S.J.,a nd Schenpuer, C.J., 1987, Salmonellosis during infection with human immunodeficiency virus, Rev. Infect. Dis., 9: 925.

Tilney, L.T., and Portnoy, d.A., 1989, Actin filaments and the growth and spread of the intracellular bacterial parasite, *L. monocytogenes*, J. Cell Biol., 109: 1597.

Turnbull, P.C., 1979, Food poisoning with special reference to *Salmonella* - its epidemiology, pathogenesis and control, Clin. Gastroenterol., 8: 663.

Udhayakumar, V., and Mathukaruppan, v.R., 1987, Protective immunity induced by outer membrane proteins of *S. typhimurium* in mice, Infect. Immun., 55: 816.

Volkman, A., and Collins, F.M., 1973, Polyarthritis associated with *Salmonella* infection in rats, Infect. Immun., 8: 814.

Vordermeier, H.M., and Kotlaski, I., 1990, Identification of antigens which stimulate T lymphocytes of *S. enteritidis* 11RX-immunized mice, Immunol. Cell. Biol., 68: 299.

Warren, R.L., Lu, D., Sizemore, D.R., Baron, L.S., and Kopecko, D.J., 1990, Method for identifying microbial antigens that stimulate specific lymphocyte responses: application to *Salmonella*, Proc. Natl. Acad. Sci. USA, 87: 9823.

oung, D.B., Kent, L., Rees, A., Lamb, J.R., and Ivanyl, J., 1986, Immunological activity of a 38-kilodalton protein purified from *M. tuberculosis*, Infect. Immun., 54: 177.
oung, D.B., Mehlert, A., Bal, V., Mendez-Samperio, P., Ivanyl, J. and Lamb, D.B., 1988, Stress proteins and the immune response to mycobacteria - antigens as virulence factors?, Ant. van Leeuweenhoek, 54: 431.
oung, R.A., 1990, Stress proteins and immunology, Ann. Rev. Immunol., 8: 401.

IMMUNITY MECHANISMS IN EXPERIMENTAL SALMONELLOSIS

Carlos E. Hormaeche[1], B. Villarreal[1], P. Mastroeni[1], G. Dougan[2] and S. N. Chatfield[3]

[1]Department of Pathology
Tennis Court Road
Cambridge CB2 1QP, UK

[2]Department of Biochemistry
[3]Medeva Group Research
Imperial College of Science, Technology and Medicine
Exhibition Road
London SW7 2AY, UK

Salmonellae are generally believed to be facultative intracellular parasites capable of growing inside professional phagocytes, although this view has been questioned (Hsu, this volume). Some salmonellae have a broad host range and can affect many species ("salmonellae of animal origin"), whereas others are more restricted and cause disease in only one or a few hosts, such as *S. typhi*, the agent of human typhoid fever, which affects only humans. The latter tend to cause invasive disease (e.g. enteric fever), whereas human infections with "animal" salmonellae - which can be invasive in animals - can be localised to the gut (salmonella gastroenteritis).

There is a need for improved vaccines for invasive salmonellosis of man and animals. Vaccines consisting of killed whole organisms administered parenterally have been in use for many years, but these are excessively reactogenic and protection is not as high as might be desired. Modern preparations using purified antigens, e.g. the Vi antigen of typhoid fever, are less reactogenic than whole organisms and are giving promising results in human trials (Levine *et al.*, 1989).

There is evidence to suggest that cell mediated immunity is an important component of immunity to invasive salmonellosis, and this is best elicited by live attenuated vaccines rather than by inactivated preparations. In humans, the Ty21a *galE S. typhi* vaccine for typhoid fever has been in use for some years. Several newer vaccine mutants (e.g. *aro* and *cya crp* mutants) are under investigation (Dougan *et al.*, this volume). Aro salmonella vaccines are effective in mice, cattle, sheep and chickens, and *S. typhi aroC aroD* vaccines are undergoing phase 1 trials in humans with promising results (Tacket *et al.*, 1992). A further development of live salmonella vaccines is their use as carriers of recombinant antigens from other pathogens (combined vaccines), which are showing great promise in experimental models (Hormaeche, 1991).

Not all live attenuated salmonellae are good vaccines. For reasons that remain totally unclear, live vaccines seem to have to be able to persist in the tissues for a limited period in order to confer high - level protection against challenge with virulent organisms. Even then, some vaccines confer better protection than others. Unfortunately, the great advances in molecular genetics which have allowed the development of live vaccines rationally attenuated by the introduction of safe, non-reverting mutations in defined loci have not been matched as yet by comparable advances in understanding the mechanisms by which salmonellae confer immunity.

There is still some disagreement regarding the relative roles of humoral *vs.* cell mediated immunity in protection. The nature of the protective immunogen(s) is uncertain, and the strain requirements for an effective vaccine are unknown. These are serious drawbacks to rational vaccine development because although several defined mechanisms of strain attenuation are being investigated, selection of candidate vaccine strains is empirical, evaluation of the immune status of a population is difficult, and the reasons for the success or failure of new candidate vaccines is often unexplainable.

The agent of human typhoid fever, *S. typhi*, will not cause invasive disease in laboratory rodents under normal conditions, and much experimental work is done on mouse typhoid, considered to have many points in common with human typhoid fever. It is with the mechanisms of immunity in this laboratory model that this brief review is mainly concerned.

MECHANISMS OPERATING IN PRIMARY SALMONELLOSIS

Attenuated salmonellae causing sublethal infections can confer solid protection from subsequent challenge with virulent organisms, and the analysis of such infections is giving many clues as to the mechanisms of immunity. Salmonellae grow in the RES, and the course of the infection can be followed by viable counts on homogenates of liver and spleen.

The early work of Frank Collins (1974) showed that a sublethal infection in laboratory mice infected intravenously proceeds in (at least) four distinct phases, schematically depicted in Fig. 1: 1) initial inactivation of a large fraction of the challenge, 2) exponential growth in the RES over the first week, 3) a plateau phase in which growth is suppressed, and 4) clearance of the organisms from the RES. With some (but not all) salmonellae, this process can leave solid immunity to rechallenge.

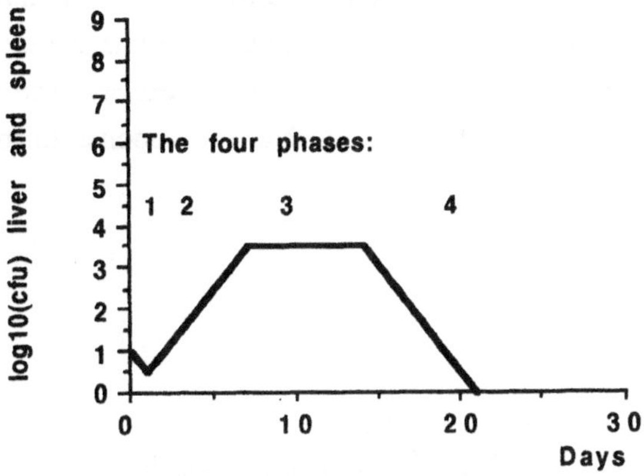

Figure 1. The four phases of a sublethal salmonella infection in mice.

Phase 1: Initial inactivation of the challenge

This is a constant finding in mice infected i/v; the mechanism is unclear. It can be further increased by treatments leading to macrophage activation. The growth phase of the salmonellae can also affect the degree of initial inactivation, log - phase organisms dropping further than stationary phase bacteria (Benjamin *et al.*, 1986). It is becoming increasingly clear that organisms growing *in vivo* are phenotypically different from those growing in culture, and it may be that there is a transition phase from the *in vitro* to the *in vivo* growth modes during which salmonellae are more susceptible to host defences (Dougan *et al.*, this volume). Antibody can also markedly increase this initial inactivation, even for serum resistant salmonellae, possibly by promoting killing by phagocytes. This effect of antibody can be sufficient to reduce a potentially lethal inoculum to a sublethal dose, and this may be one of the ways in which antibody can protect in salmonellosis (see below).

Phase 2: Exponential growth in the RES

With infections caused by invasive salmonellae, cfu counts in the RES increase exponentially in the first week. The net growth rate is roughly 0.5 - 1.5 log/day depending on the host - parasite combination, with true *in vivo* doubling times estimated at approx. 2 -

5 hours. Killing rates are also slow, the whole process appearing rather sluggish compared to growth in broth (Hormaeche 1980; Benjamin *et al.*, 1990).

Three main factors affect the course of phase 2: inoculum dose, virulence of the organism, and innate resistance of the host. Infections resulting from increasing doses are schematically depicted in Fig 2 (adapted from Collins, 1971). The course of the infection is substantially the same, but transposed to a higher level, until a lethal dose is reached. Doses just above the LD50 progress initially in the same way as sublethal infections, but phase 3 is absent: bacterial numbers continue to increase past the phase 3 inflection point of the curve until "lethal numbers" of approx. 10^8 - 10^9 are reached. The precise reason why high doses overwhelm the host defence mechanism responsible for phase 3 remains unknown.

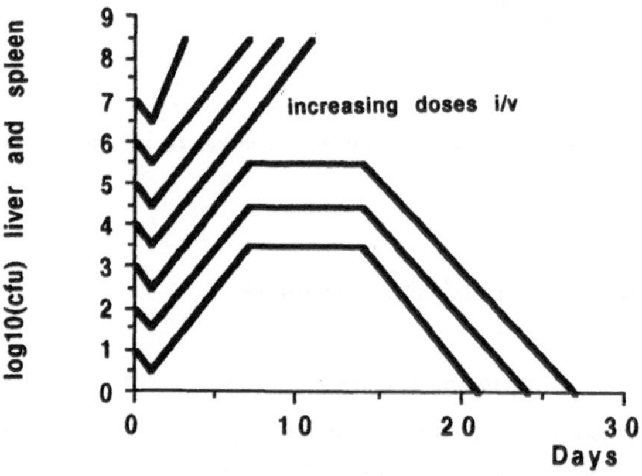

Figure 2. The course of infections resulting from increasing doses of salmonellae.

Increasing doses thus cause proportionally shorter times to death, except when doses above 10^6 - 10^7 are inoculated. With high doses, net growth rate accelerates and time to death becomes much shorter. This accelerated mortality with high doses can be mimicked using low inocula incorporated with large numbers of dead salmonellae (Hormaeche, 1990). The effect is acceleration of early net growth rate in a manner similar to that observed following administration of antimacrophage agents such as silica. Endotoxin will produce the same effect. The mechanism is unclear, but clearly indicates that the course of an infection following parenteral administration of large numbers of organisms is very different to that of an infection commencing from the more usual lower doses.

The course of infections with salmonellae of increasing virulence is depicted in Fig. 3. Whereas some non-virulent salmonellae can be cleared from the RES without any increase in cfu, more invasive organisms always "take" and increase in numbers. More virulent bacteria show faster *in vivo* net growth rates, until a degree of virulence is reached at which even low doses will grow exponentially without a plateau causing death. As with the effect of increasing doses described above, bacteria causing lethal infections appear to overwhelm the undefined host resistance mechanism early in the infection.

A host innate resistance gene with a major effect in phase 2 controls resistance to salmonellae, *Leishmania donovani*, *Mycobacterium tuberculosis* BCG and *Mycobacterium*

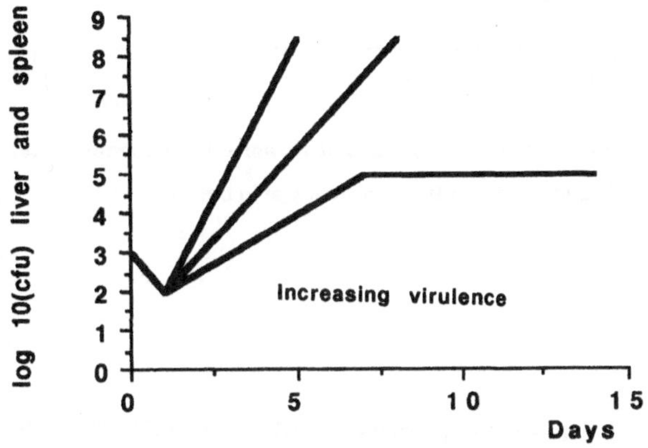

Figure 3. The course of infections resulting from salmonellae of increasing virulence.

lepraemurium. Variously designated *Ity*, *Lsh*, *Bcg* and *Inr* (Blackwell, 1989), it is situated on chromosome 1, and has resistant (^{r}) and susceptible (^{s}) alleles, resistance being dominant. *Ity*^{s} mice are more susceptible than *Ity*^{r} mice, allowing faster net growth rate of the salmonellae in the RES. With some salmonellae, the slower net growth rate in resistant mice allows the plateau to develop enabling survival (Fig. 4). The *Ity* effect appears to be expressed through macrophages. Administration of antimacrophage agents *in vivo* will accelerate net growth rate, and cultured macrophages (independently of lymphocytes) from resistant mice can deal with salmonellae more efficiently than macrophages from susceptible mice. The effect on salmonellae may be bacteriostatic or bactericidal. Several other host genes (*Lps, xid,* "the fourth gene" and others) modulate innate resistance to salmonellae in the early stages of the infection (Blackwell, 1988).

Phase 3: the plateau phase

This early suppression of exponential bacterial growth must occur for the host to survive. The mechanism is still uncertain, but it is becoming clear that this phase, which represents the host's first strong suppression of the infection, is not the result of a response in which T-cells are a major determinant. T-cell deficient mice can still mount a plateau (O'Brien and Metcalf, 1982). Experiments using whole-body irradiation, with or without shielding of the hind legs, and adoptive transfer of normal T-cell depleted bone marrow,

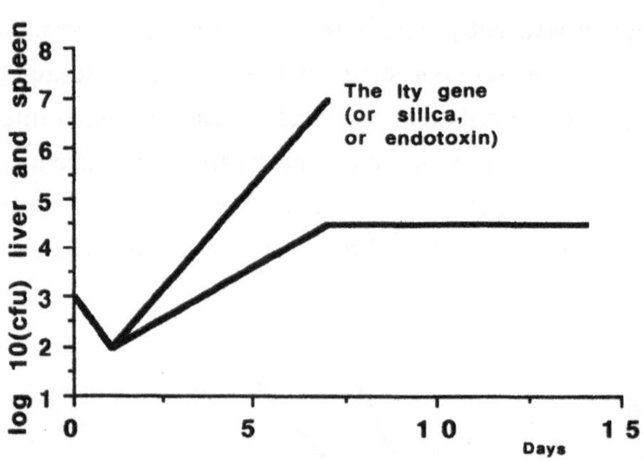

Figure 4. The course of infections in innately resistant *vs.* innately susceptible mice, similar to the effect of silica or endotoxin: aceleration of the early net growth rate in the RES.

showed that phase 3 requires the presence of radiation sensitive, non-T bone marrow cells. Adoptive transfer of normal spleen cells will not restore the plateau in irradiated mice (Hormaeche *et al*, 1990).

We further showed that two concurrent sublethal infections with tagged variants of the same organism started three days apart reach a plateau dependently of one another. This result is difficult to explain in terms of a systemic (e.g. antibody) response being the major determinant suppressing early bacterial growth, but suggests that some local effect may be operating (Maskell *et al.*, 1987).

TNFα and IFNγ are important mediators of phase 3 in salmonella infections (Mastroeni *et al.*, 1991; Muotiala and Makela, 1990). Administration of antibodies to block either TNFα or IFNγ will abolish the plateau phase, with bacteria continuing to grow unchecked. The cells responsible for this effect have not been pinpointed in salmonella models. However, results from other workers using the *Listeria* model indicate that the mechanism responsible for the early suppression of *Listeria* infections involves TNFα release by macrophages which trigger IFNγ production by NK cells (Bancroft *et. al.*, 1989). NK cells would be expected to be present in mice reconstituted with the T-cell depleted bone marrow used to restore the plateau in irradiated mice in our salmonella experiments described above. It may be that a similar mechanisms to that described for *Listeria* also obtains in salmonella infections, but this remains to be demonstrated.

We have also observed that administration of anti-TNFα antibodies will exacerbate a salmonella infection after the plateau has been established to the point where a sublethal infection will become lethal, indicating that containment of the bacterial load in a sublethal infection is a dynamic process in which TNFα is continually required (our unpublished observations). However, we could not measure sustained levels of TNFα in serum of infected mice; rather, the presence of a TNFα inhibitor was detected (Mastroeni *et al.*, 1992).

During the plateau phase, mice still carrying appreciable numbers of salmonellae in their spleens show a manifest macrophage - mediated immunosuppression towards unrelated antigens; this is discussed in detail elsewhere (Eisenstein *et al.*, this volume).

Phase 4: the clearance phase

The plateau is normally followed after a variable time by clearance of the organisms from the RES. This phase clearly requires the presence of T-cells. Athymic mice show a progressive increase in cfu count in the RES in the weeks following the initial plateau (O'Brien and Metcalf, 1982), and *in vivo* T-cell depletion impairs clearance of attenuated salmonellae from the RES (Nauciel, 1990). Host genes are important in controlling this late clearance phase, and genes both within and outside the major histocompatibility complex modulate the rate of bacterial clearance (Hormaeche *et al.*, 1985; Nauciel, 1990).

The antigens which trigger the T-cell response which mediates the clearance of the organisms in a primary infection have not been identified. We found that H-2 congenic mice on a B10 background differed in late emerging resistance to salmonellae, with mice of H-2^b (B10) and H-2^d (B10.D2) haplotypes being more susceptible than H-2^a (B10.A), H-2^k (B10.Br) and H-2^s (B10.M) mice. We found that this haplotype distribution of resistance still obtained when the mice were tested using variant strains of *S. typhimurium* expressing either the native O4 LPS determinant of *S. typhimurium* or the O9 determinant of *S. enteritidis*. This result suggested that H-2 linked differences in late emerging resistance are not the result of differences in specific immune recognition of the main O antigen, as one might perhaps have expected a different haplotype distribution of resistance when using strains of different antigenic specificity (Hormaeche *et al.*, 1985).

IMMUNITY CONFERRED BY VACCINATION

Mice can be solidly protected from invasive salmonellosis by prior immunisation. Current trends in development of live vaccines are aimed at attenuating virulent strains in such a way that the rapid growth in the RES during phase 2 is much reduced, allowing the development of the host immune response. Several ways of attenuating virulent salmonellae have been described with much attention being given to strains with mutations in genes of the aromatic pathway (Stocker *et al.*, Dougan *et al.*, Levine *et al.*, this volume). Such *aro* salmonellae grow very slowly *in vivo* due to their requirement for *p*-amino benzoic acid which is absent in mammalian tissues. We are using *aro* mutants extensively in our studies on the mechanisms of immunity in experimental salmonellosis. The total bacterial load in the RES rarely exceeds the inoculum dose and they are cleared slowly, leaving potent cellular immunity and solid protection to rechallenge. For reasons that remain unknown, strains with "overattenuating" lesions such as *purA* which persist poorly in the RES do not protect (O'Callaghan *et al.*, 1988). The strain requirements for an effective vaccine remain unknown, and selection of effective vaccines remains empirical.

Mechanisms of acquired immunity

It is generally agreed that both humoral and cellular mechanisms are involved in protection. Live vaccines are generally more effective than killed vaccines; the reasons for this difference in the protective capacity of live and killed vaccines are not entirely clear, but may be due to the ability of live vaccines to confer cell mediated immunity in addition to humoral immunity. Killed vaccines can protect innately resistant mice better than susceptible mice; the latter require immunisation with living attenuated organisms in order to be protected from oral challenge with virulent organisms (Eisenstein *et al.*, 1984).

It is unclear whether the failure of killed vaccines to protect innately susceptible mice represents a fundamental inability of these animals to respond in some way to killed organisms whereas resistant mice can respond, or whether it is simply due to the greater invasiveness of virulent organisms in susceptible mice (cf. Fig 4). It may be that killed vaccines would protect innately susceptible mice if they were challenged with salmonellae of intermediate virulence which could be partially controlled by the host in the way that resistant mice can deal with more virulent organisms, but this possibility has not been investigated.

Much work is currently being done on the mechanisms of protection conferred by live vaccines, as these will protect both innately resistant and susceptible animals from oral challenge with virulent organisms. Again, the degree of protection conferred by live vaccines is dependent on the experimental protocol. The highest degree of protection is seen in animals which are still carrying the live vaccines in the RES. Animals which are still carrying the vaccine one or two weeks after immunisation are virtually refractory to reinfection, and immunity is non-specific, such that immunisation with one serotype will protect against challenge with salmonellae of different serotypes and unrelated organisms such as *Listeria* (O'Callaghan *et al.*, 1988). However, the degree of protection displayed after the challenge has been cleared - i.e. recall of immunity to rechallenge - is not as high as in carriers, and in some models is serotype specific (Hormaeche *et al.*, 1991).

In susceptible mice vaccinated with *aroA* salmonellae, protection is greater to oral challenge than to i/v challenge. We found that innately susceptible BALB/c mice vaccinated i/v with *aroA S. typhimurium* and challenged i/v two to three months later with virulent salmonellae were protected as measured by an increase in the LD50 of vaccinated *vs.* naive mice. However, the dose-response was irregular with vaccinated mice dying over a wide dose range indicating that protection to i/v challenge was somehow only partial. In contrast, when mice similarly vaccinated i/v were challenged orally, much better protection was observed, with the LD50 increasing by three to four logs without the scattered deaths seen with i/v challenge. The reason for this difference is unclear, but could be taken to indicate that, for this particular host-parasite combination, passage of the challenge organisms through the lymph nodes in some way favours resistance (Hormaeche *et al.*, 1991).

Mediators of acquired immunity

Both antibody and cell mediated mechanisms have been shown to confer protection in different mouse models. Antibody can confer protection, but the degree of protection depends on the particular experimental model. Antibody is especially effective when mice are challenged parenterally, and protection can be O-specific. One of the effects of passively administered antibody is to increase the initial inactivation of the challenge in phase 1 of the infection. In moderately "resistant" host-parasite combinations, this can reduce a challenge which is lethal, but does not greatly exceed the LD50, to sublethal levels. This is especially true when virulent organisms are inoculated i/p in naive mice; whereas salmonellae increase rapidly in the peritoneal cavity immediately after challenge, antibody can prevent this rapid outgrowth and increase the LD50. Antibody seems to have less effect in a primary infection after it has become established and the organisms have started to increase exponentially, perhaps because the organisms are already within phagocytes and less accessible to the effect of circulating antibody.

Conversely, attempts to transfer immunity adoptively with cells alone have not uniformly met with success. As with antibody transfer, adoptive transfer of protection using immune cells has been demonstrated for parenteral challenge, and protection has been obtained using T-cell lines, B cells and T cells (Paul *et al.*, 1985; Hochadel and Keller, 1977). A role for both CD4 and CD8 T-cells has been demonstrated in adoptive transfer of immunity to *S. abortusovis* in mice challenged parenterally (Bernard *et al.*, this volume). Adoptive transfer of immunity to oral challenge has not yet been shown.

The specific recall of immunity and protection to oral challenge in susceptible mice after the live vaccine has been cleared from the RES requires the participation of T-cells; TNFα and IFNγ are also involved in expression of resistance in immune mice. Administration of anti CD4 and CD8 monoclonal antibodies seriously impairs protection in vaccinated mice; this impairment is greatest when both CD4 and CD8 monoclonals are given together, indicating that both T-cell subsets participate in immunity (Mastroeni *et al.*, this volume; Nauciel *et al.*, 1990). This is in contrast to the events occurring in the early plateau (phase 3 of the infection), in which we have been unable to demonstrate a decisive role for T-cells. Conversely, TNFα and IFNγ also participate in the specific recall of vaccine induced immunity as is the case in the plateau phase of a primary infection. Administration of antisera to either cytokine impairs recall of immunity in vaccinated mice (Mastroeni *et al.*, Nauciel *et al.*, this volume; Tite *et al.*, 1991).

However, histological examination of the spleens of immunised mice depleted of either T-cells, TNFα or IFNγ and challenged with virulent organisms shows very different pictures. Anti-TNFα antiserum leads to scarce mononuclear infiltration of the RES with little or no granuloma formation. This contrasts with the enlarged spleens and livers and the gross mononuclear infiltrate which is seen when mice are depleted of T-cells or treated with anti-IFNγ antisera. The results indicate that TNFα is somehow involved in recruitment of monocytes to the RES, whereas T-cells and IFNγ appear to be involved in the expression of antibacterial immunity by the recruited cells (Mastroeni *et al.*, this volume).

It therefore appears that the mediators of acquired immunity have points in common with those operating in the primary response, with TNFα and IFNγ playing a central role. Current evidence from our laboratory would indicate that one of the deciding factors responsible for the greatly increased resistance of vaccinated *vs.* naive mice is the additional contribution of primed T-cells.

Specificity of protection

The antigens involved in specific recall of immunity are unclear, and evidence has been presented for both protein and LPS antigens being the main determinants of immunity. If protection depends on recall of cell mediated immunity, it is *a priori* difficult to see how LPS (polysaccharide) antigens could be involved in this arm of the response as T-cells generally recognise protein antigens in association with self, rather than polysaccharide determinants. In cross-challenge experiments with *S. typhimurium* and *S. enteritidis* vaccines, using as challenge strains defined variants of these serotypes engineered to express either the *S. typhimurium* or the *S. enteritidis* O-antigen, we found evidence for a strong element of serotype (background) specificity in recall of immunity rather than immunity being defined by the main LPS O-specificity (Hormaeche *et al.*, 1991). Similar observations have been recently made following vaccination of chickens and calves (Cooper *et al.*, Lindberg *et al.*, this volume).

If specific recall of immunity following vaccination requires a cell mediated response, it should be possible to analyse the nature of this response and the antigens involved by conventional *in vitro* and *in vivo* studies of T-cell responses. There is evidence to indicate that in some way polysaccharide antigens could be involved in cell mediated responses to salmonella antigens. DTH-like responses to polysaccharide salmonella antigens have been seen in calves vaccinated with *aro* salmonellae (Robertsson *et al.*, 1982), and proliferative responses to salmonella polysaccharide antigens have been obtained using peripheral blood leukocytes from human typhoid cases and vaccinees (Murphy *et al.*, 1987).

It is however by no means clear that a proliferative response to crude salmonella antigens containing LPS can be taken as a measure of (T-cell) cellular responses. In mice, LPS is by itself a potent B-cell mitogen; this effect is not seen with human cells. However, if vaccination induces an antibody response, cell suspensions containing LPS immune B cells could produce a specific proliferative response towards LPS, a T-independent antigen. We have found that whereas lymphoid cells (including T-cell enriched populations) from mice vaccinated with *aroA* salmonellae will give an *in vitro* proliferative response to crude salmonella antigens and also to purified salmonella LPS, only the crude (protein-rich) extracts and not purified LPS will elicit the IL2/IL4 response which would be expected from a CMI response. The results suggest that the antigen responsible for the cell mediated response in the mouse model is probably protein in nature (Villarreal *et al.*, this volume). The identity of the determinant(s) responsible for the specific recall of immunity following vaccination remains, however, uncertain.

Correlates of immunity

The correlates of immunity to salmonellae following vaccination are not well defined. Mice vaccinated with both live and killed vaccines will develop circulating antibodies, but as mentioned above antibody alone is not a measure of protection. Delayed type hypersensitivity reactions have been extensively used and they are known to be elicited by vaccination with live but not killed organisms (Collins 1974). However, DTH reactivity does not correlate well with immunity in the mouse model, as animals can show immunity in the absence of DTH reactivity (Hormaeche *et al.*, 1981; Killar and Eisenstein, 1986). The very nature of the skin test can also be misleading if the antigen used contains LPS. We have previously reported that salmonella "elicitins" enriched for protein content can contain appreciable amounts of O-specific material (Hormaeche, 1979). We observed that footpad DTH reactions in response to such elicitins can consist of a cellular infiltrate which contains an abundance of PMNs, not normally to be expected in "true" DTH reactions but more reminiscent of Arthus or Shwartzman type reactions. It is difficult to correlate these complex reactions with immunity, and a better test of cellular immunity is needed.

SUMMARY AND CONCLUSIONS

Live attenuated salmonella vaccines confer excellent protection in experimental models. The mechanisms of immunity remain undefined, but it is becoming increasingly clear that cell mediated mechanisms play a fundamental role in protection. In primary responses a major determinant of the outcome of the infection depends on the interplay of the host-parasite relationship which determines the rate at which the initial infection progresses and the ability of the host to suppress exponential growth in the RES at the end of the first week of the infection, with genetic control of bacterial virulence on the one hand and of host resistance on the other playing a crucial role. Acquired immunity is maximal following immunisation with live vaccines which elicit cell mediated immunity. TNFα and IFNγ are central in control of both primary and secondary infections, with T-cell (CD4 and CD8) mediated immunity being a deciding factor in increased resistance following vaccination. The specificity of recall of immunity may be determined by protein rather than polysaccharide antigens, although the nature of the protective immunogen(s) and the correlates of protection remain to be defined.

REFERENCES

Bancroft, J., Sheehan, K. C. F., Schreiber, R. D. E., and Unanue, E. R., 1989, Tumor necrosis factor is involved in the T-cell independent pathway of macrophage activation in mice, *J. Immunol.* 143: 127.

Benjamin, W.H., Jr., Posey, B.S., and Briles, D.E., 1986, Effects of *in vitro* growth phase on the pathogenesis of *Salmonella typhimurium* in mice, *J. Gen. Microbiol.* 132: 1283.

Benjamin, W. H. Jr., Hall, P., Roberts, S.J., and Briles, D. E., 1990, The primary effect of the *Ity* locus is on the rate of growth of *Salmonella typhimurium* that are relatively protected from killing, *J. Immunol.* 144: 3143.

Blackwell, J.M.*et al.*, 1989, 27th. Forum in Immunology: The Macrophage Resistance Gene *Lsh/Ity/Bcg*, *Res. Immunol* 140: 767.

Blackwell, J. M., 1988, Bacterial infections, *in* "Genetics of Resistance to Bacterial and Parasitic Infection", D. M. Wakelin and J. M. Blackwell, eds., Taylor and Francis, London.

Collins, F.M., 1971, Mechanisms in antimicrobial immunity, *J. Reticuloendothel. Soc.*, 10: 58.

Collins, F.M., 1974, Vaccines and cell mediated immunity, *Bact. Revs.* 38: 371.

Eisenstein, T. K., Killar, L. M., and Sultzer, B. M., 1984, Immunity and infection in *Salmonella typhimurium*: mouse-strain differences in vaccine and serum-induced protection, *J. Infect. Dis.* 150: 425.

Hochadel, J. F., and Keller, K. F., 1977, Protective effects of passively transferred immune T- or B-lymphocytes in mice infected with *Salmonella typhimurium*, *J. Infect. Dis.* 135: 813.

Hormaeche, C. E., 1979, Natural resistance to *Salmonella typhimurium* in different inbred mouse strains, *Immunology* 37: 311.

Hormaeche, C.E., 1980, The *in vivo* division and death rates of *Salmonella typhimurium* in the spleen of naturally resistant and susceptible mice measured by the superinfecting phage technique of Meynell, *Immunology* 37: 329.

Hormaeche, C.E., 1990, Dead salmonellae or their endotoxin accelerate a salmonella infection in mice, *Microb. Pathogen.* 9: 213.

Hormaeche, C.E., 1991, Live attenuated salmonella vaccines and their potential as oral combined vaccines carrying heterologous antigen, *J. Immunol. Methods* 142: 113.

Hormaeche, C. E., Fahrenkrog, M. C., Pettifor, R. A., and Brock, J., 1981, Acquired immunity to *Salmonella typhimurium* and delayed (footpad) hypersensitivity in BALB/c mice, *Immunology*, 43: 569.

Hormaeche, C. E., Harrington, K. A., and Joysey, H. S., 1985, Natural resistance to salmonellae in mice:

control by genes within the major histocompatibility complex, *J. Infect. Dis. 152: 1050.*

Hormaeche, C. E., Mastroeni, P., Arena, A., Uddin, J., and Joysey, H. S., 1990, T-cells do not mediate the initial suppression of a salmonella infection in the RES, *Immunology* 70: 247.

Hormaeche, C. E., Joysey, H. S., Desilva, L., Izhar, M., and Stocker, B. A. D., 1991, Immunity conferred by Aro-salmonella live vaccines, *Microb. Pathogen.* 10: 149.

Killar, L. M., and Eisenstein, T. K., 1986, Delayed-type hypersensitivity and immunity to *Salmonella typhimurium*, *Infect. Immun.* 52: 504.

Levine, M.M., Ferreccio, C., Black, R.E., Tacket, C.O., Germanier, R., and the Chilean Typhoid Committee, 1989, Progress in vaccines against typhoid fever, *J. Infect. Dis.* 11(Supp 3):S552.

Maskell, D. J., Hormaeche, C. E., Harrington, K. A., Joysey, H. S., and Liew, F. Y., 1987, The initial suppression of a salmonella infection in the RES is mediated by a localised rather than by a systemic response, *Microb. Pathogen.* 2: 295.

Mastroeni, P., Arena, A., Costa, G. B., Liberto, M. C., Bonina, L., and Hormaeche, C. E., 1991, Serum TNFα in mouse typhoid and enhancement of the infection by anti-TNFα antibodies, *Microb. Pathogen.* 11:33.

Mastroeni, P., Villarreal, B., Demarco de Hormaeche, R., and Hormaeche, C. E., 1992, Serum TNFα inhibitor in mouse typhoid, *Microb. Pathogen.*, in press.

Muotiala, A., and Makela, P. H., 1990, The role of IFNγ in murine *Salmonella typhimurium* infection, *Microb. Pathogen.* 8: 135.

Murphy, J. R., Baqar, S., Muñoz, C., Schlesinger, L., Ferreccio, C., Lindberg, A. A., Svenson, S., Losonsky, G., Koster, F., and Levine, M. M., 1987, Characteristics of humoral and cellular immunity to *Salmonella typhi* in residents of typhoid-endemic and typhoid-free regions, *J. Infect. Dis.* 156: 1005.

Nauciel, C., 1990, Role of $CD4^+$ T cells and T-independent mechanisms in acquired resistance to *Salmonella typhimurium* infection, *J. Immunol.* 145: 1265.

Nauciel, C., Ronco, E., and Pla, M., 1990, Influence of different regions of the *H-2* complex on the rate of clearance of *Salmonella typhimurium*, *Infect. Immun.* 58: 573.

O'Brien, A. D., and Metcalf, E. S., 1982, Control of early *Salmonella typhimurium* growth in innately salmonella-resistant mice does not require functional T-lymphocytes, *J. Immunol.* 129: 1349.

O'Callaghan, D., Maskell, D., Liew, F. Y., Easmon, C. S. F., and Dougan, G., 1988, Characterization of aromatic- and purine-dependent *Salmonella typhimurium*: attenuation, persistence, and ability to induce protective immunity in BALB/c mice, *Infect. Immun.* 56: 419.

Paul, C.K., Shalala, R., Warren, R., and Smith, R., 1985, Adoptive transfer of murine host protection to salmonellosis with T-cell growth factor dependent, *Salmonella*-specific T-cell lines, *Infect. Immun.* 48: 40.

Robertsson, J. A., Svenson, S., B., and Lindberg, A. A., 1982, *Salmonella typhimurium* infection in calves: delayed specific skin reactions directed against the O-antigenic polysaccharide side chain, *Infect. Immun.* 37: 737.

Tacket, C.O., Hone, D.M., Losonsky, G.A., Guers, L., Edelman, R., and Levine, M.M., 1992, Clinical acceptability and immunogenicity of CVD908 *Salmonella typhi* vaccine strain, *Vaccine* 10: 443.

Tite, J. P., Dougan, G., and Chatfield, S. N., 1991, The involvement of tumour necrosis factor in immunity to *Salmonella typhimurium*, *J. Immunol.* 47: 3161.

HUMORAL IMMUNE RESPONSE TO *SALMONELLA TYPHI* OUTER MEMBRANE PROTEINS

Lieselotte Aron, Gustavo Faundez, Jesus Aguero,
Maria E. Fernandez-Beros, Carlos Gonzalez,
and Felipe Cabello

Department of Microbiology and Immunology
New York Medical College
Valhalla, NY, USA

INTRODUCTION

Typhoid is a febrile and bacteremic disease of intestinal origin characterized by low but persistent numbers of *Salmonella typhi* in the blood stream (Christie, 1987). The defervescence of the febrile disease in untreated typhoid coincides with the appearance of cell-mediated immunity, as manifested by the delayed type hypersensitivity phenomena of necrosis, intestinal perforation and hemorrhage (Christie, 1987; Hook, 1990). Experimental evidence obtained with *S. typhimurium* in mouse typhoid suggests that cell-mediated and humoral immunity play a role in the ability of the host to control and eradicate *Salmonella* infections (Collins and Mackaness, 1968; Saxen, et al, 1986; Brown and Hormaeche, 1989). However, the relevance of each arm of the immune response in the evolution and recovery from typhoid, as well as the *S. typhi* antigens responsible for these responses, remain uncharacterized (Edelman and Levine, 1986).

Some clinical aspects of typhoid indicate that antibodies may play a role in the evolution of the disease. An incubation time that fluctuates between 7 to 21 days (average 14) allows enough time for the synthesis of specific antibodies before the appearance of symptomatic disease (Hornick, et al., 1970; Forrest, et al., 1991). *S. typhi* can infect individuals having cross-reacting antibodies to its antigens from stimulation by normal flora and infections with other Gram negative bacteria (Aron, et al., 1992 a). This may explain in part the bactericidal effect of normal human sera for *S. typhi* (Muschel, et al., 1958; Jimenez-Lucho and Foulds, 1990). From the clinical standpoint, typhoid appears to be a rather atypical Gram-negative bacteremia (Edelman and Levine, 1986; Christie, 1987). Among the atypical characteristics found in typhoid are low numbers of bacteria in the blood stream (Watson, 1955), bacteremia with endotoxin-negative sera (Butler, et al., 1978), low frequency of septic shock (Hornick, et al., 1970), chronic bacteremia (Rocha, et al, 1971), benign bacteremia (Ferrechio, et al., 1984), low frequency of metastatic infections (Christie, et al., 1987), bactericidal activity of typhoid patients' sera

Biology of Salmonella, Edited by F. Cabello *et al.*,
Plenum Press, New York, 1993

(Nagington, 1956; Osawa and Muschel, 1964), and relatively low mortality (Christie, 1987). All of these characteristics could be explained by the presence of antibodies synthesized during the incubation period that are able to modulate the virulence of bacteria present in tissues and the blood (Christie, 1987). Moreover, immunoprotection against typhoid fever has been produced by acetone-killed whole-cell and Vi antigen vaccines suggesting a role for humoral immunity in protection and clinical evolution of typhoid fever (Edelman and Levine, 1986; Acharya, et al., 1987).

The human humoral immune response to *S. typhi* LPS and Vi antigen has been relatively well characterized in typhoid fever (Edelman and Levine, 1986; Robbins and Robbins, 1987). However, the humoral response to *S. typhi* outer membrane proteins has been less well studied (Brown and Hormaeche, et al., 1989; Fernandez-Beros, et al., 1989). While the ability of *Salmonella typhi* outer membrane proteins (OMPs) to stimulate cellular and humoral immune responses has been well established (Calderon, et al., 1986; Udhayakumar and Muthukkarupan, 1987; Matsui and Arai, 1989), the protective ability of these responses is much less characterized (Edelman and Levine, 1986). The complex nature of the Gram-negative outer membrane has hindered both isolation of purified outer membranes without contaminating LPS (Calderon, et al., 1986; Sharma, et al., 1989; Fernandez-Beros, et al., 1989) and identification of the host immune response to them (Saxen, et al., 1986). Moreover, the common presence on these bacteria of LPS and capsular antigens shielding the outer membrane structures raises questions about the ability of specific antibodies to interact with whole bacterial cells and activate antibacterial systems such as complement and phagocytosis (van der Ley, et al., 1986).

IMMUNOBLOT ANALYSIS OF HUMORAL IMMUNE RESPONSE TO *S. TYPHI* OMPs

Humoral immune responses in typhoid fever patients can be detected against *S. typhi* proteins using ELISA and immunoblot assays (Calderon, et al., 1986; Aguero, et al., 1987; Ortiz, et al., 1989; Fernandez-Beros, et al, 1989). Unfortunately, interpretation of these results has been difficult because convalescent sera contain high titers of anti-LPS antibodies (Calderon, et al., 1986). For example, Figure 1 shows that convalescent antisera react strongly with the *S. typhi* 36 kDal porin when this porin is present in the outer membranes of *S. typhi* (lane a), and that this reaction is minimal or absent when the same protein is expressed in the outer membranes of *E. coli* (lane b). This suggests that the reactivity observed when the porin is present in *S. typhi* may be due to antibodies able to recognize LPS epitopes that contaminate the porin or hybrid LPS-protein epitopes. Similar problems in analysis of human host humoral immune response against Gram-negative OMPs arise from their similar amino acid sequence and composition of these OMPs leading to the presence of cross reacting antibodies in human sera (Hofstra, et al., 1980; Singh, et al., 1992).

Figure 2 shows that convalescent sera from patients with typhoid contain antibodies able to react with iron starvation-induced proteins of *S. typhimurium* and *E. coli* indicating the presence of cross reacting antibodies against these bacteria in the sera of patients infected with *S. typhi* (Hofstra, et al., 1980; Aron, et al., 1992 a). Moreover, analysis of the immune response using immmunoblots is also problematic because the denaturation of proteins produces the loss of epitopes and generates new epitopes (Maurer and Callahan, 1980), confusing the interpretation of the relevance of OMP antibodies in host defense.

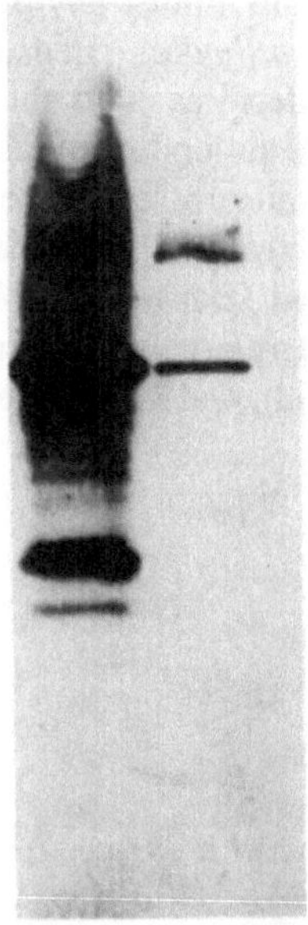

Figure 1. Immunoblot of the 36 kDa porin of *S. typhi* in *S. typhi* and *E. coli*. Immunoblot of outer membrane proteins of *S. typhi* Ty2 (lane a) and the porin-less *E. coli* UH302 expressing the *S. typhi* 36 kDa porin (lane b) separated in 12% SDS-PAGE and transferred to nitrocellulose filters, reacted with diluted typhoid fever sera and developed with anti-human IgG horseradish peroxidase conjugate (Aguero, et al., 1987). The arrow marks the position of the porin band.

RADIOIMMUNOPRECIPITATION ANALYSIS OF HUMORAL IMMUNE RESPONSE TO *S. TYPHI*

To confirm that the reactions that we and others have detected in immunoblots are really due to the presence of antibodies against OMPs, we decided to initiate the study of this immune response using radioimmunoprecipitation assays of *S. typhi* cells and cell membranes labelled with I^{125} and S^{35} methionine (Gulig, et al., 1982).

Labeling of S. typhi OMPs. Iodination of whole cells by lactoperoxidase treatment labels only OMPs exposed on the surface of the bacteria, while the metabolic label S^{35} methionine labels the total bacterial proteins (Hansen, et al., 1981; Gulig, et al., 1982; Aron, et al., 1992 a). Subsequently the cells are reacted with inactivated patients' and normal individuals' sera or their DEAE-cellulose purified IgG. Spheroplasts are produced, membrane vesicles are prepared with the French press and then separated by centrifugation in sucrose gradients (35 to 55%). Outer membrane vesicles are solubilized and immunoprecipitated with *Staphylococcus aureus* Cowan I strain or purified protein A. Precipitated proteins are separated by SDS-PAGE electrophoresis. Autoradiography identifies the immunogenic OMPs precipitated by antibodies present in the antisera.

This type of analysis is facilitated by the fact that most strains of *S. typhi* belong to a single clone (Reeves., et al., 1989; Faundez, et al., 1990), and exhibit a very well conserved pattern of OMPs on SDS-PAGE (Aron, et al., 1992 a). The study of a few strains can therefore be extrapolated to almost any *S. typhi* isolate.

Absorption of S. typhi anti-LPS antibodies. Because the presence of anti-*S. typhi* LPS in normal and convalescent sera interferes with the study of the immune response to proteins, we removed these anti-LPS antibodies by absorption to insure the specificity of our radioimmunoprecipitations. The absorptions were performed by repeated passages of the sera or the purified IgG over *S. typhi* LPS coupled to Sepharose (Aron, et al., 1992 a). Smooth *S. typhi* LPS was purified (Darveau and Hancock, 1983). Free lipid A was extracted (Bligh and Dyer, 1959). Any organic solvent remaining in the water phase was removed by several extractions with water-saturated ether and remaining ether was

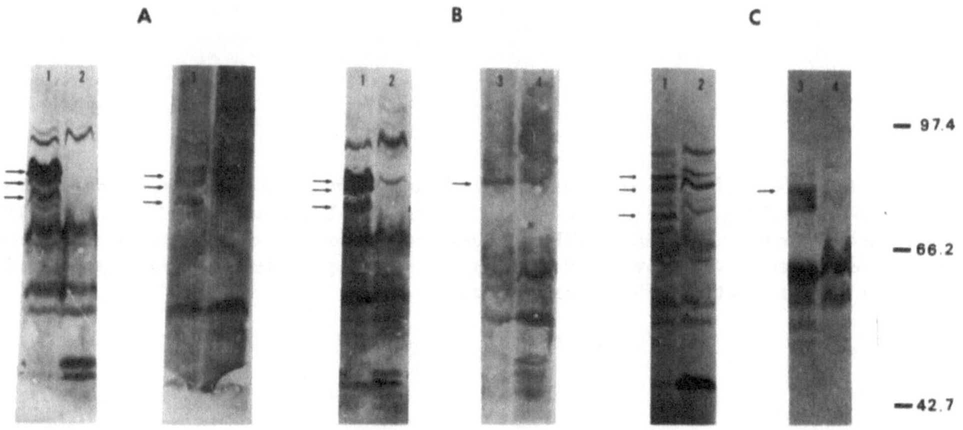

Figure 2. Immunoblot of *S. typhi, S. typhimurium* and *E. coli* outer membrane proteins. Panel A. *S. typhi*, Panel B. *S. typhimurium*. Panel C. *E. coli* FC004. 1 and 2 india ink stain of proteins transferred onto nitrocellulose paper; 3 and 4 immunoblot with typhoid fever sera. 1 and 3 cells grown in nutrient broth plus 200 μM 2,2'- dipyridil, 2 and 4 outer membrane proteins from cells grown in nutrient broth (Fernandez-Beros, et al., 1989).

evaporated. The product was then loaded onto the cathodic cup of an electrophoretic sample concentrator (ISCO) and subjected to electrodialysis (Galanos and Luderitz, 1975) at 400 V in deionized water for several hours at 4°C. After removal of the sample from the anodic cup, the pH was adjusted to 7.0 with dilute sodium hydroxide and stored at 4°C. Yields were monitored by 2-keto-3-deoxymanno-octusolonic acid (KDO) measurements (Karkhanis, et al., 1978). Samples of the product were compared to the starting material by electrophoresis on 14% acrylamide gels (Laemmli, 1970). LPS staining of gels was done (Tsai and Frasch, 1982) and protein staining of gels was done using standard methods (Weber and Osborn, 1969).

Purified *S. typhi* LPS (10 mg) was reacted with 1 g AH-Sepharose (Sigma) according to the manufacturer's instructions, using 250 mg of 1-ethyl-3-(3 dimethyl-amino propyl) carbodiimide hydrochloride (EDC, Sigma) as the coupling agent. The reaction was allowed to proceed for 20 hrs (3 hrs at room temperature, the rest at 4° C, keeping the pH at 4.9). Afterwards, 10 μl glacial acetic acid and 20 mg EDC were added to block unreacted groups on the matrix, and the pH was kept at 5.4 for the same amount of time and temperatures as above. Unbound LPS fraction was retained and the matrix washed sequentially with 0.1 M acetate-1 M NaCl, pH 4.0, and 0.05 M Tris-HCl-1 M NaCl pH 8.0 several times and finally with 0.2 M glycine, pH 2.5. The matrix was stored at pH 6.5 in 1 M NaCl and 0.05% sodium azide at 4°C. The unbound LPS fraction and subsequent washings were pooled and concentrated and washed on Centricon C-10 (Amicon), after adjusting the pH to 7.0. To rule out any artifacts, a parallel LPS sample was treated with EDC in the absence of matrix, following the same procedure as for the coupling. Binding efficiency and recovery were monitored by KDO measurements of the fractions.

Two methods were used to remove anti-*S. typhi* LPS antibodies. Purified immunoglobulins in Tris-HCl pH 8.5-0.5 M NaCl (binding buffer) were passed several times over a column of LPS-AH Sepharose at 4°C. The amounts eluted were monitored at 280 nm in a Spectronic 21 spectrophotometer. Absorption was repeated as many times as needed until no more anti-*S. typhi* LPS antibodies were absorbed. Removal was monitored by *S. typhi* LPS-immunodot blotting. Since absolute removal of anti-LPS antibodies was never achieved by this method, especially of antibodies against R_a LPS, it was combined with the following method. Nitrocellulose sheets (Bio-Rad) were incubated for 20 min in TBS (20 mM Tris-Hcl, pH 7.5, 0.25 M NaCl) with 0.1 mg/ml of polymyxin B (Sigma). Sheets were removed and allowed to dry. Purified *S. typhi* LPS (about 0.5 mg for a 10 x 10 cm sheet) or *S. typhimurium* TV119 R_a LPS or *S. minnesota* 595 R_e LPS were added dropwise in TBS to wet the sheet. Sheets were allowed to dry and subsequently blocked for 1 hr at 37°C with a blocking agent of choice in TBS: 5% skim milk (SM, Difco) + 5% bovine serum albumin (BSA, Sigma, globulin free), or 0.3% Tween 20-1% SM (Bio-Rad). Serum or immunoglobulin dilutions in blocking buffer of choice were incubated overnight at 4°C with LPS loaded strips. Protein concentrations of the absorbed immunoglobulin pools were monitored by the methods described above and adjusted to that of the original pools.

Table 1 indicates that a convalescent pool of sera from patients with typhoid had a positive titer of 1:3200 when determined by either ELISA or immunodots using *S. typhi* LPS as antigens, and that after absorption over the LPS-Sepharose column no reaction could be detected at a 1:2 dilution. Similarly, the same pooled sera when reacted in immunodots with *S. typhimurium* TV119 R_a LPS showed a titer of 1:400, and after absorption with LPS this titer went down to 1:2. Other experiments indicated that only absorption of the convalescent sera with LPS-Sepharose was able to absorb antibodies against rough and smooth *S. typhi* LPS as sera absorbed 6 times with whole *S. typhi* cell still reacted with *S. typhimurium* R_a LPS (Aron, et al., 1992 a). This may be an important point as some typhoid patients have higher titers of antibodies against the rough core of *S. typhi* LPS than against the smooth LPS.

Whole-cell and outer membrane radioimmunoprecipitation assays. The analysis of the pattern of the *S. typhi* whole- cell proteins precipitated with patients' sera indicated that this sera was able to precipitate proteins in the 80, 55, 36, 30, 18, 14 and 6.5 kDal range. The patterns obtained with ^{125}I and ^{35}S-methionine were similar except in the 80 kDa and 6.5 kDa range because these proteins appear to be poorly labeled with methionine (Aron, et al., 1992 a). Comparative analysis of autoradiography of the outer and inner membrane proteins also indicated that the proteins found in these membranes were different even when some cross contamination was evident. Further analysis of the humoral host immune response to *S. typhi* was dome by studying the ability of different antibody preparations to immunoprecipitate proteins present in the outer membrane preparations obtained from induced *S. typhi* and labeled with ^{125}I (Fig. 3A) or ^{35}S-methionine (Fig. 3B). As Figure 3A indicates, several proteins are precipitated when the immunoprecipitation was performed with IgG isolated from the convalescent serum pool (lanes e and h). These prominent protein bands have molecular weights of 83, 80, 76, 36, 30, 18.5, 15 and 6.5 KDa. The immunoprecipitations with the 76, 36, 30, 18.5, 15 and 6.5 kDa proteins

Table 1. Titer of *S. typhi* O antipolysaccharide antibodies in convalescent Ig before and after passage through an ST O-polysaccharide column

IgG preparation	Absorption of anti-LPS antibodies	ELISA titer	Dot blotting
Convalescent IgG	-	1:3200	1:3200
Convalescent IgG	+	1:2	1:2
NHS IgG	-	<1:5	ND*
NHS IgG	+	none detectable	ND

IgG purification and preparation of ST O-antigen column was done as described in the text. Dot blotting was performed using ST O-polysaccharides antigen purified as described (Aron, et al., 1992 a).
*ND, not done.

persisted, albeit somewhat fainter when performed with convalescent IgG lacking the anti-*S. typhi* LPS antibodies, while the 83 kDa band decreased several folds in intensity and the 80 kDa band dissppeared (lane e). Fig. 3A also shows a rather similar ^{125}I-protein immunoprecipitation pattern when using normal human serum immunoglobulins whether with or without anti-LPS antibodies (lanes i and f, respectively). Removal of anti-*S. typhi* LPS antibodies from the convalescent sera seemed to have only quantitative effect, except for the 69 kDal iron-starvation induced protein, which was not apparent when normal serum was used, and its precipitation seemed to be independent from the presence of anti-LPS antibodies. Normal and convalescent sera exhibited similar qualitative patterns, but large quantitative differences are apparent for the 36 kDa range proteins. Further experiments indicated that 1:1000 dilutions of the convalescent sera were able to precipitate the outer membrane proteins while the immunoprecipitation observed with the normal sera disappeared at 1:10 dilutions (Aron, et al., 1992 a).

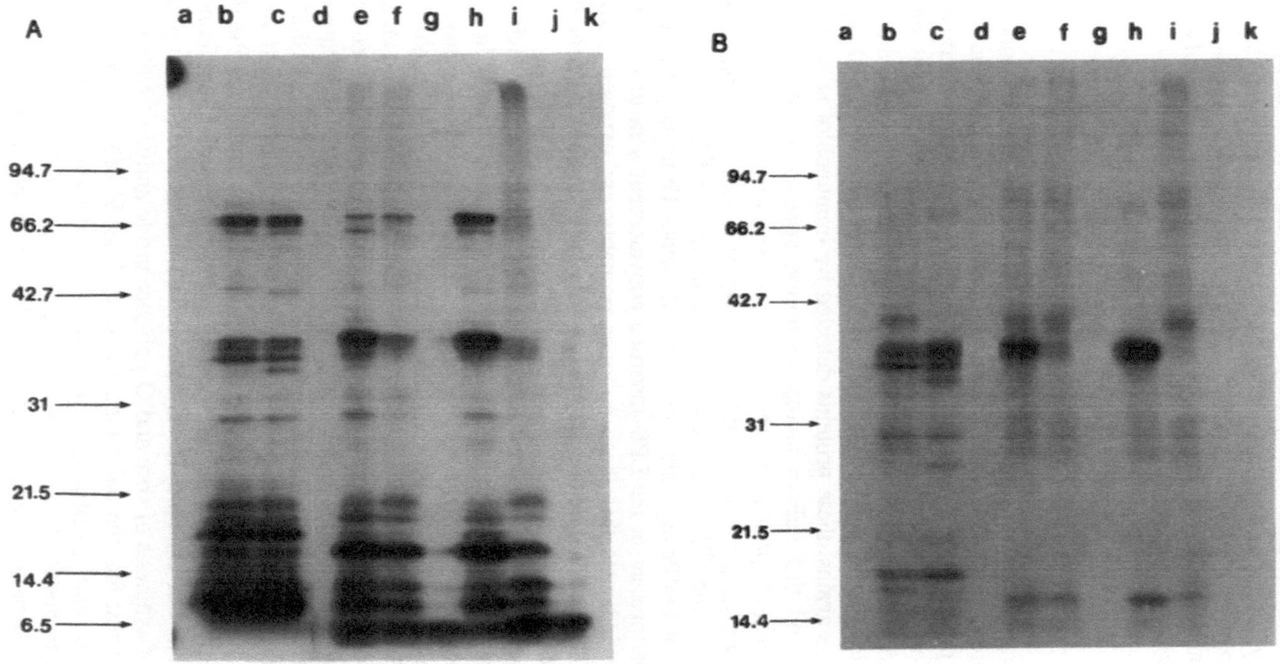

Figure 3. *S. typhi* outer membrane radioimmunoprecipitations. **A.** ^{125}-I OMPs. Lane a; protein markers (molecular weigth indicated on the left in kDa); lanes d and k: cold outer membranes; lanes b and c: ^{125}I-outer membranes solubilized directly in loading buffer or previously in solubilizing buffer respectively; lanes e and h; ^{125}I-outer membranes reacted with convalescent typhoid fever IgG pool devoid and containing anti-*S. typhi* LPS antibodies respectively; lanes f and i: ^{125}I-outer membranes reacted with the LPS-absorbed and non-absorbed normal human IgG pool respectively; lanes g and j: ^{125}I-outer membranes controls (PBS added instead of IgG) (Aron, et al., 1992 a). **B.** ^{35}-methionine OMPs. Lanes as in A.

A similar analysis performed with purified outer membrane from *S. typhi* labeled with ^{35}S methionine and a convalescent sera obtained from one patient at different times during the course of typhoid fever, indicate that the antibodies against these proteins appear early in the clinical course of the disease. They may reach a peak 2 to 3 weeks after the appearance of clinical symptoms and begin to decline after 4 weeks. Since in this specific patient, anti-LPS antibody titer remained constant over the evolution of the disease, the pattern of increasing and decreasing immunoprecipitation observed probably reflects changes in the antibody concentration against OMPs (Aron, et al., 1992 a).

DISCUSSION

Analysis of the humoral immune response to *S. typhi* OMPs by radioimmunoprecipitation indicates that there is a detectable and important humoral immune response against epitopes of these proteins in typhoid fever. The important decrease in radioimmunoprecipitation of iron starvation-induced proteins and the proteins observed after absorption of the anti-LPS antibodies indicates that these proteins are mainly precipitated by antibodies that recognize LPS epitopes. These experiments also suggest that human host response against these proteins detected by immunoblot is partially due to contaminating LPS, and that radioimmunoprecipitation appears to be an improved method to detect immunogenic proteins on the surface of Gram-negative bacteria. The identities of the patterns of immunoprecipitation observed with *S. typhi* cells and membranes labeled either with ^{125}I or ^{35}S methionine indicates that ^{125}I labeling does not alter the characteristics of OMPs as is also the case with other Gram-negative bacteria. The ability of antibodies present in the LPS-absorbed convalescent sera to interact with proteins in whole cells indicates that these epitopes probably are exposed during the infective process, and that these antibodies may play a role in the evolution of the disease and in disease prevention.

The ability of antibodies against selected *S. typhi* OMPs to interact with their targets in whole cells indicates that these proteins and antibodies behave differently in *S. typhi* than in *E. coli* and other Enterobacteriaceae, where the interaction between protein and antibodies is hindered by the shield provided by the LPS and capsular polysaccharide (van der Ley, et al., 1986; Bentley and Klebba, 1988). This behavior could also be an artifact produced by the loss of normal cellular architecture resulting in usually unexposed proteins capable of reacting with the antibodies (Hoekstra, et al., 1976). Nonetheless, experiments using *S. typhi* whole cells and immunofluorescent antibodies seem to indicate that under the conditions used in these experiments there is minimal disruption of the cell structure (Aron, et al., 1992 a). These results could also be explained by variations in the density of LPS and capsular polysaccharide of bacteria growing under different conditions (Grossman, et al., 1987). For example, it has been shown that iron starvation generates differences in the rate of synthesis of core and O-polysaccharide antigen (Nelson, et al., 1991). These results suggest that the outer membrane of *S. typhi* behaves like the outer membrane of *H. influenzae* and *P. aeruginosa* permitting the interaction of antibodies and outer membrane proteins despite the presence of capsular and LPS O-antigen layers (Gulig, et al., 1982; Mutharia and Hancock, 1983; Srikumar, et al., 1992).

Our experiments also confirm that the sera of normal individuals contain antibodies that are able to cross react with *S. typhi* OMPs (Hofstra, et al., 1980). These antibodies are probably raised against OMPs of Gram-negative bacteria constitutive of the normal intestinal flora (Hofstra, et al., 1980). We have demonstrated the existence of this cross reactivity by the ability of antibodies raised against the FepA protein of *E. coli* to recognize the FepA protein of *S. typhi* (Fernandez-Beros, et al., 1989). Nonetheless, in this case the specificity of the reactions against *S. typhi* OMPs is demonstrated by the lack of immunoprecipitation when convalescent sera are absorbed with whole *S. typhi* cells. The radioimmunoprecipitation of proteins induced by iron starvation and probably involved in iron transport confirms previous results by us that the *S. typhi* iron uptake system is fully induced in typhoid, suggesting a role for iron uptake in the pathogenesis of the disease and a role for these antibodies in host defense (Fernandez-Beros, et al., 1989). These results demonstrating the immunogenicity of the *S. typhi* OMPs, and results of other investigators that indicate that conjugates of *H. influenzae* polyribosyl ribitol phosphate with a major protein of the outer membrane of the Gram-negative *N. meningitidis* are more immunogenic than similar conjugates with other bacterial proteins (Santosham, et al. 1991), suggest that the ability of *S. typhi* OMPs to stimulate T-cells should be investigated further. If the ability of these immunogenic proteins to stimulate T-cell proliferation is confirmed, they could be the ideal carrier in *S. typhi* vaccines using the Vi antigen (Acharya, et al., 1987) or the O-antigen chain (Aron, et al., 1992 b). In summary, we have described a series of methods that allowed us to characterize the immune response to proteins of *S. typhi* in typhoid independently from the response to *S. typhi* LPS, and that could be used to characterize potentially protective antibodies to intracellular proteins, to proteins induced by *in vivo* environments, and to vaccines and foreign proteins expressed in *S. typhi* vaccines (Saddof, et al., 1988).

REFERENCES

Acharya, I.L., Lowe, C.U., Thaps, R., Gurubacharya, V.L., Shresta, M.B., Cadoz, M., Schulz, D., Armand, J., Bryla, D.A., Trollfors, B., Cramton, T., Schneerson, R., and B. Robbins, J.B., 1987, Prevention of typhoid fever in Nepal with the Vi capsular polysaccharide of *Salmonella typhi*. A preliminary report, *N. Engl. J. Med.* 317:1101.

Aguero, J., Mora, G., Mroczenski-Widley, M.J., Fernandez-Beros, M.E., Aron, L., and Cabello, F.C., 1987, Cloning, expression and characterization of 36 kdal *Salmonella typhi* porin gene in *Escherichia coli*, *Microbial Pathogenesis* 3:399.

Aron, L., Faundez, G., Gonzalez, C., Roessler, E., and Cabello, F.C., 1992 a, Radioimmunoprecipitation and identification of immunogenic surface proteins of *Salmonella typhi* in typhoid fever, accepted for publication, *Vaccine*.

Aron, L., DiFabio, J.L. and Cabello, F.C., 1992 b, *Salmonella typhi* O:9, 12-polysaccharide conjugates: Characterization and immunoreactivity with pooled and individual normal sera, paratyphoid A and B, typhoid and animal sera, submitted to *J. Clin. Microbiol.*

Bentley, A.T., and Klebba, P.E., 1988, Effect of lipopolysaccharide structure on reactivity of antiporin monoclonal antibodies with the bacterial cell surface, *J. Bacteriol.* 170:1063.

Bligh, E.J., and Dyer, W.J., 1959, A rapid method of total lipid extraction and purification, *Can. J. Biochem. Physiol.* 37:911.

Brown, A., and Hormaeche, C.E., 1989, The antibody response to *Salmonellae* in mice and humans: studies by immunoblots and ELISA, *Microbial Pathogenesis* 6:445.

Butler, T., Bell, W.R., Levin, J., et al., 1978, Typhoid fever: A study of blood coagulation, bacteremia and endotoxemia, *Arch. Inter. Med.* 138:407.

Calderon, I., Lobos, S.R., Rojas, H.A., Palominos, C., Rodriguez, L.H., and Mora, G., 1986, Antibodies to porin antigens of *Salmonella typhi* induced during typhoid fever in humans, *Infect. Immun.* 52:209.

Christie, A.B., 1987, Typhoid and paratyphoid fevers, *in*: "Infectious Diseases," A.B. Christie, ed., 4th Ed., Churchill Livingstone, New York, p. 100.

Collins, F.M., and Mackaness, G.B., 1968, Delayed hypersensitivity and Arthus reactivity in relation to host resistance in *Salmonella*-infected mice, *J. Immunol.* 101:830.

Darveau, R.P., and Hancock, R.E., 1983, Procedure for isolation of bacterial lipopolysaccharides from both smooth and rough *Pseudomonas aeruginosa* and *Salmonella typhimurium* strains, *J. Bacteriol.* 155:831.

Dubois, M., Gilles, K.A., Hamilton, J.K., Revers, P.A., and Smith, F., 1956, Colorimetric method for the determination of sugars and related substances, *Anal. Chem.* 28:350.

Edelman, R., and Levine, M.M., 1986, Summary of an international workshop on typhoid fever, *Rev. Infect. Dis.* 8:329.

Faundez, G., Aron, L., and Cabello, F.C., 1990, Chromosomal DNA, iron-transport system, outer membrane proteins and enterotoxin (Heat labile) production in *Salmonella typhi* strains, *J. Clin. Microbiol.* 28:894.

Fernandez-Beros, M.E., Gonzalez, C., McIntosh, M., and Cabello, F.C., 1989, Immune response to the iron-deprivation induced proteins of *Salmonella typhi* in typhoid fever, *Infect Immun.* 57:1271.

Ferreccio, C., Levine, M.M., Manterola, A., Rodriguez, G., Rivara, I. Prenzel, I., Black, R.E., Mancuso, T., and Bulas, D., 1984, Benign bacteremia caused by *Salmonella typhi paratyphi* in children younger than 2 years, *J. Pediat.* 104:899.

Forrest, B., Labrooy, J.T., Beyer, L., Dearlove, C.E., and Shearman, D.J.C., 1991, The humoral immune response to *Salmonella typhi* Ty21a, *J. Infect. Dis.* 163:336.

Galanos, C., and Luderitz, 1975, Electrodialysis of lipopolysaccharides and their conversion to uniform salt forms, *Eur. J. Biochem.* 54:63.

Granoff, D.M., and Munson, R.S., 1986, Prospects for prevention of *Haemophilus influenzae* type b disease by immunization, *J. Infect. Dis.* 153:448.

Grossman, N.A., Schnertz, M.A., Foulds, J., Klima, E.N., Jimenez, V., Leibe, L.L. and Joiner, K.A., 1987, Lipopolysaccharide size and distribution determine serum resistance in *Salmonella montevideo*, *J. Bacteriol.* 169:856.

Gulig, P.A., McCracken, G.H., Jr., Frisch, C.F., Johnston, K.H., and Hansen, E.C., 1982, Antibody response of infants to cell surface-exposed outer membrane proteins of *Haemophilus influenzae* type b after systemic *Haemophilus* disease, *Infect. Immun.* 37:82.

Hansen, E.J., Frisch, C.F., McDade, R.L., Jr., and Johnston, K.H., 1981, Identification of immunogenic outer membrane proteins of *Haemophilus influenzae* type b in the infant rat model system, *Infect. Immun.* 32:1084.

Hoekstra, D., Van Der Laan, J.W., De Leij, L., and Witholt, B., 1976, Release of outer membrane fragments from normal growing *Escherichia coli*, *Biochim. Biophys. Act.* 455:889.

Hofstra, J., van Tol, M.J.D., and Danbert, J., 1980, Cross reactivity of major outer membrane proteins, of *Enterobacteriaceae*, studied by crossed immunoelectrophoresis, *J. Bacteriol.* 143:328.

Hook, E.W., 1990, *Salmonella* species (including typhoid fever), *in*: " Principles and Practices of Infectious Diseases, G.L. Mandell, R.G. Douglass, Jr., J.E. Bennet, eds., Churchill Livingstone, 3rd Ed., New York, p. 1700.

Hornick, R.B., Greisman, S.E., Woodward, T.E., DuPont, H.L., Dawkins, A.T., Snyder, M.J., 1970, Typhoid fever: pathogenesis and immunological control. *N. Engl. J. Med.* 283:686, 739.

Jimenez-Lucho, V., and Foulds, J., 1990, Heterogeneity of lipopolysaccharide phenotype among *Salmonella typhi* strains, *J. Infect. Dis.* 162:763.

Karkhanis, Y.D., Zeltner, J.Y., Jackson, J.J., and Carlo, D.J., 1978, A new and improved micro-assay to determine 2-keto-3-deoxyoctonate in lipopolysaccharide of Gram-negative bacteria, *Anal. Biochem.* 85:595.

Laemmli, U.K., 1970, Cleavage of structural proteins during the assembly of the head of bacteriophage T4, *Nature (London)* 227:680.

Maurer, P.H., and Callahan, H.J., 1980, Proteins and polypeptides as antigens, *Methods Enzymol.* 70:49.

Matsui, K., and Arai, T., 1989, Specificity of *Salmonella* porin as an eliciting antigen for cell-mediated immunity (CMI) reaction in murine salmonellosis, *Microbiol. Immunol.* 33:1063.

Muschel, I.H., Chamberlin, R., and Osawa, E., 1958, Bactericidal activity of normal serum against bacterial cultures. I. Activity against *Salmonella typhi* strains, *Proc. Soc. Exp. Biol. Med.* 97:376.

Mutharia, L., and Hancock, R.E., 1983, Surface localization of Pseudomonas aeruginosa outer membrane protein F by using monoclonal antibodies, *Infect. Immun.* 42:1027.

Nagington, J., 1956, The sensitivity of *Salmonella typhi* to the bactericidal action of antibody, *Brit. J. Exp. Pathol.* 37:396.

Nelson, D., Bathgate, A.J., and Poxton, I.R., 1991, Monoclonal antibodies as probes for detecting lipopolysaccharide expression on *Escherichia coli* from different growth conditions, *J. Gen. Microbiol.* 137:2741.

Ortiz, V., Isibasi, A., Garcia-Ortigosa, E., and Kumate, J., 1989, Immunoblot detection of class-specific humoral immune response to outer membrane proteins isolated from *Salmonella typhi* in humans with typhoid fever, *J. Clin. Microbiol.* 27:1640.

Osawa, E., and Muschel, L.H., 1964, The bactericidal actions of O and Vi antibodies against *Salmonella typhosa*, *J. Immunol.* 92:281.

Reeves, M.W., Evins, G.M., and Heiba, A.A., Plikaytis, B. D., and Farmer, J.J.,III, 1989, Clonal nature of *Salmonella typhi* and its genetic relatedness to other salmonellae as shown by multilocus enzyme electrophoresis, and proposal of *Salmonella bongori* comb. nov, *J. Clin. Microbiol.* 27:313.

Robbins, J.D., and Robbins, J.B., 1984, Reexamination of the protective role of the capsular polysaccharide (Vi antigen) of *Salmonella typhi*, *J. Infect. Dis.* 150:436.

Rocha, H., Kirk, J.W., and Heaney, C.D., Jr., 1971, Prolonged *Salmonella* bacteremia in patients with *Schistosoma mansoni* infection, *Arch. Intern. Med.* 128:254.

Sadoff, J.C., Ballou, W.R., Baron, L.S., Majarian, W.R., Brey, R.N., Hockmayer, W.T., Young, J.F., Cryz, S.J., Ou, J., Lowell, G.H., and Chulay, J.D., 1988, Oral *Salmonella typhimurium* vaccine expressing circumsporozoite protein protect against malaria, *Science* 240:336.

Saxen, H, Nurminen, M., Kuusi, N., Soerison, S.B., Makela, P.H., 1986, Evidence for the importan, of O antigenic antibodies in mouse-protective *Salmonella* outer membrane (porin) antisera, *Microbial Pathogenesis* 1:433.

Santosham, M., Wolff, M., Reid, R., Hohenboken, M., Bateman, M., Goepp. J., Cortese, M., Sack, D., Hill, J., Newcomer, W., Capriotti, L., Smith, J., Owen, M., Gahagan, S., Hu, D., Kling, R., Lukacs, L., Ellis, r.W., Vella, P.P., Calandra, G., Matthews, H., and Ahonkai, V., 1991, The efficacy in navajo infants of a conjugate vaccine consisting of *Haemophilus influenzae* type B polysaccharide and *Neisseria meningitidis* outer-membrane protein complex, *N. Engl. J. Med.* 324:1767.

Sharma, P., Ganguly, N.E., Sharma, B.K., Sharma, S., and Seghal, R., 1989, Specific immunoglobulin response in mice immunized with porins and challenged with *Salmonella typhi*. *Microbial Immunol.* 33:519.

Singh, S.P., Upshaw, Y., Abdullah, T., Singh, S. R., and Klebba, P.E., 1992, Structural relatedness of enteric bacterial porins assessed with monoclonal antibodies to *Salmonella typhimurium* OmpD and OmpC, *J. Bacteriol.* 174:1965.

Srikumar, R., Chin, A.C., Vachon, V., Richardson, C.D., Ratcliffe, J.H., Saarinen, L., L., Kaythy, H., Makela, P.H., and Coulton, J.W., 1992, Monoclonal antibodies specific to porin of *Haemophilus influenzae* type B: localization of their cognate epitop[es and tests of their biological activities, *Molec. Microbiol.* 6:665.

Tsai, C.M., and Frasch, K., 1982, A sensitive silver stain for detecting lipopolysaccharides in polyacrylamide gels, *Anal. Biochem.* 119:115.

Udhayakumar, V. and Muthukkaruppan, V.R., 1987, Protective immunity induced by outer membrane proteins of *Salmonella typhimurium* in mice. Infect. Immun. 55:816.

van der Ley, P., Kuipers, O., Tommassen, J., and Lugtenberg, B., 1986, O-antigenic chains of lipopolysaccharide prevent binding of antibody molecules to an outer membrane por protein in *Enterobacteriaceae*, *Microbial Pathogenesis* 1:43.

Watson, K.C., 1955, Isolation of *Salmonella typhi* from the blood stream. *J. Lab. Clin. Med.* 46:128.

Weber, K., and Osborn, M.J., 1969, The reliability of molecular weight determination by dodecyl sulfate polyacrylamide gel electrophoresis, *J. Biol. Chem.* 244:4406.

THE ROLE OF GRANULOCYTES, NAIVE AND ACTIVATED MACROPHAGES IN THE HOST RESISTANCE AGAINST *SALMONELLA TYPHIMURIUM*

Ralph van Furth, Jaap T. van Dissel, Jan A.M. Langermans

Department of Infectious Diseases
University Hospital Leiden
PO BOX 9600
2300 RC Leiden, The Netherlands

Mononuclear phagoytes play an important role in the innate resistance of mice against infection by *Salmonella typhimurium*. The outcome of an infection is influenced by a number of factors, ranging from the handling of the microorganisms by resident macrophages during the earliest phase of infection to the appearance of specific cellular and humoral immune responses in a later phase. Control of the rate of proliferation of *S. typhimurium* in the liver and spleen of mice during the first week of infection is the earliest process known to be genetically determined (Plant and Glynn, 1976; Hormeache, 1979; Hormeache, 1980a; Skamene et al., 1982). On the basis of the differences in the early in vivo outgrowth, inbred mice can be divided into resistant and susceptible strains. In vivo studies have indicated that the control of *S. typhimurium* proliferation is due to an inherent property of macrophages (Robson and Vas, 1972; Maier and Oels, 1972; Vas, 1978; O'Brien et al., 1979; Hormaeche et al., 1980b; O'Brien and Metcalf, 1982). Our studies showed no difference in the phagocytic ability of peritoneal macrophages from resistant CBA and susceptible C57BL/10 mice for *S. typhimurium* opsonized with immune serum, but we found that the initial rate of intracellular killing of ingested *S. typhimurium* by CBA macrophages was more rapid than by macrophages from C57BL/10 mice (Van Dissel et al., 1985), which reflects the constitutive difference between the two strains of mice. However, equal rates of intracellular killing of *L. monocytogenes* were found for macrophages from the *Listeria*-susceptible CBA mice and the *Listeria*-resistant C57BL/10 mice (Van Dissel et al., 1985), which indicates that other factors do well influence the outcome of infections with these bacteria. Such a factor could be the greater number of monocyte-derived exudate macrophages in the inflammatory exudate in C57BL/10 mice than in CBA mice (Stevenson et al., 1981; Sluiter et al., 1984), which is due to a difference in the response to the factor-increasing monocytopoiesis (Sluiter et al., 1984).

There is also a contribution of granulocytes to differences in the innate susceptibility of mouse strains to infection by *S. typhimurium*. After an i.p. injection of various numbers of live *S. typhimurium* the increase in the numbers of both peritoneal granulocytes and macrophages was two to four times greater in C57BL/10 mice than in CBA mice. However, despite the larger number of phagocytes in the inflammatory exudate, the numbers of viable *S. typhimurium* in the peritoneal cavity 24 hr after injection was much higher in C57BL/mice than in CBA mice. Because the proportion of phagocytosed bacteria was similar in the two mouse strains, these findings indicate a difference in the rate of intracellular killing of *S. typhimurium* by exudate peritoneal granulocytes of the two mouse strains. This conclusion is supported by the observation that the initial rate of intracellular killing of *S. typhimurium* by exudate peritoneal granulocytes of CBA mice, after in vivo phagocytosis of the bacteria, was twice as efficiently than that by exudate peritoneal granulocytes of C57BL/10 mice. Similarly, the initial rate of intracellular killing of the ingested *S. typhimurium* by blood granulocytes of CBA mice was two times higher than that of C57BL/10 mice (Van Dissel et al., 1986). These findings are relevant with respect to the innate resistance of mice to *S. typhimurium*, particularly during the initial phase of infection when the inflammatory exudate contains predominantly granulocytes. The findings for granulocytes are specific for *S. typhimurium*, because exudate granulocytes and blood granulocytes of both mouse strains kill *L. monocytogenes* with equal efficiency (Van Dissel et al., 1986).

The differences in growth rates of *S. typhimurium* in the liver and spleen of resistant CBA and susceptible C57BL/10 mice and the rates of intracellular killing by resident peritoneal macrophages was found both for virulent and avirulent mutant strains (Van Dissel et al., 1987a). However, such differences were not found for other salmonella strains, such as S. dublin and S. heidelberg (Van Dissel et al., 1987a), and these salmonella strains were killed with equal efficiency by macrophages of CBA and C57BL/10 mice.

Since no differences between CBA and C57BL/10 mice are found in the magnitude of the respiratory burst by peritoneal macrophages and exudate peritoneal cells (mainly comprising granulocytes) upon stimulation, most likely non-oxidative bactericidal mechanisms are responsible for the difference in rates of intracellular killing of *S. typhimurium*. In mouse macrophages we found proteins with bactericidal activity against *S. typhimurium* (Hiemstra et al., submitted for publication).

Strong evidence indicates that the resident macrophage is the effector cell for the expression of the Ity (Lissner et al., 1983) and other resistance genes identical to or closely clustered around Ity (Plant et al., 1982; Stach et al., 1984; Crocker et al., 1984; Lissner et al., 1985). It is not clear whether the difference between CBA and C57BL/10 mice as to the rate of intracellular killing of ingested *S. typhimurium* by resident macrophages is due to or is supplementary to the Ity gene expression. In this context it is of interest that for peritoneal macrophages of Ity-congenic mice the efficiency of intracellular killing of various microorganisms has been found to depend on Ity expression (Lissner et al., 1985).

Activation of macrophages by T lymphocyte-derived lymphokines is thought to be essential to overcome infections with facultative intracellular pathogens (Mackaness, 1971). Mice that are infected with non-lethal numbers of BCG, Brucella spp. or *L. monocytogenes* are more resistant to infections with these bacteria, and this is accompanied by a decreased intracellular survival of the bacteria in macrophages (Mackaness, 1964; Blanden et al., 1969). These findings led to the generally accepted concept that activation of macrophages contributes to nonspecific enhancement of bactericidal activity.

Activation of macrophages results in a variety of functional and biochemical changes compared to the effect of resident macrophages. These macrophages are able to inhibit the intracellular proliferation of certain protozoa, such as Toxoplasma gondii (Murray and Cohen, 1980; Eisenhower et al., 1988), Leishmania donovani (Pappas and Nacy, 1983), and Trypanosoma cruzi (Hoff, 1975), and display both an enhanced ability to secrete reactive oxygen metabolites (Nathan et al., 1980; Murray et al., 1985) and reactive nitrogen intermediates (Nibbering et al., 1991), and increased expression of Ia antigen (Koerner et al., 1987). Murine peritoneal macrophages become activated, according to these criteria, during infection of mice with BCG followed by an i.p. booster with purified protein derivative (PPD) and during an infection with *L. monocytogenes* (Ruco and Meltzer, 1977; Van Dissel et al., 1987b).

Activated macrophages are regarded as the most important defense mechanism (Mackaness et al., 1966; Collins 1974) against *Salmonella typhimurium*. However, *L. monocytogenes*-activated peritoneal macrophages do not kill *S. typhimurium* faster than resident peritoneal macrophages, whereas these activated cells show an enhanced intracellular killing of *L. monocytogenes* (Van Dissel et al., 1987b). Peritoneal macrophages become also activated during a systemic infection with *S. typhimurium*, because they fulfil the above mentioned criteria, but these macrophages do not express an enhanced bactericidal activiy against *S. typhimurium* or *L. monocytogenes* compared to resident macrophages (Langermans et al., 1990). Consistant with these finding is that the growth rates of *S. typhimurium* in the spleen of *S. typhimurium*- and *L. monocytogenes*-activated mice and in the liver of *L. monocytogenes* activated mice are the same as in normal mice; only in the liver of *S. typhimurium*-activated mice these salmonellas do not proliferate (Langermans et al., 1990).

Since interferon γ (IFN-γ) is the most important lymphokine that activates macrophages, it has been tried to activate mice with an i.p. or i.v. injection(s) of IFN-γ. However, although resistent peritoneal macrophages were activated according to the relevant criteria discussed above, these cells do not kill phagocyted *S. typhimurium* faster in vitro and the proliferation of these bacteria in the liver and spleen of these mice was not affected either (Van Dissel et al., 1987c; Langermans et al., 1991). In vivo or in vitro activation of macrophages result in an increase of tumor necrosis factor-α (TNFα) production by these cells. Binding of secreted TNFα with anti-TNFα antibodies demonstrated that TNFα is essential for the activation of macrophages in an autocrine fashion (Langermans et al., 1992). When naive mice were treated with this antibody, *S. typhimurium* proliferated faster in the liver and spleen relative to that in control mice. This indicates that TNFα is also essential to limit the bacterial growth in non-stimulated mice.

During the first stage of an infection with *S. typhimurium*, humoral immunity does not play an essential role in the defense against the infection, because only very low levels of antibodies were observed in infected mice until day 12. In mice with high levels of antibodies after repeated immunisation with *S. typhimurium* the growth of these bacteria in the spleen and liver was similar to that in normal mice (Langermans et al., 1990).

It might well be that macrophages are not the most important cells in the elimination of Salmonellae. It has been reported that immunization with avirulent live *S. typhimurium* activates NK cells (Schafer and Eisenstein, 1988). Thus it is possible that these cells kill *S. typhimurium*-infected cells (Garcia-Penarrubia et al., 1989), which results in the release of bacteria that are subsequently opsonized by antibodies and ingested and killed by granulocytes. There is also evidence that granulocytes rather than macrophages are the predominant cells that eliminate *S. typhimurium* at primary foci of infection (Nakoneczna and Hsu, 1980; Wang et al., 1988). Thus, although during an infection with *S. typhimurium* mice become activated, there is great doubt whether immunologically activated macrophages are an important factor in the defence against a *Salmonella* infection.

REFERENCES

Blanden, R.V., Lefford, M.J., and Mackaness, G.B., 1969, The host response to Calmette-Guérin bacillus infection in mice, *J. Exp. Med.* 129:1079.

Collins, F.M., 1974, Vaccines and cell-mediated immunity, *Bacteriol. Rev.* 38:371.
Crocker, P.R., Blackwell, J.M., and Bradley, D.J., 1984, Expression of the natural resistance gene Lsh in resident liver macrophages, *Infect. Immun.* 43:1033.
Eisenhower, P., Mack, D.G., and Mcleod, R., 1988, Prevention of peroral and congeital acquisition of Toxoplasma gondii by antibody and activated macrophages, *Infect. Immun.* 56:83.
Garcia-Penarrubia, P., Koster, F.T., Kelley, R.O., McDowell, T.D., and Bankhurst, A.D., 1989, Antibacterial activity of natural killer cells, *J. Exp. Med.* 169:99.
Hiemstra, P.S., Eisenhauer, P.B., Harwig, S.S.L., Van den Barselaar, M.Th., Van Furth, R., and Lehrer, R.I., Identification of major proteins with broad spectrum antimicrobial activity in murine macrophages, (submitted for publication).
Hoff, R., 1975, Killing in vitro of Trypanosoma cruzi by macrophages from mice immunized with T. cruzi or BCG, and absence of cross-immunity on challenge in vivo, *J. Exp. Med.* 142:299.
Hormeache, C.E., 1979, Natural resistance to *Salmonella typhimurium* in different inbred strains of mice, *Immunology* 37:311.
Hormaeche, C.E., 1980a, The in vivo division and death rates of *Salmonella typhimurium* in the spleens of naturally resistant and susceptible mice measured by the superinfecting phage technique of Meynell, *Immunology* 41:973.
Hormaeche, C.E., Brock, J., and Pettifor, P., 1980b, Natural resistance to mouse typhoid:possible role for the macrophage, *in*: "Genetic control of natural resistance to infection and Malignancy", E. Skamene, P.A.L. Kongshavn, and M. Landy, ed., Academic Press, New York.
Langermans, J.A.M., Van der Hulst, M.E.B., Nibbering, P.H., and Van Furth, R., 1990, Activation of mouse peritoneal macrophages during infection with *Salmonella typhimurium* does not result in enhanced intracellular killing, *J. Immunol.* 144:4340.
Koerner, T.J., Hamilton, T.A., and Adams, D.O., 1987, Suppressed expression of surface Ia on macrophages by lipopolysaccharide: evidence for regulation at the leve of accumulation of mRNA. *J. Immunol.* 139:239.
Langermans, J.A.M., Nibbering, P.H., Van der Hulst, M.E.B., and Van Furth, R., 1991, Microbicidal activities of *Salmonella typhimurium*- and interferon-gamma-activated mouse peritoneal macrophages, *Pathobiology* 59:189.
Langermans, J.A.M., Van der Hulst, M.E.B., Nibbering, P.H., Van der Meide, P.H., and Van Furth, R., 1992, Intravenous injection of IFN-γ-induced L-arginine-dependent toxoplasmastatic activity in murine peritoneal macrophages is mediated by endogenous tumor necrosis factor-α. *J. Immunol.* 148:568.
Lissner, C.R., Swanson, R.N., and O'Brien, A.D., 1983, Genetic control of the innate resistance of mice to *Salmonella typhimurium*: expression of the Ity gene in peritoneal and splenic macrophages isolated in vitro, *J. Immunol.* 131:3006.
Lissner, C.R., Weinstein, D.L., and O'Brien, A.D., 1985, Mouse chromosome 1 Ity locus regulates microbicidal activity of isolated peritoneal macrophages against a diverse group of intracellular and extracellular bacteria, *J. Immunol.* 135:544.
Mackaness, G.B., 1964, The immunological basis of acquired cellular resistance, *J. Exp. Med.* 120:105.
Mackaness, G.B., Blanden, R.V., and Collins, F.M., 1966, Hostparasite relations in mouse typhoid. *J. Exp. Med.* 124:573.
Mackaness, G.B., 1971, Resistance to intracellular infection, *J. Infect. Dis.* 123:439.
Maier, T., and Oels, H.C., 1972, Role of the macrophage in natural resistance to Salmonellosis in mice, *Infect. Immun.* 6:438.
Murray, H.W., and Cohn, Z.A., 1980, Macrophage oxygen-dependent antimicrobial activity. III. Enhancate metabolism as an expression of macrophage activation, *J. Exp. Med.* 152:1596.
Murray, H.W., Spitalny, G.L., and Nathan, C.F., 1985, Activation of mouse peritoneal macrophages in vitro and in vivo by interferon-γ, *J. Immunol.* 134:1619.
Nakoneczna, I., and Hsu, H.S., 1980, The comparative histopathology of primary and secondary lesions in murine salmonellosis, *Br. J. Exp. Pathol.* 61:76.
Nathan, C.F., Murray, H.W., and Cohn, Z.A., 1980, The macrophage as an effector cell, *N. Engl. J. Med.* 303:622.
Nibbering, P.H., Langermans, J.A.M., Van de Gevel, J.S., Van der Hulst, M.E.B., and Van Furth, R., 1991, Nitrite production by activated murine macrophages correlates with their toxoplasmastatic activity, Ia antigen expression, and production of H_2O_2 *Immunobiology* 184:93.
O'Brien, A.D., Scher, I., and Formai, S.B., 1979, Effect of silica on the innate resistance of inbred mice to *Salmonella typhimurium* infection, *Infect. Immun.* 25:513.

O'Brien, A.D., and Metcalf, E.S., 1982, Control of early *Salmonella typhimurium* growth in innately *Salmonella*-resistant mice does not require functional T lymphocytes, *J. Immunol.* 129:1349.

Pappas, M.P., and Nacy, C.A., 1983, Antileishmanial activities of macrophages from C3H/HeN and C3H/HeJ mice treated with Mycobacterium bovis strain BCG, *Cell. Immunol.* 80:217.

Plant, J., and Glynn, A.A., 1976, Genetics of resistance to infection with *Salmonella typhimurium* in mice, *J. Infect. Dis.* 133:72.

Plant, J.E., Blackwell, J.M., O'Brien, A.D., Bradley, D.J., and Glynn, A.A., 1982, Are the Lsh and Ity disease genes at one locus on mouse chromosome 1?, *Nature* 297:510.

Robson, H.G., and Vas, S.I., 1972, Resistance of inbred mice to *Salmonella typhimurium*, *J. Infect. Dis.* 126:378.

Ruco, L.P., and Meltzer, M.S., 1977, Macrophage activation for tumor cytotoxicity: induction of tumoricidal macrophages by PPD in BCG-immune mice, *Cell. Immunol.* 32:203.

Schafer, R., and Eisenstein, T.K., 1988, Induction of natural killer cell activity by a *Salmonella* vaccin. *FASEB J.* 2:A677.

Skamene, E., Gros, P., Forget, A., Kongshavn, P.A.L., St.Charles, C., and Taylor, B.A., 1982, Genetic regulation of resistance to intracellular pathogens, *Nature* 297:506.

Sluiter, W., Elzenga-Claasen, I., Van der Voort van der Kley-van Andel, A., and Van Furth, R., 1984, Differences in the response of inbred mouse strains to the factor increasing monocytopoiesis, *J. Exp. Med.* 159:524.

Stach, J.L., Gros, P., Forget, A., and Skamene, E., 1984, Phenotypic expression of genetically-controlled natural resistance to Mycobacterium bovis (BCG), *J. Immunol.* 132:888.

Stevenson, M.M., Kongshavn, P.A.L., and Skamene, E., 1981, Genetic linkage of resistance to *Listeria monocytogenes* with macrophage inflammatory responses, *J. Immunol.* 127:402.

Van Dissel, J.T., Leijh, P.C.J., and Van Furth, R., 1972, Differences in initial rate of intracellular killing of *Salmonella typhimurium* by resident peritoneal macrophages from various mouse strains, *J. Immunol.* 134:3404.

Van Dissel, J.T., Stikkelbroeck, J.J.M., Sluiter, W., Leijh, P.C.J., and Van Furth, R., 1986, Differences in initial rate of intracellular killing of *Salmonella* typhimurium by granulocytes of *Salmonella*-susceptible C57BL/10 mice and *Salmonella*-resistant CBA mice, *J. Immunol.* 136:1074.

Van Dissel, J.T., Stikkelbroeck, J.J.M., Michel, B.C., Leijh, P.C.J., and Van Furth, R., 1987a, *Salmonella typhimurium*-specific difference in rate of intracellular killing by resident peritoneal macrophages from *Salmonella*-resistant CBA and *Salmonella*-susceptible C57BL/10 mice, *J. Immunol.* 138:4428.

Van Dissel, J.T., Stikkelbroeck, J.M., Van den Barselaar, M. Th., Sluiter, W., Leijh, P.C.J., and Van Furth, R., 1987b, Divergent changes in antimicrobial activity after immunologic activation of mouse peritoneal macrophages, *J. Immunol.* 139:1665.

Van Dissel, J.T., Stikkelbroeck, J.J.M., Michel, B.C., Van den Barselaar, M.Th., Leijh, P.C.J., and Van Furth, R., 1987c, Inability of recombinant interferon-" to activate the antibacterial activity of mouse peritoneal macrophages against *Listeria monocytogenes* and *Salmonella typhimurium*, *J. Immunol.* 139:1673.

Vas, S.I., 1978, Genetic control of defenses against intracellular pathogens, *in*: "Infection Immunity, and Genetics", H. Friedman, T.J. Linna, and J.E. Prier, eds., University Park Press, Baltimore

Wang, X., Lin, F., Hsu, H.S., Mumaw, V.R., and Nakoneczna, I., 1988, Electronmicroscopic studies on the location of *Salmonella* proliferation in murine spleen, *J. Med. Microbiol.* 25:41.

ROLE OF GAMMA INTERFERON AND TUMOR NECROSIS FACTOR IN EARLY RESISTANCE TO MURINE SALMONELLOSIS

Charles Nauciel, Florence Espinasse-Maes, and Peggy Matsiota-Bernard
Laboratoire de Microbiologie, Faculté de Médecine de Paris-Ouest
Université Paris 5, 92380 Garches, France

INTRODUCTION

It is generally assumed that *Salmonella typhimurium* is a facultative intracellular pathogen although this point is controversial (Hsu, 1989). This bacterium induces in mice a disease similar to human typhoid fever. After an intravenous (i.v.) challenge with a virulent strain the bacteria replicate essentially in the spleen and liver. During the first few days of infection natural resistance is controlled by several genes, particularly *Ity* which regulates the growth rate of *S. typhimurium* in resident macrophages (O'Brien, 1986; Benjamin et al., 1990). In sublethal infections acquired resistance is expressed by an inhibition of bacterial growth in the organs, characterized by a plateau phase (Hormaeche et al., 1990) and later by progressive clearance. Moreover, infected animals develop increased resistance to reinfection, which appears within the first few days of infection (Nauciel et al., 1985). Resistance to reinfection is non-specific (i.e. directed against other intracellular pathogens) as long as mice are harboring bacteria from the primary inoculum (Blanden et al., 1966; Nauciel et al., 1985).

The relative importance of cell-mediated and humoral immunity in acquired resistance to *Salmonella typhimurium* has been the subject of controversy (Collins, 1974; Eisenstein and Sultzer, 1983). Previous studies have shown that the clearance of bacteria, which begins after about 2 weeks, is regulated by the H-2 complex (Hormaeche et al., 1985; Nauciel et al., 1988) and mediated by T cells (Nauciel, 1990). Athymic nude mice fail to clear bacteria and so do euthymic mice depleted of $CD4^+$ T cells by monoclonal antibodies (mAb). $CD4^+$ T cells

Biology of Salmonella, Edited by F. Cabello *et al.*,
Plenum Press, New York, 1993

are also involved in resistance to reinfection in the late phase of primary infection (Nauciel, 1990).

In contrast, depletion of $CD4^+$ or $CD8^+$ T cells in vivo does not impair the host response during the first 2 weeks, despite evidence for increased resistance (Nauciel, 1990). This suggests that T cell-independent mechanisms of resistance are involved at this stage. The aim of the present study was to determine whether cytokines known to activate macrophages, such as gamma-interferon (IFN-γ) and tumor necrosis factor-α (TNF-α), could play a role in early resistance to infection by *S. typhimurium*

MATERIALS AND METHODS

Bacteria

Virulent *S. typhimurium* C5, its temperature-sensitive avirulent mutant C5TS (Hormaeche et al., 1981) and a virulent strain of *Listeria monocytogenes* were grown overnight in tryptic soy broth at 37°C (or 30°C for C5TS).

Infection of Mice

Genetically resistant (*Ity*^{r}) female CBA mice purchased from Iffa Credo (L'Arbresles, France) were used between 6 and 8 weeks of age. Mice were inoculated i.v. with 0.2 ml of an appropriate dilution of bacteria in saline. Groups of 3 to 4 mice were killed at various times. Spleens were homogenized in 2 ml of distilled water. Samples of the homogenates and 10-fold serial dilutions in saline were plated onto tryptic soy agar. Colonies were counted after overnight incubation at 37°C.

Anti-Cytokines Antibodies

Hybridoma R4-6A2 (Spitalny and Havell, 1984) producing rat IgG1 against murine IFN-γ was kindly provided by G. Milon (Institut Pasteur, Paris). The hybridoma was grown intraperitoneally in pristane-primed Swiss nude mice. Antibodies were partially purified from ascites by 50% ammonium sulfate precipitation. The preparation contained 17 mg of protein per ml and had a neutralizing titer of 1/10,000 against 10 U/ml of recombinant mouse IFN-γ (Holland Biotechnology bv, Leiden, The Netherlands). An appropriate dilution of this preparation was administrated intraperitoneally. Ascites containing mAb against an irrelevant antigen (penicilloyl group) was used as a control.

Antibodies against TNF-α were obtained by immunizing a rabbit with 10 μg of recombinant murine TNF-α (Genenzyme, Boston, Mass.) in Freund's complete adjuvant.

The rabbit was boosted 1 and 2 months later with the same amount of TNF-α in incomplete adjuvant, and bled 1 week after each boost. The serum was precipitated with 33% ammonium sulfate and dialyzed against phosphate buffered saline. One milliliter of this preparation neutralized 10^5 units of murine TNF-α. TNF-α activity was assayed on L-929 cells as described by Fish and Gifford (1983). Anti-TNF-α antibodies were administrated by i.v. route. Normal rabbit serum precipitated with 33% ammonium sulfate was used as a control.

T-Cell Subset Depletion

The mAb produced by hybridomas GK1.5 (anti-CD4) and H35-17.2 (anti-CD8) were partially purified from ascites as previously described (Nauciel, 1990). C5-infected CBA mice were injected intraperitoneally on day 0 with 2 mg of anti-CD4 mAb, or i.v. with 400 μg of anti-CD8 mAb. Control mice received 0.2 ml of ascites containing an irrelevant mAb.

The percentage of $CD4^+$ and $CD8^+$ T cells in the spleen was determined in a flow cytometer (Facscan, Becton Dickinson, Mountain View, Calif.) as previously described (Nauciel, 1990).

Statistical Analysis

Student's unpaired *t* test was used to determine the significance of differences between control and experimental groups.

RESULTS

Effect of anti-IFN-γ mAb and anti-TNF-α Antibodies on Primary Infection by *S. typhimurium*

CBA mice were infected i.v. with a sublethal dose (10^3 CFU) of *S. typhimurium* C5 and treated with 400 μg of anti-IFN-γ mAb, 0.2 ml of anti-TNF-α or control antibodies. Spleen counts increased progressively in control mice until day 7 and then reached a plateau. In mice treated with anti-IFN-γ or anti-TNF-α antibodies, spleen counts were not different from control values until day 4 but continued to increase rapidly over the following days (Fig. 1) and animals died between day 7 and 9.

The response to anti-IFN-γ mAb was dose-dependent (Fig. 2). As little as 25 μg of mAb induced a significant increase in spleen counts at day 7.

In order to determine whether IFN-γ was produced by T cells during the early phase of infection, the effect of *in vivo* T cell depletion was compared to the effect of anti-IFN-γ mAb administration. Administration of anti-CD4+ or anti-CD8+ mAb induced a depletion of more than 90% of the corresponding T cell subset at day 7, as shown by flow cytometric analysis, but did not significantly modify spleen counts at that time (data not shown). In mice depleted of both T cell subpopulations spleen counts were significantly higher than in controls ($P<0.01$) but significantly lower than in mice treated with anti-IFN-γ mAb ($P<0.01$) (Fig. 3). These results suggest that during week 1 of infection IFN-γ is produced, at least in

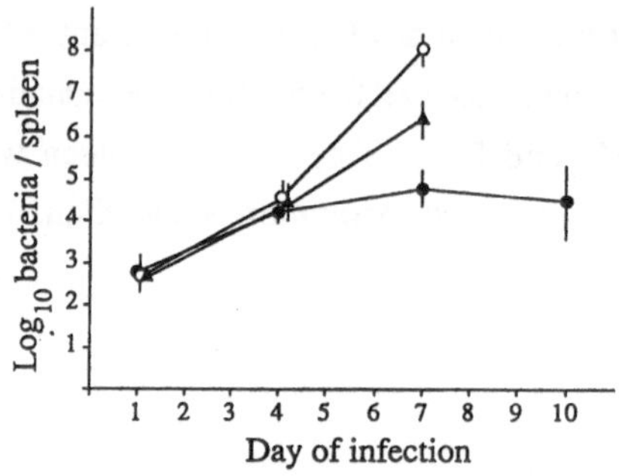

Figure 1. Time course of infection in mice inoculated with 10^3 CFU of *S. typhimurium* C5 and treated at day 0 with anti-IFN-γ (O), anti-TNF-α (▲) or control antibodies (●). Data are the geometric mean number of bacteria recovered from the spleens of three mice per group ± SD. Reproduced from *Infection and Immunity* (Nauciel and Espinasse-Maes, 1992) with permission of the publisher

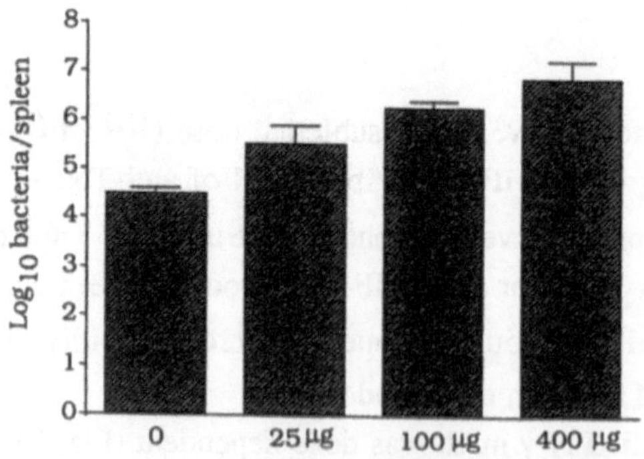

Figure 2. Dose response to anti-IFN-γ mAb of mice infected with 10^3 CFU of *S. typhimurium* C5. The indicated doses of mAb were injected on day 0 and spleen counts carried out on day 7. Reproduced from *Infection and Immunity* (Nauciel and Espinasse-Maes, 1992) with permission of the publisher.

part, by cells other than T lymphocytes. In addition, anti-IFN-γ and anti-TNF-α antibodies showed a synergistic interaction (Fig. 3).

Effect of anti-IFN-γ and anti-TNF-α Antibodies on Resistance to Reinfection.

Mice immunized with an avirulent strain of *S. typhimurium* exhibit increased resistance to a secondary challenge with a virulent strain of the same species or with another intracellular pathogen such as *L. monocytogenes*. This increased resistance is detectable at day 2 postinfection and its maximum level is reached at day 8 (Nauciel et al., 1985). In order to analyze the role of IFN-γ and TNF-α in this increase of resistance, CBA mice were infected with 10^6 CFU of the avirulent *S. typhimurium* C5TS strain and treated on day 6 with 400 μg

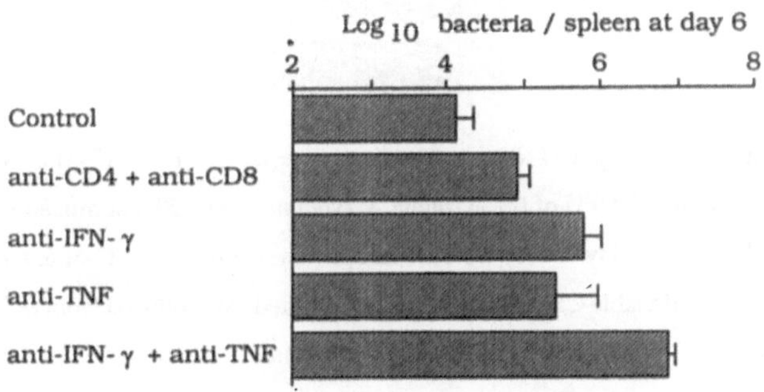

Figure 3. Effects of T cell depletion, anti-IFN-γ and anti-TNF-α antibodies on resistance to primary infection with *S. typhimurium*. CBA mice were infected with 10^3 CFU of *S. typhimurium* C5 and treated 1 h later with indicated antibodies. Results are the geometric mean number ± SD of bacteria recovered on day 6 from the spleens of four mice per group.

of anti-IFN-γ mAb and/or 0.2 ml of anti-TNF-α antibodies. Controls received irrelevant antibodies. On day 7 mice were challenged i.v. with either 10^5 CFU of *S. typhimurium* C5 or 10^5 CFU of *L. monocytogenes* and the number of challenge organisms per spleen was determined 3 days later. In mice challenged with *S. typhimurium* C5, anti-IFN-γ and anti-TNF-α antibodies given separately induced a slight but significant ($P < 0.05$) increase in spleen counts, while simultaneous administration had a more marked effect ($P < 0.001$) (Fig.4). However, the protective effect induced by the avirulent strain of *S. typhimurium* was not totally abrogated by the treatment.

In mice challenged with *L. monocytogenes*, anti-IFN-γ and anti-TNF-α antibodies, again induced a slight but significant ($P < 0.01$) increase in spleen counts when given separately. The simultaneous administration of the two antibodies totally abrogated the non-specific resistance to *L. monocytogenes* (Fig. 5).

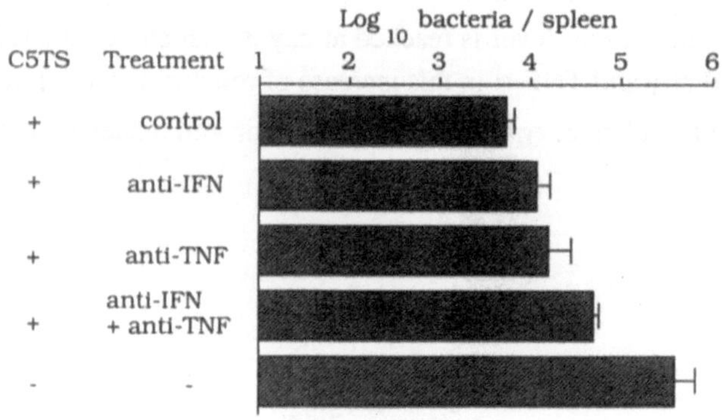

Figure 4. Effect of anti-IFN-γ and anti-TNF-α antibodies on resistance to reinfection with *S. typhimurium*. CBA mice were infected with 10^6 CFU of the avirulent *S. typhimurium* C5TS strain, and treated on day 6 with the indicated antibodies. On day 7, they were challenged i.v. with 10^5 CFU of *S. typhimurium* C5. Three days later, the number of viable C5 per spleen was determined. Reproduced from *Infection and Immunity* (Nauciel and Espinasse-Maes, 1992) with permission of the publisher.

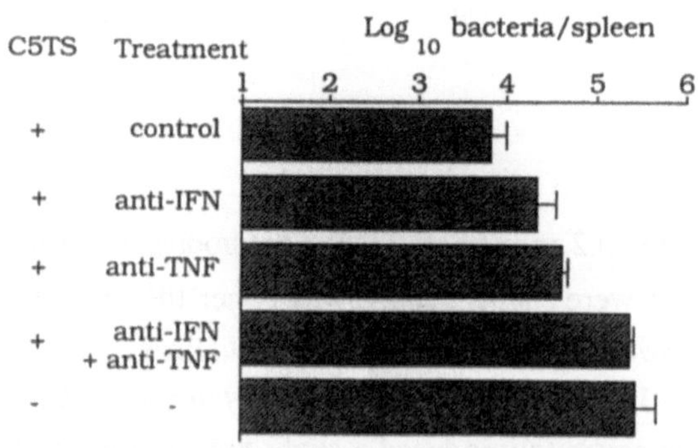

Figure 5. Effect of anti-IFN-γ and anti-TNF antibodies on nonspecific resistance to reinfection. CBA mice were infected with 10^6 CFU of *S. typhimurium* C5TS and treated on day 6 with the indicated antibodies. On day 7, they were challenged i.v. with 10^5 CFU of *L. monocytogenes*. Three days later, the number of viable *L. monocytogenes* per spleen was determined. Reproduced from *Infection and Immunity* (Nauciel and Espinasse-Maes, 1992) with permission of the publisher.

DISCUSSION

Our results show that IFN-γ and TNF-α play a critical role in the early phase of *S. typhimurium* infection. When antibodies against one of these cytokines were given at the onset of a primary infection with a sublethal dose, the bacteria continued to grow exponentially in the organs until the animals' death, whereas bacterial growth was inhibited at the end of the first week in controls. The increase in resistance which develops at the end of the first week in normal animals appears to be mediated by activated macrophages (Killar and Eisenstein, 1985; Hormaeche et al., 1990). It is likely that IFN-γ and TNF-α play a role in macrophage activation.

Other studies have shown that exogenous IFN-γ increases resistance to *S. typhimurium* infection (Matsumura et al., 1990) and bactericidal activity of macrophages against this bacterium (Kagaya et al., 1989). Van Dissel et al. (1987) have, however, reported conflicting results. Anti-IFN-γ antibodies have also been shown to decrease resistance to *S. typhimurium* (Muotiala and Makela, 1990) and *L. monocytogenes* (Buchmeier and Schreiber, 1985) infection. The cellular origin of IFN-γ in the early phase of *S. typhimurium* infection is not known. Depletion of both $CD4^+$ and $CD8^+$ T cells *in vivo* slightly depressed resistance to *S. typhimurium* infection, but to a lesser extent than anti-IFN-γ mAb. This suggests that the major source of IFN-γ is not T cells at this stage of infection. It has been shown that natural killer cells can produce IFN-γ in *L. monocytogenes* infection (Bancroft et al., 1989; Dunn and North, 1991),but further studies are needed to determine whether this also occurs in *S. typhimurium* infection.

The role of TNF-α in resistance to infection has been demonstrated in various experimental models. Exogenous TNF-α can increase resistance to *S. typhimurium* infection (Nakano et al., 1990). Anti-TNF-α antibodies depress resistance to *L. monocytogenes* (Havell, 1987), *Chlamydia trachomatis* (Williams et al., 1990) and *Mycobacterium bovis* BCG (Kindler et al., 1989) infections. Mastroeni et al.(1991) and Tite et al. (1991) have also reported independently that anti-TNF-α antibodies decrease resistance to *S. typhimurium* infection. TNF-α can activate the bactericidal activity of macrophages (Bermudez and Young, 1998; Denis, 1991) and play a role in the development of granulomas (Kindler et al., 1989).

Our data also show that IFN-γ and TNF-α participate in resistance to reinfection at week 1 postinfection. A recent report by Tite et al. (1991) has also shown that the administration of anti-TNF-α antibodies to mice immunized with an attenuated strain of *S. typhimurium* abrogates resistance to reinfection with a virulent strain. Interestingly anti-TNF-α antibodies also abrogate resistance to reinfection in the late phase of infection (Tite et al., 1991) when acquired resistance is T-dependent (Nauciel, 1990).

IFN-γ and TNF-α interacted synergistically to increase resistance to *S. typhimurium* infection. Such a synergistic interaction has been reported in several studies (Liew et al., 1990; Nakane et al., 1989). IFN-γ increases the number of TNF-α receptors (Aggarwall et al., 1985) and the transcription of TNF-α mRNA (Collart et al. 1986) in macrophages.

Moreover, TNF-α has been shown to be necessary for the production of IFN-γ by natural killer cells (Bancroft et al., 1989).

Taken together, our results show that IFN-γ and TNF-α are produced in the early phase of *S. typhimurium* infection. The production of these cytokines seems to be an early mechanism of nonspecific resistance which is of critical importance for the survival of the host before the development of the specific immune response.

REFERENCES

Aggarwall, B.B., Eesallu, T.E., and Hass, P.E., 1985, Characterization of receptors for human tumor necrosis factor and their regulation by γ-interferon, *Nature* (London), 318:665

Bancroft, G.J., Sheehan, K.C.F., Schreiber, R.D., and Unanue, E.R., 1989, Tumor necrosis factor is involved in the T cell-independent pathway of macrophage activation in *scid* mice, *J. Immunol.* 143:127.

Benjamin Jr., W.H., Hall, P., Roberts, S.J., and Briles, D.E., 1990, The primary effect of the *Ity* locus is on the rate of growth of *Salmonella typhimurium* that are relatively protected from killing, *J. Immunol.* 144:3143.

Bermudez, L.E.M., and Young, L.S., 1988, Tumor necrosis factor, alone or in combination with IL-2, but not IFN-γ, is associated with macrophage killing of *Mycobaterium avium* complex, *J. Immunol.* 140:3006.

Blanden, R.V., Mackaness, G.B., and Collins, F.M., 1966, Mechanisms of acquired resistance in mouse typhoid, *J. Exp. Med.* 124:585.

Buchmeier, N.A., and Schreiber, R.D., 1985, Requirement of endogenous interferon-γ production for resolution of *Listeria monocytogenes* infection, *Proc. Natl. Acad. Sci.* USA. 82:7404.

Collart, M.A., Belin, D., Vassali, J.D., de Kossodo, S., and Vassali, P., 1986, γ interferon enhances macrophage transcription of the tumor necrosis factor/cachectin, interleukin 1, and urokinase genes, which are controlled by short-lived repressors, *J. Exp. Med.* 164:2113.

Collins, F.M., 1974, Vaccines and cell-mediated immunity, *Bact. Rev.* 38:371.

Denis, M., 1991, Modulation of *Mycobacterium lepraemurium* growth in murine macrophages : beneficial effect of tumor necrosis factor alpha and granulocyte-macrophage colony-stimulating factor, *Infect. Immun.* 59:705.

Dunn, P.L., and North, R.J., 1991, Early gamma interferon production by natural killer cells is important in defense against murine listeriosis, *Infect. Immun.* 59:2892.

Eisenstein, T.K., and Sultzer, B.M., 1983, Immunity to Salmonella infection, *Adv. Exp. Med. Biol.* 162:261.

Fish, H., and Gifford, G.E., 1983, *In vitro* production of rabbit macrophage tumor cell cytotoxin, *Int. J. Cancer* 32:105.

Havell, E.A., 1987, Production of tumor necrosis factor during murine listeriosis, *J. Immunol.* 139:4225.

Hormaeche, C.E., Harrington, K.A., and Joysey, H.S., 1985, Natural resistance to Salmonellae in mice : control by genes within the major histocompatibility complex, *J. Infect. Dis.* 152:1050.

Hormaeche, C.E., Mastroeni, P., Arena, A., Uddin, J., and Joysey, H.S., 1990, T cells do not mediate the initial suppression of a salmonella infection in the RES, *Immunology* 70:247.

Hormaeche, C.E., Pettifor, R.A., and Brock, J., 1981, The fate of temperature-sensitive Salmonella mutants *in vivo* in naturally resistant and susceptible mice, *Immunology* 42:569.

Hsu, H.S., 1989, Pathogenesis and immunity in murine salmonellosis, *Microbiol. Rev.* 53:390.

Kagaya, K., Watanabe, K., and Fukazawa, Y., 1989, Capacity of recombinant gamma-interferon to activate macrophages for *Salmonella*-killing activity, *Infect. Immun.* 57:609 .

Killar, L.M., and Eisenstein, T.K., 1985, Immunity to *Salmonella typhimurium* infection in C3H/HeJ and C3H/HeNCrBR mice : studies with an aromatic-dependent live *S. typhimurium* strain as a vaccine,. *Infect. Immun.* 47:605.

Kindler, V., Sappino, A.P., Grau, G.E., Piguet, P.F., and Vassali, P., 1989, The inducing role of tumor necrosis factor in the development of bactericidal granulomas during BCG infection, *Cell* 56:731.

Liew, F.Y., Li, Y., and Millott, S., 1990, Tumor necrosis factor-α synergizes with IFN-γ in mediating killing of *Leishmania major* through the induction of nitric oxide, *J. Immunol.* 145:4306.

Mastroeni, P., Arena, A., Costa, G.B., Liberto, M.C., Bonina, L., and Hormaeche, C.E., 1991, Serum TNF-α in mouse typhoid and enhancement of a *Salmonella* infection by anti-TNF-α antibodies, *Microb. Pathog.* 11:33.

Matsumura, H., Onozuka, K., Terada, Y., Nakano, Y., and Nakano, M., 1990, Effect of murine recombinant interferon-γ in the protection of mice against Salmonella, *Int. J. Immunopharmac.*12:49 (1990).

Muotiala, A., and Makela, P.H., 1990, The role of IFN-γ in murine *Salmonella typhimurium* infection, *Microb. Pathog.* 8:135.

Nakane, A., Minagawa, T., Kohanawa, M., Chen, Y., Sato, H., Moriyama, M., and Tsuruoka, N., 1989, Interactions between endogenous gamma interferon and tumor necrosis factor in host resistance against primary and secondary *Listeria monocytogenes* infections, *Infect. Immun.* 57:3331.

Nakano, Y., Onozuka, K., Terada, Y., Shinomiya, H., and Nakano, M., 1990, Protective effect of recombinant tumor necrosis factor-α in murine salmonellosis, *J. Immunol.* 144:1935.

Nauciel, C., 1990, Role of $CD4^+$ T cells and T-independent mechanisms in acquired resistance to *Salmonella typhimurium* infection, *J. Immunol.* 145:1265.

Nauciel, C., and Espinasse-Maes, F., 1992, Role of gamma interferon and tumor necrosis factor alpha in resistance to *Salmonella typhimurium* infection, *Infect. Immun.* 60:450.

Nauciel, C., Ronco, E., Guenet, J.L., and Pla, M., 1988, Role of *H-2* and non-*H-2* genes in control of bacterial clearance from the spleen in *Salmonella typhimurium*-infected mice, *Infect. Immun.* 56:2407.

Nauciel, C., Vildé, D., and Ronco, E., 1985, Host response to infection with a temperature-sensitive mutant of *Salmonella typhimurium* in a susceptible and a resistant strain of mice, *Infect. Immun.* 49: 523.

O'Brien, A.D., 1986, Influence of host genes on resistance of inbred mice to lethal infection with *Salmonella typhimurium*, *Curr. Top. Microbiol. Immunol.* 124:37.

Spitalny, G.L., and Havell, E.A., 1984, Monoclonal antibody to murine gamma interferon inhibits lymphokine-induced antiviral and macrophage tumoricidal activities, *J. Exp. Med.* 159:1560.

Tite, J.P., Dougan, G., and Chatfield, S.N., 1991, The involvement of tumor necrosis factor in immunity to *Salmonella* infection, *J. Immunol.* 147:3161.

Van Dissel, J.T., Stikkelbroeck, J.J.M., Michel, B.C., Van den Barselaar, M.T., Leijh, P.C.J., and Van Furth, R., 1987, Inability of recombinant interferon-γ to activate the antibacterial activity of mouse peritoneal macrophages against *Listeria monocytogenes* and *Salmonella typhimurium*, *J. Immunol.* 139:1673.

Williams, D.M., Magee, D.M., Bonewald, L.F., Smith, J.G., Bleicker, C.A., Byrne, G.I., and Schachter, J., 1990, A role *in vivo* for tumor necrosis factor alpha in host defense against *Chlamydia trachomatis*, *Infect. Immun.* 58:1572.

IMMUNITY AND IMMUNOSUPPRESSION INDUCED BY ATTENUATED *SALMONELLA*: THE ROLE OF NITRIC OXIDE

T. K. Eisenstein, B. K. Al-Ramadi, D. Huang, R. Schafer, L. Killar, J.-C. Lee, and J. J. Meissler, Jr.

Department of Microbiology and Immunology
Temple University School of Medicine
Philadelphia, PA 19140

INTRODUCTION

Our laboratory has been investigating mechanisms of immunity to Salmonella infection for many years. Previous work showed that nonviable vaccines, in particular acetone-killed and dried cells (AKC) of *Salmonella typhimurium*, were capable of giving high levels of protection in mice which were genetically inherently resistant to Salmonella infection, but protected very poorly in mice which were innately hypersusceptible to Salmonella infection (Eisenstein et al., 1984*b*). Extension of this observation showed that passive high-titered serum raised to AKC could protect inherently resistant but not inherently hypersusceptible mice (Eisenstein et al., 1984*b*). Studies initiated nearly 10 years ago tested the protective capacity of an aroA⁻ mutant of *S. typhimurium* derived by Hoiseth and Stocker as a prototype for oral, attenuated vaccines against typhoid (Hoiseth and Stocker, 1981). We confirmed Stocker's findings that SL3235, a *Salmonella* mutant with an aroA⁻ lesion, was highly protective in mice against virulent *Salmonella* challenge and showed that it was very effective in both genetically hypersusceptible and resistant mouse strains in the C3H lineage. In the course of our investigations, however, we made a paradoxical observation; namely, that the attenuated *Salmonella* induced strong parameters of immunosuppression while it was conferring protection (Eisenstein et al., 1984*a*). Our investigations over the last several years have probed the suppressive phenomenon to discover the mechanisms involved. Our findings lead to the conclusion that the attenuated *Salmonella* are powerful macrophage-activating agents and that activated macrophages may be responsible for both protection and immunosuppression.

Protection Experiments

The standard immunization and challenge procedure which we adopted to evaluate various vaccines was to immunize with the live aroA⁻ strain SL3235 on day 0

Biology of Salmonella, Edited by F. Cabello *et al.*,
Plenum Press, New York, 1993

Table 1. Protection induced by SL3235 in C3H/HeJ and C3HeB/FeJ mice.

	Survival[2] (alive/total)	
Day of challenge[1]	C3H/HeJ	C3HeB/FeJ
7	6/6[3]	N.D.
21	8/11[4]	12/12[5]

[1]Mice previously immunized i.p. with SL3235.
[2]Mice challenged i.p. with *S. typhimurium* W118-2. LD_{50} = 1 cell.
[3]Challenge dose = 1,290 LD_{50}s.
[4]Challenge dose = 9,000 LD_{50}s.
[5]Challenge dose = 12,000 LD_{50}s.

and challenge with virulent *S. typhimurium* W118-2 21 days later. As shown in Table 1, strain SL3235 protected against approximately 10,000 LD_{50} doses of attenuated *Salmonella* when the immunization and challenge were intraperitoneal (Eisenstein et al., 1984*a*). Other experiments designed to monitor the time of onset of protection showed that substantial levels of protection were also evident 7 days postchallenge against greater than 1,000 LD_{50} doses (Killar and Eisenstein, 1985). (Higher doses were not tested at 7 days, so the strength of the protection at that time might have been even higher.) It is evident that SL3235 confers high levels of protection at 7 and 21 days. In C3H/HeJ mice, we found that the protection was long lasting, as animals were resistant to challenge doses of at least 1,000 LD_{50} as long as 6 months after vaccination (longer times were not tested) (Killar and Eisenstein, 1985). In addition, it was found that SL3235 induced cross-protection against challenge with *Listeria monocytogenes* that was evident as early as 3 days after vaccination, strong for the first 2 weeks, waning by 21 days, and gone by 30 days. The temporal pattern of cross-protection by one intracellular pathogen against challenge with a second, fit the pattern of "cellular immunity" described by Mackaness, and thought to be attributable to transiently activated macrophages (Mackaness, 1971). Direct evidence for induction of activated macrophages was the observation that peritoneal macrophages harvested from animals 7 days after vaccination with SL3235 inhibited replication of *Leishmania major* (Schafer et al., 1988). Additional evidence for activated macrophages was obtained by investigating the tumoricidal capacity of cells harvested from SL3235-immunized mice. It was found that SL3235 was capable of inducing tumoricidal macrophages in C3H/HeJ mice, as well as in C3HeB/FeJ mice, 7 to 10 days postvaccination (Table 2) (Schafer et al., 1988). It is noteworthy that BCG is not capable of activating C3H/HeJ macrophages to tumoricidal activity, owing to a genetic defect in macrophage activation in this strain. The macrophage-activating capacity of SL3235 was strong enough to overcome this defect. AKC of SL3235 could prime C3H/HeJ mice, but could not fully activate them to tumoricidal activity (Schafer et al., 1988). SL3235 was also tested as a therapeutic agent against a plasmacytoma in ICR mice and found to significantly increase survival (Schafer et al., 1986). We also found that splenic fibronectin-adherent cells harvested 7 days after SL3235 (assumed to be splenic macrophages) could transfer resistance to *Salmonella* (Killar and Eisenstein, 1985).

Table 2. Induction of tumoricidal macrophages by SL3235 and BCG.

	CPM ± S.D. (% cytotoxicity)[2]	
Stimulating agent[1]	C3H/HeJ	C3HeB/FeJ
Saline	1%	1%
BCG	5%	57%
SL3235	40%	51%

[1]Injection of BCG given 12 days previously and SL3235 given 10 days prior to harvest of adherent peritoneal cells.

[2] ^{51}Cr release at 16 hr against P815 mastocytoma cells at an effector-to-target ratio of 10:1.

Table 3. Beneficial effects of attenuated *Salmonella.*

Protection against virulent *Salmonella* - 3 days to 6 months
Protection against *Listeria* - up to 21 days
Transfer of protection with splenic macrophages - 7 days
Tumoricidal peritoneal macrophages - 7 to 10 days
Leishmaniacidal peritoneal macrophages - 7 to 10 days
Therapeutic tumor regression in vivo

In summary, these studies showed that SL3235 induced high levels of protection against challenge with virulent *Salmonella* and was protective against *Listeria*. The vaccine induced activated macrophages which were protective, microbicidal and tumoricidal (Table 3).

Immunosuppression

We wished to test whether SL3235 could overcome the defect in responsiveness to the lipid A portion of lipopolysaccharide (LPS) in C3H/HeJ mice, since SL3235 overcame the macrophage activation defect of this strain. Spleen cells were harvested from C3H/HeJ mice 21 days after administration of SL3235 and stimulated in vitro with phenol-water-extracted LPS. As a control, concanavalin A (ConA) was also used. It was discovered that not only was the defect in LPS nonresponsiveness not reversed, but the responses to ConA were also suppressed (Eisenstein et al., 1984*a*). Expanded studies of mitogenic suppression (Lee et al., 1985) showed that responses to a panel of B and T cell mitogens were suppressed by 3 days postadministration of SL3235, and suppression was maximal by 7 days postvaccine injection (Fig. 1). Removal of adherent macrophages by two rounds of plastic adherence followed by passage over a Sephadex G10 column partially relieved mitogenic suppression. Selective depletion of B or T lymphocytes did not abrogate suppression. These observations suggested that macrophages mediated the suppression. The suppression was not inhibitable by prostaglandins (Lee et al., 1985).

To more directly assess the immunological significance of what appeared to be an immunosuppressive phenomenon, a new paradigm was used. Mice were vaccinated on day 0 intraperitoneally with SL3235, and 7 days later they were injected with sheep red blood cells intravenously. Four days later spleens of individual animals were removed, and the number of plaque-forming cells (PFCs) determined (Cunningham and Szenberg, 1968). Suppression of the PFC response was observed and it was found to be dependent on the dose of SL3235 (Fig. 2) (Al-Ramadi et al., 1992*a*). Similar experiments were carried out using tetanus toxoid as the immunogen. This soluble antigen did not result in a primary immune response, so the protocol was modified to include a booster dose. On day 0 animals received SL3235,

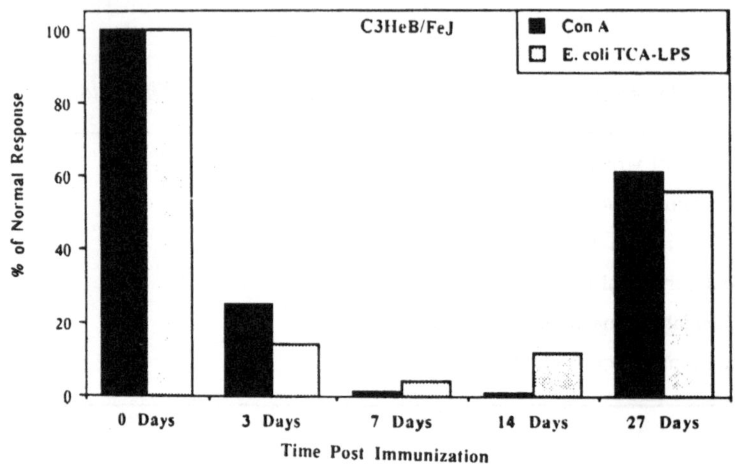

Figure 1. Suppression of mitogen responses in C3HeB/FeJ mice after SL3235 administration.

on day 7 they received the primary immunizing dose of tetanus toxoid, on day 14 they were boosted, and on day 21 individual sera were obtained and assayed for IgG anti-tetanus toxoid antibody titer by ELISA. Animals which had received SL3235 7 days prior to the primary dose of tetanus toxoid showed approximately a 10-fold lower titer to tetanus toxoid than controls (Fig. 3) (Al-Ramadi et al., 1992*a*).

To dissect the mechanism of suppression, an in vitro paradigm was developed. Animals were immunized on day 0 with SL3235, and 7 days later their spleen cells were removed and placed in Mishell-Dutton cultures in vitro, using sheep red blood cells as the immunogen. Five days later spleen cells were harvested, and the number of PFCs determined. It was shown that vaccination with SL3235 7 days prior to harvest of spleen cells, inhibited the primary in vitro PFC response to sheep red blood cells in a dose-dependent manner (Fig. 2).

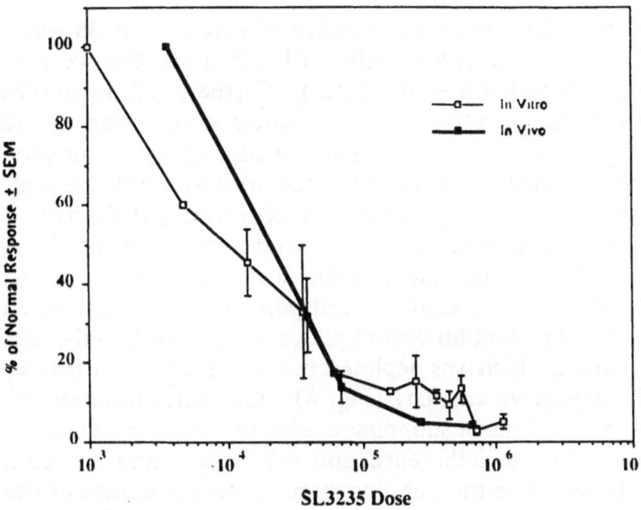

Figure 2. Immunization with SL3235 inhibits both the in vivo and in vitro plaque-forming cells responses to SRBC in a dose-dependent fashion in C3HeB/FeJ mice. [Al-Ramadi et al., 1992*a*. Copyright 1992, *Microbial Pathogenesis*]

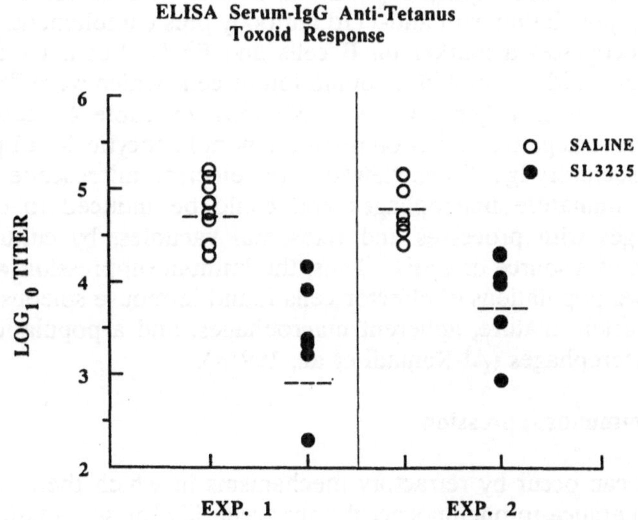

Figure 3. SL3235 inhibits the secondary in vivo IgG response to tetanus toxoid. Tetanus toxoid antibody responses were measured by ELISA, using toxoid-coated plates. [Al-Ramadi et al., 1992*a*. Copyright 1992, *Microbial Pathogenesis*]

Cells Mediating Immunosuppression

It was found that the immunosuppressive effects of SL3235 were not reversed by irradiation of the immune spleen cells with 500 rads, but were reversed when 3,000 rads was used (Al-Ramadi et al., 1992*a*). Further, cell metabolism was necessary for the suppression as mitomycin C inhibited suppression (Al-Ramadi et al., 1992*a*). To distinguish between suppression mediated by suppressor T cells and macrophages, immune spleen cells were treated with anti-Thy 1.2 and complement. No significant abrogation of suppression was observed (Al-Ramadi et al., 1991*a*). When immune spleen cells were subjected to one round of plastic adherence and passage over a G10 column, suppression was abrogated approximately 60% (Al-Ramadi et al., 1992*a*). The resulting cell population was further fractionated by panning over anti-immunoglobulin-coated plates to remove B cells. Surprisingly, the nonadherent population, which was depleted of mature adherent macrophages and of B cells, regained suppressive capacity (Fig. 4). Our interpretation of this result is that there are two populations of suppressor cells, the mature adherent macrophages, which were removed by the adherence and G10 steps, and a second population, which was concentrated after the panning step. By examination of the postpanning population, using two-color immunofluorescence in which Mac1 expression was monitored with phycoerythrin and phagocytosis was assessed using fluorescent latex beads, it was found that the postpanning population was Mac1 positive but not phagocytic (Al-Ramadi et al., 1991*a*). Light microscopic studies showed that the sequential purification steps led to a decrease in the number of cells with nonspecific esterase activity, so that the postpanning population was esterase negative (Al-Ramadi et al., 1991*a*). The postpanning population contained approximately 12% lymphocytes, 46% monocytes, and 42% polymorphonuclear leukocytes (PMNs) by differential count. In subsequent experiments PMNs were depleted by treatment of the postpanning population with anti-J11d antibody plus complement. This monoclonal antibody recognizes a marker on B cells and PMNs but not on monocytes. Treatment with anti-J11d resulted in a population of cells which were 75% mononuclear, 12% PMNs, and 13% lymphocytes. As shown in Table 4, these cells were nonadherent, esterase negative, radiation resistant, nonphagocytic, Mac1 positive, Thy 1.2 negative, and secretory Ig^-. When viewed in the electron microscope they had the characteristics of immature macrophages and could be induced to develop into typical macrophages with processes and lysosomal vacuoles, by culture in L-cell condition medium as a source of CSF1. Thus, the immunosuppression was shown to be mediated by two populations of effector cells found in mouse spleens 7 days after SL3235 administration; mature, adherent macrophages, and a population of immature, precursor macrophages (Al-Ramadi et al., 1991*a*).

Mechanisms of Immunosuppression

Suppression can occur by refractory mechanisms in which there is a response failure, such as tolerance to pneumococcal polysaccharide, or suppression can result from an active mechanism mediated by suppressor T cells or macrophages. In order to show whether or not the diminished number of PFCs observed after SL3235 administration was due to lack of responsiveness or to active suppression, coculture experiments were carried out. In these experiments, small numbers of immune cells were cultured with normal spleen cells. It was found that the immune cells could suppress normal cells. In order to assess more rigorously the capacity of immune cells to suppress normal cultures, an assay was developed using "transwell" plates in which normal spleen cells were placed in the bottom wells of a 24-well plate, and

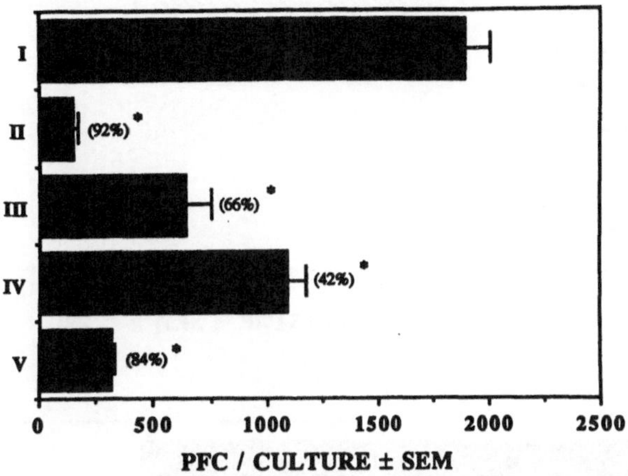

Figure 4. Changes in suppressive capacity of immune splenocytes subjected to different depletion procedures. Cells (5 x 10^6) from the indicated group were cocultured with 1 x 10^7 normal cells and PFCs were determined on day 5 of culture. Group *I*, normal cells; group *II*, whole immune spleen cells; group *III*, immune cells after 2-hr round of plastic adherence; group *IV*, as for group *III* plus passage over Sephadex G-10 column; group *V*, as for group *IV* plus panning on anti-mouse Ig-coated plates. Results are the means ± SEM of triplicate cultures. The numbers in parentheses indicate % suppression of the PFC responses in comparison with the control group (group *I*). Asterisks denote significant differences from control ($p < 0.005$). The SL3235 dose was 4.4 x 10^5 per mouse. Background PFCs (with no SRBC added) were typically <10% of experimental groups' responses, and have been omitted for clarity.

Table 4. Characteristics of suppressor cells.

1. Depletion of T lymphocytes has no effect on the suppressive capacity of immune spleen cells, suggesting that T cells are not involved in the induction of suppression.

2. Depletion of adherent macrophages results in a partial alleviation of suppression.

3. Sequential depletion leads to the enrichment of a highly suppressive population exhibiting the following characteristics:
   - $Mac1^+$, $Thy1.2^-$, $J11d^-$, and sIg^-
   - Non adherent
   - Non phagocytic
   - Radiation resistant
   - NS esterase negative

4. In the presence of L cell-conditioned medium, a rich source of CSF-1, but not Con A-supernatant, these cells mature into typical macrophages within 72 hr of culture.

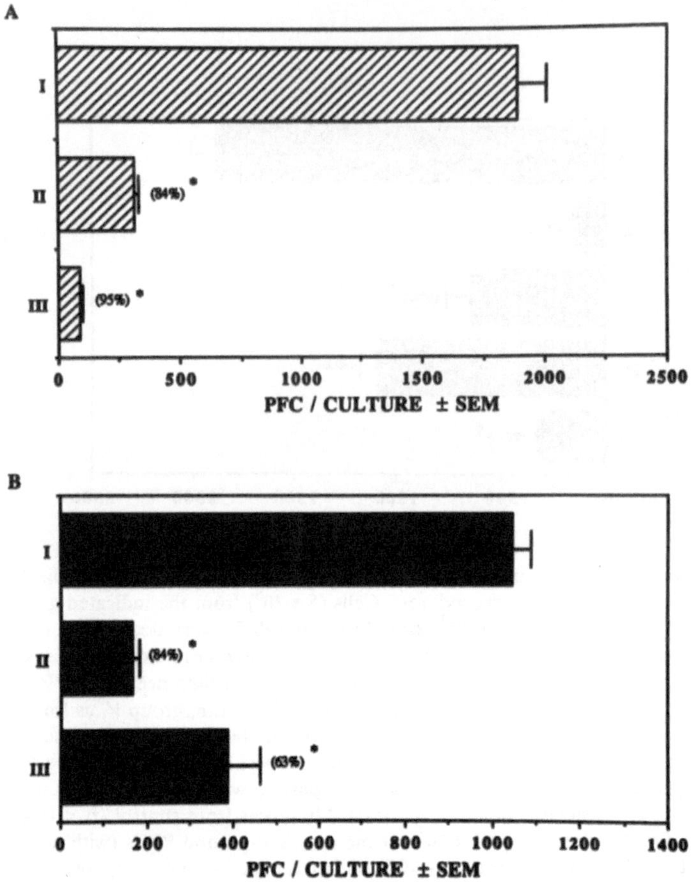

Figure 5. Effect of J11d plus C treatment on the suppressive capacity of purified immune splenocytes. Spleen cells (5 x 10^6) from the indicated group were cocultured, either directly (*A*) or in transwell plates (*B*), with 1 x 10^7 normal cells. Group *I*, normal cells; group *II*, immune cells after plastic adherence, G-10, and panning depletion steps; group *III*, as for group *II* plus J11d + C treatment. Data represent the means ± SEM of triplicate cultures per group. The numbers in parentheses indicate % suppression of the PFC responses in comparison with the control group (group *I*). Asterisks denote significant differences from control ($p < 0.001$). [Al-Ramadi et al., 1991*a*. Copyright 1991, *Journal of Immunology*]

immune spleen cells were placed in the upper chamber (Al-Ramadi et al., 1991*a*). The chambers were separated by a 0.4-μ membrane. After incubation for 5 days in vitro, the normal cells were harvested from the bottom wells, and the number of PFCs determined. It was found that immune cells placed in the top chambers were able to inhibit the normal PFC responses in the bottom chambers, suggesting that a factor was produced by the immune cells which could pass through the membrane. As shown in Fig. 5, both whole immune cell cultures, the postpanning population, and the post-J11d population were able to inhibit in cocultures and across a membrane in transwell plates. These experiments indicated that suppression was active and that the precursor macrophage population was capable of being suppressive. Studies using immune cells from C3HeB/FeJ (H-2^k) mice and normal cells, from Balb/c (H-2^d) or C57BL/6 (H-2^b) mice, showed that the suppression was not H2 restricted (Al-Ramadi et al., 1991*b*).

What Is the Suppressor Factor?

Experiments were carried out to determine the identity of the suppressor factor. As was the case for mitogenic suppression, the suppression of the PFC response was not reversed by treatment of immune cell cultures with indomethacin, suggesting that prostaglandins were not mediating the suppression. Treatment of immune cell cultures with catalase also did not reverse suppression, ruling out peroxide as the suppressor factor. Suppression did not appear to be due to the oxidation state of the cultures, as addition of 2-mercaptoethanol did not reverse suppression. Anti-TGFβ did not reverse suppression. Further, we had observed that suppressed cultures were deficient in interleukin-2 (IL-2), and that the reduction in amount of IL-2 in cultures was proportional to the vaccinating dose of SL3235 (Al-Ramadi et al., 1991*b*). Addition of IL-2 did not restore responsiveness. Table 5 summarizes the factors which were tested and found not to mediate the suppression.

In the course of control experiments using IL-2, it was found that interleukin-4 (IL-4) could restore responsiveness to suppressed cultures (Al-Ramadi et al., 1991*b*). Anti-interferon γ (αIFNγ) was also found to reverse the immunosuppression (Al-Ramadi et al., 1991*b*; Al-Ramadi et al., 1992*b*).

Table 5. Factors which do not reverse of suppression.

Indomethacin
Catalase
Interleukin-2
2-Mercaptoethanol
Anti-TGFβ

Nitric Oxide as the Suppressor Factor

Two reports were published in 1991 providing evidence that nitric oxide (NO) could cause suppression of mitogenic responses in rats and mice (Albina et al., 1991; Mills, 1991). L-arginine is converted to NO by nitric oxide synthase. The role of NO can be probed using N-monomethyl L-arginine (NMLA), a competitive inhibitor of L-arginine. As shown in Table 6, spleen cells from SL3235-immunized mice produce significant quantities of NO, measured as nitrite, 7 days after immunization, and NO production is reversed by NMLA, IL-4, and αIFNγ (Al-Ramadi et al., 1992*b*). Further, NMLA, IL-4 and αIFNγ reverse suppression (Al-Ramadi et al., 1992*b*) (Table 7). (The concentrations of additives needed to reverse nitrite levels and PFC numbers are different because the media for the two assays differ, and because RBCs in the PFC assay quench NO.) These studies indicate that the major suppressor factor induced by attenuated *Salmonella* is NO. This conclusion agrees with other unpublished observations from our laboratory, namely, that the suppressor factor is not found in the supernatants of immune spleen cells. As NO is labile and is quickly converted into nitrite ion, the failure to find NO in supernatants is consistent with the suppressor factor's being nitric oxide.

Table 6. Nitrite production by spleen cells of mice immunized 7 days previously with SL3235.

Additive[1]	Nitrite, $\mu M/10^7$ cells[2]
None	27.7 ± 6.8
IL-4	1.6 ± 0.3
Anti-IFNγ	6.5 ± 0.7
NMLA	3.5 ± 0.8

[1]Concentrations of additives: IL-4, 100 units/ml; anti-IFNγ, 10 μg/ml; NMLA, 2 mM. (Spleen cells from control mice produce undetectable quantities of nitrite.)
[2]Nitrite measured in culture supernatants 48 hr after culture initiation.

Table 7. Reversal of immunosuppression induced by SL3235.

	PFC/culture	
Addition[1]	Normal	Immune
None	1977 ± 346	170 ± 8
IL-4	813 ± 190	951 ± 31
Anti-IFNγ	1201 ± 141	965 ± 154
NMLA	1402 ± 190	1079 ± 266

[1]Concentrations of additives: IL-4, 10 units/ml; anti-IFNγ, 1 μg/ml; NMLA, 0.05 mM.

DISCUSSION

We have shown that a strain of *Salmonella* with an aroA⁻ lesion can induce high levels of protection against challenge with virulent *Salmonella*. Concomitant with protection, we found immunosuppression to nonsalmonella antigens and to mitogens. The suppressor factor appears to be NO and suppression can be reversed by inhibitors of the pathway of NO, by IL-4, and by αIFNγ. We interpret the cytokine regulation to be occurring by pathways illustrated in Fig. 6. Induction of NO requires macrophage activation. It is well established that INF-γ is one of the major factors upregulating the inducible NO synthase of macrophages (Ding et al., 1988). IL-4 is known to have the capacity to downregulate various macrophage activities (Hart et al., 1989; Lehn et al., 1989).

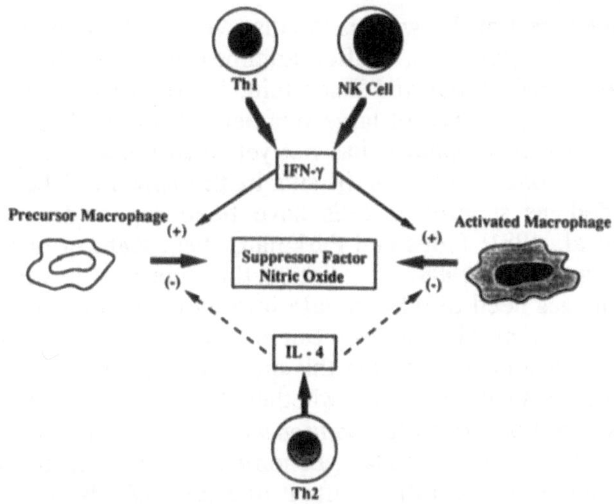

Figure 6. Model for mechanism of cytokine regulation of macrophage nitric oxide production and immunosuppression.

It is our hypothesis that immunosuppression occurs because production of excess NO by activated macrophages has toxicity for the host cells which surround the macrophages. NO has been shown to inhibit iron sulphur-containing enzymes in mitochondria of eukaryotic cells (Stuehr et al., 1989). Overproduction of NO by large numbers of activated macrophages might metabolically poison lymphocytes in the vicinity. In our PFC and mitogen cultures, we find that the lymphocytes do not stain with trypan blue, indicating that they are not dead even though they are non-functional. Further, we have found that immune cells can be added as late as the third day of a Mishell-Dutton culture and still give significant suppression (Al-Ramadi et al., 1991*b*). These observations would be consistent with metabolic inhibition of the functional activity of lymphocytes by NO.

Our studies have shown that attenuated *Salmonella* result in activation of peritoneal macrophages to be leishmanicidal and tumoricidal and in induction of suppressor macrophages in the spleen. We have yet to prove that suppressive capacity, and tumoricidal and microbicidal capacities, reside in the same cell. Studies to prove that suppressor macrophages are tumoricidal, and that tumoricidal macrophages are suppressive are currently in progress. However, it can certainly be speculated that both activities might go hand in hand. NO was discovered as the major mechanism by which activated macrophages inhibit tumor cells (Hibbs et al., 1987).

A model in which a macrophage' that produces NO has both effector and suppressor capacities would help to explain why inflammatory responses mediated by the cellular arm of the immune system are frequently a double-edged sword, resulting in antimicrobial and antitumor activity on one hand, but in tissue destruction and even anergy on the other hand. Examination of the dose-response curves for induction of NO indicate a threshold response when SL3235 exceed 10^5 (Al-Ramadi et al., 1992*b*). Perhaps in the face of a dangerous systemic microbial burden, NO synthase is activated. Its bystander toxicity to nearby lymphocytes is tolerated by the host in the hopes of controlling potentially fatal infection or neoplasia, in much the same way near-lethal doses of chemotherapy are used to treat cancer patients, in spite of their potential to damage proliferating normal cells in the bone marrow.

An important aspect of this work is the discovery that a population of precursor macrophages can be suppressive. After administration of SL3235, considerable splenomegaly is observed (Killar and Eisenstein, 1985). At least part of the splenomegaly is due to the appearance of large numbers of macrophage precursors. The origin of the precursor macrophages has not yet been determined; that is, whether they arise from the bone marrow or in situ in the spleen. The latter possibility cannot be excluded, as suppressor cells have been reported in neonatal mouse spleens (Piquet et al., 1981; Jadus and Parkman, 1986; Kato et al., 1985) and adult bone marrow (Noga et al., 1988). An interesting question is whether or not the precursor macrophages need to differentiate into more mature macrophages before they can be suppressive, or whether they can exhibit suppression in their less differentiated state. Experiments to examine this question are currently in progress.

An important practical question is whether the suppression induced by SL3235 is peculiar to this strain of attenuated *Salmonella* (SL3235) or whether other *Salmonella* are capable of inducing a similar phenomenon. Nauciel and coworkers have reported that a temperature-sensitive mutant of *Salmonella* is also immunosuppressive (Deschenes et al., 1986), which supports the hypothesis that suppression is not unique to aroA$^-$ strains. We plan to test other strains of *Salmonella* in the near future.

From a practical point of view these studies raise the question as to whether administration of live, attenuated *Salmonella* as vaccines could induce a window of immunosuppression in the recipients. This question cannot be addressed adequately at the present time. The experiments reported in this paper were carried out using interperitoneal administration of SL3235. Preliminary unpublished studies suggest that if oral doses are used, which are large enough to result in organisms becoming systemic and reaching the spleen, then splenic suppression may also result. Whether the doses used in humans to induce protection will also induce suppression will have to be determined. Furthermore, what effect oral administration will have on secretory immune function is yet to be assessed.

ACKNOWLEDGMENTS

This work was supported by grant AI15613 from the National Institutes of Health. Thanks to Mr. Greg Harvey for expert editing and typing of this manuscript.

REFERENCES

Albina, J.E., Abate, J.A., and Henry, W.L., Jr., 1991, Nitric oxide production is required for murine resident peritoneal macrophages to suppress mitogen-stimulated T cell proliferation. Role of IFN-γ in the induction of the nitric oxide-synthesizing pathway, J. Immunol., 147: 144.

Al-Ramadi, B.K., Brodkin, M.A., Mosser, D.M., and Eisenstein, T.K., 1991*a*, Immunosuppression induced by attenuated *Salmonella*. Evidence for mediation by macrophage precursors, J. Immunol., 146: 2737.

Al-Ramadi, B.K., Chen, Y.-W., Meissler, J.J., Jr., and Eisenstein, T.K., 1991*b*, Immunosuppression induced by attenuated *Salmonella*. Reversal by IL-4, J. Immunol., 147: 1954.

Al-Ramadi, B.K., Greene, J.M., Meissler, J.J., Jr., and Eisenstein, T.K., 1992*a*, Immunosuppression induced by attenuated *Salmonella*: effect of LPS responsiveness on development of suppression, Microb. Pathogen., 12: 267.

Al-Ramadi, B.K., Meissler, J.J., Huang, D., and Eisenstein, T.K., 1992*b*, Immunosuppression induced by nitric oxide and its inhibition by interleukin-4, Eur. J. Immunol., in press.

Cunningham, A., and Szenberg, A., 1968, Further improvements in the plaque technique for detecting single antibody-forming cells, Immunology, 14: 599.

Deschenes, M., Guenounou, M., Ronco, E., Vacheron, F., and Nauciel, C., 1986, Impairment of lymphocyte proliferative responses and interleukin-2 production in susceptible (C57BL/6) mice infected with *Salmonella typhimurium*, Immunology, 58: 225.
Ding, A., Nathan, C.F., and Stuehr, D.J., 1988, Release of reactive nitrogen intermediates and reactive oxygen intermediates from mouse peritoneal macrophages: comparison of activating cytokines and evidence for independent production, J. Immunol., 141: 2407.
Eisenstein, T.K., Killar, L.M., Stocker, B.A.D., and Sultzer, B.M., 1984*a*, Cellular immunity induced by avirulent *Salmonella* in LPS-defective C3H/HeJ mice, J. Immunol., 133: 958.
Eisenstein, T.K., Killar, L.M., and Sultzer, B.M., 1984*b*, Immunity to infection with *Salmonella typhimurium:* mouse-strain differences in vaccine- and serum-mediated protection, J. Inf. Dis., 150: 425.
Hart, P.H., Vitti, G.F., Burgess, D.R., Whitty, G.A., Piccoli D.S., and Hamilton, J.A., 1989, Potential antiinflammatory effects of interleukin-4: suppression of human monocyte tumor necrosis factor alpha, interleukin-1, and prostaglandin E2, Proc. Natl. Acad. Sci. USA, 86: 3803.
Hibbs, J.B., Jr., Taintor, R.R., and Vavrin, Z., 1987, Macrophage cytotoxicity: role for L-arginine deiminase and imino nitrogen oxidation to nitrite, Science, 235: 473.
Hoiseth, S.K., and Stocker, B.A.D., 1981, Aromatic-dependent *Salmonella typhimurium* are non-virulent and effective as live vaccines, Nature (Lond.), 291: 238.
Jadus, M.R., and Parkman, R., 1986, The selective growth of murine newborn-derived suppressor cells and their probable mode of action, J. Immunol., 136: 783.
Kato, K., Yamamoto, K.-I., and Kimura, T., 1985, Migration of natural suppressor cells from bone marrow to peritoneal cavity by live BCG, J. Immunol., 135: 3661.
Killar, L.M., and Eisenstein, T.K., 1985, Immunity to *Salmonella typhimurium* infection in C3H/HeJ and C3H/HeNCr1BR mice: studies with an aromatic-dependent live *S. typhimurium* strain as a vaccine, Infect. Immun., 47: 605.
Lee, J.-C., Gibson, C.W., and Eisenstein, T.K., 1985, Macrophage-mediated mitogenic suppression induced in mice of the C3H lineage by a vaccine strain of *Salmonella typhimurium*, Cell. Immunol., 91: 75.
Lehn, M., Weiser, W.Y., Engelhorn, S., Gillis, S., and Remold, H.G., 1989, IL-4 inhibits H_2O_2 production and antileishmanial capacity of human cultured monocytes mediated by IFN-γ, J. Immunol., 143: 3020.
Mackaness, G.B., 1971, Resistance to intracellular infection, J. Inf. Dis., 123: 434.
Mills, C.D., 1991, Molecular basis of "suppressor" macrophages. Arginine metabolism via the nitric oxide synthetase pathway, J. Immunol., 146: 2719.
Noga, S.J., Wagner, J.E., Horwitz, L.R., Donnenberg, A.D., Santos, G.W., and Hess, A.D., 1988, Characterization of the natural suppressor cell population in adult rat bone marrow, J. Leukocyte Biol., 43: 279.
Piquet, P.-F., Irle, C., and Vassalli, P., 1981, Immunosuppressor cells from newborn mouse spleen are macrophages differentiating in vitro from monoblastic precursors, Eur. J. Immunol., 11: 56.
Schafer, R., Largen, M.T., Nacy, C.A., Dalal, N., Havas, H.F., and Eisenstein, T.K., 1986, An attenuated strain of *Salmonella typhimurium* as an activator of C3H/HeJ and P/J macrophages and anti-tumor agent, Intl. Symp. Immunolog. Adjuvants and Modulators of Non-Specific Resistance to Microbial Infections, Columbia, Md., Abstract #34.
Schafer, R., Nacy, C.A., and Eisenstein, T.K., 1988, Induction of activated macrophages in C3H/HeJ mice by avirulent *Salmonella*, J. Immunol., 140: 1638.
Stuehr, D.J., and Nathan, C.F., 1989, Nitric oxide. A Macrophage product responsible for cytostasis and respiratory inhibition in tumor target cells, J. Exp. Med., 169: 1543.

ROLE OF NATURAL RESISTANCE AND IMMUNE RESPONSE IN THE EFFICACY OF ANTIBIOTIC TREATMENT FOR SYSTEMIC SALMONELLOSIS

Letterio Bonina, and Pasquale Mastroeni
Microbiology Institute, Medical School, Messina University
Piazza xx Settembre, 4
98100 Messina, Italy

INTRODUCTION

Salmonella infections are still not declining and cause much morbidity and mortality. For this reason we have looked closely at the action of antibiotics on the host-parasite relationship.

The ability of a bacterial population to settle in the host and to lead to an infectious disease depends on the virulence of the infecting agent as well as on the host's natural resistance and immune response. Antibiotics are able to influence the microorganism and the host in different ways. Although the focus of therapy is the infection, we now recognise that the host not only has to deal with the causative organism but also, directly and indirectly, with the antibiotics administered.

Salmonella typhimurium is a facultative intracellular bacterium; it multiplies within macrophages during the early phase of infection in mice (Hormaeche, 1980; Harrington and Hormaeche, 1986); it is therefore important for antibacterial agents to penetrate macrophages to be effective in salmonella infection.

Penetration and accumulation of an antimicrobial agent within eucaryotic cells are important determinants of its activity against

Biology of Salmonella, Edited by F. Cabello *et al.*,
Plenum Press, New York, 1993

intracellular bacteria (Easmon and Crane, 1986). Having penetrated into cells, however, an antibiotic must also reach the bacteria which are often, but not always, localised in vacuoles. The drug must then remain stable and capable of expressing its activity in the physico-chemical conditions prevailing at this specific site. Moreover, the bacteria must be in a metabolic state that renders them sensitive to the drug. Inability to meet one or several of these conditions will result in the drug being largely inactive *in vivo.*

Innate resistance to salmonellae in mice is largely controlled by a single gene or gene cluster: designated *Ity* (Plant and Glynn, 1974, 1979), which is expressed in macrophages and regulates the *in vivo* bacterial net growth rate in the early, acute phase of the infection (Hormaeche et al. 1983; Lissner et al. 1983). In addition, genes within the H-2 complex influence the development of resistance later in the infection (Nauciel et al 1900). Nevertheless, survival of the host requires that the early exponential bacterial growth in the RES be suppressed leading to a plateau phase and a carrier state of variable duration. Recent results indicate that the host response causing the early plateau and suppression of bacterial growth at the end of the first week of the infection, which is essential for survival, does not require T cells (Hormaeche et al. 1990).

In this report we will address the early phase of infection, studying the effects of different antibiotics both *in vivo* and *in vitro.*

The efficacy of ampicillin therapy in mouse typhoid, as judged by the severity of relapse following its cessation, has previously been shown to be influenced by the duration of treatment, the virulence of the infecting organism, and the genetic constitution of the mice (Maskel and Hormaeche, 1985). The resistant Ity^r phenotype is advantageous, because Ity^r mice are much less prone to relapse than Ity^s mice. The resistant phenotype may be beneficial because it allows time for the development of an effective immune response after the second out growth of bacteria following cessation of therapy.

In this report we have used innately susceptible BALB/c (Ity^s) and resistant CBA (Ity^r) mice, to investigate the efficacy of one monocyclic (aztreonam) and three bicyclic (ampicillin, cefazolin, ceftazidime) beta-lactam antibiotics in controlling systemic salmonella infection when given for brief or prolonged periods.

The *in vitro* approach was carried out on cultured liver macrophages (Kupffer cells) from BALB/c and CBA mice, infected with a virulent *S. typhimurium,* in presence or absence of the different antibiotics. Survival of the target cells was controlled over short period of time.

MATERIALS AND METHODS

Micro-organisms *S. typhimurium* strain M525 has been described (Hormaeche, 1979). Micro-organisms were grown overnight in 100 ml of nutrient broth (Difco, Detroit, Michigan, USA) at 37°C , sedimented by centrifugation, washed once and re suspended in PBS pH 7.2, the concentration determined by nephelometry and diluted accordingly for inoculation. The *in vitro* study was carried out with the virulent strain C5 of *S. typhimurium*.

Animals. BALB/c and CBA mice were purchased from Charles River, Como, Italy. Animals of either sex four weeks old were used.

Inoculation of animals. Groups of approximately 60 animals received c. 10^4 *S. typhimurium* M525 intravenously in a lateral tail vein in 0.1 ml PBS. The number of viable inoculated organisms was checked by viable counts on nutrient agar (Difco, Michigan, USA). The exact number of mice and the inoculum dose for each experiment are given in the figure legends.

Antibiotics. The monobactam aztreonam (Squibb, Rome, Italy) and the bicyclic ampicillin (Sigma, St. Louis, MO, USA), cefazolin (Chemil, Chemioterapici, Milan, Italy) and ceftazidime (Glaxo, Verona, Italy), were dissolved in sterile saline.

MIC determination. The MIC of the antibiotics for the *Salmonella typhimurium* strains C5 and M525 used were determined by the broth dilution test by standard methods (Washington and Sutter,1980). The value obtained were 0.125 mg/l for aztreonam and ceftazidime, and 32 mg/l for ampicillin and cefazolin.

Administration of antibiotics. All antibiotics were administered as a single daily ip dose in 0.1 ml. The dose was 10 mg for aztreonam and ceftazidime and 50 mg for ampicillin and cefazolin In all experiments, treatment was started on day 3, and stopped on day 5 for half the mice; the remainder were treated up to or beyond day 12.

Enumeration of organisms in liver and spleen homogenates. On successive days after inoculation, groups of three mice were killed by cervical dislocation before their daily dose of antibiotic. The livers and spleens were homogenised separately in distilled water and 0.1 ml aliquots of suitable dilutions were plated in duplicate on nutrient agar. Results are expressed as geometric mean $\log_{10}$ cfu (colony forming units) per whole organ.

Preliminary experiments were performed to exclude the possibility that bacterial counts could be influenced by carry-over of antibiotic in organ homogenates. Mice were given a dose of antibiotics twice as high as that used for treatment of the salmonella infection and killed 20 h later. The livers and spleens were homogenised as above, and 1 ml of homogenate was added to pour plates in TSB containing approximately $2x10^2$ *S. typhimurium*. The viable counts did not differ from those performed with homogenates from untreated mice. It is therefore highly unlikely that carry-over could have influenced the present data.

Relapse. Relapse is defined as a sustained and progressive increase in the bacterial load in the liver and spleen following cessation of antibiotic therapy leading to death of the mice .

Kupffer cell cultures. These were performed as described (Harrington and Hormaeche, 1986). Briefly, mice were killed, and the liver perfused with HBSS containing 0.05% collagenase. The liver was then removed and finely chopped in presence of collagenase plus 100 µg/ml DNAse. The cells were resuspended in 35% isotonic bovine serum albumin and then overlaid with Ca/Mg- HBSS and centrifuged at 600 g for 15 min. The interface cells were collected, and resuspended in RPMI 1640 supplemented with 10% fetal calf serum. Cells were placed to adhere in culture dishes. Approximately 90% of BALB/c and CBA monolayer cells were phagocytic, according to *Candida albicans* and latex tests.

Infection of monolayers. The Kupffer cells, after 24 h in culture, were infected with salmonellae, at a multiplicity of infection 10, opsonized with 0.5% heat-inactivated rabbit anti-salmonella antiserum just before addition to the monolayer. The cells were washed after 2 h and fresh medium without antibiotic was added in the control, and plus 50 µg/ml of aztreonam or cefazolin to control the effects of these antibiotics on cell survival.

RESULTS

Kinetics of infection in untreated control mice

Infection of susceptible BALB/c mice with *S. typhimurium* M525 was followed by exponential bacterial growth at approximately 1 log a day; no mice survived beyond day 6

The infection in resistant CBA mice progressed more slowly, with higher counts in livers than in spleens.

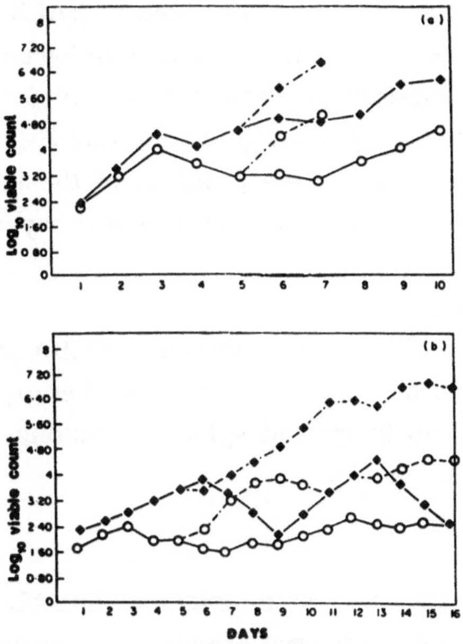

Figure 1. Cefazolin treatment. (a) 67 BALB/c mice were injected iv with $6.2x10^4$ M525. (b) 61 CBA mice were injected iv with $7.2x10^4$ M525. Squares: livers; circles: spleens. Dotted lines indicate the counts after cessation of treatment. (From J. Antimicr. Chemother. 1990, 25:813)

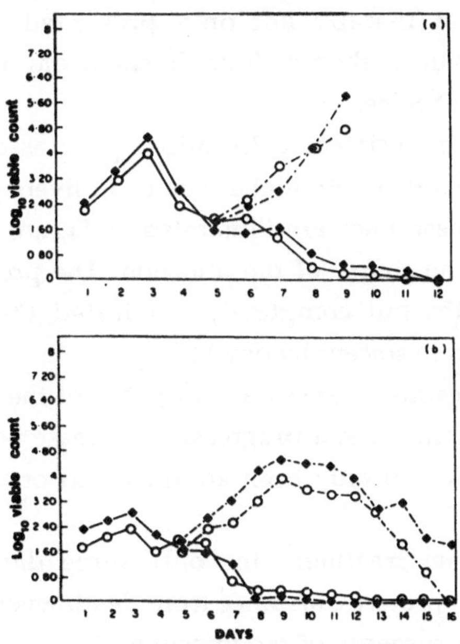

Figure 2. Aztreonam treatment. (a) 68 BALB/c mice were injected iv with $7x10^4$ M525. (b) 63 CBA mice were injected with $6.8x10^4$ M525. Squares: livers; circles: spleens. Dotted lines indicate the counts after cessation of treatment. (From: J. Antimicr. Chemother. 1990, 25:813).

Cefazolin in susceptible mice

Figure 1a illustrates the results obtained in BALB/c mice. Treatment commencing on day 3 caused an initial decrease in bacterial numbers which was only transitory. Cessation of treatment on day 5 was followed by a marked relapse, with no mice surviving beyond day 7. However, treatment until day 10 could not suppress the progress of the infection, with liver counts in treated mice increasing ; no mice survived beyond day 10.

Cefazolin in resistant mice

Figure 1b illustrates the results obtained in CBA mice. A suppression of bacterial growth was observed. Cessation of therapy caused a marked increase in counts in both livers and spleens. Prolonged therapy did not reverse bacterial growth.

Aztreonam treatments

Figure 2a illustrates the results obtained in BALB/c mice. Infection was followed by exponential bacterial growth at approximately 1 log a day; no control mice survived beyond day 6. Treatment with aztreonam started on day 3 reversed the course of the infection. Its cessation on day 5 caused a relapse which killed all mice in this group by day 9.

Prolongation of therapy not only prevented further deaths but completely eradicated the pathogens from livers by day 12.

Aztreonam in resistant mice.

The infection in resistant CBA mice progressed more slowly, at approximately 0.5 log a day with higher counts in livers than in spleens. The results obtained in these mice are illustrated in Figure 2b. Treatment with aztreonam reversed the course of the infection. The prolonged therapy not only prevented deaths but completely eradicated the salmonellae from livers by day 12 and from spleens by day 11.

Cessation of treatment caused a resurgence of the infection. However, this resurgence was transitory, a progressive decrease began, in spite of the fact that the antibiotic had only been administered on days 3-5; no deaths were observed.

Thus, aztreonam treatment for only three days is ineffective in BALB/c mice, but can protect CBA mice from death, even though it does not prevent a transitory resurgence of the infection.

Effect of antibiotics on infected Kupffer cells

Preliminary experiments showed that infection of Kupffer cells with

S. typhimurium C5 was very deleterious for BALB/c (susceptible) cells, with a high proportion of the cells being lost from the monolayer, whereas CBA (resistant) macrophages were not lost to the same extent. The effect of antibiotics on macrophage survival following infection was therefore studied using the highly labile BALB/c cells, comparing the survival of treated versus untreated cultures. The results of cefazolin and aztreonam are illustrated in Figure 3 and 4 respectively.

A striking difference was observed : whereas cefazolin had no effect, with the monolayer decaying as for controls, aztreonam virtually blocked macrophage loss with the monolayers surviving essentially as for uninfected controls. Thus aztreonam has a profound effect on survival of infected macrophages, paralleling its effect *in vivo*

DISCUSSION

Different antibiotics vary markedly in their efficacy in the eradication of salmonellae in the mouse typhoid model, with aztreonam being clearly superior to all others tested, in that it not only reversed the infection but can actually eradicate the organism from the tissues if given early in the infection (Bonina et al. 1990) We present data which further show that the increased efficacy of aztreonam *in vivo* is paralleled by its efficacy in promoting the survival of infected Kupffer cells, strongly suggesting that the efficacy of this antibiotic is due to its interactions with this particular host cell.

In vivo cefazolin treatment was only able to delay the infection. Bacterial growth in the RES was slowed but not stopped; susceptible mice undergoing continued treatment died in ten days; resistant mice survived better, but bacteria continued to increase throughout the experiment. Cefazolin was therefore ineffective.

Ceftazidime was as effective as ampicillin (data not shown); the results were comparable for both antibiotics, even though the MIC of ceftazidime is lower than that of ampicillin for this organism. Nevertheless, no clearance was observed with both antibiotics.

The present results show that aztreonam is effective for the treatment of mouse typhoid when given early in the infection. The most striking result was obtained with prolonged treatment, which not only reversed bacterial growth in the RES and prevented death in all mice, but actually eradicated the pathogen after nine days of treatment in both resistant and susceptible mice.

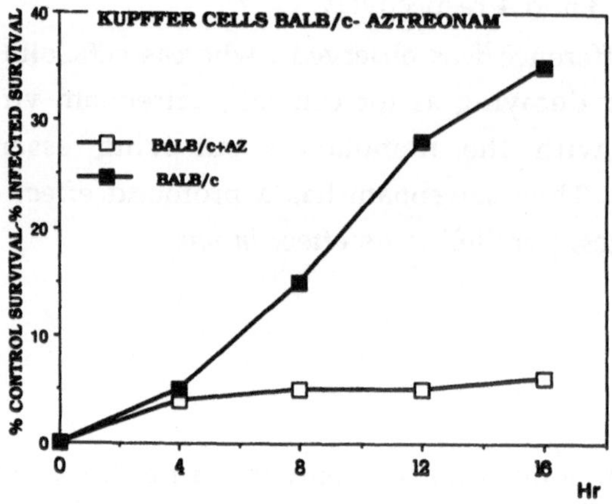

Figure 3. Cefazolin effect on Kupffer cell monolayers after infection with *S. typhimurium* C5. Results are expressed as the difference in percent survival of control minus infected monolayers. Open squares: Infected cells without antibiotic; closed squares: infected cells plus cefazolin.

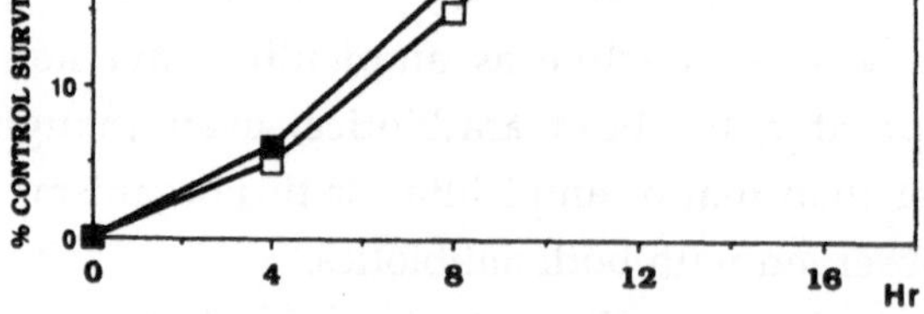

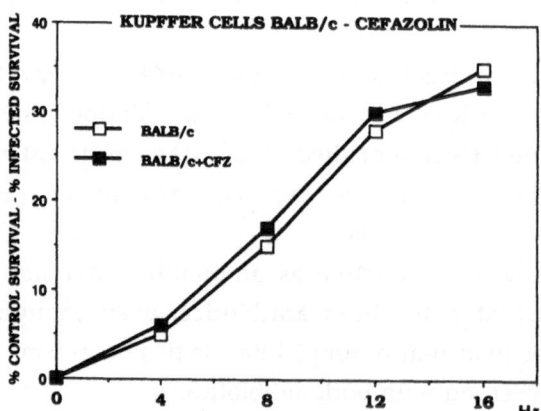

Figure 4. Aztreonam effect on Kupffer cell monolayers after infection with *S. typhimurium* C5. Results are expressed as the difference in percent survival of control minus infected monolayers. Closed squares: Infected cells without antibiotic; Open squares: Infected cells plus aztreonam.

Host resistance was very important in the response to treatment; overall, resistant CBA mice responded better to treatment, with less severe relapses and fewer deaths as compared to susceptible BALB/c mice. It is noteworthy in this respect that treatment with aztreonam for only three days in CBA mice was followed by only a transitory increase in bacterial counts, which then decreased in the absence of antibiotic treatment.

Thus, the mouse typhoid model is important since it may provide important information in the field of antibiotic influence on host-parasite relationships. We were able to demonstrate that for beta-lactam antibiotics the MIC is no guide as to the *in vivo* effectiveness versus *S.typhimurium*; moreover, survival alone provides an incomplete assessment of the efficacy of treatment. In a kinetics study previously reported (Bonina et al. 1990) we were in fact able to show that mice treated with cefazolin which survived to the end of the observation period were actually undergoing a slow but progressive infection.

The reasons for the striking superiority of aztreonam over the other three antibiotics is not clear, but must represent some fundamental difference between the *in vivo* pharmacokinetics of this antibiotic and the other three employed.

We have shown a positive stimulatory effect of aztreonam on phagocytic cells; moreover, unlike other beta-lactam antibiotics, aztreonam can penetrate intracellularly particularly in professional phagocytic. Intracellular distribution of the monobactam is dose dependent and energy-requiring process (Mastroeni et al. 1984). These properties could explain not only the *in vivo* results, but also the *in vitro* data on Kupffer cells. Differences in Kupffer cells from naturally resistant or susceptible mice in their response to infection with salmonella have been demonstrated (Harrington and Hormaeche, 1986). The survival of monolayers gave more reliable results than bacterial counts. The results illustrated in Figure 3 show that cefazolin has no effect *in vitro,* paralleling the *in vivo* findings. The results obtained with aztreonam are striking, since *Ity*^{s} Kupffer cells behave as *Ity*^{r} cells in presence of the monobactam; in fact the rate of survival share the same percentage as in the control liver macrophages from CBA mice (data not shown).

If the molecular basis of the undetlying mechanisms responsible for the effectiveness of aztreonam can be elucidated, it could provide a powerful tool for the treatment of typhoid fever.

REFERENCES

Bonina, L., Carbone, M., Matera, G., Teti, G., Joysey, S.H., Hormaeche, C.E., Mastroeni, P.

1990: Beta-lactam antibiotics (aztreonam, ampicillin, cefazolin, and ceftazidime) in the control and eradication of *Salmonella typhimurium* in naturally rresistant and susceptible mice. J. Antimicr. Chemother. 25: 813.

Easmon, C.S.F., Crane, J.P,.1985, Uptake of ciprofloxacin by human neutrophils. J. Antimicr. Chemother. 16: 67.

Harrington, K.A., Hormaeche, C.E.,1986, Expression of the innate resistance gene Ity in mouse Kupffer cells infected with *Salmonella tyuphimurium* in vitro. Microbial Pathogenesis. 1:299.

Hormaeche, C.E.,1979, The natural resistance of radiation chimeras to *S. typhimurium* C5. Immunology, 37: 329.

Hormaeche, C.E., 1980, The "in vivo" division and death rates of *Salmonella typhimurium* in the spleens of naturally resistant and susceptible mice measured by the superinfecting phage technique of Meynell. Immunology, 5:973.

Hormaeche, C.E., Maskell, D.J., Harrington, K., Joysey, H., Brock, J.,1983, Mechanisms of natural resistance to mouse typhoid. Bull. Eur. Physiopath. Resp., 19: 137

Hormaeche, C.E., Maskell, D.J., Harrington, K.A., Joysey, H., and Brock, J. 1985, Natural resistance to salmonellae in mice: control by genes within the major histocompatibility complex. J. Infect. Dis. 152:1050.

Lissner, C.R., Swanson, R.N., O'Brien, A.D. 1983, Genetic control of the innate resistance of mice to *Salmonella typhimurium*: expression of the Ity gene in peritoneal and splenic macrophages isolated in vitro. Journal of Immunology, 131: 3006.

Maskell, D.J. , Hormaeche, C.E.1985 Relapse following cessation of antibiotic therapy for mouse typhoid in resistant and susceptible mice infected with salmonellae of different virulence. J. Infect. Dis., 152:1044.

Mastroeni, P., Bonina, L., Leonardi, M.S., Carbone, M., Berlinghieri, M.C.1984, Study of aztreonam intracellular kinetics and its relationships with different cell types. In Proceedings of the IV Mediterranean Congress of Chemotherapy, Rhodos, Greece, 345.

Nauciel, C., Ronco, E. Guenet, J.L., and Pla, M. 1988, Role of H-2 and non-H-2 genes in control of bacterial clearance from the spleen in *Salmonella typhimurium* infected mice. Infect. Immun. 56:2407.

Plant, J., Glynn, A.A. 1974, Natural resistance to salmonella infection. Delayed hypersensitivity and Ir genes in different strains of mice. Nature, 248: 345.

Plant, J., Glynn, A.A., 1979, Locating salmonella resistance gene on mouse chromosome 1. Clinical Exper. Immunol. 37: 1.

Washington, J.A., Sutter, V.L.,1980, Dilution susceptibility test: agar and macro-broth dilution procedures. In Manual of Clinical Microbiology, 3rd edn (Lennette, E.H., Balows, A., Hausler, W.J., Truant, J.P. Eds), pp 453-458. ASM, Washington, DC.

MUCOSAL IMMUNITY: THE ROLE OF SECRETORY IMMUNOGLOBULIN A IN PROTECTION AGAINST THE INVASIVE PATHOGEN *SALMONELLA TYPHIMURIUM*

James M. Slauch[1], Michael J. Mahan[1], Pierre Michetti[3,2], Marian R. Neutra[2], and John J. Mekalanos[1]

[1] Dept. of Microbiology and Molecular Genetics, and [2] GI Cell Biology Laboratory, Children's Hospital, and Department of Pediatrics, Harvard Medical School, Boston, MA 02115, [3] Division of Gastroenterology, Chuv CH-1011 Chuv-Lausanne, Switzerland

INTRODUCTION

Many pathogens gain entry into a host organism by crossing the epithelia of the digestive, respiratory or genital tracts. The main defense against this entry is the mucosal immune system. In the gut mucosa, antigen sampling sites contain organized mucosal lymphoid tissue including lymphoid follicles. These cellular assemblies, which form large aggregates in Peyer's patches, sample lumenal antigens, resulting in the stimulation of both T cells and B lymphoblasts, committed to IgA synthesis. This leads to the production of secretory antibodies of the IgA isotype. Secretory IgA (sIgA) antibodies are thought to act by immune exclusion. That is, sIgA prevents the pathogen from contacting the mucosal surface by agglutination, entrapment of immune complexes in the mucus, and clearance by peristalsis (Brandtzaeg, 1989; Childers, et al., 1989; Mestecky, 1988).

We have recently developed methods for the production of IgA hybridomas by fusing Peyer's patch lymphocytes to myeloma cells (Weltzin et al., 1989). Subcutaneous injection of these hybridoma cells into BALB/c mice results in the formation of tumors that secrete dimeric and polymeric IgA in the serum. This serum IgA is delivered to mucosal surfaces by the normal transepithelial transport system (Winner et al., 1991). The hybridoma "backpack" tumor method provides a continuous source of monoclonal sIgA. The level of the monoclonal IgA in the intestinal lumen has been shown to be proportional to the serum level of the antibody (Winner et al., 1991; Michetti et al., 1992). However, it has also been shown that the total concentration of IgA in the intestinal lumen does not increase in this animal model (Apter et al., 1991). This suggests that the concentration of IgA in the intestinal lumen is limited by transport of the IgA from the serum via the receptor mediated transepithelial transport system. Polymeric IgA receptors on the basolateral surfaces of mucosal epithelial cells bind IgA and transcytose the antibodies to the apical surface where a cell-associated protease releases the antibody in association with the ectodomain of the receptor, called secretory component, which protects the secreted antibody from proteolytic cleavage (Underdown and Dorrington, 1974; Mostov et al., 1980).

Using this backpack tumor model, we have shown that mucosal delivery of monoclonal sIgA can confer protection against an oral challenge of *Vibrio cholerae* in a neonate mouse model (Winner et al., 1991). However, *V. cholerae* does not penetrate the mucosal epithelium and does not result in systemic infection. Thus, it is not known whether this "backpack tumor" delivery system can confer mucosal protection against an oral challenge of an invasive pathogen such as *Salmonella typhimurium.*

S. typhimurium invades the intestinal mucosa of Peyer's patches by transepithelial transport across M cells. The organisms are released basolaterally and phagocytized by macrophages (Kohbata et al., 1986) in which they survive and are disseminated in several host tissues resulting in systemic infection (Hohmann et al., 1978). We have shown that *in vivo* delivery of a single monoclonal sIgA is sufficient to confer protection against an oral challenge of *S. typhimurium.* This IgA did not confer any protection against intra-peritoneal challenge, suggesting that the protection was localized to the mucosal surface. The details of this study are presented in Michetti et al. (1992). In this report, we also present preliminary characterization of the antigen to which the monoclonal antibody was directed. This antigen is a surface carbohydrate that dominated the mucosal immune response induced by oral immunization with *S. typhimurium.*

RESULTS

Isolation and Characterization of Monoclonal sIgA Directed Against Salmonella Surface Antigens

Monoclonal antibodies (mAbs) were obtained by immunizing mice with attenuated *phoP* strains of *S. typhimurium* (Miller et al., 1989; Miller and Mekalanos, 1990). Such mutant strains colonize the intestinal epithelium, providing sufficient antigen to stimulate an immune response, but are avirulent because of a macrophage survival defect (Miller and Mekalanos, 1990). This allows one to deliver a large amount of antigen without killing the mouse. Two weeks after immunization, Peyer's patch lymphocytes were fused to myeloma cells, generating immunoglobulin-secreting hybridomas (Weltzin et al., 1989).

A total of 46 Salmonella specific hybridomas were identified by ELISA analysis, using crude lysates of *S. typhimurium* as the immobilized antigen. Twelve IgA-producing hybridomas were selected for further characterization. All 12 IgAs produced by these hybridoma cells agglutinated *in vitro* grown *S. typhimurium* cells, indicating that the mAbs are directed against surface antigens. One of these hybridoma subclones, Sal4, was chosen for further study and was shown to be capable of forming hybridoma tumors when injected subcutaneously into the backs of BALB/c mice.

Sal4 IgA Recognizes an Externally Exposed Carbohydrate Epitope of *S. typhimurium*

The ability of Sal4 to agglutinate *S. typhimurium* cells *in vitro* suggests that it is directed against a surface exposed epitope. To characterize this epitope further, we probed Western blots of whole *S. typhimurium* cell lysates with Sal4 antibody. Figure 1 (lane 1) shows that the Sal4 antibody reacted with a broad band of material from 80 kD to >200 kD rather than with a single distinct band. To determine if this epitope had a carbohydrate component, we treated the nitrocellulose blots with periodate, which cleaves vicinal hydroxyl bonds and disrupts most carbohydrate epitopes, and subsequently probed with the

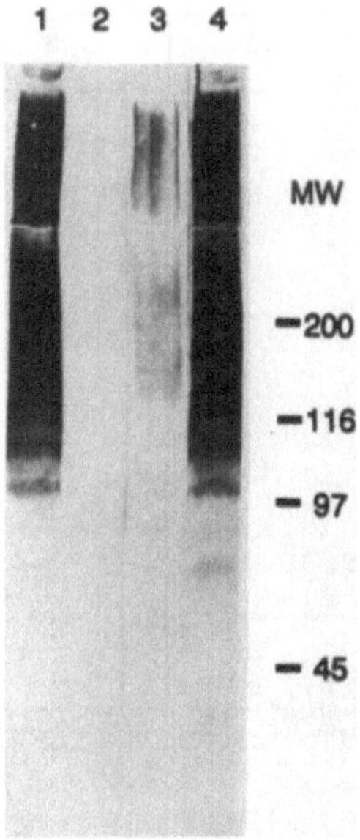

Figure 1. Immunoblot assay of monoclonal IgA antibody Sal4. *S. typhimurium* lysates were separated by SDS-PAGE, transferred to nitrocellulose, and probed with Sal4 IgA. Lane 1, wild type strain MT110. Lane 2, mutant strain MT114. Lane 3, wild type strain MT110: western blot pretreated with periodic acid at pH 4.5. Lane 4, wild type strain MT110: control blot incubated in pH 4.5 buffer without periodic acid. From Michetti et al. (1992) with permission. Copyright 1992 ASM.

Sal4 antibody (Woodward et al., 1985). Figure 1 (lanes 3 and 4) show that the epitope is very sensitive to periodate treatment, indicating that it is carbohydrate in nature. Taken together with the ability to agglutinate *S. typhimurium* cells, these results indicate that Sal4 IgA is directed against an externally exposed carbohydrate epitope.

Isolation of a Fully-Virulent Mutant That is Not Recognized by the Sal4 Antibody

The fact that the Sal4 IgA agglutinates *S. typhimurium* cells allowed us to enrich for mutants that no longer produce the epitope and therefore are no longer recognized by the monoclonal antibody. Briefly, a "pool" of insertion mutants of *S. typhimurium* was grown overnight in rich medium and 0.1 ml of Sal4 IgA was used to agglutinate 0.1 ml of this pool in a well of a microtiter dish. After a 2 hr incubation, a small aliquot of cells that remained in suspension were removed and again grown overnight in rich medium. This procedure was repeated 6 times and the final suspension was plated out on rich medium to obtain single clones that were then tested for agglutination with the Sal4 antibody. In one particular enrichment, for example, 56% (27/48) of the individual colonies tested were composed of cells that were not agglutinated by the Sal4 antibody. Using this enrichment procedure, 3 independent insertion mutants were isolated using 3 different transposons. In each case, the non-agglutination phenotype was inextricably linked to the insertion mutation by bacteriophage P22 transduction. One of the mutants, termed MT114, was used for *in vivo* experiments.

The strain MT114 contains an insertion mutation in a gene whose product is required for production of the Sal4 epitope. To confirm this, we performed a western blot in which we probed crude cell lysates of MT114 with Sal4 antibody. Figure 1 (lane 2) indicates that Sal4 antibody does not react with any external or internal components of MT114.

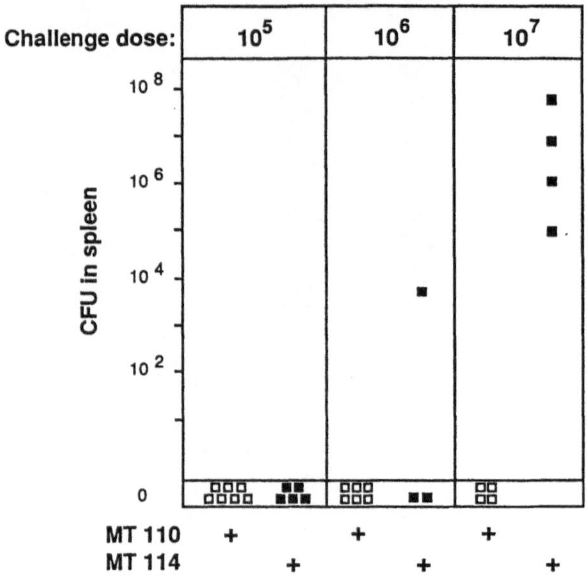

Figure 2. Systemic infection after oral challenge. *S. typhimurium* recovered from spleens of mice bearing Sal4 IgA hybridoma backpack tumors, 5 days after oral challenge with the wild type strain MT110 (open squares) or the mutant MT114 that is not recognized by the Sal4 antibody (solid squares). Total colony forming units (CFU) in spleens of individual mice are shown for each of 3 challenge doses. Separate experiments established that an oral challenge of 2 $x10^6$ MT110 represents the approximate LD_{50} in control BALB/c mice bearing hybridoma tumors secreting irrelevant IgA antibodies. From Michetti et al. (1992) with permission. Copyright 1992 ASM.

To determine if the insertion mutation in MT114 resulted in a virulence defect, we challenged mice both orally and intraperitoneally (i.p.) with either MT114 or the wild type strain, MT110. The oral and i.p. dose required to kill 50% of the animals (LD_{50}) for both strains was 2.0 X 10^6 organisms orally and <20 organisms i.p. This result indicates that, although MT114 lacks the epitope recognized by Sal4, it is a fully virulent mutant.

Sal4 IgA Protects BALB/c Mice from Systemic Infection After an Oral Challenge of *S. typhimurium*

To test the role of secretory IgA in protection against systemic infection after an oral challenge of *S. typhimurium*, Sal4 IgA was delivered to mucosal surfaces from subcutaneous hybridoma tumors on the backs of BALB/c mice as described previously (Winner et al., 1991). The tumor-bearing mice were then challenged orally with *S. typhimurium*. The bacterium normally crosses the intestinal epithelium and, after causing a transient bacteremia, infects the reticulo-endothelial system. A characteristic result of this systemic spread is high numbers of bacterial cells in target organs such as the liver and the spleen (Hohmann et al., 1978). Therefore, we monitored bacterial CFU in the spleen as a measurement of systemic infection. *S. typhimurium* is very efficient in establishing a systemic infection once it has crossed the mucosal barrier, as evidenced by the extremely low i.p. LD_{50} of less than 20 organisms. Thus, our criterion for protection against systemic infection after an oral challenge of *S. typhimurium* was a sterile spleen.

The sera of tumor-bearing mice were tested by ELISA analysis to ensure that the hybridoma tumors were producing levels of anti-*S. typhimurium* antibody detectable at 1:1000 dilution. Those tumor bearing mice that had sufficient levels of serum IgA were orally challenged with either wild type, MT110, or the virulent mutant, MT114, which no longer produces the Sal4 epitope. Figure 2 shows that tumor-bearing mice that were orally challenged with 10^7 cells of MT110 were protected from systemic infection. However, tumor-bearing mice were not protected when challenged with the control strain, MT114, which does not produce the Sal4 epitope. Taken together, these results indicate that secretion of a single monoclonal IgA onto mucosal surfaces via the physiological transport mechanism is sufficient to confer protection against an oral challenge with the invasive pathogen, *S. typhimurium*. Moreover, this protection is dependent on specific recognition of the bacteria by the monoclonal IgA.

Sal4 IgA Does Not Prevent Systemic Infection After an I.P. Challenge of *S. typhimurium*

We wanted to determine at which stage of the infection cycle the protection occurred. Normal blood levels of IgA are low. However, since the tumor-bearing mice produce relatively high levels of Sal4 IgA in their serum, the protection observed after oral challenge could have occurred systemically. To test this hypothesis, tumor-bearing mice were challenged i.p. with either wild type, MT110, or the agglutination deficient mutant, MT114. Figure 3 shows that bacterial CFU were recovered from all mice challenged with as low as 20 organisms of either strain. The numbers of MT110 cells recovered from the spleen of infected mice were reduced compared to MT114, indicating that serum IgA may have formed some immune complexes that somewhat slowed the infection process. Nevertheless, significant systemic protection against an i.p. challenge of *S. typhimurium* did not occur in tumor-bearing mice. Since the protection was observed only after oral challenge, the protection must have occurred at the intestinal mucosa.

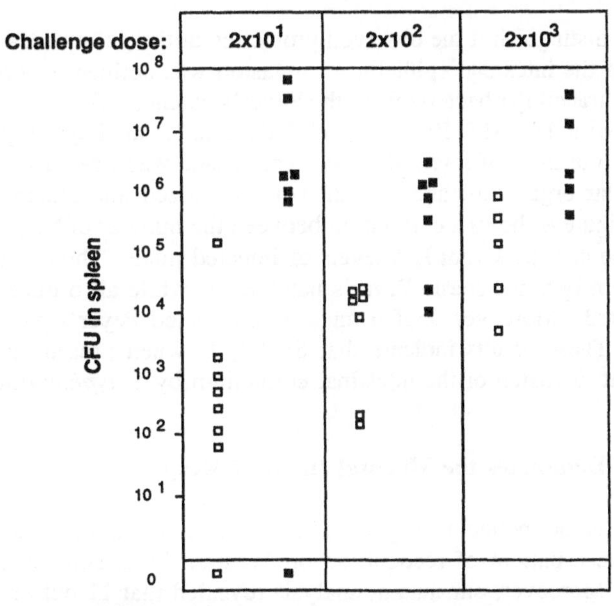

Figure 3. Systemic infection after i.p. challenge. *S. typhimurium* recovered from spleens of mice bearing Sal4 hybridoma backpack tumors, 3 days after i.p. challenge with the wild type strain MT110 (open squares) or the mutant MT114 that is not recognized by the Sal4 antibody (solid squares). Total colony forming units (CFU) in spleens of individual mice are shown for each of 3 challenge doses. From Michetti et al. (1992) with permission. Copyright 1992 ASM.

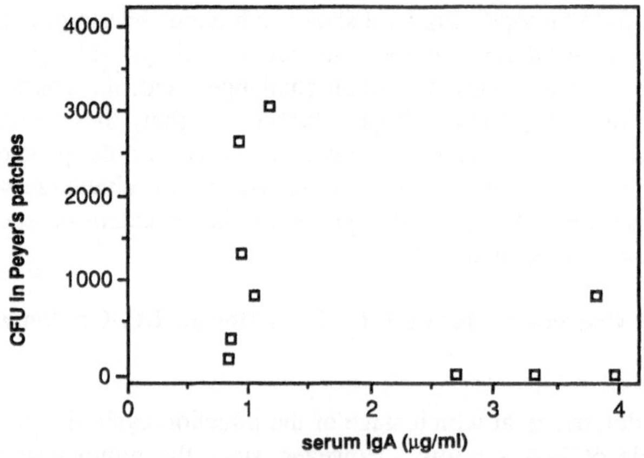

Figure 4. Infection of Peyer's patch mucosa after oral challenge. *S. typhimurium* recovered from Peyer's patch tissue of mice bearing Sal4 IgA hybridoma backpack tumors 2 days after oral challenge with 10^7 MT110 cells. Total colony forming units (CFU) in pooled Peyer's patch tissue from individual mice is plotted against the concentration of Sal4 IgA in serum. From Michetti et al. (1992) with permission. Copyright 1992 ASM.

Sal4 IgA Can Prevent Infection of Peyer's Patch Cells After Oral Challenge with *S. typhimurium*

We wanted to distinguish if the observed protection at the gut mucosa occurred before or after invasion of the intestinal epithelium. Invasion was defined experimentally as the ability to recover intracellular bacteria from the Peyer's patches. Tumor-bearing mice were orally challenged with 10^7 MT110 cells, and 2 days later, all Peyer's patch tissue was removed from the intestines of infected mice. This tissue was treated with gentamicin to kill any extracellular organisms, and was then homogenized and plated to detect viable *S. typhimurium.* Figure 4 shows a correlation between the number of bacteria in the pooled Peyer's patch tissue and the serum IgA levels of infected mice. Three out of 4 mice with high levels of serum IgA had sterile Peyer's patch cells, while all 6 mice with low levels of IgA were infected. Moreover, 2 of 6 mice with infected Peyer's patch cells also had infected spleens. These results indicate that Sal4 IgA, when present at levels above 2 μg/ml, can prevent invasion of the intestinal epithelium by *S. typhimurium.*

The Sal4 Antigen Dominates the Mucosal Immune Response

In characterizing the monoclonal IgA antibodies generated in this study, we noted that many of them gave a pattern of recognition on Western blots similar to that shown in Figure 1 for Sal4. Moreover, our mutant analysis revealed that 11 out of 12 of the mAbs chosen for study were directed against the same surface carbohydrate antigen; i.e., the mutants that were no longer agglutinated by Sal4, were also not agglutinated by other mAbs. These 12 hybridomas were derived from Peyer's patch cells from 2 different groups

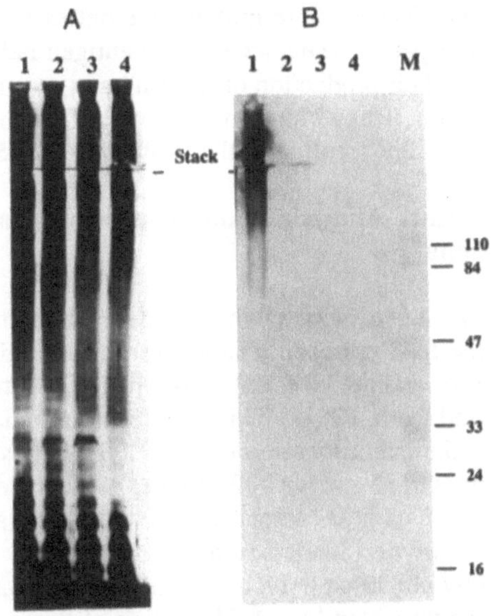

Figure 5. Sal4 antigen is immunologically distinct from LPS. LPS preparations from wild type MT110 (lane 1) and the three insertion mutants that do not produce the Sal4 epitope (lanes 2, 3 and 4) separated by SDS-PAGE. Panel A. The gel was silver stained. Panel B. The gel was used for a Western blot. The material was transferred to nitrocellulose and probed with Sal4 antibody.

of mice immunized with different attenuated strains of *S. typhimurium*. These results suggest that this carbohydrate antigen dominated the mucosal immune response.

Since the antigen recognized by Sal4 is a surface polysaccharide (as discussed above), we presumed that the Sal4 antibody was directed against lipopolysaccharide (LPS). As an initial test of this hypothesis, we isolated LPS (Silverman and Benson, 1987) from both the wild type and the insertion mutants that no longer express the Sal4 epitope, and ran these preparations on 10% polyacrylamide gels. One gel was silver stained to visualize the LPS, while the second gel was used in a Western blot and probed with Sal4 antibody. Figure 5 shows that the mutants that do not produce the Sal4 epitope have no apparent defect in LPS synthesis as visualized with silver stain. Moreover, in the Western blot, the Sal4 antibody does not seem to recognize the entire LPS ladder, but rather recognizes material that remains largely in the stacking gel. These results suggest that the epitope recognized by the Sal4 mAb is not simply LPS.

We have explored further the relationship between LPS and the Sal4 antigen. Mutants that contain a deletion of the entire galactose operon are defective in the synthesis of O-antigen and part of the outer core of LPS (Fukasawa and Nikaido, 1960). These *gal* mutants are also defective in the synthesis of the Sal4 epitope, indicating that this surface carbohydrate has a galactose component. The bacteriophage Felix-O uses LPS as a receptor. We have isolated isogenic mutants defective in LPS synthesis by selecting resistance to this bacteriophage. We were interested in mutations that affect LPS synthesis in genes other than *galE*, so we subsequently checked to ensure these Felix-O resistant strains remained *gal*$^+$ (Subbaiah and Stocker, 1964; Wilinson et al., 1972). As monitored visually by silver stained SDS-PAGE, these mutants are defective in some aspect of LPS synthesis in that they do no display the characteristic O-antigen ladder. These mutants are also 10 to 100 fold reduced in their production of the Sal4 antigen, and this residual activity maybe because the mutants are "leaky" (data not shown). Taken together, these results suggest a relationship between LPS (or at least the synthesis of LPS) and the Sal4 epitope.

Genetic Mapping and Sequence Analysis of the Insertion Mutations that Prevent Synthesis of the Sal4 Epitope

As described above, we have isolated three independent insertion mutations that prevent the expression of the Sal4 epitope. These insertion mutations have been mapped to the Salmonella chromosome and are 10% linked (by phage P22 transduction) to *glpQTA* at minute 45.5 (Sanderson and Roth, 1988). This map location indicates that the mutations are not genetically linked to the major loci involved in LPS synthesis, including *rfb*, which is responsible for the synthesis of O-antigen (42 minutes). However, these mutations may be related to *oafA*, a locus involved in O-factor 5 acetylation that maps in a similar position on the *S. typhimurium* chromosome (Sanderson and Roth, 1988). Indeed, although all of the strains are agglutinated by undiluted polyclonal anti-O-factor 4,5 antibody (Difco), our insertion mutants were not agglutinated by a 1:64 dilution of this antibody. This titer was, however, sufficient to agglutinate both the wild type and the Felix-O resistant (LPS-specific) mutants described above. Although the three insertions confer the same phenotype, they are not all tightly linked to one another; one of the insertion mutations is only 10% linked by phage P22 transduction to the remaining 2 insertion mutations, which are 95% linked to each other. These results suggest that either one relatively large locus or at least 2 linked loci are involved in the synthesis of the Sal4 epitope. Therefore, if there is a relationship to *oafA*, this locus is more complex than previously realized.

One of these insertion mutations has been cloned and the chromosomal segment surrounding the insertion has been sequenced. The DNA sequence reveals an open reading frame corresponding to 367 amino acids. The amino acid sequence of the putative protein shows similarity to 2 proteins found in Rhizobium species (Devereux et al., 1984). There is a 54.6% similarity to ExoZ, involved in exopolysaccharide synthesis in *Rhizobium meliloti* (Buendia et al., 1991) and a 51.7% similarity to NodX, which confers host specificity to *Rhizobium leguminosarum* biovar *viciae* strain TOM (Davis et al., 1988).

SUMMARY

The mucosal immune system is thought to provide an early line of defense against both invasive and noninvasive pathogens. Here we have examined the role of secretory IgA in mucosal protection and have shown that a single monoclonal IgA, delivered in a physiological fashion to mucosal surfaces, is capable of preventing infection of the invasive pathogen *S. typhimurium*. This protection is directly dependent on specific recognition of the bacteria by the monoclonal IgA and protection takes place at the mucosal surface, presumably by immune exclusion.

The mucosal immune response to oral *S. typhimurium* was dominated by antibodies directed against a single carbohydrate moiety on the surface of the bacterial cells. Preliminary characterization of this surface epitope suggests that it is related to lipopolysaccharide (LPS), in that mutations that affect the synthesis of LPS affect the synthesis of the Sal4 epitope. However, western blot analysis shows that the Sal4 antibody is not recognizing LPS per se, but rather, recognizes material that remains largely in the stacking gel. Also, insertion mutations that prevent expression of the Sal4 epitope show no genetic linkage to the major genes responsible for LPS production, and these insertion mutations have no obvious effect on LPS synthesis. Thus, if the Sal4 antigen is LPS, then recognition by the antibody requires a structure that is only realized in a subset of the LPS molecules. This structure must be dependent on the locus or loci defined by the insertion mutations. This suggests that this locus is involved in modification of the LPS, perhaps O-acetylation. If true, it is interesting that this modified form of LPS dominates the mucosal immune response. We are exploring further the relationship between LPS and the Sal4 epitope.

ACKNOWLEDGEMENTS

We would like to thank Helen Mäkelä and Bruce Stocker for helpful suggestions, and Don Wilbur and other members of the animal research facility at the School of Public Health, Harvard Medical School, for their excellent assistance with mice.

This work was supported by NIH research grants HD17557 and DK21505 (M.R.N.) and AI18045 (J.J.M.), NIH Center grant DK34854 (to the Harvard Digestive Diseases Center), research grant FNS32.30011.90 from the Swiss National Science Foundation (P.M.), National Research Service Award AI08245 (M.J.M.), and Damon Runyon-Walter Winchell Cancer Research Fund Fellowship DRG-1016 (J.M.S.).

REFERENCES

Brandtzaeg, P., 1989, Overview of the mucosal system, *Curr. Top. Microbiol. Immunol.* 146:13.

Buendia, A.M., Enenkel, B., Köplin, R., Niehaus, K., Arnold, W. and Pühler, A., 1991, The Rhizobium meliloti *exoZ/exoB* fragment of megaplasmid 2: ExoB functions as a UDP-glucose 4-epimerase and ExoZ shows homology to NodX of *Rhizobium leguminosarum* biovar *viciae* strain TOM, *Mol. Microbiol.* 5:1519.

Childers, N.K., Bruce, M.G. and McGhee, J.R., 1989, Molecular mechanisms of immunoglobulin A defense, *Annu. Rev. Microbiol.* 43:503.

Davis, E.O., Evans, I.J. and Johnston, A.W.B., 1988, Identification of *nodX*, a gene that allows *Rhizobium leguminosarum* biovar *viciae* strain TOM to nodulate Afghanistan peas, *Mol. Gen. Genet.* 212:531.

Devereux, J., Haeberli, P. and Smithies, O., 1984, A comprehensive set of sequence analysis programs for the VAX, *Nucleic Acids Res.* 12:387.

Fukasawa, T. and Nikaido, H., 1960, Formation of phage receptors induced by galactose in a galactose-sensitive strain of *Salmonella, Virology* 11:508.

Hohmann, A.W., Schmidt, G. and Rowley, D., 1978, Intestinal colonization and virulence of *Salmonella* in mice, *Infect. Immun.* 22:763.

Kohbata, S., Yokobata, H. and Yabuuchi, E., 1986, Cytopathogenic effect of *Salmonella typhi* GIFU 10007 on M cells of murine ileal Peyer's patches in ligated ileal loops: an ultrastructural study, *Microbiol. Immunol.* 30:1225.

Mestecky, J., 1988, Immunobiology of IgA, *Am. J. Kidney Dis.* 12:378.

Michetti, P., Mahan, M.J., Slauch, J.M., Mekalanos, J.J. and Neutra, M.R., 1992, Monoclonal secretory immunoglobulin A protects mice against oral challenge with the invasive pathogen *Salmonella typhimurium, Infect. Immun.* 60:1786.

Miller, S.I., Kukral, A.M. and Mekalanos, J.J., 1989, A two-component regulatory system (*phoP phoQ*) controls *Salmonella typhimurium* virulence, *Proc. Natl. Acad. Sci. USA* 86:5054.

Miller, S.I. and Mekalanos, J.J., 1990, Constitutive expression of the *phoP* regulon attenuates *Salmonella* virulence and survival within macrophages, *Journal of Bacteriology* 172:2485.

Mostov, K.E., Kraehenbuhl, J.P. and Blobel, G., 1980, Receptor mediated transcellular transport of immunoglobulin: synthesis of secretory component as multiple and larger transmembrane forms, *Proc. Natl. Acad. Sci. USA* 77:7257.

Sanderson, K.E. and Roth, J.R., 1988, Linkage map of *Salmonella typhimurium*, edition VII, *Microbiological Reviews* 52:485.

Silverman, J.A. and Benson, S.A., 1987, Bacteriophage K20 requires both the OmpF porin and lipopolysaccharide for receptor function, *Journal of Bacteriology* 169:4830.

Subbaiah, T.V. and Stocker, B.A.D., 1964, Rough mutants of *Salmonella typhimurium* (I) genetics, *Nature* 201:1298.

Underdown, B.J. and Dorrington, K.J., 1974, Studies on the structural and conformational basis for the relative resistance of serum and secretory immunoglobulin A to proteolysis, *J. Immunol.* 112:949.

Weltzin, R.A., Lucia Jandris, P., Michetti, P., Fields, B.N., Kraehenbuhl, J.P. and Neutra, M.R., 1989, Binding and transepithelial transport of immunoglobulins by intestinal M cells: demonstration using monoclonal IgA antibodies against enteric viral proteins, *J. Cell Biol.* 108:1673.

Wilkinson, R.G., Gemski, P. and Stocker, B.A.D., 1972, Non-smooth mutants of *Salmonella typhimurium*: differentiation by phage sensitivity and genetic mapping, *J. Gen. Microbiol.* 70:527.

Winner, L.S.,III, Weltzin, R.A., Mekalanos, J.J., Kraehenbuhl, J.P. and Neutra, M.R., 1991, New model for analysis of mucosal immunity: intestinal secretion of specific monoclonal immunoglobulin A from hybridoma tumors protects against Vibrio cholerae infection, *Infect. Immun.* 59:977.

Woodward, M.P., Young, W.W.,Jr. and Bloodgood, R.A., 1985, Detection of monoclonal antibodies specific for carbohydrate epitopes using periodate oxydation, *J. Immunol. Methods* 78:143.

T CELL FIBRONECTIN, DELAYED HYPERSENSITIVITY AND HUMAN DISEASE

Henry P. Godfrey

Department of Experimental Pathology
New York Medical College
Valhalla, NY 10595

INTRODUCTION

Extracellular matrix molecules such as fibronectin (FN) are commonly thought of as structural elements of tissue, and as such, to have limited involvement in the pathogenesis of infectious disease. In point of fact, FN is involved in multiple aspects of the host-parasite relationship. On the host side, a special FN, T cell FN, is produced as a lymphokine by antigen-activated T lymphocytes. T cell FN is associated with initiation of the tissue-damaging delayed hypersensitivity reactions seen in typhoid fever and tuberculosis (Godfrey, 1990). On the parasite side, FN-binding proteins are found in the outer membranes of many bacteria, including *Salmonella typhi* and *Mycobacterium tuberculosis*, and may be involved in bacterial adhesion to and/or infection of host cells (Van de Water et al., 1983; Froman et al., 1984; Baloda et al., 1988, 1991; Barnes et al., 1989; George and Falkingham, 1989; Ljungh et al., 1990; Wiker et al., 1990; Visai et al., 1991). Bacterial FN-binding proteins are sometimes major secretory products (Wiker et al., 1986; De Bruyn et al., 1987; Abou-Zeid et al., 1988). For example, FN-binding proteins of the antigen 85 (Ag85) complex are both major mycobacterial secretory proteins and strong immunogens in human beings infected with *M. tuberculosis* (Kaplan and Chase, 1980; Wiker et al., 1986; De Bruyn et al., 1987; Abou-Zeid et al., 1988; Turneer et al., 1988; Sada et al., 1990). Interactions between FN and bacterial FN-binding proteins can also modulate expression of delayed hypersensitivity in sensitized hosts (Godfrey et al., 1992).

Before summarizing the properties and functional activities of T cell FN, the structure of FN will be briefly reviewed. The role of T cell fibronectin in initiating delayed hypersensitivity reactions will be examined. Finally, the activity of FN-binding secretory proteins in modifying expression of delayed hypersensitivity will be described. The implications of these findings for studies on *S. typhi* and pathogenesis of typhoid fever will be briefly explored.

Biology of Salmonella, Edited by F. Cabello *et al.*.
Plenum Press, New York, 1993

GENERAL PROPERTIES OF FN

The FN are a family of closely related high molecular weight glycoproteins found in plasma and tissues (Hynes, 1989). They bind specifically to fibrin, Gram-positive and Gram-negative bacteria, gelatin (collagen), heparin and eukaryotic cells. Fibronectins are involved in cell adhesion and motility, regulation of cell

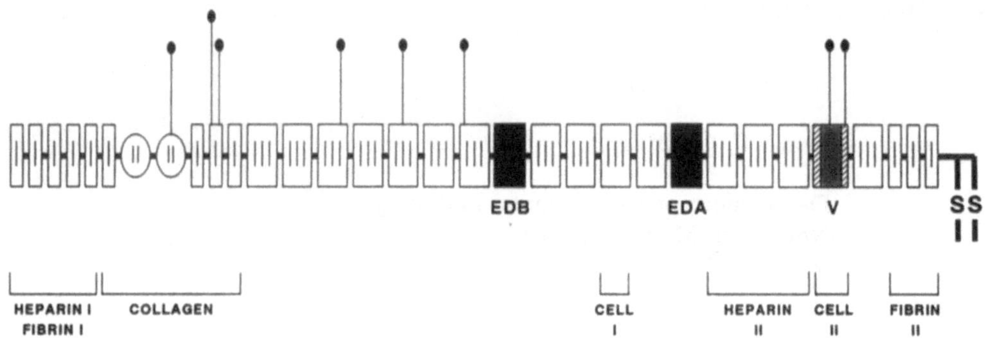

Figure 1. Schematic representation of FN monomeric subunit showing location of type I, II, and III sequence homology units (labeled I, II, III), type III extra domains A and B (EDA, EDB) (filled rectangles), the alternatively spliced variable (V) region (hatched), functional domains, and glycosylation sites (—●) (Hynes, 1989). EDA and EDB can either be included or excluded. The V region can be partially or completely included, or completely excluded. Reprinted with permission from Godfrey (1990).

morphology, wound healing, phagocytic function and inflammation (Godfrey, 1990). Their exact role(s) in many of these processes is frequently unclear. Molecularly, FN are disulfide-bonded, heterodimeric glycoproteins with molecular weights of 400-550 kDa (Hynes, 1989; Petersen et al., 1989) containing 4-12% carbohydrate depending on their origin (Fig. 1) (Fukuda et al., 1982; Zhu et al., 1984). The role of the carbohydrate in FN function is unclear. Recent studies suggest that it can modify FN function by affecting protein conformation (Zhu et al., 1990). Each FN monomer is composed of structurally and functionally distinct globular domains connected by short, flexible, proteolytically sensitive segments (Mosher, 1984; Akiyama and Yamada, 1987; Petersen et al., 1989). Domains are relatively resistant to proteolytic attack and can be individually isolated after partial proteolysis of FN. Various binding activities of FN (fibrin, heparin, gelatin, cell) have been localized to one or more domains in the molecule. In those cases where multiple domains for the same ligand are present, each domain has distinct affinities for ligand and requires different conditions for optimal binding.

The FN are produced by alternative splicing of a primary transcript from the single complex FN gene in the genome (Tamkun et al., 1984; Kornblihtt et al., 1985). For this reason, much of the molecule is constant regardless of the cell source, and individual members of this family have numerous epitopes and properties in common (Godfrey, 1990). Cellular FN contain extra domains EDA and/or EDB (Hynes, 1989;

Petersen et al., 1989). These extra domains are not present in plasma FN, and their functional role is unknown. As a result of the existence of these extra domains and the variable region, there are 20 possible variations in each chain of the dimer, and 190 possible variations in FN dimers (Godfrey, 1990). It is not known if all mathematically possible FN variants actually exist, and whether they differ in biological properties.

T CELL FN

T cell FN is produced as a lymphokine by mitogen and antigen activated human, murine and guinea pig T cells, and appears to play an important role in the initiation of *in vivo* DH reactions. It can be quantitated by immunoassay in culture supernatants of mitogen stimulated cloned T cells using appropriate anti-FN antibodies (Godfrey et al., 1988, 1989a). T cell FN can also be demonstrated in the cytoplasm of activated cloned T cells (Godfrey et al., 1988). It is not present immunohistochemically or by flow cytometry on the surface membranes of these cells.

T cell FN appears to be a unique FN, biochemically different from other FN. The properties of T cell FN are summarized in Table 1. T cell FN differs from plasma FN in that it contains EDA (Godfrey et al., 1988, 1989a, b, 1990), a hallmark of cellular FN. T cell FN differs from other cellular FN in that it has a lower M_r (Godfrey et al., 1988, 1989a, 1990) and a smaller mRNA than other cellular FN. T cell FN is otherwise similar to other FN. Like other FN, T cell FN is a disulfide-bonded heterodimer with properties of a typical FN: pI, 6.1; rapid loss of activity at temperatures above 60-65 °C; stable between pH 3 and 9; lability to reduction and proteolytic enzymes; reversible binding to fibrin, heparin, gelatin, and cells (monocytes/macrophages). T cell FN shares many epitopes with other FN and cross reacts with polyclonal and monoclonal antibodies raised to plasma and cellular FN.

T cell FN is a cellular FN. It contains EDA, an extra domain present only in cellular FN and lacking in plasma FN (Godfrey et al., 1988, 1989a, 1990). The M_r of T cell FN is slightly smaller than or similar to that of plasma FN by both immunoblotting and chromatography. This is surprising since T cell FN contains an extra domain lacking in plasma FN. Molecules with the functional properties of T cell FN cannot, however, be generated from purified high M_r lymphoid cell FN by partial proteolysis with trypsin or thrombin (Godfrey, 1990; Godfrey et al., 1990). The small molecular mass of T cell FN is confirmed by studies of mRNA coding for its protein. T cell FN mRNA from mitogen activated human peripheral blood mononuclear cells is only 5800-6000 bases long, rather than the 6300 bases of smooth muscle FN (Z.-H. Feng and H.P. Godfrey, unpublished observations).

Two *in vitro* biological activities have been associated with T cell FN, mononuclear phagocyte agglutination and matrix-driven translocation. The monocyte agglutination activity of T cell FN co-purifies with the well studied lymphokine, macrophage agglutination factor or MAggF, on HPLC (Godfrey et al., 1989a, 1990). Monocyte agglutination is mediated by the cell-binding and gelatin-binding domains (Godfrey et al., 1989b). The other activity of T cell FN, matrix-driven translocation, is mediated by the amino-terminal heparin-binding and gelatin-binding domains (Godfrey et al., 1989b). T cell FN is 10^4 to 10^6 times more potent than other FN in these activities, and is active at femtomolar concentrations (Godfrey et al., 1989a, b). Different mechanisms are involved in these two activities. T cell FN agglutinates monocytes or macrophages by interacting with multiple classes of cell surface integrin receptors and fucose receptors (Godfrey, 1990; Donson et al., 1991). It translocates monocytes and neutrophils in the generic physical process of matrix-driven translocation by interacting with heparin-like surface molecules (Godfrey et al., 1989b).

Table 1. Selected Properties of T cell and Plasma Fibronectins

	T cell FN	Plasma FN
Electrophoretic mobility	ß	ß
pI	6.1	6.1
M_r (unreduced)	400-420 kDa	450-470 kDa
(reduced)	210-220 kDa	230-250 kDa
Activity labile to:		
Reduction	+	+
Heat (>60-65 °C, 30 min)	+	+
pH <3, pH >9	+	+
Proteolytic enzymes	+	+
Hyaluronidase	-	-
Fucosidase	+	ND
Functional domains		
Heparin-binding	+	+
Fibrin-binding	+	+
Gelatin-binding	+	+
Cell-binding	+	+
Immunochemical domains		
NH_2-terminal heparin-binding	+	+
Gelatin-binding	+	+
Cell-binding	+	+
Extra domain A	+	-

Data from Godfrey and Purohit, 1982a, b; Godfrey et al., 1984, 1986, 1988, 1989a, b, 1990; Donson et al., 1991.

T CELL FN AND INITIATION OF DELAYED HYPERSENSITIVITY REACTIONS

Several lines of evidence point to T cell FN involvement in initiating DH reactions *in vivo* (reviewed in Godfrey, 1990). Antigen-induced elicitation of T cell FN from sensitized polyclonal T cells is closely correlated with delayed skin reactivity to the same antigen both in man and laboratory animals. T cell FN is rapidly produced in two to four hours following T cell activation (Godfrey et al., 1988; Godfrey, 1990). There is a close association between the ability of several cloned murine $CD4^+$ and $CD8^+$ T cell lines of various antigenic specificities such as the cloned T cell lines described above to produce T cell FN *in vitro* after antigenic stimulation and their ability to transfer delayed hypersensitivity reactions *in vivo*. Monoclonal anti-T cell FN antibodies neutralized T cell FN agglutination activity *in vitro* in a dose-dependent manner (S. Mandy and H.P. Godfrey, unpublished observations). These antibodies also significantly inhibited the expression of delayed hypersensitivity to purified protein derivative of tuberculin (PPD) (Figure 2). The decrease in visible erythema caused by the antibodies was not associated with a decrease in superficial or deep dermal infiltrating cells at the site of the reaction. This suggests that T cell FN acts on the vascular rather than the cellular component of delayed hypersensitivity.

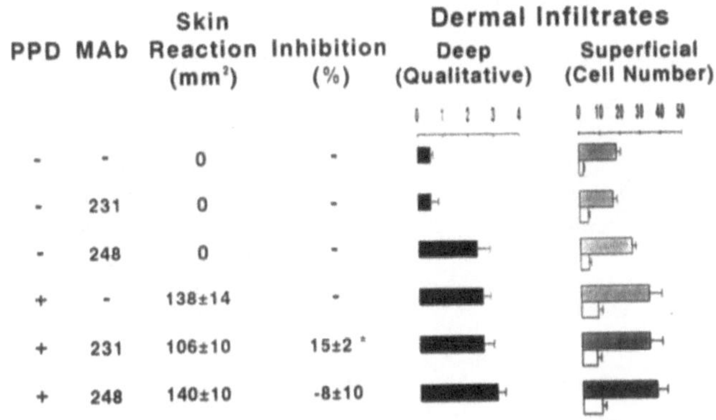

PPD
MAb
Skin Reaction (mm²)
Inhibition (%)
Dermal Infiltrates
Deep (Qualitative)
Superficial (Cell Number)
- - 0 -
- 231 0 -
- 248 0 -
+ - 138±14 -
+ 231 106±10 15±2 *
+ 248 140±10 -8±10

Figure 2. Inhibition of PPD-elicited delayed hypersensitivity reactions in guinea pigs sensitized 13 weeks previously with Freund's complete adjuvant (FCA) by monoclonal anti-T cell FN antibodies. Animals (n=6) skin tested with 0.1 ml phosphate buffered saline, or with 90 ng monoclonal antibody (clone 231.28 or 248.13), 500 ng PPD or 500 ng PPD mixed with monoclonal antibody in 0.1 ml phosphate buffered saline, and reactions measured at 24 h. Skin reactions and relative inhibition are shown as means ± SE. Histologic evaluation of skin biopsies are shown as mean qualitative scores ± SE for deep dermal cellular infiltrates (solid bars) and as mean cell number/5 oil fields ± SE for superficial dermal infiltrates of mononuclear cells (shaded bars) and basophils (open bars). See Godfrey et al. (1992) for details of methods. *, mean differs significantly from value obtained with PPD alone ($P<0.05$, paired *t*-test).

T CELL FN AND INTRACELLULAR PARASITISM

Infection with facultative intracellular bacteria such as *S. typhi* or *M. tuberculosis* leads to production of specific antibodies and cellular immune responses (Collins and Mackaness, 1968, 1970; Kaplan and Chase, 1980; Orme and Collins, 1984; Orme, 1988). These cellular immune responses include specific bactericidal activity and delayed hypersensitivity, and are involved in protective immunity and tissue damage. In both typhoid fever and tuberculosis, cellular immune responses play an important role in pathogenesis and recovery. Despite the obvious differences in typhoid fever and tuberculosis, the basic mechanisms of host cellular immunity are likely to be similar whatever the inciting organism. The rapidity and ease of culture of *S. typhi* are advantageous for studying mechanisms of bactericidal responses. On the other hand, the ease of induction of cellular immune responses to non-viable *M. tuberculosis* is advantageous for studying mechanisms of tissue damaging aspects of cellular immunity.

Protective immunity to *Salmonella* and *Mycobacteria* requires immunization with living bacteria (Collins and Mackaness, 1970; Orme and Collins, 1984; Orme, 1988). This implies that secreted bacterial proteins may be important for development of protective host immunity (Abou-Zeid et al., 1988; Andersen et al., 1991a, b). One such group of secreted proteins are the FN-binding proteins of the mycobacterial antigen 85 (Ag85) complex (Wiker et al., 1986; De Bruyn et al., 1987; Abou-Zeid et al., 1988). These 30-32 kDa proteins can account for up to 30% of secreted proteins in culture and are strong immunogens in man (Huygen et al., 1988; Havlir et al., 1991). Ag85 binds equally well to all FN, including T cell FN (Godfrey et al., 1992). However, the individual components differ in their ability to inactivate purified T cell FN *in vitro*. The reasons for this are not known.

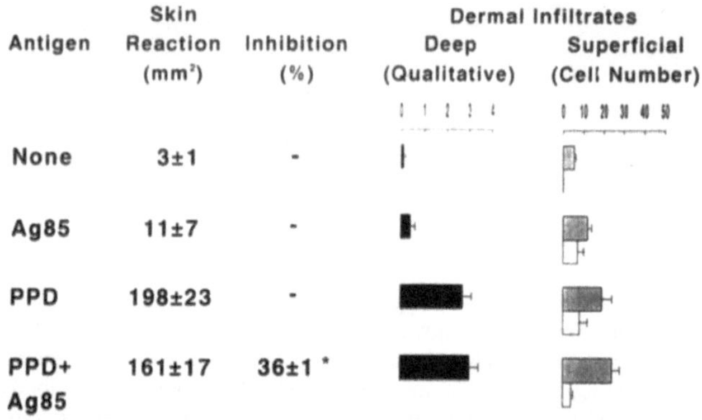

Antigen
Skin Reaction (mm²)
Inhibition (%)
Dermal Infiltrates
Deep (Qualitative)
Superficial (Cell Number)
None
3±1
-
Ag85
11±7
-
PPD
198±23
-
PPD+ Ag85
161±17
36±1 *

Figure 3. Inhibition of PPD-elicited delayed hypersensitivity reactions in guinea pigs sensitized 13 weeks previously with FCA by BCG Ag85 complex. Animals (n=9) skin tested with 0.1 ml phosphate buffered saline, or with 100 ng Ag85 complex (De Bruyn et al., 1987), 500 ng PPD or 500 ng PPD mixed with 100 ng Ag85 in 0.1 ml phosphate buffered saline, and reactions measured at 24 h. Skin reactions and relative inhibition are shown as means ± SE. Histologic evaluation of skin biopsies are shown as mean qualitative scores ± SE for deep dermal cellular infiltrates (solid bars) and as mean cell number/5 oil fields ± SE for superficial dermal infiltrates of mononuclear cells (shaded bars) and basophils (open bars). See Godfrey et al. (1992) for details of methods. *, mean differs significantly from value obtained with PPD alone ($P<0.05$, paired *t*-test).

The inhibitory action of Ag85 components on monocyte agglutination activity of human T cell FN *in vitro* prompted an examination of their ability to modulate expression of delayed hypersensitivity *in vivo*. Animals sensitized with FCA containing mixed strains of human heat-killed *M. tuberculosis* showed strong delayed hypersensitivity to PPD and to another mycobacterial protein antigen, hsp65, but only weak delayed hypersensitivity to Ag85 (Figure 3, and Godfrey et al., 1992). Sensitized guinea pigs had significant anti-Ag85 titers (700-900) at the time of skin testing, but skin testing was not associated with any significant increases in Ag85 antibody titer two weeks later (Godfrey et al., 1992). Even though 100 ng Ag85 complex elicited little delayed hypersensitivity by itself, delayed reactions elicited by mixtures of Ag85 complex and PPD were significantly inhibited in a dose dependent manner when compared to those elicited by PPD alone (Figure 3). The individual purified Ag85 components did not vary significantly in their inhibition of expression of delayed hypersensitivity (Godfrey et al., 1992). Inhibition by Ag85 was limited to the vascular component of delayed hypersensitivity (Figure 3), and significant inhibition of delayed erythema was not associated with any qualitative or quantitative decrease in numbers of infiltrating dermal cells. Inhibition of expression of delayed hypersensitivity by Ag85 was not mycobacterial antigen specific. Ag85 also inhibited expression of delayed hypersensitivity to a hapten-protein conjugate (Godfrey et al., 1992). The inhibition of expression of delayed hypersensitivity by Ag85 was similar to that observed with monoclonal anti-T cell FN antibodies, and suggests that Ag85 interferes with T cell FN activity *in vivo*. Inhibition of expression of delayed hypersensitivity by Ag85 proteins might occur during mycobacterial infections because of Ag85 interaction with T cell FN, and be responsible for some of the clinical anergy seen in this disease.

The involvement of FN and mycobacterial FN-binding proteins in expression of delayed hypersensitivity suggests that similar studies on *Salmonella* might yield important insights into the pathogenesis of typhoid fever. *M. tuberculosis* has several cell-associated, FN-binding proteins, including the Ag85 complex and p55 (Barnes et al., 1989; George and Falkingham, 1989; Wiker et al., 1990; Abou-Zeid et al., 1991; Harboe et al., 1992). FN-binding protein(s) of *S. typhi* presumably exist, since *S. typhi* bind to FN (Baloda et al., 1988, 1991). These proteins have not been studied in any detail. *E. coli* appears to have two such proteins, one of which binds to the amino-terminal domain of FN and a second which has not been fully characterized (Froman et al., 1984; Visai et al., 1991). FN-binding proteins of *S. typhi*, like those of *E. coli*, might be involved in adhesion and/or infection of eukaryotic cells.

Mycobacterial FN-binding proteins are major bacterial secretory products. It would be reasonable to examine *S. typhi* culture media for proteins of similar activity. The rapid growth rate and simpler chemistry of *S. typhi* compared to *M. tuberculosis* would facilitate such biochemical analysis. In view of the activity of secreted mycobacterial FN-binding proteins in modulating the expression of delayed hypersensitivity, it would also be reasonable to evaluate secreted FN-binding proteins of *S. typhi* for their ability to modulate anti-salmonellal cellular immunity. The strong human immunogenicity of mycobacterial FN-binding proteins suggests that studies of the human immunogenicity of any newly discovered salmonellal FN-binding proteins could have important ramifications for our understanding of infection immunity in salmonellosis.

CONCLUSION

T cell FN is a unique cellular FN produced as a lymphokine by mitogen and antigen activated human, murine and guinea pig T cells. It is biochemically different and markedly more potent *in vitro* than other cellular FN. T cell FN appears to play an important role in initiating delayed hypersensitivity reactions *in vivo*. Secreted mycobacterial FN-binding proteins can bind T cell FN and inhibit its activity *in vitro*, as well as modify expression of delayed hypersensitivity to mycobacterial and other antigens *in vivo*. Modulation of T cell FN activity by secreted FN-binding proteins could prove to be important in typhoid fever. For these reasons, studies on cell-associated and secreted FN-binding proteins of *Salmonella* are likely to yield key information on the pathogenesis of typhoid fever.

Acknowledgements This work was supported by grant CA 34141 from the National Cancer Institute, United States Public Health Service, Department of Health and Human Services.

REFERENCES

Abou-Zeid, C., Smith, I., Grange, J.M., Ratliff, T.L., Steele, J., and Rook, G.A.W., 1988, The secreted antigens of *Mycobacterium tuberculosis* and their relationship to those recognized by the available antibodies, *J. Gen. Microbiol.* 134:531.

Abou-Zeid, C., Garbe, T., Lathigra, R., Wiker, H.G., Harboe, M., Rook, G.A.W., and Young, D.B., 1991, Genetic and immunologic analysis of *Mycobacterium tuberculosis* fibronectin-binding proteins, *Infect. Immun.* 59:2712.

Akiyama, S.K., and Yamada, K.M., 1987, Fibronectin, *Adv. Enzymol.* 59:1.

Andersen, P., Askgaard, D., Ljungqvist, L., Bennedsen, J., and Heron, I., 1991a, Proteins released from *Mycobacterium tuberculosis* during growth, *Infect. Immun.* 59:1905.

Andersen, P., Askgard, D., Ljungqvist, L., Bentzon, M.W., and Heron, I., 1991b, T-cell proliferative response to antigens secreted by *Mycobacterium tuberculosis*, *Infect. Immun.* 59:1558.

Baloda, S.B., Faris, A., and Krovacek, K., 1988, Cell-surface properties of enterotoxigenic and cytotoxic *Salmonella enteriditis* and *Salmonella typhimurium*: studies on hemagglutination, cell-surface hydrophobicity, attachment to human intestinal cells and fibronectin-binding, *Microbiol. Immunol.* 32:447.

Baloda, S.B., Dyal, R., Gonzalez, E.A., Blanco, J., Hajdu, L., and Mansson, I., 1991, Fibronectin binding by *Salmonella* strains - Evaluation of a particle agglutination assay, *J. Clin. Microbiol.* 29:2824.

Barnes, P.F., Mehra, V., Hirschfield, G.R., Fong, S.J., Abou-Zeid, C., Rook, G.A.W., Hunter, S.W., Brennan, P.J., and Modlin, R.L., 1989, Characterization of T cell antigens associated with the cell wall protein-peptidoglycan complex of *Mycobacterium tuberculosis*, *J. Immunol.* 143:2656.

Collins, F.M., and Mackaness, G.B., 1968, Delayed hypersensitivity and Arthus reactivity in relation to host resistance in *Salmonella*-infected mice, *J. Immunol.* 101:830.

Collins, F.M., and Mackaness, G.B., 1970, The relationship of delayed hypersensitivity to acquired antituberculous immunity. I. Tuberculin sensitivity and resistance to reinfection in BCG-vaccinated mice, *Cell. Immunol.* 1:253.

De Bruyn, J., Huygen, K., Bosmans, R., Fauville, M., Lippens, R., Van Vooren, J.P., Falmagne, P., Wiker, H.G., Harboe, M., and Turneer, M., 1987, Purification, characterization and identification of a 32 kDa protein antigen of *Mycobacterium bovis* BCG, *Microb. Pathog.* 2:351.

Donson, J., Mandy, K., Feng, Z.-H., Mandy, S., Brown, E.J., and Godfrey, H.P., 1991, Role of monocyte fucose-receptors in T cell fibronectin activity, *Immunology* 74:473.

Froman, L., Switalski, L.M., Faris, A., Wadstrom, T., and Höök, M., 1984, Binding of *Escherichia coli* to fibronectin. A mechanism of tissue adherence, *J. Biol. Chem.* 259:14899.

Fukuda, M., Levery, S.B., and Hakamori, S., 1982, Carbohydrate structure of hamster plasma fibronectin. Evidence for chemical diversity between cellular and plasma fibronectin, *J. Biol. Chem.* 257:6856.

George, K.L., and Falkingham, J.O., III, 1989, Identification of cytoplasmic membrane protein antigens of *Mycobacterium avium*, *M. intracellulare*, and *M. scrofulaceum*, *Can. J. Microbiol.* 35:529.

Godfrey, H.P., 1990, T cell fibronectin: an unexpected inflammatory lymphokine, *Lymphokine Res.* 9:435.

Godfrey, H.P., and Purohit, A., 1982a, Properties of guinea-pig macrophage agglutination factor, *Immunology* 46:507.

Godfrey, H.P., and Purohit, A., 1982b, Reversible binding of a guinea-pig lymphokine to gelatin and fibrinogen: possible relationship of macrophage agglutination factor and fibronectin, *Immunology* 46:515.

Godfrey, H.P., Angadi, C.V., Wolstencroft, R.A., and Bianco, C., 1984, Localization of macrophage agglutination factor activity to the gelatin-binding domain of fibronectin, *J. Immunol.* 133:1417.

Godfrey, H.P., Angadi, C.V., Haak-Frendscho, M., and Kaplan, A.P., 1986, Concurrent production of macrophage agglutination factor and factor VII by antigen-stimulated human peripheral blood mononuclear cells, *Immunology* 57:77.

Godfrey, H.P., Canfield, L.S., Kindler, H.L., Angadi, C.V., Tomasek, J.J., and Goodman, J.W., 1988, Production of a fibronectin-associated lymphokine by cloned mouse T cells, *J. Immunol.* 141:1508.

Godfrey, H.P., Canfield, L.S., Haak-Frendscho, M., Melancon-Kaplan, J., Brown, E.J., and Kaplan, A.P., 1989a, Relationship of human macrophage agglutination factor to other fibronectins, *Immunology* 67:321.

Godfrey, H.P., Frenz, D.A., Canfield, L.S., Akiyama, S.K., and Newman, S.A., 1989b, Non-chemotactic translocation of phagocytic cells mediated by a fibronectin-related human lymphokine, *J. Immunol.* 143:3691.
Godfrey, H.P., Canfield, L.S., Angadi, C.V., Zagachin, L.M., Kielpinski, G.G., and Colvin, R.B., 1990, Characterization of lymphokine fibronectin from guinea pig lymphoid cell culture supernatants, *Immunobiology* 180:109.
Godfrey, H.P., Feng, Z.-H., Mandy, S., Mandy, K., Huygen, K., DeBruyn, J., Abou-Zeid, C., Wiker, H.G., Nagai, S., and Tasaka, H., 1992, Modulation of expression of delayed hypersensitivity by mycobacterial antigen 85 fibronectin-binding proteins, *Infect. Immun.* 60:2522.
Harboe, M., Wiker, H.G., and Nagai, S., 1992, Protein antigens of mycobacteria studied by quantitative immunologic techniques, *Clin. Infect. Dis.* 14:313.
Havlir, D.V., Wallis, R.S., Boom, W.H., Daniel, T.M., Chervenak, K., and Ellner, J.J., 1991, Human immune response to *Mycobacterium tuberculosis* antigens, *Infect. Immun.* 59:665.
Huygen, K., Van Vooren, J.P., Turneer, M., Bosmans, R., Dierckx, P., and De Bruyn, J., 1988, Specific lymphoproliferation, gamma interferon production, and serum immunoglobulin G directed against a purified 32 kDa mycobacterial protein antigen (P32) in patients with active tuberculosis, *Scand. J. Immunol.* 27:187.
Hynes, R.O., 1989, "Fibronectins," Springer Verlag, New York.
Kaplan, M.H., and Chase, M.W., 1980, Antibodies to mycobacteria in human tuberculosis. II. Response to nine defined mycobacterial antigens with evidence for an antibody common to tuberculosis and leprosy patients with antigens from BCG, *J. Infect. Dis.* 142:835.
Kornblihtt, A.R., Umezawa, K., Vibe-Petersen, K., and Baralle, F.E., 1985, Primary structure of human fibronectin: differential splicing may generate at least 10 polypeptides from a single gene, *EMBO J.* 4:1755.
Ljungh, A., Emody, L., Steinruck, H., Sullivan, P., West, B., Zetterberg, E., and Wadstrom, T., 1990, Fibronectin, vitronectin, and collagen binding to Escherichia coli of intestinal and extraintestinal origin, *Zbl. Bakt. Int. J. Med. Microbiol.* 274:126.
Mosher, D.F., 1984, Physiology of fibronectin, *Annu. Rev. Med.* 35:561.
Orme, I.M., 1988, Characteristics and specificity of acquired immunologic memory to *Mycobacterium tuberculosis* infection, *J. Immunol.* 140:3589.
Orme, I.M., and Collins, F.M., 1984, Adoptive protection of the *Mycobacterium tuberculosis*-infected lung: dissociation between cells that passively transfer protective immunity and those that transfer delayed-type hypersensitivity to tuberculin, *Cell. Immunol.* 84:113.
Petersen, T.E., Skorstengaard, K., and Vibe-Pedersen, K., 1989, Primary structure of fibronectin, *in*: "Fibronectin," D.F. Mosher, ed., Academic Press, New York.
Sada, E.D., Ferguson, L.E., and Daniel, T.M., 1990, An ELISA for the serodiagnosis of tuberculosis using a 30,000-Da native antigen of *Mycobacterium tuberculosis*, *J. Infect. Dis.* 162:928.
Tamkun, J.W., Schwarzbauer, J.E., and Hynes, R.O., 1984, A single rat fibronectin gene generates three different mRNAs by aloternative splicing of a complex exon, *Proc. Natl. Acad. Sci. USA.* 81:5140.
Turneer, M., Van Vooren, J.P., De Bruyn, J., Serruys, E., Dierckx, P., and Yernault, J.C., 1988, Humoral immune response in human tuberculosis: immunoglobulins G, A, and M directed against the purified P32 protein antigen of Mycobacterium bovis bacillus Calmette-Guerin, *J. Clin. Microbiol.* 26:1714.
Van de Water, L., Destree, A.T., and Hynes, R.O., 1983, Fibronectin binds to some bacteria but does not promote their uptake by phagocytic cells, *Science* 220:201.
Visai, L., Bozzini, S., Petersen, T.E., Speciale, L., and Speziale, P., 1991, Binding sites in fibronectin for an enterotoxigenic strain of E coli B342289c, *FEBS Lett.* 290:111.
Wiker, H.G., Harboe, M., and Lea, T.E., 1986, Purification and characterization of two protein antigens from the heterogeneous BCG85 complex in *Mycobacterium bovis* BCG, *Int. Arch. Allergy Appl. Immunol.* 81:298.
Wiker, H.G., Harboe, M., Nagai, S., and Bennedsen, J., 1990, Quantitative and qualitative studies on the major extracellular antigen of *Mycobacterium tuberculosis* H37Rv and *Mycobacterium bovis* BCG, *Am. Rev. Resp. Dis.* 141:830.
Zhu, B.C., Fisher, S.F., Pande, H., Calaycay, J., Shively, J.E., and Laine, R.A., 1984, Human placental (fetal) fibronectin: increased glycosylation and higher protease resistance than plasma fibronectin. Presence of polylactosamine glycopeptides and properties of a 44-kilodalton chymotryptic collagen-binding domain: difference from human plasma fibronectin, *J. Biol. Chem.* 259:3962.

Zhu, B.C., Laine, R.A., and Barkley, M.D., 1990, Intrinsic tryptophan fluorescence measurements suggest that polylactosaminylglycosylation affects the protein conformation of the gelatin-binding domain from human placental fibronectin, *Eur. J. Biochem.* 189:509.

ATTENUATION OF *SALMONELLA* BY AUXOTROPHY

Bruce A.D. Stocker

Department of Microbiology and Immunology
Stanford University School of Medicine
Stanford, CA 94305-5402

INTRODUCTION

It seems obvious that for pathogenic bacteria which cause disease by extensive multiplication in the tissues of the infected animal (host) the bacteria must acquire from the host all the nutrients they need for growth and consequently that mutations causing new requirements, that is new auxotrophic characters, will interfere with multiplication in host tissues, and so cause reduced virulence or non-virulence, if the required metabolite or metabolites are not present in the relevant "compartment" of the host or are present but only at concentrations insufficient for growth of the auxotrophic organism. Most *Salmonella* serotypes had been shown to be prototrophic by the mid 1940's, at which time methods for inducing bacterial mutations and for the detection and isolation of auxotrophic mutants from mutagen-treated cultures first became available. Not much later three reports (Bacon, Burrows and Yates, 1950 *a, b,* 1951) described a very thorough and extensive investigation of the effect on virulence (for the mouse) of mutations causing auxotrophy in a strain of *S. typhi*. These reports are, I think, the earliest study of the effect of auxotrophy on bacterial virulence; they revealed a large fraction of what we now know, more than 40 years later, on this subject, in particular that most mutations causing requirement for a single amino acid or for a single vitamin have little or no effect on virulence. The success of this investigation is the more striking in that the infection model used, that is the interaction of *S. typhi* and mice, is a highly artificial one. The typhoid organism is highly specialized for infection of man and does not multiply significantly when administered to mice (or other non-primates) unless either a very large number of bacteria are injected into the peritoneal cavity or the inoculum includes an adjuvant, usually hog gastric mucin, 5%, which in some not-understood way facilitates bacterial multiplication or impedes killing of the bacteria, by host phagocytes or otherwise. Bacon and his colleagues used the former method, and injected ca. 5 x 10^7 colony-forming units (CFU), without adjuvant, into the peritoneal cavity. This dose resulted in about 80% mortality within two or three days, for their virulent starting strain. Another complication of use of *S. typhi* for such studies is the auxotrophic character of most strains, including strain Ty2, the standard strain used for production of (killed) vaccine, chosen by Bacon

Biology of Salmonella, Edited by F. Cabello *et al.*,
Plenum Press, New York, 1993

and his colleagues. Strain Ty2 had been shown to require tryptophan, but to give rise to prototrophic descendants when "trained" by serial passage in defined medium (or, as we should now interpret the observations, to produce *trp+* mutants). A prototrophic derivative of strain Ty2 was exposed to different mutagenic treatments and the survivors were tested for nutritional character, in some experiments after penicillin treatment to enrich for auxotrophs. Characterized auxotrophic mutants, of independent origin, were tested in mice. Auxotrophs of only three classes, of many tested, consistently caused attenuation.

Several later investigations confirmed the attenuating effect of some auxotrophic characters and showed that, in general, what was true for *S. typhi*, tested by injection of large inocula in mice, held true also for much smaller inocula of *S. typhimurium* given to mice by various routes and for a mouse-virulent strain of *Klebsiella pneumoniae*. In recent years the use as live vaccine of strains of *S. typhimurium, S. typhi* and *S. dublin* attenuated by auxotrophic character has been the subject of investigation in many laboratories. My contribution to this workshop deals not with such applications but with the basic phenomenon of attenuation by auxotrophy, and discusses both the original study and several later publications from various laboratories including my own, and also describes some as yet unpublished results from my laboratory.

THE EARLIEST INVESTIGATION

The 97 stable or reasonably stable auxotrophic mutants of *S. typhi* identified by Bacon and his colleagues fell in 25 classes according to nutritional character. In a screening test the number of mice dying by three days after inoculation of 10 mice with ca. 5×10^7 CFU of a mutant was compared with the number of deaths in 10 mice similarly inoculated with the parent strain, tested at the same time. These tests showed that the mutants of nearly all the classes were about as virulent as the prototrophic parent. Further tests, using graded inocula, showed attenuation of: all of four mutants requiring a purine (i.e. growing if provided with either adenine, guanine, xanthine or hypoxanthine) or a purine plus thiamine; the one mutant with a requirement for guanine or xanthine; all of three mutants growing only if provided with aspartic acid; and the only mutant with a need for *p*-aminobenzoic acid (pAB). Probit analysis indicated virulence in the range 0.005 to 0.2, relative to that of the parent strain, for all these nine mutants. In addition one of the ten histidine-requiring mutants and one of the nine cystine-requiring mutants proved less virulent than the parent strain. However for these two auxotrophs reversion to histidine-independence or cystine-independence did not restore virulence, an indication that attenuation resulted from coincidental mutation, not from auxotrophy. Three mutants responding to cytosine or uracil were not detectably reduced in virulence. Non-dependent (revertant) variants were obtained from two of the purine-requiring, from one aspartate-requiring and from the pAB-requiring mutant; in each case the revertant was about as virulent as its prototrophic ancestor, strain Ty22. This showed that attenuation of these mutants resulted from the mutation causing auxotrophy, not from a coincidental mutation.

A further proof that non-virulence resulted from the new growth-factor requirement was the virulence-enhancing effect of administration of the required metabolite to the experimentally infected host. Mice given hypoxanthine, i.p., as a fine suspension, or given pAB suspended in oil, by the same route, died from inocula of respectively, a purine-requiring or the pAB-requiring mutant which were tolerated by mice not given hypoxanthine or pAB. Partial restoration of virulence of an aspartate-requiring mutant was observed in mice given aspartic acid, 6 mg, in oil by injection. The fluid obtained by washing out the peritoneal cavity of a normal mouse added to liquid defined medium allowed growth from ca. 200 CFU to more than 10^9 CFU, for several auxotrophic mutants of not-reduced virulence, and of one aspartate-requiring mutant but did not support any substantial multiplication of either of two purine or purine-plus-thiamine mutants or of the pAB-requiring mutant.

The main observations made by Bacon and his colleagues may be summarized thus: Only three of the 20 or so auxotrophic classes tested were consistently associated with reduced virulence, those involving purine requirement, aspartate requirement or pAB requirement. For at least one mutant of each of these classes reversion to non-dependence restored virulence and for at least one representative of each class administration of the required metabolite to the experimentally infected animal restored virulence, at least in part. The metabolites required by mutants of reduced virulence, with the partial exception of aspartic acid, were not available in normal peritoneal fluid to an extent sufficient to allow extensive multiplication of reduced-virulence mutants.

LATER INVESTIGATIONS

I consider below the main results obtained in later investigations, including those from my own laboratory, for each of the classes of auxotroph of reduced virulence identified by Bacon and his colleagues (1950 *a, b,* 1951) and for additional classes, not encountered by them. In the infection model used by Bacon, that is i.p. inoculation of mice with a very large number of *Salmonella typhi*, death results from extracellular multiplication of the bacteria within the peritoneal cavity; by contrast in typhoid fever in man and in experimental models of the human disease, including *S. typhimurium* infection in mice, all or nearly all multiplication is believed to occur within host cells. It is therefore noteworthy that auxotrophic characters found to prevent extracellular multiplication of *S. typhi* have later been found to attenuate *S. typhi* for man and *S. typhimurium* and other serotypes for mice, presumably by preventing intracellular multiplication.

ATTENUATION BY REQUIREMENT FOR SUBSTANCES WHICH ARE METABOLITES OF THE HOST

Auxotrophic characters which reduce bacterial virulence can be divided into two classes according to the role of the required substance in the physiology of the vertebrate host. It may be a normal metabolite of the host, commonly a small molecule which is either manufactured by host cells or acquired by the host from the exterior and used as a building block for the synthesis of a co-enzyme or of a macromolecular compound, e.g. purines and pyrimidines as precursors of DNA or RNA. Other attenuating auxotrophic characters make bacteria dependent on an external supply of a substance which is a normal metabolite of the bacteria, but not of the host (but such substances if absorbed from the gut may be handled like other xenobiotics, that is detoxified by conjugation, before being excreted).

Requirement for a Purine

Table 1 lists bacterial genes whose mutation causes requirement for a substance which is a normal vertebrate metabolite and is such that the mutation results in partial or complete loss of virulence, at least if the mutation causes complete dependence, i.e. is non-leaky.

Table 1. Effect of Auxotrophy on *Salmonella* virulence.

Mutated gene(s)	Relevant Requirement[1]	Effect on virulence[2]	Live-vaccine suitability[3]
(a) Requirement is a host metabolite			
purF, C, HD, G	Any purine	Reduced	Effective, not safe
guaA, B	Guanine	Reduced	Not tested
purA, B	Adenine	Non-virulent	Safe, poor efficacy
thyA	Thymine	Reduced	Effective, not safe
pyrC	Cytosine or Uracil	Reduced	Not tested
hemA	5-aminolevulinic acid	Non-virulent	Effective, safe
"asp", ppc	Aspartate; citrate[4]	Reduced	Not tested
(b) Requirement is not a host metabolite			
pab	*p*-aminobenzoate	Non-virulent	Effective, safe
aroA, C, D	*p*-aminobenzoate and enterobactin	Non-virulent	Effective, safe
ent	2,3 dihydroxy-Genzoate, or enterobactin	Reduced (?)	Not safe
dap, asd	*m*-diaminopimelate	Non-virulent or reduced	Safe (?); non-effective or poorly effective

[1]In general requirement for purines or pyrimidines is satisfied by provision either of base or of nucleoside.

[2]Virulence as indicated by parenteral administration to mice. Non-virulent, LD50 of auxotroph at least 10^5 larger than that of non-auxotrophic parent or revertant. Reduced, LD50 of auxotroph at least 10-fold larger than that of related prototroph, but not consistently 10^5 larger.

[3]As indicated by result of challenge of mice given auxotroph as live-vaccine by administration of a virulent strain of the same or a related serotype, or by inference from trial of strains of similar phenotype.

[4]*ppc* mutants of *S. typhimurium* tested on defined medium with glucose as energy source grow if provided with aspartate or citrate. An *"asp"* mutant of *S. typhi* is satisfied only by aspartate.

The first three classes of mutation listed block steps in one or other of the three segments of the *de novo* synthesis pathway leading to the purine mononucleotides, adenosine monophosphate (AMP) and guanosine monophosphate (GMP), required as source of the adenine and guanine nucleotides of DNA and RNA and with many other functions in cellular metabolism. The first segment, or common pathway, comprises several steps leading to inosine monophosphate (IMP), which is the precursor of both AMP and GMP. A complete block in any of the earlier steps of this segment results in a requirement both for thiamine (vitamin B1) and for a purine. The purine requirement can be satisfied by provision of either hypoxanthine, adenine or guanine (or xanthine), as free base or as nucleoside. The later steps of the common pathway cause similar purine requirement but without need for thiamine. The use of adenine as a precursor of GMP, or of guanine as precursor of AMP, results from the activity of salvage-pathway enzymes, for conversion of AMP or GMP back to IMP, the branch-point compound. IMP is converted to GMP via xanthine mononucleotide (XMP) by two steps, requiring the products of genes *guaB* and *guaA*. Loss of function of either gene blocks conversion of IMP to GMP and causes a purine requirement satisfied only by provision of guanine (or guanosine), or of xanthine, in the case of *guaB* mutants. In the first investigation of the effect on virulence of new auxotrophic characters in *Salmonella typhi* all of the four mutants of the first class, with a purine requirement satisfied by any purine, were found to be of reduced virulence, as determined by probit analysis of dose/mortality data, by a factor of 0.05 to 0.003. In a much later investigation in my laboratory on the effect on virulence of mutation to purine auxotrophy a mouse-virulent strain of Vi-positive *S. dublin* and two mouse-virulent strains of *S. typhimurium* were given different blocks in purine biosynthesis, by transducing in alleles inactivated by insertion of transposon Tn10 (McFarland and Stocker, 1987). Transductants with any of several blocks in the common segment of the pathway, and those with blocks at either step in the conversion of IMP to GMP were of considerably reduced virulence, as shown by survival of some mice given large inocula, by the i.p. route. However in these experiments, in which only a small number of mice were used for each strain-allele combination, there were always some deaths from i.p. inocula of less than 10^6 CFU. This was taken to mean that blocks either in the common pathway or in its GMP branch cause only partial attenuation, and that this sort of auxotrophic character, at least when used by itself, would not be appropriate for construction of a live-vaccine strain for trial in humans.

Inactivation of gene *purA* or *purB* blocks the conversion of IMP to AMP and causes purine requirement satisfied only by provision of adenine (or adenosine). No mutants of this sort were encountered amongst the 97 auxotrophic mutants, of independent origin, of *S. typhi* tested (Bacon *et al.*, 1951*a*), from a total of more than 250,000 colonies screened. However, in a somewhat later investigation Furness and Rowley (1956) found that all of three independently-isolated adenine-requiring mutants of *S. typhimurium* were effectively completely attenuated, in that i.p. inocula of more than ten million bacteria did not cause deaths. Correspondingly genetically susceptible BALB/c mice inoculated i.p. with the *purA* or *purB* derivatives of two mouse-virulent *S. typhimurium* strains were unaffected by the largest doses tested, i.e. 10^7 or more CFU (McFarland and Stocker, 1987).

The virtually complete attenuation of several mouse-virulent strains by adenine requirement resulting from transposon insertion or deletion at *purA* or *purB* suggested that this requirement would cause loss of virulence of *S. typhi* for man. Furthermore, mice given a single inoculum of at least 2.5 x 10^5 CFU of the *purA* derivatives of two strains of *S. typhimurium* all survived challenge, 30 or 58 days later, with a 250 LD50 dose of a virulent strain of the same species. It thus seemed likely that an irreversible genetic lesion preventing synthesis of AMP from IMP would similarly attenuate *S. typhi* without causing loss of protecting ability. A *purA* allele, earlier identified by genetic analysis as an extensive deletion, was therefore introduced, by co-transduction with an adjacent silent Tn10 insertion, into an *S. typhi* strain, of phage type A, which had been given a deletion of aroA (Edwards and Stocker, 1988). [The reason for doing this was the necessity for two independent attenuating mutations, each irreversible, in a candidate live-vaccine strain to be submitted to licensing authorities for trial in humans, even though the dangers against which a second mutation guards, that is correction of one defect by some hypothetical unknown mechanism or by recombination with a wild-type strain in the gut lumen, are perhaps so small that they could with impunity be ignored]. The resulting *aroA*(deletion) *purA*(deletion) strain, 541Ty, and its Vi-negative mutant, 543Ty, given by mouth to volunteers in a trial at the Center for Vaccine Development, University of Maryland School of Medicine, in doses up to 2 x 10^{10} CFU, caused no ill-effects in them and nearly all of those taking the larger dose showed evidence of a cellular immune response; however there was no or very little serological response. More extensive trials (O'Callaghan *et al.*, 1988; Sigwart *et al.*, 1989) of *purA* and *purA aroA* strains of *S.typhimurium* or *S. dublin* as live vaccines in mice showed that the number of bacteria recoverable from the liver and spleen one day after administration by injection or feeding is a much smaller fraction of the number inoculated if the bacteria have a *purA* mutation in addition to *aroA* character and that the *purA* and double mutants have much less ability to protect against a later challenge than has the live-vaccine strain with only the aromatic requirement.

The cause of the extreme attenuation of all of several independently-derived adenine-dependent strains, contrasting with the generally incomplete attenuation of strains with blocks in the common pathway or in the GMP branch, is not known. One factor may be the conversion of most assimilated adenosine to inosine, which *purA* or *purB* bacteria cannot use as precursor of AMP.

Requirement for a Pyrimidine

The virulence of *Salmonella* mutants requiring pyrimidine bases or nucleosides has not been as extensively investigated as that of mutants with blocks in the *de novo* purine biosynthesis pathway. Loss of function of gene *thyA*, specifying thymidylate synthase, needed for conversion of deoxyuridine monophosphate (dUMP) to deoxythymidine monophosphate (dTMP) causes resistance to trimethoprim and other inhibitors of dihydrofolate reductase; mutants requiring thymine or thymidine can therefore be obtained in the laboratory by selection for resistance to trimethoprim and they are also encountered amongst strains selected *in vivo*, in patients treated with trimethoprim or related compounds or in animals given such agents as food additives. Smith and Tucker (1976) tested thymine-requiring strains of *S. typhimurium* isolated from the feces of an experimentally infected chicken fed a diet laced with trimethoprim and sulphadiazine, comparing its virulence with that of a thymine-independent mutant derived from it. They similarly compared the virulence for day-old chicks of *thyA* mutants selected as trimethoprim-resistant in several other *Salmonella* serotypes. An oral dose of 5 x 10^8 CFU given to day-old chicks resulted in fatal infections in from 13% to 100% of them, according to serotype used, when the wild-type strains were tested, and for each strain the *thyA* variant caused fewer deaths than the parent strain, e.g. 3% as against 100% for *S. gallinarum*, a natural pathogen of poultry. A similar reduction in virulence, by a factor of 0.1 to 0.01, was seen when chicks were challenged by the subcutaneous (s.c.) route. The two *S. typhimurium* strains studied were of relatively low virulence for the mice used, but here also the thymine requiring strains were less virulent than their thymine-independent parents. For instance a s.c. dose of 5 x 10^8 CFU caused deaths in 15 of 18 mice given one of the non-exacting forms, compared to only 4 of 18 deaths in those given its thymine-requiring mutant. Aqueous extracts of ground chick organs added to simple defined medium allowed growth of thymine-requiring strains. In a later investigation in my laboratory (C.Spurdon and Stocker, unpublished) the strain of *S. typhimurium* used, SL3201, and mouse line, C57Bl, were such that an i.p. inoculum of less than 10 CFU resulted in fatal infections, with mice dying six to fourteen days later. By contrast nearly all mice given 800 CFU of a thymine-requiring derivative of strain SL3201 survived to day 22 but most of those given 8 x 10^4 CFU succumbed in three to nine days. Cultures taken at autopsy yielded only thymine-requiring *Salmonella*. It thus appears that thymine requirement reduces the virulence of *Salmonella*, both for day-old chicks and for mice, but not to an extent which would suggest possible usefulness as attenuating character for construction of a safe live-vaccine strain.

Blocks at any of several points in the pathway for the *de novo* biosynthesis of the pyrimidine mononucleotides, deoxyuridine monophosphate (dUMP) and deoxycytidine monophosphate (dCMP) and so of dTMP, cause requirement satisfied by provision of uracil or cytosine, or their nucleosides. As noted above three such mutants of *S. typhi* were not detectably attenuated when tested by i.p. inoculation of mice in the original investigation. In the course of a recent experiment a fully virulent strain of *S. typhimurium* was made *pyrC*::Tn10, as a step in the construction of a strain for use for another purpose; the resulting pyrimidine-requiring strain tested in Balb/C mice was somewhat attenuated (Carsiotis, Stocker and O'Brien, unpublished).

Requirement for Heme

The list of auxotrophic characters associated with reduction of virulence of *S. typhimurium* was recently extended by the report (Benjamin et al., 1991) that an inactivated *hemA* allele, introduced into a virulent *S. typhimurium* strain by co-transduction with a silent Tn10 insertion, caused extreme attenuation, with LD50 values, i.v., increased from about 10 CFU to more than 10^7 CFU, as tested in Ity-s mice. Gene *hemA* is involved in the biosynthesis pathway leading to heme, needed for production of the cytochromes and of catalase and also for biosynthesis of siderоheme, needed for reduction of sulfate and of nitrite, and of vitamin B_{12} (Elliott, 1989). Loss of *hemA* function prevents synthesis of 5-amino levulinic acid, a precursor of heme in both bacteria and mammals. The *hemA* gene of *S. typhimurium* has been cloned and sequenced but the role of the protein which it specifies is not known. Mutants lacking *hemA* function cannot carry out oxidative respiration unless supplied with 5-amino levulinic acid and fail to grow under aerobic conditions if glycerol is present. Presumably the non-availability of 5-amino levulinic acid at the required concentration to *Salmonella* in the supposedly aerobic environment of bacteria located within macrophage-like cells is involved in the attenuation of *hemA* strains. A considerable fraction of the number of CFU injected could be recovered from the liver and spleen for up to two weeks after inoculation of *hemA* bacteria into mice; it therefore seems that the attenuation of the *hemA* strain is not due to extensive killing by some oxygen-involved mechanism because of defect in catalase production. The excellent protection observed in mice vaccinated with a live *hemA* strain suggests that this character may prove of service for construction of live vaccine strains.

Requirement for Aspartic Acid

The auxotrophic mutants of *S. typhi* tested in the original investigation included three which grew on defined medium only if it was supplemented with aspartic acid; all three of them had virulence ca. 0.1 relative to the parent strain. A prototrophic revertant of the only one from which revertants could be obtained had about the same virulence as its prototrophic ancestor and injection of aspartate increased the mortality resulting from injection of one of the mutants into mice. The nature of their mutation (assuming it to be the same in all three) is not clear, because three different transaminases can convert oxaloacetate to aspartate, the only reaction specific for its formation. The three mutants were noted (Bacon et al., 1951) to require 10- to 25-fold more aspartic acid as supplement in order to obtain full growth in defined medium, as compared to the requirements of other single-amino-acid mutants tested. Such a phenotype would be expected if the metabolic defect produced by the mutation was in the re-supply of 4-carbon skeletons to the Krebs (tricarboxylic-acid) cycle, to replace those withdrawn from the cycle for synthesis of various amino acids (aspartate, and from it asparagine, methionine, lysine, threonine and isoleucine; glutamate, glutamine, arginine and proline, from alpha-oxoglutarate). In bacteria such as *E. coli* this resupply is effected by the condensation of phosphoenolpyruvate with carbon dioxide to form alpha-ketoglutarate, by action of phosphoenolpyruvate carboxylase. The one available aspartate-responding mutant, 8505, examined in my laboratory was deficient of activity for this enzyme (R.F. Brown and Stocker, unpublished). However, our tentative interpretation of strain 8505 as having a mutation of the gene, *ppc*, which specifies this enzyme, could not be confirmed by transductional crosses to *ppc*::Tn10 mutants of *S. typhimurium*. (These latter mutants, identified in the laboratory of Dr John Roth, University of Utah, as Tn10-induced auxotrophs responding to aspartic acid, to our surprise grew on the simple defined medium used at Stanford without the need for supplementation with aspartic acid or any other amino acid. The difference between results here and those at University of Utah was found to result from our use of defined medium containing sodium citrate, 0.4g/l, and glycerol, 5 ml/l, whereas the Roth laboratory medium had glucose but no citrate or glycerol. Citrate is a compound of the Kreb's cycle and can be taken up by *S. typhimurium* and most other serotypes, since they can use it as sole carbon and energy source. However, it would not be available to *S. typhi* which is known to lack the mechanism for uptake of citrate. This would explain why strain 8505 behaved as an asparte auxotroph on our defined medium despite its citrate content). A virulent strain of *S. typhi* made deficient of phosphoenolpyruvate carboxylase by transducing into it a *ppc*::Tn10 allele had LD50 of less than 25,000 CFU when tested in mice by i.p. injection with hog gastric mucin. Thus, the aspartate requirement resulting from this defect did not cause any substantial loss of virulence, as tested by this technique. Furthermore, this defect introduced into *S. typhimurium* would not be likely to attenuate, since citrate and perhaps other intermediates of the Kreb cycle would be available to the bacteria from mammalian plasma or cytosol.

ATTENUATION BY REQUIREMENT FOR SUBSTANCES WHICH ARE NOT METABOLITES OF THE HOST

Requirement for p-Aminobenzoic Acid or for Aromatic Compounds

A pAB-requiring mutant of *S. typhi* was found to be of reduced virulence by Bacon *et al.* (1950 *b*, 1951) and this requirement was shown to cause the attenuation both by the virulence of a pAB-independent revertant and by the increased mortality observed when a continuous supply of pAB was provided to the experimentally infected mice by the injection of pAB suspended in oil or by pAB in their water. A later investigation in my laboratory (Brown and Stocker, 1987) showed that this strain, 20063, had LD50 of ca. 10^7 CFU when given i.p. to mice suspended in hog gastric mucin, 5%, whereas its pAB-independent ancestor, strain Ty2, and a pAB-independent transductant derived from 20063 had LD50 of less than 200 CFU by this test (the result expected for Vi-positive *S. typhi*). So far as I have been able to ascertain from the literature the only proven role of pAB in bacteria and plants is as precursor of folic acid (which consists of a pAB molecule with a pteridine molecule attached at one end and one or more glutamic acid residues at the other end). Vertebrates and other higher animals acquire folic acid preformed from their diet and cannot make it themselves; PAB administered experimentally is subject to rapid detoxification by conjugation with glucuronic acid, acetic acid and glycine, and excreted by the kidneys. Most bacteria are unable to take up folic acid from the exterior, perhaps because of the large size of the molecule; they instead synthesize it from pAB, which latter is derived from chorismic acid, the final product of the common aromatic biosynthesis pathway. Inactivation of any one of the three *pab* genes concerned in the conversion of chorismic acid to pAB results in requirement for exogenous pAB. Normal human plasma is expected to contain small amounts of pAB, derived from dietary components and perhaps also from the products of metabolism of the normal flora of the cecum and large intestine; bioassay indeed indicates the presence of traces of pAB, or perhaps of its conjugates, since the availability of the latter to bacteria as pAB source has not been investigated. (Many *Plasmodium* species are pAB auxotrophs, as shown both by their requirement for it for *in vitro* propagation in red cells and by failure of progression of experimental malaria in animals taking a pAB-free diet.) However it seems unlikely that pAB will penetrate from plasma into the interior of macrophage-like cells, so that the non-virulence of a *S. typhi* strain Ty22 made *pab* is not surprising. Transductional analysis identified the *pab* mutation of strain 20063 as affecting gene *pabB* (Brown and Stocker, 1987). A *pabB* mutation, apparently a point mutation but not detectably leaky, transduced into a virulent strain of *S. typhimurium* resulted in attenuation similar to that caused by *aroA*(deletion) mutation, with LD50 for Ity-s mice, i.p. route, greater than 10^6 CFU (Stocker, unpublished). A limited trial showed the constructed *pabB* strain tested as live vaccine to be about as effective as an *aroA* derivative of the same parent strain.

Plants and prokaryotes make all their aromatic metabolites, including the three major aromatic amino acids, tryptophan, phenylalanine and tyrosine, from chorismic acid, the product of the aromatic biosynthesis pathway. In the summer of 1977 it occurred to me that a complete block in any of the steps of the *aro* pathway would make *S. typhimurium*, etc., dependent on provision of essential aromatic compounds from the exterior and that two such metabolites, pAB, for synthesis of folic acid, and 2,3-dihydroxybenzoic acid, for synthesis of the iron-capturing compound enterobactin, are not mammalian metabolites and so unlikely to be present in mammalian tissues. The availability of *S. typhimurium* strains with transposon Tn10, which confers resistance to tetracycline, inserted into different *aro* genes made it possible to test the prediction that blocking aromatic biosynthesis by introduction of such an inactivated allele would result in loss of virulence (Hoiseth and Stocker, 1981). The non-virulence of *Salmonella*, of several serotypes, made aromatic-dependent in this way and prevented from reversion by transposon-generated secondary mutations and their usefulness both as live vaccines and as live-vaccine presenters of heterologous antigens and epitopes has been extensively reported and will not be further considered here. It was at first thought that their pAB requirement and their inability to make enterobactin could each, by itself, explain the non-virulence of such *aro* strains. However, it now seems that the latter inability is not likely to be of importance, since making *S. typhimurium* deficient only in ability to make enterobactin from chorismic acid via dihydroxybenzoic acid appeared to have no significant effect on virulence in two investigations (Benjamin et al., 1985; S.K. Hoiseth and Stocker, unpublished) though Yancey et al. (1979) using different strains found an *ent* mutant to be partially attenuated.

Bacteria of some genera resemble fungi by the possession of a catabolic pathway allowing the use of quinic acid and related compounds as carbon and energy source. Two of the enzymes of this catabolic pathway catalyse the reactions which are run by the products of genes *aroD* and *aroE* of the biosynthetic pathway. It seems that bacteria of a species possessing both pathways would not become aromatic-dependent by inactivation of gene *aroD* or *aroE* of the biosynthetic pathway. Inactivation of any of genes *aroA*, *aroC* and *aroD* results in apparently the same, almost complete, extent of attenuation of *S. typhimurium*, *S. dublin*, etc. The same may well hold for the virulence of *S. typhi* for man but the candidate live-vaccine strains so far tested in volunteers all had either two different *aro*(deletion) mutations or an *aroA*(deletion) and an independent attenuating mutation, *purA*(deletion). Though inactivation of any of the above-mentined three *aro* genes causes apparently identical loss of virulence in *S. typhimurium* this does not hold for gene *aroE*; two different transposon insertions at this locus transferred into mouse-virulent *S. typhimurium* caused only incomplete loss of virulence (C. Hormaeche and Stocker, unpublished). In one case this was explicable by incomplete blockage of aromatic biosynthesis, since the *aroE* strain grew, though only very slowly, on defined medium supplemented with other amino acids but without tryptophan (perhaps because of transposon insertion in the promotor region, rather than in the coding part of the gene). In the other case the allele used, *aroE552*::Tn10, appears to cause complete dependence on the provision of aromatic metabolites, so that incomplete attenuation remains to be explained.

Requirement for *m*-Diaminopimelic Acid

The *meso*diaminopimeleic acid (DAP) component of the murein making up the bacterial cell wall is derived from aspartic acid by a pathway whose several steps are catalyzed by enzymes specified by *dap* genes, which in *E. coli* and *Salmonella* are scattered around the chromosome. The DAP product of this pathway serves not only for synthesis of murein but also for conversion to lysine, by a single enzymic reaction controlled by gene *lysA*. Thus, complete loss of function of any of the *dap* genes prevents synthesis both of murein and of lysine. Such mutants require the provision of DAP for growth. If incubated in medium lacking DAP but otherwise suitable for growth, including presence of lysine, they lyse, within an hour or so, presumably because the cell wall made when DAP is not available is ruptured. DAP is not a metabolite of higher animals, which acquire lysine from their diet. For this reason DAP is not expected to be present in host tissues and *dap* mutants of *Salmonella* should be non-virulent, as has proved to be the case. In an investigation in my laboratory of the possible usefulness of DAP requirement as an additional attenuating character in an *aroA* live-vaccine strain a transposon-inactivated allele, *dap-140*::Tn10, was transduced into a fully virulent strain of *S. typhimurium*, SL1344, and into an *aroA* live-vaccine derivative of it, SL3261. A tetracycline-sensitive mutant of each *dap*::Tn10 strain was then tested by i.p. inoculation of BALB/c mice, one dose of 2 x 10^5 CFU. One group of mice given the double mutant had been started on drinking water containing DAP, 1g/l, two days previously. The results (S.K. Hoiseth and Stocker, unpublished) showed that *dap* mutation alone caused loss of virulence, for none of the mice given the *aro*+ *dap* strain died or showed signs of illness. As the LD50, i.p., of the wild-type parent strain is less than 20 CFU the *dap* character attenuates by a factor of at least 10^4, as judged by LD50 values. We did not test for persistence of live bacteria in the liver and spleen but the results of challenge, 28 days, by i.p. inoculation of 3 x 10^5 CFU of the virulent parent strain, SL1344, showed that the *dap* strain, attenuated only by its DAP requirement, and the *dap aroA* strain, with two attenuating characters, had no or minimal live-vaccine efficacy for mice not given DAP in their drinking water (2 of 7 and 0 of 7 mice surviving challenge). The *dap aroA* strain given to mice consuming DAP at the time of live-vaccine administration was perhaps rather more protective, 3 of 8 mice surviving challenge and the time to death of the remainder significantly prolonged. It seems likely that the non-virulence of the *aro+ dap* strain results from lysis and that the failure of the double mutant strain to confer substantial protection against later challenge even in mice taking DAP in their water supply likewise results from lysis because of failure of the DAP (presumably present in the plasma of these mice) to penetrate into the macrophage-like cells where the bacteria are thought to reside. The result was disappointing in that we did not obtain the hoped-for result, that is good survival of the double mutant despite DAP deprivation because of failure to grow as a result of starvation for aromatic compounds; at the time this experiment was made we were not aware of the ability of strain SL3261 and other *aroA* live-vaccine strains to multiply for several generations after injection into mice. (Strain SL3261, *aroA* but *dap+*, was not included as live vaccine in this experiment but from previous experience would certainly have given complete protection against the challenge dose used).

Clarke and Gyles (1987) similarly constructed a tetracycline-sensitive derivative of a calf-virulent strain of *S. typhimurium* made *dap*::Tn10 by transduction. Five one-week-old calves were given the *dap* strain as live vaccine, one oral dose of ca. 2×10^{10} CFU followed by 5×10^{8} CFU, s.c., at age two and age three weeks. The oral dose produced no ill-effects and the subcutaneous doses caused only some local inflammation. Challenge, on day 24, by feeding 7×10^{10} CFU of the virulent parent strain caused high fever and severe diarrhea in 5 non-vaccinated control calves and these calves all died, 23 to 72 hours after challenge. The vaccinated calves were less affected by the challenge; two succumbed, 113 and 130 hours after challenge but the other three made a full recovery from several days of fever and diarrhea. Thus *dap* character causes attenuation of *S. typhimurium* for calves, challenged by feeding, and for mice, challenged by injection but in both systems when used as live vaccine gave no or only poor protection.

The pathway for biosynthesis of DAP and lysine starts from aspartate semialdehyde, as also does the pathway leading to threonine and methionine. The conversion of aspartate-6-phosphate to aspartate semialdehyde is effected by the enzyme coded by gene *asd*. In consequence *asd* mutants require threonine and methionine, as well as DAP and lyse if incubated in media lacking DAP but otherwise suitable for growth. Curtiss and his colleagues (1990) have made use of the cloned and sequenced *asd* gene for construction of a balanced lethal system in which a live-vaccine strain carries a deleted *asd* gene but can grow in the absence of DAP, for instance in mammalian tissues, so long as it maintains a plasmid constructed so as to include a wild-type *asd+* gene. This ensures maintenance by live-vaccine strains of plasmids causing production of heterologous antigens, etc., both during laboratory culture and in the tissues of vaccinated animals, in the sense that nearly all viable bacteria recoverable from the liver and spleen of such animals still have the plasmid. However, it appears to me doubtful whether this contributes to live-vaccine efficacy in respect of response to the passenger antigen. Death of bacteria losing the plasmid would not be expected to have any effect on the number of bacteria retaining it, since live-vaccine bacteria are presumably not competing with each other for nutrients when they are located in different cells of the vaccinated animal.

CONCLUSION

The compounds required by auxotrophic mutants of *Salmonella* so far identified as being of reduced virulence because of auxotrophy include some which are metabolites of the vertebrate host and others which are not, except that they may, like other xenobiotics, serve as substrates for detoxifying mechanisms when they gain entrance to the blood stream. In both cases it seems probable that non-virulence or reduced virulence results from the absence or insufficient concentration of the required metabolite in the host compartment in which the bacteria locate and where they would multiply, to cause the disease, if they were prototrophic. However, the cytosol or even plasma concentrations of most relevant metabolites are not known, so that this simplistic explanation cannot at present be tested. Some attenuating auxotrophic characters cause a rapid decline in the number of viable bacteria in host organs, instead of survival for at least several days or limited multiplication; this decline may result from a known cause, that is lysis of *dap* mutants by deprivation of diaminopimelic acid, or from a not known mechanisms, in the case of *purA* or *purB* mutants, unable to procure sufficient adenine or adenosine. However, some auxotrophs of reduced virulence persist for days or weeks with little or no multiplication or with multiplication by up to a thousand-fold, for *aro* or *pab* mutants, which require p-aminobenzoic acid, presumably because its product, folic acid, functions only catalytically and because most of the biosynthetic reactions for which it is a co-factor lead to products which are available, at least to some extent, from the host environment.

REFERENCES

Bacon, G.A, Burrows, T.W., and Yates, M., 1950a, The effects of biochemical mutation on the virulence of *Bacterium typhosum*: the induction and isolation of mutants, *Brit.J. Exp. Path.* 31:703.

Bacon, G.A., Burrows, T.W., and Yates, M., 1950b, The effects of biochemical mutation on the virulence of *Bacterium typhosum*: the virulence of mutants, *Brit. J. Exp. Path.* 31:714.

Bacon, G.A., Burrows, T.W., and Yates, M., 1951, The effects of biochemical mutation on the virulence of *Bacterium typhosum*: the loss of virulence of certain mutants, *Brit. J. Exp. Path.* 32:85.

Benjamin, W.H., Jr., Turnbough, C.L., Jr., Posey, B.S., and Briles, D.E., 1985, The ability of *Salmonella typhimurium* to produce the siderophore enterobactin is not a virulence factor in mouse typhoid, *Infect. Immun.*, 50:392.

Benjamin, W.H., Jr. Hall, T., and Briles, D.E., 1991, A *hemA* mutation renders *Salmonella typhimurium* avirulent in mice, yet capable of eliciting protection against intravenous infection with *S. typhimurium, Microb. Pathogen.* 14:289.

Brown, R.E. and Stocker, B.A.D., 1987, *Salmonella typhi* 205aTy, a strain with two attenuating auxotrophic characters, for use in laboratory teaching, *Infect. Immun.* 55:892.

Clarke, R. C. and Gyles, C.L., 1987, Vaccination of calves with a diaminaopimelic acid mutant of *Salmonella typhimurium, Can. J. Vet. Res.*,51:32.

Curtiss, R., Jr., Galan, J.E., Nakayama, K., and Kelly, S.M., 1990, Stabilization of recombinant avirulent vaccine strains *in vivo, Res. Microbiol.* 141:797.

Edwards, M.F., and Stocker, B.A.D., 1988, Construction of *aroA his pur* strains of *Salmonella typhi*, *J. Bacteriol.* 170:3991.

Elliott, T., 1989, Cloning, genetic characterization, and nucleotide sequence of *hemA-prfA* operon of *Salmonella typhimurium*, *J. Bacteriol.*, 171:3948.

Furness, G., and Rowley, D., 1956, Transduction of virulence within the species *Salmonella typhimurium*, *J. Gen Microbiol.* 15:140.

Hoiseth, S.K., and Stocker, B.A.D., 1981, Aromatic-dependent *Salmonella typhimurium* are non-virulent and are effective as live vaccines, *Nature, Lond.* 291:238.

MacFarland, W.C., and Stocker, B.A.D., 1987, Effect of different purine auxotrophic mutations on mouse-virulence of a Vi-positive strain of *Salmonella dublin* and of two strains of *Salmonella typhimurium*, *Microb. Pathogen.* 3:129.

O'Callaghan, D., Maskell, D., Liew, F.Y., Easmon, C.S.F., and Dougan, G., 1988, Characterization of aromatic- and purine-dependent *Salmonella typhimurium* attenuation, persistence, and ability to induce protective immunity in BALB/c mice, *Infect. Immun.* 56:419.

Sigwart, D.F., Stocker, B.A.D., and Clements, J.D., 1989, Effect of a *purA* mutation on efficacy of *Salmonella* live-vaccine vectors, *Infect. Immun.* 57:1858.

Smith, H.W., and Tucker, J.F., 1976, The virulence of trimethoprim-resistant thymine-requiring strains of *Salmonella*, *J. Hyg., Camb.* 76:97.

Yancey, R.J., Breeding, S.A., and Lankford, C.E., 1979, Enterochelin (enteobactin): virulence factor for *Salmonella typhimurium*, *Infect. Immun.* 24:174.

THE GENETICS OF SALMONELLA AND VACCINE DEVELOPMENT

Gordon Dougan[1], Mark Roberts[2], Gillian Douce[1], Patricia Londono[1], Carlos Hormaeche[3], Julia Harrison[3] and Steven Chatfield[2]

[1]Department of Biochemistry, Imperial College of Science, Technology and Medicine, Wolfson Laboratories, London SW7 2AY.
[2]Vaccine Research Unit, Medeva Group Research, Department of Biochemistry, Imperial College of Science, Technology and Medicine, Wolfson Laboratories, London SW7 2AY.
[3] Department of Pathology, University of Cambridge, Tennis Court Road, Cambridge CB2 1QP

ABSTRACT

The advent of molecular biological techniques coupled with the availability of *in vivo* and *in vitro* models for studying virulence has significantly contributed to our understanding of salmonella pathogenesis. Genes and gene clusters have been identified which are known to be required for *in vivo* growth and survival but other genes that are necessary for the expression of full virulence by pathogenic salmonellae have still to be identified. *Salmonella* strains harbouring fully defined mutations have been constructed and are under evaluation in systems such as the murine typhoid model. Some of these attenuated derivatives are excellent candidate vaccine strains, eliciting potent local and systemic responses in the host. Fully genetically defined strains are now being evaluated in humans as candidate oral typhoid vaccines. Because such live vaccine strains are capable of eliciting potent immune responses they have been considered as carriers for delivering heterologous antigens to the mammalian immune system. Problems have been encountered with this approach, a major obstacle being the instability of expression of foreign antigens from recombinant plasmids. We are tackling this by utilising promotors to direct the expression of heterologous sequences that are tightly regulated but switch on in particular environments within host tissue We have demonstrated the utility of such an approach by using the *nirB* promotor to control the expression of heterologous antigens and have shown that this may be a generally applicable approach to obtaining the stable *in vivo* expression of heterologous antigens in salmonella vaccine strains.

INTRODUCTION

Salmonella species are a major cause of enteric infections in man and domestic animals. These sophisticated pathogens can cause a variety of diseases and syndromes, ranging from systemic infections such as typhoid to limiting infections of the gut such as gastroenteritis. Progress has been slow in elucidating the virulence mechanisms employed by these bacteria. Recent advances in our knowledge have stemmed from the application of molecular techniques to study fully virulent salmonella strains and their derivatives. Alongside the application of advanced genetic manipulation systems, the availability of convenient *in vitro* and *in vivo* models for studying genetically manipulated strains has proved invaluable. The murine model has been extensively used for studying salmonella virulence. Many mouse strains are highly susceptible to infections by different *Salmonella* serotypes including *S. typhimurium, S. dublin* and *S. enteritidis* (Collins, 1974; Hormaeche, 1979). This model provides an excellent system for the study of invasive, systemic infections which resemble typhoid, and has been used successfully for the study of both naturally and acquired resistance to salmonellosis. It has also facilitated the development of genetically defined attenuated strains of *Salmonella* that are being considered for the development of oral vaccines. However the murine model can not be readily used as a model for gastroenteritis or diarrhoeal disease. Studies on these types of infection have been much more limited and confined mainly to *in vitro* experiments. For example it has been shown that there is a correlation between the ability of strains of *S. typhimurium* to cause gastroenteritis in humans and invasiveness for HEp-2 cells growing in tissue-culture medium (Douce *et al.*, 1991). The ability to adhere to, invade and transcytose eukaryotic cells *in vitro* has also been employed extensively to characterise genetically defined salmonella mutants or banks of mutagenised strains for defects in these phenotypic traits (Fields *et al.*, 1986; Finlay *et al.*, 1988*a*). Using the types of approaches mentioned above a rapidly increasing number of salmonella genes are being identified which are required to establish infection or cause disease.

SALMONELLA GENES AND MOUSE VIRULENCE

Evidence from animal studies suggests that pathogenic salmonella strains are able to penetrate the mucosal barriers of the intestine, enter host tissues and invade and grow within both professional (Fields *et al.*, 1986) and non-professional phagocytic cells (Takeuchi, 1967; Finlay *et al.*, 1988*b*). Genetic techniques have been used to identify the bacterial genes involved in these processes. The importance of particular genes in virulence has been demonstrated using mutants of pathogenic salmonellae which are attenuated in the mouse model when compared to the virulent parent. More recently specific genetic approaches have been used to introduce mutations into individual genes or gene clusters which simplifies the interpretation of the data. Many genes have now been shown to play a role in salmonella pathogenesis. This was highlighted by the work of Fields *et al.* (1986) who produced mutants of *S. typhimurium*, using Tn*10* transposon mutagenesis, which were unable to survive in macrophages growing in cell culture. These mutants were found to be attenuated when administered to mice intraperitoneally. Insertions were mapped to several different genes controlling properties as diverse as serum sensitivity, motility and surface hydrophobicity. Additionally, the type of infection model utilised or other experimental variations has led to the generation of apparently contradictory data on the importance of some individual genes or genetic determinants in virulence. An example of this may be work on the role of cistrons involved in flagella or fimbrial biosynthesis in virulence (Carsiotis *et al.*, 1984; Lockman and Curtiss, 1990). The genetic background of a particular strain can also have a profound influence on the apparent role of a particular gene in virulence (Benjamin *et al.*, 1991).

Mutations known to attenuate virulent salmonella strains can be arbitrarily classified into groups on the basis of there general properties; examples are described below.

Metabolic genes encode enzymes which are involved in the synthesis of key metabolic compound essential for bacterial growth. For such mutations to lead to attenuation either a compound, which can not be synthesized without the functional gene, must be in short supply *in vivo* or the gene must have a previously unrecognised secondary function. Mutations in genes involved in the chorismate biosynthetic pathway, the route for biosynthesis of aromatic compounds in the bacterial cell, are highly attenuating probably because of the limiting supply of aromatic compounds such as para- aminobenzoic acid in tissues of the animal host (Hosieth and Stocker, 1981). The functional position of a particular gene in a metabolic pathway can effect the influence of mutations on growth *in vivo*. *aro* mutations which effect the chorismate pathway usually produce a similar level of attenuation whereas *pur* mutations which effect purine biosynthesis can give rise to greatly differing levels of attenuation (Dougan *et al.*, 1988; O'Callaghan *et al.*, 1988). *purA* mutations are much more attenuated than *purE* mutations. This observation can be partly explained by the branched nature of the purine biosynthetic pathway (McFarland and Stocker, 1987).

Pathogens encounter rapidly changing environmental conditions during the infection process and regulatory genes which are vital for bacterial adaptation are likely to play key roles *in vivo*. Control of gene expression in response to environmental stimuli can be considered at different levels. Response to specific environmental stimuli results in the activation of sensory and regulatory proteins which controls activation or repression of transcription from different promoters. The idea that several virulence genes are co-ordinately regulated by a master control gene as the bacteria moves through the various environments *in vivo* is an attractive one (Mekalanos, 1991). Examples of this type of control may include the *crp/cya* system which regulates the expression of many genes via cyclic adenosine monophosphate and the *ompR/envZ* system which plays a key role in regulating outer membrane protein expression; salmonella strains with mutations in either system are attenuated *in vivo* (Curtiss and Kelly, 1987; Dorman *et al.*, 1989). The second mechanism of control in response to changing environmental conditions involves changes in the physical state of the DNA. Mutations which alter the levels of DNA supercoiling can influence bacterial virulence and the expression of individual virulence determinants (Galan and Curtiss, 1990).

Study of the role of regulatory genes *in vivo* has indirectly provided a new approach to the identification of virulence genes; this is most clearly illustrated in the case of the *pho* regulon. This regulon was initially identified by several groups as a two component regulatory system that was essential for salmonella virulence (Miller *et al.*, 1989; Miller and Mekalanos, 1990; Miller, 1991). The *pho* regulon allows bacteria to respond to changes in phosphate levels and other environmental conditions such as low pH. In terms of pathogenicity the *pho* system plays a key role in controlling responses to the intracellular environment of the macrophage, regulating the expression of genes which play a direct role in protecting the bacterial cell from macrophage killing mechanisms (Miller, 1991). Several *pho* regulated genes have now been identified some of which are activated and others which are repressed by the regulon (Miller *et al.*, 1989; Pukkinen and Miller, 1991). Many but not all of the genes regulated by this regulon appear to play a role *in vivo* as strains carrying mutations in these are attenuated.

Identification of classical virulence determinants such as toxins or adhesins has proved to be more difficult in *Salmonella* compared to some other pathogens. Biochemical approaches have failed to clearly define toxins produced by *Salmonella* and the role of those that have been reported *in vivo* at this point remain undefined (Finkelstein *et al.*, 1983). Novel genes which are required for the expression of virulence are now being identified on a regular basis although their role in the pathogenic process in most instances remains undefined. We have identified mutations in the gene *htrA* which attenuate *Salmonella* in the mouse model (Johnson *et al.*, 1991). In *Escherichia coli* the HtrA gene product is a periplasmically located serine protease whose expression is increased in response to environmental stresses, including heat-shock (Johnson *et al.*, 1991). In *Salmonella* this protein may be required for intracellular survival as mutant strains show an increased susceptibility to oxidative stress. Defined *htrA* mutants of *S. typhimurium* are excellent oral vaccines in mice (Chatfield *et al.*, 1992*a*). It has been known for many years that mutations that effect salmonella lipopolysaccaride (LPS) biosynthesis can affect virulence (Germanier and Furer, 1971;1975). The role of LPS in pathogenesis is likely in itself to be multifactorial since endotoxin has such a major effect on immune cells.

Pre-screening using simple *in vitro* assays has proved an effective method for enriching for mutations which are likely to be required *in vivo*. Cultured mammalian cells have been used to identify invasion genes from a number of intracellular pathogens (Galan and Curtiss, 1989; Isberg, 1989). The use of pre-screening using *in vitro* grown mammalian cells has proved to be a particularly effective method for identifying attenuated variants. Various cells have been utilised including macrophages (Fields *et al.*, 1986), and epithelial cells (Finlay *et al.*, 1988*b*). The approach has been to identify bacterial mutants which can not enter these cells by the utilisation of antibiotics which kill extracellular bacteria such as gentamicin. Several loci, some of which encode multiple genes, have now been identified on the salmonella chromosome which are required for entry into cultured cells and these are described elsewhere in this book. Interestingly mutations in these invasins do not always attenuate suggesting *Salmonella* may have evolved several ways for entering cells (Finlay *et al.*, 1988*a*). As studies progress we are likely to identify more genes required in someway for virulence although defining the exact role *in vivo* of these genes is likely to prove much more difficult. A good illustration of this problem is the present work on the genes encoded by the virulence-associated plasmid found in many pathogenic *Salmonella* isolates (Jones *et al.*, 1982).

SALMONELLA VACCINES

Initial interest in salmonella vaccines centered on efforts to design vaccines against salmonellae themselves. Many studies indicated that live oral vaccines based on attenuated strains of *Salmonella* were particularly promising (Germanier and Furer, 1975; Hoiseth and Stocker, 1981). Efforts to develop practical salmonella vaccines for man have been limited in the main to *S. typhi* since the use of a vaccine against food-poisoning has been considered impractical. Preliminary work leading to an *S. typhi* vaccine must proceed using mouse adapted non-typhi salmonella strains for practical reasons. Since we now have identified many attenuating lesions in *Salmonella* there are many candidate genes for testing in real vaccine strains. How then can we choose the best candidates ?

For safety reasons it is desirable to include at least two attenuating mutations in a single vaccine strain to minimise the potential for reversion to virulence. Different mutations or combinations of mutations can have a dramatically different effect on virulence and ultimately on the immunogenicity of a vaccine strain (Dougan *et al.*, 1988; O'Callaghan *et al.*, 1988). Preliminary testing in a model system can give a strong indication of how a particular combination of attenuating lesions might effect the immunogenicity of a strain. This was certainly true of studies using auxotrophic mutants of *S.typhimurium* in the murine model where the immunisation results obtained with various combinations of *aro* and *pur* mutations closely parallel the results currently being obtained in humans (Levine *et al.*, 1987; Tacket *et al.*, 1992). However this is not always be the case. *S. typhi* strains harbouring mutations in *galE* differ in their behaviour in mice and humans, retaining virulence in man and being attenuated in mice (Hone *et al.*, 1988).

The variety of genetically defined *S. typhi* vaccine strains tested in humans is very limited. The results of these studies are summarised elsewhere in this book. To date only *aro*, *aro/pur* and *cya/crp* mutants of *S. typhi* have been tested. These strains were constructed using modern genetic procedures which minimise the potential for introducing foreign DNA or undefined mutations. In the case of double *aro* mutants we utilised cloned genes and fully defined deletions to construct *S. typhi* mutants (Hone *et al.*, 1991; Chatfield *et al.*, 1992*b*). These strains are easier to quality control since the nature of the attenuating lesion is fully defined. Extensive trials will be required before we know if any of these strains are likely to form the basis of useful oral typhoid vaccines.

SALMONELLA AS CARRIERS OF FOREIGN ANTIGENS

Salmonella are known to be able to survive inside immune cells and this may be the basis of the potent immunogenicity of some salmonella vaccine strains. This factor coupled with their ability to be delivered orally has made *Salmonella* attractive as potential carriers of foreign antigens to the immune system. This topic has been extensively reviewed recently elsewhere (Chatfield *et al.*, 1989). Here we will discuss our recent efforts to improve the stability of foreign gene expression in these hybrid vaccine strains. In the murine model *Salmonella* has been used to deliver antigens from a variety of pathogens to the immune system and in some cases protection against both the *Salmonella* strain and the pathogen have been obtained. We have utilised a region of the tetanus toxin protein expressed in *Salmonella* to develop a candidate oral tetanus vaccine (Fairweather *et al.*, 1990; Chatfield *et al.*, 1992*b*). Tetanus is caused by intoxication with the tetanus neurotoxin produced by *Clostridium tetani.* The molecular weight of the toxin is 150,000 daltons. A 50,000 dalton region of the toxin derived from the carboxl-terminal end of the protein referred to as Fragment C is known to be non-toxic and to be able to induce protective immunity, if injected into mice, against tetanus toxin challenge. Fragment C has been expressed in *Escherichia. coli* from the tac promoter which is under the control of the lac repressor protein (Fairweather *et al.*, 1990). Since salmonellae do not express the lac repressor, recombinant plasmids introduced into salmonella vaccine strains,which direct the expression of Fragment C from tac, do so in a constitutive manner. This high level constitutive expression results in plasmid instability and loss upon subculture of the ability to express Fragment C.

Plasmid instability is a general problem when vaccinating animals with hybrid salmonella vaccine strains since the plasmid is rapidly lost *in vivo* due to the absence of antibiotic selection during the period of self-limiting replication that usually follows shortly after the vaccine strain has entered the host tissues. Loss of the plasmid directing the expression of the recombinant protein can severely reduce the immune response to the foreign antigen. Many conventional promoters modified and developed for foreign gene expression in vitro are unsuitable for use *in vivo* as chemical or thermal induction can not easily be effected. In view of this we have initiated a search for promoters which will turn on when the vaccine strain enters the host tissue so that the foreign gene will be expressed when the bacteria encounter the host immune system.

An example of this approach is the use of the *nirB* promoter to direct the expression of Fragment C. Expression from the *nirB* promoter is regulated by environmental signals such as the level of oxygen in the growth environment (Chatfield *et al.*, 1992*c*). Expression from *nirB* occurs preferentially at low oxygen concentrations. Salmonella vaccine strains harbouring the *nirB*- Fragment C plasmid can be grown to prepare vaccine inocula under aerobic conditions were Fragment C expression is repressed and this inocula can be used to orally vaccinate mice.

Mice which receive such a vaccine respond by producing consistently higher levels of protective antibodies to fragment C than mice vaccinated with similar strains where Fragment C expression is directed by tac. Further the *nir B*-Fragment C plasmid is much more stable in vivo than the tac-fragment C plasmid. Thus the use of an in vivo inducible promoter may be a useful route towards developing better hybrid salmonella vaccines.

CONCLUSIONS

The use of modern genetic approaches particularly coupled with the use of the murine salmonellosis model is proving to be of particular value for identifying genes required for the full expression of virulence by pathogenic salmonella strains. Already several categories of genes have been identified although so called classical virulence genes are proving more elusive. Genetic studies are clearly helping the development of salmonella vaccines and a number of well defined candidate oral typhoid vaccines are being tested. The identification of the true role of these genes *in vivo* is proving more difficult to define and will require much more work in the future.

REFERENCES

Benjamin Jr, W.H., Yother, J., Hall, P. and Briles, D.E., 1991, The *Salmonella typhimurium* locus *mviA* regulates virulence in *Ity*^{s} but not *Ity*^{r} mice : functional *mviA* results in avirulence and mutant *mviA* results in virulence, *J. Exp. Med.* 174: 1073.

Carsiotis, M., Weinstein, D.L., Karch, H., Holder, I.A. and O'Brien, A.D., 1984, Flagella of *Salmonella typhimurium* are a virulence factor in infected C57BL/6J mice, *Infect. Immun.* 46: 814.

Chatfield, S.N., Strahan, K., Pickard, D., Charles, I.G., Hormaeche, C.E. and Dougan, G., 1992*a*, Evaluation of *Salmonella typhimurium* strains harbouring defined mutations in *htrA* and *aroA* in the murine salmonellosis model, *Microb. Pathog.* 12:145.

Chatfield, S.N., Fairweather, N.F., Charles, I., Pickard, D., Levine, M., Hone, D., Posada, M., Strugnell, R.A. and Dougan, G., 1992*b*, Construction of a genetically defined *Salmonella typhi* Ty2 *aroA aroC* mutant for the engineering of a candidate typhoid-tetanus vaccine, *Vaccine*. 10: 53.

Chatfield, S.N., Charles, I.G., Makoff, A.J., Oxer, M.D., Dougan, G., Pickard, D., Slater, D. and Fairweather, N., 1992*c*, Use of the *nirB* promoter to direct the stable expression of heterologous antigens in *Salmonella* oral vaccine strains: Development of a single-dose oral tetanus vaccine, *Bio/Tech*. 10:888.

Chatfield, S.N., Strugnell, R.A. and Dougan, G., 1989, Live salmonellae as vaccines and carriers of foreign antigenic determinants, *Vaccine*. 7:495.

Collins, F.M., 1974, Vaccines and cell mediated immunity, *Bact. Revs*. 38:371.

Curtiss, R. and Kelly, S.M., 1987, *Salmonella typhimurium* deletion mutants lacking adenylate cyclase and cyclic AMP receptor protein are avirulent and immunogenic, *Infect. Immun.* 57:2136.

Dorman, C.J., Chatfield, S.N., Higgins, C.F., Hayward, C. and Dougan, G., 1989, Characterisation of porin and *ompR* mutants of a virulent strain of *Salmonella typhimurium : ompR* mutants are attenuated in vivo, *Infect. Immun.* 57 : 2136.

Douce, G. R., Amin, I.I. and Stephen, J., 1991, Invasion of HEp-2 cells by strains of *Salmonella typhimurium* of different virulence in relation to gastroenteritis, *J. Med. Micro.* 35:347.

Dougan, G., Chatfield, S.N., Pickard, D., Bester, J., O'Callaghan, D. and Maskell, D., 1988, Construction and characterisation of vaccine strains of *Salmonella* harbouring mutations in two different *aro* genes, *J. Infect. Dis.* 158:1329.

Fairweather, N.F., Chatfield, S.N., Makoff, A., Strugnell, R., Bester, J., Maskell, D.J. and Dougan, G., 1990, Oral vaccination of mice against tetanus by use of a live attenuated *Salmonella* carrier, *Infect. Immun.* 58:1323.

Fields, P.I., Swanson, R.V., Haidaris, D.G. and Heffron, F., 1986, Mutants of *Salmonella typhimurium* that cannot survive within the macrophage are avirulent, *P.N.A.S., USA*. 83:5189.

Finkelstein.R.A., Mechlewiz, B.A., McDonald, R.J. and Finkelstein, M.B., 1983, Isolation and characterisation of a cholera-related enterotoxin from *Salmonella typhimurium,* FEMS Micro. Letts. 17:239.

Finlay, B.B., Stanbach, M.N., Francis, C.L., Stocker, B.A.D., Chatfield, S.N., Dougan, G. and Falkow, S., 1988*a*, Identification and characterisation of Tn*phoA* mutants of *Salmonella* which are unable to pass through a polarised MDCK epithelial monolayer, *Mol. Microbiol.* 2:757.

Finlay, B.B., Gumbiner, B. and Falkow, S., 1988*b*, Penetration of Salmonella through a polarised MDCK epithelial cell monolayer, *J. Cell Biol.* 107:221.

Galan, J.E. and Curtiss, R., 1990, Expression of *Salmonella typhimurium* genes required for invasion is regulated by changes in DNA supercoiling, *Infect Immun.* 58:1879.

Galan, J.E. and Curtiss, R., 1989, Cloning and molecular characterisation of genes whose products allow *Salmonella typhimurium* to penetrate tissue culture cells, *P.N.A.S.,USA*. 86:6383.

Germanier, R. and Furer, E., 1975, Isolation and characterisation of a *galE* mutant Ty21a of *Salmonella typhi* : a candidate strain for a live oral typhoid vaccine, *J. Infect. Dis.* 131:553.

Germanier, R. and Furer, E., 1971, Immunity in experimental salmonellosis. II Basis of the avirulence and protective capacity of *galE* mutants of *Salmonella typhimurium, Infect. Immun.* 4:663.

Hone, D.M., Harris, A.M., Chatfield, S.N., Dougan, G. and Levine, M.M., 1991, Construction of a genetically defined double *aro* mutant of *Salmonella typhi* , *Vaccine.* 9:810.

Hone, D.M., Attridge, S.R., Forrest, B., Morrona, R., Daniels, D., LaBrooy, J.T., Bartholomeus, R.C.A., Shearman, D.J.C. and Hackett, J., 1988, A *galE* (Vi negative) mutant of *Salmonella typhi* Ty2 retains virulence in humans, *Infect. Immun.* 56:1326.

Hormaeche, C.E., 1979, Natural resistance to *Salmonella typhimurium* in different inbred mouse strains, *Immunol.* 37:311.

Hosieth, S.K. and Stocker, B.A.D., 1981, Aromatic-dependent *Salmonella typhimurium* are non-virulent and effective as live vaccines, *Nature* . 291:238.

Isberg, R.R., 1989, Mammalian cell adhesion functions and cellular penetration of enteropathogenic *Yersinia enterocolitica, Mol. Microbiol.* 3:1449.

Johnson, K., Charles, I., Dougan, G., Pickard, D., O'Goara, P., Costa, G., Ali, T., Miller, I. and Hormaeche, C., 1991, The role of a stress-response protein in *Salmonella typhimurium* virulence, *Mol. Microbiol.* 5:401.

Jones, G.W., Rabert, D.K., Svinarich, D.M. and Whifield, H.J., 1982, Association of adhesiveness, invasive and virulence phenotypes of *Salmonella typhimurium* with an autonomous 60 megadalton plasmid, *Infect. Immun.* 38:476.

Levine, M.M., Herrington, D., Murphy, J.R., Morris, J.G., Losonsky, G., Tall, B., Lindberg, A.A., Svenson, S., Baqar, S., Edwards, M.F. and Stocker, B.A.D., 1987, Safety, infectivity, immunogenicity and in vivo stability of two attenuated auxotrophic mutant strains of *Salmonella typhi*, 541Ty and 543Ty, as live oral vaccines in humans, *J. Clin. Invest.* 79:888.

Lockman, H.A. and Curtiss, R., 1990, *Salmonella typhimurium* mutants lacking flagella remain virulent in BALB/c mice, *Infect. Immun.* 58:137.

McFarland, W.C., and Stocker, B.A.D., 1987, Effect of different purine auxotrophic mutations on mouse virulence of a Vi-positive *Salmonella dublin* and of two strains of *Salmonella typhimurium, Microb. Pathog.* 3:129.

Mekalanos, J.J., 1991, Environmental signals controlling expression of virulence determinants in bacteria, *J. Bacteriol.* 174:1.

Miller, S.I., 1991, PhoP/PhoQ: macrophage-specific modulators of Salmonella virulence? *Mol. Microbiol.* 5:2073.

Miller, S.I. and Mekalanos, J.J., 1990, Constitutive expression of the *phoP* regulon attenuates *Salmonella* virulence and survival within macrophages, *J. Bacteriol.* 172:2485.

Miller, S.I., Kukral, A.M. and Mekalanos, J.J., 1989, A two component regulatory system (*phoP phoQ*) controls *Salmonella typhimurium* virulence, *P.N.A.S., USA.* 86:5054.

O'Callaghan, D., Maskell, D., Liew, F.Y., Easmon, C.S.F. and Dougan, G., 1988, Characterisation of aromatic and purine-dependent *Salmonella typhimurium* ; attenuation, persistence and ability to induce protective immunity in BALB/c mice, *Infect. Immun.* 56:419.

Pukkinen, W.S. and Miller, S.I., 1991, A *Salmonella typhimurium* virulence protein is similar to a *Yersinia enterocolitica* invasion protein and bacteriophage lambda outer membrane protein, *J. Bacteriol.* 173:86.

Tacket, C.O., Hone, D.M., Losonsky, G.A., Guers, L., Edelman, R. and Levine, M.M., 1992, Clinical acceptability and immunogenicity of CVD908 *Salmonella typhi* vaccine strain, *Infect. Immun.* 60:536.

Takeuchi, A., 1967, Electron microscopic studies of experimental *Salmonella* infection. I. Penetration into the intestinal epithelium by *Salmonella typhimurium, Amer. J. Path.* 50:109.

SALMONELLA STRAINS WITH BOTH ANTIGEN O4 AND O9: CHARACTERIZATION OF THEIR LIPOPOLYSACCHARIDES AND USE AS IMMUNOGENS

Alf A. Lindberg,[1] Andrej Weintraub,[1] Thomas Segall,[3] and Bruce A.D. Stocker[2]

[1] Karolinska Institute, Department of Clinical Bacteriology, Huddinge Hospital, S-141 86 Huddinge, Sweden
[2] Department of Microbiology and Immunology, Stanford University School of Medicine, Stanford, California 94305-5402, USA
[1,3] The National Veterinary Institute, S-750 07 Uppsala, Sweden

INTRODUCTION

Salmonella serotypes are grouped by O antigen, i.e., the antigenic character of the polysaccharide component of their surface lipopolysaccharide (LPS). Many species including *Salmonella typhimurium*, fall into O-group B, with O-antigen factor 4; others, including *S. typhi*, *S. enteritidis* and *S. dublin* fall into group D with O-antigen factor 9. The O polysaccharides of group B and D strains are polymers of an oligosaccharide repeat unit, which in each of these O groups is an α-D-mannose1→2-α-L-rhamnose1→3-α-D-galactose trisaccharide. In serogroup B bacteria, abequose (3,6-dideoxygalactose) and in serogroup D tyvelose (3,6-dideoxymannose) are α1,3-linked to D-mannose (Hellerquist et al., 1968; Hellerquist et al., 1969). The basic O repeat units of group B and D strains are identical. In many group B strains, the abequose is substituted with an acetyl group in position 2, determining the O-factor 5; no corresponding factor occurs in group D strains. Other O factors may be present or absent in strains of either group. A glucosyl branch units on the galactose results in either O-antigen 1 if the glucose is α1,6 linked or 12_2 when the linkage is α1,4 (Lindberg and Le Minor, 1984). All the genetical information to build the repeating unit of the *Salmonella* group B and

D are comprised in the *rfb* gene cluster located near the *his* operon, at ca 42' on the *S. typhimurium* linkage map (Mäkelä and Stocker, 1984; Sanderson and Roth, 1988). The complete *rfb* gene cluster of a group B *Salmonella* strain (B*rfb*) may be replaced by the corresponding D*rfb* gene cluster of a group D strain. As a consequence, an original group B strain may be converted to a group D strain.

Aromatic-dependent strain of *S. typhimurium* and *S. dublin* used as live vaccines in mice (Stocker at al., 1983) and in calves (Robertsson et al., 1983) give substantial protection against challenge with virulent strains of the corresponding serotypes. Only limited cross-protection between serogroups B and D can be seen shortly after the vaccination. A *Salmonella* live-vaccine able to protect against both O4 and O9 strains would be therefore of interest. From nature, however, no *Salmonella* strains possessing both O4 and O9 antigens have been isolated. We therefore constructed the following hybrid-strains: *S. typhimurium* (serogroup B) with an additional *rfb* gene cluster from *S. enteritidis* (as in SL5313); *S. enteritidis* (serogroup D) with an additional *rfb* gene cluster from *S. typhimurium* (as in SL5396) and *S. dublin* (serogroup D) with an additional *rfb* gene cluster from *S. typhimurium* (as in SL7103) (Johnson et al., 1992; Lindberg et al., 1992).

RESULTS AND DISCUSSION

Serological studies

In slide agglutination tests with O-factor-specific rabbit antisera and mouse monoclonal antibodies, strain SL5313 reacted with the anti-O1, anti-O4, anti-O5 and anti-O9 and both SL5396 and SL7103 strains reacted with the anti-O4, and anti-O9 reagents.

The antigenic character was stable as shown by tests of single colonies after 25 serial transfers in broth and on nutrient agar and all strains expressed both the O4 and O9 epitopes as shown by indirect immunofluorescence using O4- and O9-specific mouse monoclonal and rabbit polyclonal antisera (Kaufman, 1966; Carlin et al., 1987; Svenungsson and Lindberg, 1977) (Table 1).

The simultaneous expression of both the *S. dublin* O9 and *S. typhimurium* O4 O-antigens in strains SL5315, SL5396 and SL7103 was studied using immunofluorescence with a mixture of rabbit anti α-Tyvelose1→3-α-D-Mannose-bovine-serum-albumin (O9-specific) serum and an O4-specific mouse monoclonal antibody (MAST4-2), or in other tests with a mixture of rabbit anti α-Abequose1→3-α-D-Mannose-bovine-serum-albumin (O4-specific) antiserum and an O9-specific mouse monoclonal antibody (MASE9-1). Using fluoresceinthiocyanate-conjugated anti-mouse and tetramethyl-rhodamin-isothiocyanate-conjugated anti-rabbit antisera we could observe the simultaneous expression of the O9 and O4 epitopes in the bacteria (Table 1). Only the hybrid strains i.e SL5313, SL5396 and SL7103 reacted with both O4 and O9 specific antibodies. In contrast, the *S. typhimurium* SH4809 and SL1479 control strains reacted only with the O4 specific antibodies. The *S. enteritidis* SH1262 and *S. dublin* SL5631 control strains reacted only with the O9 specific antibodies. These results show that the bacterial cells of each hybrid strain carry both O4 and O9 epitopes.

Table 1. Immunofluorescence study using rabbit antibodies against synthetic disaccharides: α-Abequose1→3-α-D-Mannose (O4 specific) and α-Tyvelose1→3-α-D-Mannose (O9 specific) in combination with mouse monoclonal antibodies MAST 4-2 (O4 specific) and MASE 9-1 (O9 specific). A mixture of anti-rabbit tetramethyl-rhodamin-isothiocyanate (TRITC) and anti-mouse fluoresceinthiocyanate-conjugate (FITC) was used to detect the bound primary antibodies.

Strains	Rabbit-anti AM-BSA[1] + MASE 9-1[2]		Rabbit-anti TM-BSA[3] + MAST 4-2[4]	
	TRITC[5]	FITC[6]	TRITC	FITC
SH4809 (O4,5,12)	+	-	-	+
SH1262 (O9,12)	-	+	+	-
SL5631 (O9,12)	-	+	+	-
SL1479 (O4,12)	+	-	-	+
SL5313 (O1,4,5,9,12)	+	+	+	+
SL5396 (O4,9,12)	+	+	+	+
SL7103 (O4,9,12)	+	+	+	+

[1]Anti α-Abequose1→3-α-D-Mannose-bovine-serum-albumin (O4-specific)
[2]O9-specific mouse monoclonal antibody
[3]Anti α-Tyvelose1→3-α-D-Mannose-bovine-serum-albumin (O9-specific)
[4]O4-specific mouse monoclonal antibody
[5]Anti-rabbit Tetramethyl-rhodamin-isothiocyanate-conjugate
[6]Anti-mouse Fluoresceinthiocyanate-conjugate

Chemical studies

The LPS from strains SL5313, SL5396 and SL7103 were prepared (Westphal et al. 1952) and subjected to qualitative and quantitative sugar analysis using gas liquid chromatography of alditol acetate derivatives (Sawardeker et al., 1965) (Table 2). All three LPS preparations contained the dideoxyhexosyls abequose and tyvelose, in ratios of 1.2:1 (SL7103), 1:1.5 (SL5313) and 1:2.5 (SL5396). The relative amounts of rhamnose, mannose and galactose were approximately 1:1:1. The dideoxyhexosyls were in non-stoichiometric amounts to mannose. This is expected since no special precautions were taken to protect the acid-labile and volatile dideoxyhexosyls during hydrolysis and work-up of the peracetylated alditols.

The relatively large amount of glucose in strain SL5313 can be attributed to the O1 antigenic specificity of this strain which is caused by 1,6 glucosylation of galactose in the repeating unit (Lindberg and Le Minor, 1984), reflecting the lysogeny of this strain with the converting P22 phage.

Table 2. Neutral carbohydrates in *Salmonella typhimurium* SH4809 and SL1479, *S.enteritidis* SH1262, *S. dublin* SL5631 and hybrid *Salmonella* strains SL5313, SL5396 and SL7103 lipopolysaccharides.

Sugar[a]	Mol% in LPS of:						
	SH4809	SH1262	SL1479	SL5361	SL5313	SL5396	SL7103
Abequose	9		11		4	4	6
Tyvelose		9		10	6	10	5
Rhamnose	29	30	28	27	27	29	29
Mannose	27	26	27	28	24	25	27
Galactose	30	31	30	32	26	27	29
Glucose	5	4	4	3	13	5	4

[a] Determined by GLC as alditol acetates

Methylation analysis (Hakomori, 1964) also documented that the methyl ethers expected for the sugars of the repeating units of serogroup B and D were present in the O-polysaccharides from the SL5313, SL5396 and SL7103 (data not shown).

Immunochemical studies

The question whether individual O polysaccharide chains of the hybrid *Salmonella* SL5313 and SL5396 strains include both abequose-containing and tyvelose-containing repeating units was subsequently addressed by sandwich EIA tests (Weintraub et al., 1992).

EIA microtiter plates were coated either with purified anti-AM-BSA (O4-specific) or purified anti-TM-BSA (O9-specific) rabbit IgG antibodies. The plates were incubated with varying amounts of lipid-free polysaccharide fractions from SL5313, SL5396, SH4809 or SH1262. Bound PS was detected after incubation with mouse monoclonal antibodies MAST O4 or MASE O9 followed by incubation with a rabbit anti-mouse immunoglobulin-alkaline-phosphatase conjugate.

The O4-specific anti AM-BSA antibody had bound the PS of the two hybrid strains and that of the control *S. typhimurium* strain since all three were detected by the subsequently added anti-O4 monoclonal antibody (data not shown). Only the bound PS of the two hybrid strains were detected by the added anti-O9-specific monoclonal antibody (Table 3).

The O9-specific anti TM-BSA antibody had bound the PS of the two hybrid strains and that of the control *S. enteritidis* strain since all three were detected by the subsequently added anti-O9 monoclonal antibody (data not shown). Only the bound PS of the two hybrid strains were detected by the added anti-O4-specific monoclonal antibody (Table 3).

Since PS from each hybrid strain, bound either by anti-O4 or anti-O9 was detectable by the anti-O9 or by the anti-O4 antibody the chains of each must contain both abequose and tyvelose repeat units.

This assay convincingly showed that polysaccharide chains from either SL5313 or SL5396 could bind both O4 and O9-specific antibodies,

Table 3. Sandwich EIA with anti-AM antibodies or anti-TM antibodies coated to the solid phase. After incubation with polysaccharides (10μg) isolated from *Salmonella typhimurium* SH4809, *Salmonella enteritidis* SH1262, *Salmonella* SL5313 or *Salmonella* SL5396. Bound polysaccharide was detected with either O4 or O9-specific mouse monoclonal antibody.

Polysaccharide	Rabbit-anti AM-BSA + MASE 9-1	Rabbit-anti TM-BSA + MAST 4-2
SH4809 (O4,5,12)	0.05[1]	0.03
SH1262 (O9,12)	0.04	0.04
SL5313 (O1,4,5,9,12)	0.85	1.1
SL5396 (O4,9,12)	1.1	0.88

[1]Absorbance value at 405 nm after 100 minutes

which neither polysaccharide from *S. typhimurium* SH4809 nor that from *S. enteritidis* SH1262 could (Table 3).

Thus an individual O-polysaccharide chain in either hybrid strain can contain both O4- and O9-specific repeating units but the assay does not tell us anything about the distribution of the different repeating units in a single chain. It is not unlikely that each hybrid may have some pure O4- or pure O9-polysaccharide chains.

The protective efficacy of the *Salmonella* hybrid strains was subsequently tested in mice and calves.

Mouse immunization studies

Immunization of NMRI mice with the *Salmonella* SL7103 hybrid strain expressing both the O4 and O9 repeating units resulted in elicitation of O4 and O9 specific antibodies (Lindberg et al., 1992). To see if mice immunized with hybrid *Salmonella* SL7301 strain also were protected against virulent strains, the animals were challenged with *S. typhimurium* SVA44 and *S. dublin* SVA47. Groups of 10 mice were vaccinated intraperitoneally with ~$2x10^5$ live bacteria, a dose which caused no visible ill-effect, given in a volume of 0.2 ml, on days 0, 7 and 14. The strains used as vaccines were *S. dublin* SL5631, *aroA* (O9, 12), *S. typhimurium* SL1479, *aroA* (O4,12) and *S. dublin* SL7103, *aroA* (O9,12; O4,12). The mice were challenged intraperitoneally on day 28 with graded doses, 10^1 to 10^7, of either of the virulent strains *S. dublin* SVA47 or *S. typhimurium* SVA44 (Table 3).

Mice vaccinated with *S. dublin* SL5631 were protected against challenge with its virulent parent strain *S. dublin* SVA47: the LD_{50} dose was $1x10^6$ cfu SVA47 for vaccinated mice as compared to <10 cfu SVA47 in the non-vaccinated control group (Table 4).

Immunization with *S. dublin* SL5631 did not, however, result in any significant protection against challenge with *S. typhimurium* SVA44: the LD_{50} dose in vaccinated and non-vaccinated mice was ~10 and <10 cfu SVA44, respectively (Table 4).

Mice immunized with live *S. typhimurium* SL1479 were protected against challenge with virulent *S. typhimurium* SVA44; LD_{50} dose $1x10^7$ in vaccinated and <10 cfu in non-vaccinated mice (Table 4). However, no significant protection was elicited against challenge with *S. dublin* SVA47, LD_{50} dose ~15 cfu, compared to <10 cfu for non-vaccinated mice.

For immunization with the hybrid *S. dublin* strain SL7103, which expresses the O4,O9,12 antigens, groups of 20 mice were used. Ten mice were challenged with S. *dublin* and 10 with *S. typhimurium.* The vaccination elicited protection against each challenge : The LD_{50} dose was 1.6 x 10^4 for *S. dublin* SVA47 and 1 x 10^5 for the *S. typhimurium* SVA44 challenge (Table 4). This means that the immunization increased the LD_{50} dose >10,000-fold for each of the two virulent challenge strains. Compared to the protection elicited by either auxotrophic *aroD* vaccine (*S. dublin* SL5631 or *S. typhimurium* SL1479) the protection elicited by the hybrid SL7103 was approximately 70-fold lower for the *S. dublin* challenge and 100-fold lower for the *S. typhimurium* challenge (Table 4).

NMRI mice immunized with the live *S. dublin* SL5631 vaccine strain were protected against challenge with the wild-type *S. dublin* SVA47 (Table 4). Compared to non-immunized mice the vaccination increased the LD_{50} dose 100,000-fold. However no, or only a marginal, protection was seen against challenge with *S.typhimurium* SVA44. Likewise immunization with live *S. typhimurium* SL1479 increased the LD_{50} for challenge with *S. typhimurium* SVA44 about 1,000,000-fold compared to non-immunized mice and gave almost no protection against *S. dublin* SVA47 (Table 4).

The O-antigen specificity suggests that the protection seen was a consequence of elicited antibodies specific for the O4 and O9 epitopes in respectively *S. typhimurium* SL1479 and *S. dublin* SL5631. The fact that both strains share the O12 epitope, the determinant(s) of which are found

Table 4. Protection against intraperitoneal challenge with virulent *Salmonella dublin* SVA47 or *Salmonella typhimurium* SVA44 in mice immunized with live attenuated *S. dublin*, *S. typhimurium* or *S. dublin/S. typhimurium* hybrid vaccine strains.

Vaccine[a]	Challenge[b]	LD_{50}[c]
S. dublin SL5631	None	>8.0
S. typhimurium SL1479	None	>8.0
Salmonella SL7103	None	>8.0
None	*S. dublin* SVA47	<1.0
None	*S. typhimurium* SVA44	<1.0
S. dublin SL5631	*S. dublin* SVA47	6.0
S. dublin SL5631	*S. typhimurium* SVA44	1.0
S. typhimurium SL1479	*S. dublin* SVA47	1.2
S. typhimurium SL1479	*S. typhimurium* SVA44	7.0
Salmonella SL7103	*S. dublin* SVA47	4.2
Salmonella SL7103	*S. typhimurium* SVA44	5.0

[a] Live vaccine in doses of ca $2x10^5$ cfu was given on day 0, 7 and 14

[b] The live challenge was given intraperitonally on day 28

[c] Mice were observed for up to 30 days. The LD_{50} values were calculated according to Reed & Muench, 1938

in the common α-D-mannose1→2-α-L-rhamnose1→3-α-D-galactose trisaccharide, was apparently not enough to elicit a demonstrable cross-protection. We demonstrated in studies of the protective effect of mouse monoclonal antibodies, specific for O-antigen epitopes and passively administered to NMRI mice, that both IgG and IgM antibodies which were O9-specific were protective but that only the IgM subclass of O12-specific antibodies showed a significant protection (Carlin et al., 1987). We surmise that the poor cross-protective effect seen in this study after *S. dublin* vaccination and *S. typhimurium* challenge and *vice versa* was because little, or no, anti O12-specific antibodies were generated following immunization with either *S. typhimurium* SL1479 or *S. dublin* SL5631.

In this investigation the protective efficacy of each of the two monospecific, O4 or O9, live vaccines was almost entirely O-specific. The live vaccines used here gave excellent protection against death or evident disease in mice observed for 30 days after challenge. The ability of live vaccines to reduce deaths from *Salmonella* challenge has been seen in many investigations and thought to result from persistence of live bacteria in host tissues - but the mechanism of protection is not known (Collins, 1970; O'Callaghan et al., 1988; Smith et al., 1984). The O-specificity of protection seen in our experiments may thus reflect the ability of *Salmonella* O polysaccharide to cause T-cell responses, as also indicated by delayed-type hypersensitivity to endotoxin-free O polysaccharide in calves immunized by live-vaccine administration (Lindberg and Robertsson, 1983). It should be observed, however, that the elicited immunity failed to eradicate the challenge bacteria.

The stable hybrid *S. dublin/S. typhimurium* SL7103, expressing equal amounts of the O4 and O9 epitopes elicits protective immunity against challenge with either a virulent *S.typhimurium* or virulent *S.dublin* (Table 4). This confirms the important role of LPS in the pathogenesis of salmonellosis (Mäkelä et al., 1990), and convincingly demonstrates that an anti-LPS antibody response effectively protects against an intraperitoneal challenge in the experimental mouse typhoid model.

Calf immunization studies

The SL7103 hybrid strain was next used as an oral live vaccine in 5- to 8-weeks old calves. Also in orally vaccinated and orally challenged calves the host defence has been shown to be more or less strain-specific so that *S. typhimurium* immunized calves were protected against *S. typhimurium* but not *S. dublin*, and vice versa.

Strain SL7103 was given to 18 calves in three doses on days 0, 7 and 14: c:a $2x10^9$ on day 0, c:a $1x10^{10}$ day 7 and c:a $1x10^{11}$ on day 14. The calves tolerated the vaccine, were feeding well and gained weight. Six of the 18 calves showed a mild and transient diarrhoea during the immunization period lasting 2 to 3 days. No diarrhoea was seen after the third vaccination. Slight increases of the rectal temperature were also seen: maxima on days 2 (mean 0.5°) and 9 (mean 0.2°) and lasting for about 4 days. The calves excreted the vaccine strain in high numbers (4 to 7 x 10^5 cfu/g) the first day after each vaccination. The excretion fell rapidly and the mean numbers were <100 cfu/g 3 to 4 days after each dose. Only one calf was found shedding the SL7103 vaccine strain at the time for challenge on day 28.

Six calves were orally challenged with virulent *S. dublin* SVA47 in a dose of 1 x 10^{10} cfu which corresponds to approximately 1000 LD_{50} doses. None of the calves became ill, but marked transient increases of the rectal temperature to a maximal mean of 40.6° on day 30. The temperatures had returned to the normal level on day 33. *S. dublin* SVA47 was not recovered from any of the blood cultures taken each day on days 29 to 31. Four of the calves had loose, mucoid stools on day 29, one on day 29 and 30, and one from day 29 to 32. *S. dublin* SVA47 was recovered from feces of all calves: from a maximal mean value of 2 x 10^6 cfu/g on day 29 the numbers fell rapidly to a mean of <100/g on day 34. At the time for sacrifice on day 49 SVA47 was no longer recovered from any calf. The protective efficacy of the hybrid SL7103 vaccine strain was equal to that of its *S. dublin aroA* parent SL5631 (Segall and Lindberg, 1991).

Another six calves were challenged with *S. typhimurium* SVA44. The dose chosen 3 x 10^9 cfu (c:a 10.000 LD_{50}) was approximately the same which when given to calves immunized orally with the *aroA* auxotrophic strain SL1479 caused no salmonellosis (Robertsson et al., 1983). However, the calves immunized with the hybrid strain SL7103 developed a severe hemorrhagic diarrhea and had to be sacrificed within 48 hrs after challenge.

A third group of six calves immunized with SL7103 was challenged with 5 x 10^8 cfu (corresponding to c:a 1.000 LD_{50}) of *S. typhimurium* SVA44. All calves became severely affected with significantly increased rectal temperatures ($p<0.01$), positive bloodcultures in 5 of 6 calves, and an enteritis which initially was catarrhal but changed to hemorrhagic within 3 days after challenge. All calves were sacrificed within 7 days after challenge. At necropsy the calves showed a purulent necrotizing panenteritis and atrophy of intestinal lymphatic tissues. Each calf was cultured from 29 different locations and *S. typhimurium* SVA44 was recovered in more than 90% of the specimens.

All calves immunized with SL7103 responded with significantly increased serum antibody titers against the LPSs as estimated by ELISA: against *S. dublin* LPS the IgM titers were significantly ($p<0.01$ and $p<0.001$) elevated on days 21 and 28, and against *S. typhimurium* already from day 14 ($p<0.01$ and $p<0.001$). The anti-LPS antibody responses were of the same magnitude and showed the same kinetics as in calves vaccinated with the parent *S. dublin* SL5631 strain (Segall & Lindberg 1991).

In a separate experiment calves orally vaccinated with 3 doses of *S. dublin* SL5631 showed high anti-LPS IgA titer responses which lasted well over day 28.

It is evident that vaccination with a *S. dublin* hybrid strain which through insertion of the *rfb* gene cluster of *S. typhimurium* expresses the *S. typhimurium* O-antigenic polysaccharide chain did not elicit a significant protective efficacy against *S. typhimurium* challenge. The immunization caused significant serum and probably local intestinal, anti-*S. typhimurium* LPS responses but this host defence was unable to control the invasion of and multiplication in the tissues of *S. typhimurium* SVA44. We feel safe in concluding that anti-LPS immunity alone will not protect the calf against high oral challenge doses.

SUMMARY

Hybrid *Salmonella* strains, which express the O-antigen of serogroups B and D, can be constructed by insertion of the chromosomal O-antigen-synthesis specifying *rfb* locus of serogroup B bacteria in a serogroup D recipient and vice versa. Such strains were shown to express the two O-antigens in approximately equal amounts in the same bacterium. Immunochemical studies also suggest that a single O polysaccharide chain can contain repeating tetrasaccharide units from both serogroups. One *aroA* attenuated hybrid strain, *S. dublin* SL7301, with the *rfb* gene cluster from *S. typhimurium*, elicited a significant protection ($p<0.001$)

against challenge with either virulent *S. dublin* or virulent *S. typhimurium* in mice intraperitoneally vaccinated with live SL7301 and intraperitoneally challenged. However, calves orally vaccinated with live SL7301 were significantly protected only against oral challenge with *S. dublin* ($p<0.001$) but not against challenge with *S. typhimurium*. Since significant and equal *S. typhimurium* and *S. dublin* serum anti LPS titers had been elicited our results suggest that anti LPS immunity alone is insufficient in an oral challenge model to protect against high challenge doses.

ACKNOWLEDGEMENTS

Work reported herein was supported by the Swedish Medical Research Council (grant no 16x-656) and the National Swedish Agricultural Marketing Board (grant no 852001).

REFERENCES

Carlin, N.I., Svenson, S.B., and Lindberg, A.A., 1987, Role of monoclonal O-antigen antibody epitope specificity and isotype in protection against experimental mouse typhoid, *Microb. Pathog.* 2:171.

Collins, F.M., 1970, Immunity to enteric infection in mice, *Infect. Immun.* 1:243.

Hakomori, S., 1964, A rapid permethylation of glycolipids and polysaccharides catalyzed by methylsulfinyl carbanion in dimethylsulfonoxide, *J. Biochem.* (Tokyo) 55:205.

Hellerquist, C.G., Lindberg, B., Svensson, S., Holme, T., and Lindberg, A.A., 1968, Structural studies on the O-specific side chains of the cell wall lipopolysaccharide from *Salmonella typhimurium* 395MS, *Carbohydr. Res.* 8:43.

Hellerquist, C.G., Lindberg, B., Svensson, S., Holme, T., and Lindberg, A.A., 1969, Structural studies on the O-specific side chains of the cell wall lipopolysaccharides from *Salmonella typhi* and *S. enteritidis*, *Acta Chem. Scand.* 23:1588.

Johnson, B.N., Weintraub, A., Lindberg, A.A., and Stocker, B.A.D., 1992, Construction of Salmonella strains with both antigen O4 (of group B) and antigen O9 (of group D), *J. Bacteriol.* 174:1911.

Kaufman, F. 1966, The bacteriology of enterobacteriaceae. Munksgaard, Copenhagen.

Lindberg, A.A., and Le Minor, L., 1984 Serology of *Salmonella*, *in*: "Methods in Microbiology," - Vol 15. T. Bergan ed, Academic Press, London and New York.

Lindberg, A.A., and Robertsson, J.Å., 1983, *Salmonella typhimurium* infection in calves: Cell-mediated and humoral reactions before and after challenge with live virulent bacteria in calves given live or inactivated vaccines, *Infect. Immun.* 41:751.

Lindberg, A.A., Segall, T., Weintraub, A., and Stocker, B.A.D., Submitted, Antibody response and protection against challenge in mice vaccinated intraperitneally with live *aroA Salmonella dublin*, *S. typhimurium* or an $O4^{+}O9^{+}$ hybrid *S. dublin* strain.

Mäkelä, P.H., and Stocker, B.A.D., 1984, Genetics of lipopolysaccharide, *in*: "Handbook of endotoxin, vol. 1: chemistry of endotoxin," E. Th. Rietschel, ed., Elsevier Science Publishing, Inc., New York.

Mäkelä, P.H., Hovi, M., Saxén, H., Moutiala, A., and Rhen, M., 1990, Role of LPS in the pathogenesis of salmonellosis. *in:* "Cellular and molecular aspects of endotoxin reactions", A. Nowotny, J.J. Spitzer, and E.J. Ziegler eds., Elsevier Science Publishers B.V.

O'Callaghan, D., Maskell, D., Liew, F.Y., Easmon, C.F.S., and Dougan, G., 1988, Characterization of aromatic- and purine dependent *Salmonella typhimurium*: Attenuation, percistence, and ability to induce protective immunity in BALB/c mice, *Infect. Immun.* 56:419.

Reed, L.J., and Muench, H., 1938, A simple method of estimating fifty per cent endpoints, 27:493.

Robertsson, J.Å., Lindberg, A.A., Hoiseth, S., and Stocker, B.A.D., 1983, *Salmonella typhimurium* infection in calves: protection and survival of virulent challenge bacteria after immunization with live or inactivated vaccines, *Infect. Immun.* 41:742.

Sanderson, K.E., and Roth, J.R. 1988, Linkage map of *Salmonella typhimurium*, edition VII, *Microbiol. Rev.* 52:485.

Sawardeker, J.S., Sloneker, J.H.,and Jeanes, A., 1965, Quantitative determination of monosaccharides as their alditol acetates by gas-liquid chromatography, *Anal. Chem.* 37:1602.

Segall, T., and Lindberg, A.A., 1991, *Salmonella dublin* experimental infection in calves: Protection after oral immunization with an auxotrophic *aroA* live vaccine, *J. Vet. Med.* B38:142.
Smith, B.P., Reina-Guerre, M., Stocker, B.A.D., Hoiseth, K.S., and . Johnson, H., 1984, Aromatic-dependent *Salmonella dublin* as a parenteral modified live vaccine for calves, *Am. J. Vet. Res.* 45:2231.
Stocker, B.A.D., Hoiseth, S.K., and Smith, B.P., 1983, Aromatic-dependent *Salmonella* sp as live vaccine in mice and calves, *Devel. Biol. Stand.* 53:47,
Svenungsson, B., and Lindberg, A.A., 1977, Synthetic disaccharide-protein antigens for production of specific O4 and O9 antisera for immunofluorescence diagnosis of *Salmonella*,.*Med. Microbiol. Immunol.* 163:1.
Weintraub, A., Johnson, B.N., Stocker, B.A.D., and Lindberg, A.A., 1992, Structural and immunochemical studies of the lipopolysaccharides of *Salmonella* strains with both antigen O4 and antigen O9, *J. Bacteriol.* 174:1916.
Westphal, O., Lüderitz, O., and Bister, F., 1952, Über die Extraktion von Bakterien mit Phenol/Wasser, *Z. Naturforsch.* 7:148.

ATTENUATED SALMONELLA TYPHI AS LIVE ORAL VACCINES TO PREVENT TYPHOID FEVER AND AS CARRIER VACCINES TO EXPRESS FOREIGN ANTIGENS

Myron M Levine,[1] David M. Hone,[1] Carol O. Tacket,[1] Marcelo Sztein,[1] Genevieve Losonsky,[1] James P. Nataro,[1] Cesar Gonzalez,[1] Gordon Dougan,[2] Stephen Chatfield,[2] Stanley Cryz,[3] Roy Curtiss,[4] and Sandra Kelley[4]

[1]The Center for Vaccine Development, University of Maryland School of Medicine, Baltimore, MD 21201, U.S.A.

[2]The Department of Physiological Biochemistry, Imperial College of Science and Medicine, London SW7, United Kingdom

[3]The Swiss Serum and Vaccine Institute, Berne, Switzerland

[4]The Department of Biology, Washington University, St. Louis, Missouri

INTRODUCTION

Attenuated strains of *Salmonella typhi* can serve as non-reactogenic, effective live oral vaccines to prevent typhoid fever (Levine et al, 1989) and as "live carrier vaccines" to express foreign antigens and deliver them to the human immune system (Formal et al, 1981; Black et al, 1987; Herrington et al, 1990; Forrest et al, 1989; Tacket et al, 1990). Strain Ty21a, developed by Germanier and coworkers (1975) by chemical mutagenesis, is a prototype live vaccine that has established the feasibility of these concepts. While Ty21a is impressively well-tolerated (Wahdan et al, 1980; Levine et al, 1987), confers significant protection against typhoid vaccine (Wahdan et al, 1980; Levine et al, 1987; Wahdan et al, 1982; Levine et al, 1990) and can serve as a carrier vaccine (Formal et al, 1981; Black et al, 1987; Herrington et al, 1990; Forrest et al, 1989; Tacket et al, 1990), it suffers from serious drawbacks which limit its utility. Foremost among the deficiencies of Ty21a are the fact that three or four spaced doses must be administered in order to stimulate protective immune responses (Levine et al, 1989) and that this mutagenized strain is difficult to manipulate genetically in attempting to utilize it as a carrier. Consequently, considerable research effort is being expended by different groups of investigators to engineer new attenuated strains of *S. typhi* that are comparably well-tolerated as Ty21a yet markedly more immunogenic.

Strain Ty21a as a Live Oral Typhoid Vaccine

The efficacy of Ty21a in preventing typhoid fever was first demonstrated in volunteers who were protected from experimental challenge (Gillman et al, 1977). A total of six field trials of efficacy were carried out in typhoid-endemic areas (one in Alexandria, Egypt, four in Santiago, Chile and one in Indonesia) (Levine et al, 1987; Wahdan et al, 1982; Levine et al, 1990; Ferreccio et al, 1989; Black et al, 1990; Simanjantuk et al, 1991). These trials

Biology of Salmonella, Edited by F. Cabello *et al.*,
Plenum Press, New York, 1993

emphasized the critical role played by different formulations and immunization schedules in affecting the efficacy of Ty21a. The optimal regimen was found to consist of three doses (every other day interval) of vaccine administered as a liquid suspension (after reconstitution of a lyophilate).

The field trials with Ty21a in Chile showed that if a vaccine "take" could be elicited, relatively long-term protection could be achieved. For example, three doses of vaccine in enteric-coated capsules provided 63% protection for at least six years in the Area Occidente field trial in Santiago, while in another trial (Area Sur Oriente/Area Norte), vaccine administered in a liquid suspension conferred 77% protection for at least four years. Ty21a is currently licensed in many countries as an oral vaccine to prevent typhoid fever. Based on the practicality of oral administration of Ty21a, its lack of reactogenicity and its efficacy (which is at least as good as that conferred by the reactogenic heat-inactivated, phenol-preserved parenteral whole cell vaccine), Ty21a enjoys considerable popularity.

Correlates of Protection

Two immunologic measurements have been identified that correlate with the protection of Ty21a as observed in field trials. These include serum IgG antibody to *S. typhi* O antigen measured by ELISA (Levine et al, 1989) and quantitation of the number gut-derived of IgA O antibody secreting cells in peripheral blood, measured by ELISPOT (Kantele, 1990; Kantele and Makela, 1991). While it is not believed that these particular immune responses actually mediate the protection conferred by Ty21a, they serve as helpful correlates for comparing immunization schedules and formulations.

Ty21a as a Carrier Vaccine

Several years ago, Ty21a was the only choice available if one desired to construct a *S. typhi*-based live vector vaccine. Accordingly, Formal et al (1981) introduced into Ty21a a modified plasmid from *Shigella sonnei* that encodes production of the *S. sonnei* O antigen, while Australian investigators (Forrest et al, 1989) introduced into Ty21a cloned genes that allow expression of the O antigen of *Vibrio cholerae* O1 serotype Inaba. The Ty21a-*S. sonnei* hybrid vaccine was immunogenic in volunteers and some lots conferred significant protection against experimental challenge (Black et al, 1987). However, because of lot-to-lot variability in the protective efficacy (Herrington et al, 1990) against experimental challenge, field trials of efficacy were not undertaken. The Ty21a-*V. cholerae* O1 Inaba construct was poorly immunogenic and did not provide significant protection against cholera in experimental challenges but did significantly ameliorate the severity of illness (Tacket et al, 1990).

Phase 1 Clinical Studies with Double *aro* and *cya,crp* Mutants of *S. typhi*

Two approaches to attenuating *S. typhimurium* have been particularly successful in animal models and in veterinary studies. One approach is based on mutating genes involved in the biosynthesis of aromatic amino acids (Hoiseth and Stocker, 1981). Such mutants become auxotrophic for para-aminobenzoic acid and 2,3 dihydroxybenzoic acid, substrates that are not available in mammalian tissues; the *Salmonella* cannot sustain proliferation and are therefore attenuated. Mutations in *aro*A, *aro*C or *aro*D each independently attenuate *S. typhimurium* for mice and protect against challenge with virulent organisms. Calves and sheep immunized with aro mutants of *S. typhimurium* have exhibited a high level of protection against challenge with the homologous serotype of *Salmonella.*

In the second approach, attenuation is based on mutations in the genes encoding adenylate cyclase (*cya*) and the cyclic AMP receptor protein (*crp*) (Curtiss and Kelly, 1987). These two genes comprise a global regulatory system that affects the transcription of many other genes. Inactivation of the system attenuates *S. typhimurium* for mice. Moreover, mice immunized with *cya,crp* mutants manifest a high level of protection against experimental challenge.

Edwards and Stocker (1988) constructed a prototype attenuated *S. typhi* vaccine strain harboring a deletion in *aro*A and another in *pur*A. This vaccine strain was well-tolerated in clinical trials but was poorly immunogenic (Levine et al, 1987), probably because of the effect of the mutation in *pur*A.

Hone et al (1991) constructed two double mutants (*aro*C, *aro*D) of *S. typhi* in two different genetic backgrounds, including CVD 908 (derived from wild type strain Ty2, the wild type parent of Ty21a) and CVD 906 (derived from strain ISP 1820, a minimally-passaged wild type strain recently isolated from the blood of a Chilean schoolchild with uncomplicated typhoid fever). Kelly and Curtiss constructed a *cya,crp* mutant of *S. typhi* (strain X3927) in the identical Ty2 background as used by Hone et al (1991).

X3927, CVD 906 and CVD 908 candidate live oral vaccine strains were tested for safety and immunogenicity in a randomized, double-blind trial where subjects received a single dose containing 5 x 10^4 or 5 x 10^5 CFU with buffer (Tacket et al, 1992). Approximately 50% of recipients manifested significant rises in serum IgG O antibody and in gut-derived trafficking IgA-class O antibody secreting cells detected in peripheral blood. These results demonstrate a high level of immunogenicity for all three vaccines. However, recipients of two vaccines, CVD 906 and X3927, manifested febrile reactions with temperature elevations $\geq$ 39.5 C (albeit without accompanying toxemia). Because of these febrile adverse reactions, further clinical trials with vaccine candidate strains CVD 906 and X3927 were discontinued and it was concluded that an additional independent attenuating mutation would have to be introduced into each of these vaccine strains. Variants of X3927 and CVD 906 that have been modified by introduction of additional attenuating mutations have been constructed and are entering Phase 1 clinical trials.

While strain CVD 906 and X3927 caused febrile reactions in volunteers when fed in a dose of 5 x 10^4 or 5 x 10^5 CFU, no volunteers who received vaccine candidate strain CVD 908 suffered febrile reactions. Accordingly, additional volunteers were fed this strain at a dose of 5 x 10^7 CFU (Tacket et al, in press). Of a total of 12 subjects who received CVD 908 at this dosage level with buffer, none experienced febrile reactions or other notable adverse reactions and excellent immune responses were observed. Gut-derived, trafficking IgA O antibody-secreting cells were detected in the peripheral blood of all 12 vaccinees who received a single 5 x 10^7 CFU dose of CVD 908. Based on the encouraging clinical and immunologic responses observed in Phase 1 clinical trials of CVD 908, plans are underway for this vaccine strain to proceed to Phase 2 clinical trials. If the Phase 2 studies proceed without incident, CVD 908 will be an attractive strain to use as a live vector vaccine to express an array of foreign antigens and to determine whether protective immune responses can be generated in human subjects who receive such hybrid vaccines.

REFERENCES

Black, R.E., Levine, M.M., Clements, M.L., Losonsky, G., Herrington, D., Berman, S., and Formal, S.B., 1987, Prevention of shigellosis by a Salmonella typhi-shigella sonnei bivalent vaccine, J. Infect. Dis., 155: 1260.

Black, R.E., Levine, M.M., Ferreccio, C., Clements, M.L., Lanata, C., Rooney, J., Germanier, R., and the Chilean Typhoid Committee, 1990, Efficacy of one or two doses of Ty21a Salmonella typhi vaccine in enteric-coated capsules in a controlled field trial, Vaccine, 8: 81.

Curtiss, R. III, Kelly, S.M., 1987, Salmonella typhimurium deletion mutants lacking adenylate cyclase and cyclic AMP receptor protein are avirulent and immunogenic, Infect. Immun., 55: 3035.

Edwards, M.F., Stocker, B.A.D., 1988, Construction of aroA his pur strains of Salmonella typhi, J. Bacteriol., 170: 3991.

Ferreccio, C., Levine, M.A., Rodriguez, H., et al., 1989, Comparative efficacy of two, three, or four doses of Ty21a live oral typhiod vaccine in enteric coated capsules, J. Infect. Dis., 159: 766.

Formal, S.B., Baron, L.S., Kopecko, D.J., et al. 1981, Construction of a potential bivalent vaccine strain: introduction of Shigella sonnei form I antigen gives into the galE Salmonella typhi Ty21a typhoid vaccine strain, Infect. Immun., 34: 746.

Forrest, B.D., LaBrooy, J., Attridge, S.R., et al., 1989, A candidate live oral typhoid/cholera hybrid vaccine is immunogenic in humans, J. Infect. Dis. 159:145.

Germanier, R., Furer, E., 1975, Isolation and characterization of gal E mutant Ty21a of Salmonella typhi: a candidate strain for a live oral typhoid vaccine, J. Infect. Dis., 141: 553.

Gilman, R.H., Hornick, R.B., Woodward, W.E., et al., 1977, Immunity in typhoid fever: evaluation of Ty21a - an epimeraseless mutant of S. typhi as a live oral vaccine, J. Infect. Dis., 136: 717.

Herrington, D.A., Van De Verg, L., Formal, S.B., Hale, T.L., Tall, B.D., Cryz, S.J., Tramont, E.C., Levine, M.M., 1990, Studies in volunteers to evaluate candidate Shigella vaccines: further experience with a bivalent Salmonella typhi-Shigella sonnei vaccine and protection conferred by previous Shigella sonnei disease, Vaccine, 8: 353.

Hoiseth, S., and Stocker, B.A.D., 1981, Aromatic-dependent Salmonella typhimurium are non-virulent and effective as live vaccines, Nature, 29: 238.

Hone, D.M., Harris, A.M., Chatfield, S., Dougan, G., Levine, M.M. 1991, Construction of genetically-defined double aro mutants of Salmonella typhi, Vaccine, 9: 810.

Kantele, A., 1990, Antibody-secreting cells in the evaluation of the immunogenicity of an oral vaccine, Vaccine, 8: 321.

Kantele, A., and Makela, H., 1991, Different profiles of the human immune response to primary and secondary immunization with an oral Salmonella typhi Ty21a vaccine, Vaccine, 9: 423.

Levine, M.M., Ferreccio, C., Black, R.E., et al. 1987, Large-scale field trial of Ty21a live oral typhoid vaccine in enteric-coated capsule formulation, Lancet, I: 1049.

Levine, M.M., Ferreccio, C., Black, R.E., Tacket, C.O., Germanier, R., Chilean Typhoid Committee, 1989, Progress in vaccines against typhoid fever, Rev. Infect. Dis., 11: (supplement 3):S552.

Levine, M.M., Ferreccio, C., Cryz, S., Ortiz, E., 1990, Comparison of enteric-coated capsules and liquid formulation of Ty21a typhoid vaccine in a randomized controlled field trial, Lancet, 336: 891.

Levine, M.M., Herrington, D., Murphy, J.R., et al. 1987, Safety, infectivity, immunogenicity, and in vivo stability of two attenuated auxotrophic mutant strains of Salmonella typhi, 541Ty and 543Ty, as live oral vaccines in man, J. Clin. Invest., 79: 888.

Levine, M.M., Taylor, D.N., Ferreccio, C., 1989, Typhoid vaccines come of age, Pediat. Infect. Dis. J., 8: 374.

Simanjantuk, C., Paleologo, F.P., Punjabi, N.H., Darmogiwoto, Soeprawato, Totosudirjo, H., Haryanto, P., Suprijanto, E., Witham, N.D., Hoffman, S.L., 1991, Oral immunisation against typhoid fever in Indonesia with Ty21a vaccine, Lancet, 338: 1055.

Tacket, C.O., Forrest, B., Morono, R., Attridge, S.R., LaBrooy, J., Tall, B.D., Reymann, M., Rowley, D., Levine, M.M., 1990, Safety, Immunogenicity, and efficacy against cholera challenge in man of a typhoid-cholera hybrid vaccine derived from S. typhi Ty21a, Infect. Immun., 58: 1620.

Tacket, C.O., Hone, D.M., Curtiss, R. III, Kelly, S.M., Losonsky, G., Guers, L., Harris, A.M., Edelman, R., Levine, M.M., 1992, Comparison of the safety and immunogenicity of aroCaroD and cyacrp Salmonella typhi strains in adult volunteers, Infect. Immun., 60: 536.

Tacket, C.O., Hone, D.M., Losonsky, G., Guers, L., Edelman, R., Levine, M.M., In press, Clinical acceptability and immunogenicity of CVD 908 Salmonella typhi vaccine strain, Vaccine.

Wahdan, M.H., Serie, C., Germanier, R., Lackany, A., Cerisier, Y., Guerin, N., Sallam, S., Geoffroy, P., Sadek el Tantawi, A., Guesry, P., 1980, A controlled field trial of live oral typhoid vaccine Ty21a, Bull WHO, 58: 469.

Wahdan, M.H., Serie, C., Cerisier, Y., Sallam, S., Germanier, R., 1982, A controlled field trial of live Salmonella typhi strain Ty21a oral vaccine against typhoid: three year results, J. Infect. Dis., 145:292.

HYBRID HEPATITIS B VIRUS CORE/PRE-S PARTICLES: POSITION EFFECTS ON IMMUNOGENICITY OF HETEROLOGOUS EPITOPES AND EXPRESSION IN AVIRULENT *SALMONELLAE* FOR ORAL VACCINATION

Florian Schödel[1*], Darrell Peterson[2], Janice Hughes[3], David R. Milich[3]

[1]Max-Planck-Institut für Biochemie, D-8033 Martinsried;[2]Department of Biochemistry, Virginia Commonwealth University, Richmond, VA;[3] The Scripps Research Institute, La Jolla, CA

*Corresponding Author, Current Address: Department of Bacterial Diseases, Walter Reed Army Institute of Research, Washington, DC 20307-5100

ABSTRACT

This paper reviews data on the use of hepatitis B virus (HBV) core (HBcAg) particles as a carrier moiety for B-cell epitopes of the HBV envelope proteins (Schödel et al., 1990a; b, 1991 and 1992a;b). Virus neutralizing epitopes of the HBV pre-S region were inserted at the N-terminus, the N-terminus through a precore linker sequence, the C-terminus and an internal position of HBcAg by genetic engineering in *E. coli* . The hybrid HBc/pre-S proteins were purified and their antigenicity and immunogenicity analyzed. All purified HBc/pre-S particles were particulate. Pre-S epitopes inserted at the N-terminus through a precore polylinker, the truncated C-terminus and at the internal position between HBcAg amino acids 75 and 81 were accessible on the particle surface. N-terminal fusions required the presence of the linker sequence to become surface accessible and immunogenic. Fusions to the N- and C-termini of HBcAg did not interfere with HBcAg antigenicity and immunogenicity. In contrast, insertion at the internal site abrogated recognition of HBcAg by 5 of 6 monoclonal antibodies and diminished recognition by human polyclonal anti-HBc antibodies as well as HBcAg immunogenicity. A pre-S(2) sequence fused to the C-terminus of HBcAg was surface accessible and weakly immunogenic. Pre-S(1) sequences fused to the N-terminus through a precore linker were surface accessible and highly immunogenic. The same sequence fused to the core methionine was not surface accessible or immunogenic. Insertion of the same pre-S(1) sequence at an internal position of HBcAg resulted in the most efficient anti-pre-S(1) antibody response. Hybrid HBc/pre-S particles were also expressed in avirulent *aroA* or *Δcya Δcrp Salmonella typhimurium* and *S. dublin*. Oral immunisation of mice with *Salmonella* expressing C-terminally fused hybrid HBc/pre-S(2) particles resulted in high titered serum anti HBc antibodies and low titered anti-pre-S(2) antibodies. Oral immunisation with *Δcya Δcrp S. typhimurium* expressing internally fused HBc/pre-S hybrid elicited high titered serum anti-pre-S(1) antibodies.

INTRODUCTION

Despite the existence of safe and efficient hepatitis B virus vaccines these are not

Biology of Salmonella, Edited by F. Cabello *et al.*,
Plenum Press, New York, 1993

available due to economic constraints for the majority of the world population. Infection with hepatitis B virus (HBV) continues to be a major global public health problem with over 200 million chronic carriers and a high morbidity and mortality. The eradication of HBV, which has no described reservoir outside of the human population should be possible if cheap and efficient vaccines become available. The ideal vaccine would be safe, induce longlasting protective immunity without side effects after a single ideally oral dose and have a high degree of stability under field conditions. We have therefore developed approaches towards oral recombinant bacteria based vaccines. The major surface antigen of HBV, HBsAg, is a highly conformation dependent particulate immunogen including a lipid bilayer membrane which currently cannot be expressed in an immunologically meaningful way in prokaryotes. It was therefore necessary to find alternative immunogens. The nucleocapsid or core antigen (HBcAg) of hepatitis B virus (HBV) is a 183 amino acid, 21kDa protein that spontaneously assembles to form particles even when the core gene alone is expressed in prokaryotes (for a review on HBV see Schödel et al., 1990c). It is a potent T-cell and B-cell immunogen and is capable of eliciting T-cell independent as well as T-cell dependent immune responses (Milich and McLachlan, 1986: Milich et al., 1987a;b). It has therefore been suggested and used as a carrier moiety to enhance the immunogenicity of heterologous sequences fused either by chemical coupling or by gene fusion (Milich et al., 1987a; Clarke et al., 1987; Stahl and Murray, 1989; Borisova et al., 1989; Moriarty et al., 1990; Schödel et al., 1990a;b,1991; Clarke et al., 1991; Schödel et al., 1992). In addition, immunization with HBcAg and WHcAg elicits protective immune responses in chimpanzees against HBV (Murray et al., 1984; Iwarson, 1985; Murray et al.,1987) and in woodchucks against WHV challenge (Roos et al., 1989; Schödel et al., 1992c; for a review on animal hepadnaviruses see Schödel et al., 1989). While the mechanism of protection is not defined, it has been demonstrated in the murine model that HBcAg specific T-helper cells can cooperate with envelope specific B cells to produce anti-HBs (Milich et al., 1987b). Woodchucks immunized with core antigen and protected against WHV also display a secondary type of anti-WHs response after WHV challenge (Schödel et al., 1992b). Peptide sequences of the large and middle hepatitis B virus envelope proteins (pre-S-(1) and pre-S(2)) capable of mediating protective immunity in chimpanzees or induce virus neutralizing antibodies have been identified (Milich et al., 1986; Itoh et al., 1986; Neurath et al., 1986; Emini et al., 1989; Thornton et al., 1989). We have used HBcAg as a carrier for virus-neutralizing epitopes of the HBV pre-S region (epitopes reviewed in Schödel et al., 1990c). By using HBcAg as a carrier for HBV envelope epitopes the inherent protective capacity of HBcAg would be exploited and the induction of virus neutralizing antibodies is linked to the induction of core specific T-cell help. To study the influence of heterologous epitope position on antigenicity and immunogenicity of hybrid proteins we inserted B-celle sites of the pre-S(1) and the pre-S(2) region of the HBV surface antigens into HBcAg at the N-terminus, at the N-terminus through a precore linker sequence, at an internal position or at the truncated C-terminus by genetic engineering of expression vectors in *E.coli* (see Schödel et al., 1992a).

Salmonella spp. can be rendered avirulent by defined genetic manipulations while retaining invasiveness across the intestinal epithelium after oral uptake (for reviews see Curtiss, 1991 and Schödel, 1992 a+b). Such avirulent *Salmonella spp*. are also promising candidate carrier strains to immunize against antigens from other pathogens by the oral route. Many genes encoding viral envelope or capsid antigens cannot be expressed in an immunogenic form in prokaryotes or their expression is toxic to the cells. For hybrid *Salmonella*-viral oral vacines eliciting virus neutralizing antibody responses genes have to be identified that can be stably maintained in the prokaryotic host. In addition their translation products have to be immunogenic when synthesized in the *Salmonella* background. Hybrid HBc particles are highly immunogenic and can be expressed in prokaryotes. We have therefore expressed some hybrid HBcAg/pre-S fusion genes in avirulent *Salmonella spp*. and analyzed the recombinant strains as oral immunogens (Schödel; et al., 1990b; 1991). In this report we shall review our data on the effects of position on immunogenicity of heterologous epitopes in HBcAg and on the immunogenicity of avirulent *Δcya ΔcrpS. typhimurium* strains expressing hybrid HBc/pre-S genes in mice (Schödel et al. 1990b; 1991, 1992a).

EXPERIMENTAL AND RESULTS

Antigenicity of Hybrid HBc/pre-S Particles

We have inserted antibody binding sites of the HBV pre-S(1) (amino acids 27-53 or 12-49) and pre-S(2) (amino acids 133-143) region into HBcAg at the gene level using genetic engineering of *E. coli* expression vectors. The heterologous epitopes were either inserted at the amine terminus of HBcAg [designated (27-53)-HBc], the amine terminus through a linker sequence derived from the precore sequence [(12-47)-PC-HBc], at an internal site between HBcAg amino acids 75 and 81 [HBc-(27-53)-HBc-(133-143] or at the truncated C-terminus of HBcAg after amino acid 156 [HBc-(133-143)]. The numbers in the protein designations signify pre-S amino acid positions (for the structure of hybrids see Schödel et al., 1992a;b).The hybrid HBc/pre-S proteins were purified and analyzed by electron micrography (Schödel et al., 1992a; and F. Schödel, D. Peterson, D.R. Milich, unpublished results). All hybrid HBcAg/pre-S particles studied were able to form particles. The surface accessibility of HBcAg antigenic determinants as well as the inserted heterologous epitopes was analyzed using monoclonal and polyclonal anti-HBc and anti-pre-S antibodies under nondenaturing ELISA conditions. All N-terminally and C-terminally modified HBcAg hybrids were recognized by HBc specific antibodies similar to wildtype HBcAg. Particles with an internal insertion between HBcAg amino acid positions 75 and 81 showed a much reduced binding to polyclonal anti-HBc antibodies and were only recognized by 1 out of 6 HBc specific monoclonal antibodies (Schödel et al., 1992a). Pre-S(1) and pre-S(2) sequences fused to the amine terminus through a precore linker sequence, to the truncated C-terminus or at an internal position were surface accessible on the hybrid particles (Schödel et al., 1992a). Only a pre-S(1) sequence (amino acids 27-53) fused to the very amine terminus of HBcAg was not surface accessible but detectable under denaturing Western blot conditions. To ascertain that the lack of surface accessibility of this N-terminally fused pre-S(1) sequence was not peculiar to this inserted sequence we inserted the precore linker sequence RWLWG designated PC in the hybrid (12-47)-PC-HBc between pre-S(1) amino acid 53 and HBcAg amino acid 1 to create hybrid (27-53)-PC-HBc (Schödel et al., 1992 b). The insertion of the five amino acids RWLWG was fully sufficient to render the pre-S(1) sequence surface accessible in nondenaturing capture ELISA assays. The lack of surface accessibility of the same sequence fused to the very N-terminus of HBcAg suggests that the N-terminus of native HBcAg is buried inside the particles.

Immunogenicity of Hybrid HBc/pre-S Particles

To determine and compare the immunogenicities of HBcAg and the inserted pre-S sequences of the hybrid HBc/pre-S particles BALB/c mice were immunized with equal doses of the five hybrids particles. The pre-S(2) sequence 133-143 fused to the C- terminus of HBcAg induced high-titred anti-HBc and low titered anti-pre-S(2) antibodies (Schödel et al., 1990b; 1992a). Pre-S(1) sequences 12-47 or 27-53 fused to the HBcAg N-terminus through a polylinker induced high-titred anti-HBc as well as high titred anti-pre-S(1) antibodies, mainly directed against the C-terminal amino acids 32-53 of the inserted sequence. The same pre-S(1) sequence 27-53 fused N-terminal of the core methionine was predictably not immunogenic as it was not surface accessible in the assembled particle. However, these particles retained HBcAg immunogenicity. The pre-S(1) 27-53 sequence inserted between HBcAg amino acids 75 and 81 induced an extremely high antibody response against the pre-S(1) insert that is comparable to anti-HBc titres achieved after immunization with native HBcAg particles. The enhanced immunogenicity of pre-S(1) 27-53 inserted at this internal position of HBcAg is not restricted to BALB/c mice, it was also demonstrated in B10 and B10.S mice and in outbred rabbits (Schödel et al., 1992a; and F. Schödel, unpublished data). Particles with an internal pre-S(1) insertion elicited relatively low primary anti-HBc titers compared with native HBc or other chimeric particles indicating that a major immunogenic region of HBcAg was removed by the deletion of residues 76 to 80.

Immunogenicity of recombinant *Salmonella spp.* expressing hybrid HBc/pre-S genes.

Genes coding for HBc-(133-143) and HBc-(27-53)-HBc-(133-143) could be stably expressed in *Salmonella typhimurium* and *S. dublin* strains rendered avirulent by deletions in the *aroA* (Hoiseth and Stocker, 1981; Smith et al., 1984; kindly provided by Bruce Stocker) or the *cya* and *crp* genes (Curtiss and Kelly, 1987; kindly provided by Roy Curtiss) to levels ranging *in vitro* from approximately 1% to 3% of total protein content after overnight culture (Schödel et al., 1990b; 1991; and unpublished data). Recombinant *ΔaroA* or *Δcya Δcrp* salmonellae synthesizing HBc-(133-143) elicited high titered anti-HBc and low titered anti-pre-S(2) antibody responses when fed to BALB/c or B10.S mice (Schödel et al., 1990b). Antibody titers were more heterogeneous after oral than after i.p. immunisation. After i.p. immunisation all mice seroconverted after a single immunisation.The rate of seroconversion after oral immunisation was dependent on the carriers strain, with some carier strains all mice seroconverted after a single oral immunisation (Schödel et al., 1990b). A single oral dose of as little as 5 x 10^6 CFU of χ4064(pFS14PS2) was sufficient for high titered anti-HBc seroconversion within a fortnight in B10.S mice (table 1).

Table 1. Serum Ig anti-HBc titers in B10.S mice fed χ4064(pFS14PS2).

CFU[1]	anti-HBc titer (1/)[2]
5x10^9	10,240
	10,240
	40,960
5x10^8	10,240
	160
5x10^7	20,480
	640
	10,240
5x10^6	10,240
	2,560
	2,560

[1]Female B10.S mice were fed a single dose (indicated as CFU) of a χ4064(pFS14PS2) overnight culture serially diluted in PBS. CFU were verified by plating dilutions on LB/ampicillin agar plates and counting bacterial colonies.

[2]Serum Ig anti-HBc antibody titres in individual mice two weeks after the immunisation are indicated as the reciprocal serum dilution yielding an OD_{492} 4 x above that of preimmune sera in ELISA.

The plasmid pNS27-53PS2 coding for HBc-(27-53)-HBc-(133-143) was also stably expressed in various avirulent *Salmonella* strains (Schödel et al., 1991). Recombinant *Δcya Δcrp S. typhimurium* synthesizing HBc-(27-53)-HBc-(133-143) [χ4064(pNS27-53)] elicited high titered anti-pre-S antibodies when fed to BALB/c mice (Schödel et al., 1991). In further studies it was demonstrated that a single oral immunisation with χ4064(pNS27-53) elicited anti-pre-S(1) antibodies in all BALB/c mice immunized (F. Schödel, unpublished data). The kinetics of the immune response were however different from those observed after immunisation with the same *S. typhimurium* strain synthesizing HBc-(133-143). While the latter strain elicited anti-HBc antibodies within a fortnight after oral immunisation, seroconversion to anti-pre-S(1) IgG could only be detected in all animals at one month after a single oral immunisation with the former strain.

DISCUSSION

The hepatitis B virus core antigen has been used as a carrier moiety for B-cell epitopes from a variety of pathogens (e.g. Clarke et al., 1987; 1991; Moriarty et al., 1989; Stahl and Murray, 1989). We have concentrated on its use for the induction of antibodies against HBV envelope proteins and have exploited the possibility to stably express it in prokaryotes to generate oral vaccines based on recombinant avirulent *Salmonella spp*.

To determine which position in HBcAg would render heterologous epitopes most immunogenic we have made insertions of HBV pre-S epitopes at the N-terminus, the N-terminus through a precore linker sequence, an internal position between amino acids 75 and 81 and at the truncated C-terminus. A fusion to the very N-terminus was not surface accessible and not immunogenic, while interposition of 4 precore amino acids and an arginine rendered the sequence surface accessible and immunogenic (Schödel et al., 1992a; b). Most fusions described in the literature have been made to the N-terminus of HBcAg through similar linker sequence. The historical reason is the existence of a naturally occurring restriction site in the HBc gene at this position (Clarke et al., 1987). These fusions were described as N-terminal fusions to HBcAg. It may be more appropriate to postulate that fusions to precore or linker amino acids render heterologous sequences surface accessible and immunogenic in hybrid HBcAg particles.

While C-terminal fusions are surface exposed but not highly immunogenic, fusion of heterologous sequences to the N-terminus through a precore linker sequence renders them highly immunogenic. Both N-terminal and C-terminal fusions did not interfere with native HBcAg antigenicity and immunogenicity.

Internal insertions within an immunodominant region of HBcAg abrogated most of the native HBcAg antigenicity and immunogenicity. Internal insertion resulted in the most drastically enhanced immunogenicity for the guest epitopes (for a discussion see Schödel et al., 1992a).

T-cell help in hybrid HBc/pre-S particles stems predominantly from HBcAg T-cell sites (Schödel et al., 1992 a)). If one attempts to use HBcAg particles as carriers to immunize against non-HBV pathogens it may be desirable to include Th cell sites of the respective pathogen for the potential generation of T-cell recall memory. It should be noted, however, that the position of a heterologous T-cell epitope within HBcAg may also critically affect its immunogenicity (Schödel et al., 1992a). HBcAg is capable of eliciting T-cell dependent as well as T-cell independent antibody responses (Milich and McLachlan, 1986). An internally inserted pre-S(1) sequence was highly immunogenic in euthymic mice but failed to elicit antibodies in nude mice (Schödel et al., 1992a). Insertion at the internal position of HBcAg therefore does not render heterologous epitopes T-cell independent and raises the interesting question of a potential sequence specificity required for T-cell independence.

Both C-terminally and internally fused hybrid HBc/pre-S particles were stably expressed to high levels in avirulent *S. typhimurium, S. dublin* and *S. typhi* vaccine strains, either from conventional pBR322 based plasmids or from ASD complementing plasmids (Nakayama et al., 1988), which do not carry an antibiotic resistance marker. (Schödel et al., 1990b, 1991 and F. Schödel, S. Kelly, R. Curtiss, manuscript in preparation). While homogeneous high anti-HBc antibody titres were observed in all mice with all strains after a single ip immunisation the seroconversion rate of BALB/c mice fed *aroA Salmonella dublin* and *S. typhimurium* or *Δcya Δcrp S. typhimurium* (χ4064, Curtiss and Kelly, 1987) strains expressing hybrid HBc genes coding for C-terminally truncated particles with a pre-S(2) extension (pFS14PS2) was dependent on the carriers strain (Schödel et al., 1990b; and F. Schödel, unpublished results). The ratio of antibody titres against HBc and pre-S(2) closely reflected the parenteral immunogenicity of these particles: the mice developed high titered anti-HBc serum antibodies and low titered anti-pre-S(2) antibodies. The minimum oral dose for anti-HBc seroconversion required of χ4064(pFS14PS2) is approximately 10^6 CFU. Hybrid HBc-(27-53)HBc-(133-143) when synthesized in avirulent *S. typhimurium* in contrast were more toxic to the cariers strains. One of the recombinant *S. typhimurium* (χ4064(pNS27-53)PS2) strains induced relatively high titered anti-pre-S antibodies after oral immunization of mice, but the kinetic of the immune response was delayed in comparison with the same strain expressing HBc-(133-143), possibly a reflection of the impaired growth characteristics of χ4064(pNS27-53)PS2. While all mice fed 2 x 10^9 CFU χ4064(pNS27-53)PS2 seroconverted to IgG anti-pre-S(1) at 4 weeks after a single dose, again the titers achieved were quite heterogeneous (F. Schödel, unpublished results).

Methods that would increase the stability of the expression unit, for example chromosomal integration, and/or the use of promoters that are activated within host tissues could potentially relieve some of this heterogeneity in as far as it may be owed to insufficient levels of expression at the adequate site or loss of plamids. Another factor for the heterogeneity of immune responses may however be variations in the numbers of recombinant *Salmonellae* reaching gut associated lymphoid tissues or deeper tissues after oral delivery within animals. Hybrid hepatitis B virus nucleocapsid antigens which are synthesized in the cytoplasm of prokaryotes appear to be the only epitope presentation system so far described that is capable of inducing high titered antibody responses against the carried epitope(s) when administered in live avirulent *Salmonellae* by the oral route.

Acknowledgement

Partly supported by grants of the Wilhelm-Sander Stiftung and the Walter-Schulz Stiftung to FS. The expert technical help of Ursula Morgenroth is acknowledged. We thank Bruce Stocker and Roy Curtiss for *Salmonella* strains.

REFERENCES

Borisova, G.P., Berzins, I., Pushko, P.M., Pumpen, P., Grem, E., Tsibinogen, V.V., Loseva, V., Ose, V., Ulrich, R., Siakkou, H., Rosenthal, H.A. Recombinant core particles of hepatitis B virus exposing foreign antigenic determinants on their surface. *FEBS Letters*, 1989, **259**, 121-124.

Clarke, B. E., Newton, S.E., Carroll, A.R., Francis, M.J., Appleyard, G., Syred, D., Highfield, P.E., Rowlands, D.J., Brown, F. Improved immunogenicity of a peptide epitope after fusion to hepatitis B core protein. 1987, *Nature*, **330**, 381-384.

Clarke, B.R., Carroll, A.R., Brown, A.L., Jon, J., Parry, N.R., Rud, E.W., Francis, M.J., Rowlands, D.J. Expression and immunological analysis of hepatitis-B core fusion particles carrying internal heterologous sequences. In: *Vaccines 91*. (Eds. Brown, F., Chanock, R. M., Ginsberg, H. S., and Lerner, R. A.). Cold Spring Harbor Laboratory, Cold Spring Harbor, New York, 1991, p. 313-318.

Curtiss, R., and Kelly, S. M. *Salmonella typhimurium* deletion mutants lacking adenylate cyclase and cyclic AMP receptor protein are avirulent and immunogenic. *Infect. Immun.* , 1987, **55**, 3035-3043.

Curtiss, R. Attenuated *Salmonella* strains as live vectors for the expression of foreign antigens. In: New generation vaccines. (Woodrow, G. C., Levine, M. M.). Marcel Dekker, New York 1990, pp. 161-188.

Emini, E., V. Larson, J. Eichberg, P. Conrad, V. M. Garsky, D. R. Lee, R. W. Ellis, W. J. Miller, C. A. Anderson and J. L. Gerin. 1989. Protective effect of a synthetic peptide comprising the complete preS2 region of the hepatitis B virus surface protein. *J. Med. Virol.* **28**: 7-11.

Hoiseth, S. K., Stocker, B.A.D. Aromatic-dependent *Salmonella typhimurium* are non-virulent and effective as live vaccines. *Nature*, 1981, **291**, 238-239.

Iwarson, S., Tabor, E., Thomas, H. C., Snoy, P., and Gerety, R. J. Protection against hepatitis B virus infection by immunization with hepatitis B c-antigen. *Gastroenterol.*, 1985, **88**, 763-767.

Milich, D. R., and McLachlan, A. The nucleocapsid of hepatitis B virus is both a T-cell-dependent and a T-cell-independent antigen. *Science*, 1986, **234**, 1398-1401.

Milich, D. R., A. McLachlan, F. V. Chisari, F. V. Nakamura, and T. Thornton. Two distinct but overlapping antibody binding sites in the pre-S(2) region of HBsAg localized within 11 continuous residues. *J. Immunol.* 1986, **137**, 2703-2710.

Milich, D. R., McLachlan, A., Moriarty, A., and Thornton, G. B. Immune response to hepatitis B virus core antigen (HBcAg): localization of T cell recognition sites within HBcAg/HBeAg. *J. Immunol.*, 1987a, **139**, 1223-1231.

Milich, D. R., McLachlan, A., Thornton, G. B., and Hughes, J. L. Antibody production to the nucleocapsid and envelope of the hepatitis B virus primed by a single synthetic T cell site. *Nature*, 1987b, **329**, 547-549.

Milich, D.R., McLachlan, A., Hughes, J.E., Jones, J.E., Stahl, S., Wingfield, P., Thornton, G.B. Characterization of the hepatitis B virus nucleocapsid as an immunologic carrier moiety. In: *Vaccines 89*. (Eds. Brown, F., Chanock, R. M., Ginsberg, H. S., and Lerner, R. A.). Cold Spring Harbor Laboratory, Cold Spring Harbor, New York, 1989, p. 37-42.

Moriarty, A.M., McGee, J.S., Winslow, B., Inman, D., Leturcq, J., Thornton, Hughes, J.L., Milich, D.R. Expression of HIV *gag* and *env* B-cell epitopes on the surface of HBV core particles and analysis of the immune responses generated to those epitopes. In: *Vaccines 90*. (Eds. Brown, F., Chanock, R. M., Ginsberg, H. S., and Lerner, R. A.). Cold Spring Harbor Laboratory, Cold Spring Harbor, New York, 1990, p. 225-229.

Murray, K., Bruce, S. A., Hinnen, A., Wingfield, P., van Erd, P. M. C. A., de Reus, A., and Schellekens, H. Hepatitis B virus antigens made in microbial cells immunize against viral infection. *EMBO J.*, 1984, **3**, 645-650.

Murray, K., Bruce, S. A., Wingfield, P., van Erd, P. M. C. A., de Reus, A., and Schellekens, H. Protective immunization against hepatitis B virus infection by immunization with an internal antigen of the virus. *J. Med. Virol.*, 1987, **23**, 101-107.

Nakayama, K., S. M. Kelly, and R. Curtiss, III. Construction of an *asd*+ expression-cloning vector: stable maintenance and high level expression of cloned genes in a *Salmonella* vaccines strain. *Biotechnol.* 1988, **6**:693-697.

Neurath, A. R., S. B. H. Kent, K. Parker, A. M. Prince, N. Strick, B. Brotman, and P. Sproul. Antibodies to a synthetic peptide from the preS 120-145 region of the hepatitis B virus are virus-neutralizing. *Vaccine*, 1986, **4**, 35-37.

Roos, S., Fuchs, K., and Roggendorf, M. Protection of woodchucks from infection with woodchuck hepatitis virus by immunization with recombinant core protein. *J. Gen. Virol.*, 1989, **70**, 2087-2095.

Smith, B.P., Reina-Guerra, M., Stocker, B. A.D., Hoiseth, S.K., Johnson, E. Aromatic-dependent *Salmonella dublin* as a modified live vaccine for calves. *Am. J. Vet. Res.* , 1984, **45**, 2231-2235.

Schödel, F.. Prospects for oral vaccination using recombinant bacteria expressing viral epitopes. *Adv.Virus Res.*, 1992a, **41**:409-446.

Schödel, F. Recombinant avirulent *salmonella* as oral vaccine carriers. *Infection*, 1992b, **1**, 1-12.

Schödel, F., Sprengel, R., Weimer, T., Fernholz, D., Schneider, R., and Will, H. Animal hepatitis B viruses. *Adv. Viral Oncol.*, 1989, **8**, 73-102.

Schödel, F., T. Weimer, H. Will, D. R. Milich. Recombinant hepatitis B virus (HBV) core particles carrying immunodominant B-cell epitopes of the HBV pre-S(2) region. In: *Vaccines 90*, Edts. Fred Brown, Robert M. Chanock, Harold S. Ginsberg, Richard A. Lerner. Cold Spring Harbor Laboratories, 1990a, pp. 193-198.

Schödel, F., D. R. Milich, and H. Will. Hepatitis B virus nucleocapsid/pre-S2 fusion proteins expressed in attenuated *Salmonella* for oral immunisation. *J. Immunol.* 1990b, **145**: 4317-4321.

Schödel, F., T. Weimer, H. Will. HBV: Molecular Biology and Immunology. *Biotest Bull.*, 1990c, **4**: 63-83

Schödel, F., D. R. Milich, H. Will. Hybrid hepatitis B virus core/pre-S particles expressed in attenuated *salmonellae* for oral immunisation. In: *Vaccines 91*, Edts. Fred Brown, Robert M. Chanock, Harold S. Ginsberg, Richard A. Lerner. Cold Spring Harbor Laboratories, 1991, 319-325.

Schödel, F., A. M. Moriarty, D. L. Peterson, J. Zheng, J. L. Hughes, H. Will, D. J. Leturcq, J. S. McGee, D. R. Milich. The position of heterologous epitopes inserted in hepatitis B virus core particles determines their immunogenicity. *J.Virol.*, 1992a, **66**, 106-114.

Schödel. F., D. Peterson, J. Hughes, D. R. Milich. Avirulent Salmonellae expressing hybrid HBc/pre-S particles for oral vaccination. *Vaccine*, 1992b, in press.

Schödel. F., G. Neckermann, D. Peterson, K. Fuchs, S. Fuller, H. Will, M. Roggendorf. Immunization with recombinant woodchuck hepatitis virus nucleocapsid antigen or hepatitis B virus nucleocapsid antigen protects woodchucks from woodchuck hepatitis virus infection. *Vaccine*, 1992c, in press.

Thornton, G. B., A. M. Moriarty, D. R. Milich, J. W. Eichberg, R. H. Purcell, and J. L. Gerin. Protection of chimpanzees from hepatitis B virus infection after immunization with synthetic peptides: identification of protective epitopes in the preS region. In: *Vaccines 89*, Edts. Fred Brown, Robert M. Chanock, Harold S. Ginsberg, Richard A. Lerner. Cold Spring Harbor Laboratories, 1989, pp. 467-471.

SAFETY AND IMMUNOGENICITY OF *SALMONELLA TYPHI* Ty21a LIQUID FORMULATION VACCINE IN 4 TO 6 YEAR OLD THAI CHILDREN

Thavichai Olanratmanee,[1] Myron M. Levine,[2] Genny Losonsky,[2]
Usa Thisyakorn,[3] and Stanley J. Cryz, Jr.[4]

[1]Chachoengsao Hospital, Ministry of Public Health, Thailand
[2]Center for Vaccine Development, Baltimore, MD, USA
[3]Department of Pediatrics, Chulalongkorn University, Bangkok, Thailand
[4]Swiss Serum and Vaccine Institute, Berne, Switzerland

ABSTRACT

The safety and immunogenicity of the *Salmonella typhi* Ty21a vaccine in liquid form was evaluated in 170 Thai children 4-6 years of age. Subjects were randomized to receive either 3 doses of vaccine or placebo on alternate days in a double-blind fashion. Three mild adverse reactions were noted, all in the placebo group. Of the vaccine recipients, 69 of 83 (83%) responded with a significant rise in serum anti-*S. typhi* LPS antibody compared to 11 of 76 (14%) placebo recipients (P < 0.001).

INTRODUCTION

Typhoid fever remains a serious public health problem in many developing areas of the world (Thisyakorn et al., 1987; Institute of Medicine, New Vaccine Development: Establishing Priorities, Vol. II, 1986). The attenuated *S. typhi* Ty21a vaccine strain confers significant protection against typhoid fever without the attendant adverse reactions associated with the use of the parenteral vaccine (Levine et al., 1989; Levine et al., 1990; Simanjuntak et al., 1991). A liquid Ty21a vaccine formulation has recently been found to provide greater protection than the currently available enteric-coated capsules in Chile and Indonesia (Levine et al., 1990; Simanjuntak et al., 1991). The vast majority of the study populations were school-aged children (>6 years of age) and young adults. It is becoming increasing apparent that there is a high incidence of *S. typhi* infection in children 2 to 5 years of age. Unfortunately, comparatively little is known about the safety and immunogenicity of the Ty21a vaccine in this high risk age group (Murphy et al., 1991).

Because of (i) the recent recognition that the incidence of *S. typhi* infection in young children is increasing; (ii) the superiority of the liquid versus the enteric-coated capsule vaccine formulation; (iii) the frequent difficulty that very young children have in swallowing capsules; and (iv) the desire to vaccinate at the earliest age possible to gain the maximum public health benefit, we conducted a randomized, double-blind placebo-controlled trial to evaluate the safety and immunogenicity of the liquid Ty21a vaccine formulation in Thai children 4 to 6 years of age.

METHODS

The study was conducted from May through June 1991, at 5 day-care centers in Chachoengsao Province, Thailand. The trial design and consent procedure were approved by the Thai Ministry of health. Written informed consent was obtained from the parent or guardian of each child. A total of 170 healthy children were enrolled.

The vaccine and placebo were packaged in identically-appearing aluminum foil sachets bearing a letter code. Each dose of vaccine contained 3 x 10^9 colony forming units (CFU) of *S. typhi* Ty21a, while each placebo packet contained 2 x 10^9 heat-killed *Escherichia coli* K12. Aspartame (25 mg) had been added to each as a sweetening agent. Each buffer pack was composed of 1.3 g of sodium bicarbonate and 0.8 g of ascorbic acid. The vaccine or placebo was prepared as follows. The contents of the buffer pack were dissolved in 50 ml of water after which the contents of a coded packet containing either the vaccine or placebo was added and gently mixed. The suspension was ingested within 10 minutes of preparation. Upon reconstitution, the Ty21a bacteria remain viable for at least 30 minutes. Vaccine packets returned to the Swiss Serum and Vaccine Institute after the immunization program was completed contained $\geq$ 2.5 x 10^9 CFU of Ty21a per packet.

Prior to participation in the study, each child was examined by a health care worker. Only healthy children with no history of typhoid fever were enrolled. The vaccine was administered under conditions designed to facilitate its rapid passage through the stomach and into the proximal small intestine. Therefore, at the time of vaccination, children had not ingested food for at least 1 hour. After vaccination, children were instructed to refrain from eating for 1.5 hours and drinking any liquids for 15 minutes. The children were observed by a physician for 1.5 hours post-vaccination for immediate-type reactions. Parents were given an adverse reaction report sheet to complete after each vaccination. This sheet listed such reactions as diarrhea, nausea, malaise, vomiting, and fever. Parents were encouraged to contact members of the vaccination team if any reactions were noted. Three doses of vaccine or placebo were given on alternate days. A venous blood sample was collected immediately before and 21 days after vaccination and the serum collected. Serum tubes were labeled with the subject's number and the date and frozen at -20°C. The parents and children were interviewed 21 days after immunization to ascertain vaccine acceptability.

Anti-*S. typhi* lipopolysaccharide (LPS) antibody levels were measured using an enzyme-linked immunosorbent assay (ELISA) (Levine et al., 1987; Murphy et al., 1987). Paired serum samples were run in parallel on the same plate. All samples were initially screened at a final dilution of 1:100. Serum samples from subjects who possessed high antibody levels in their baseline sample [optical density (O.D.) $\geq$ 0.75] were serially diluted (2-fold starting at 1:100 dilution) and re-assayed. A significant rise in antibody level (seroconversion) was defined as a net increase of 0.15 O.D. units above pre-immunization values (Levine et al., 1987).

Significance between groups as relates to seroconversion was determined by chi-square analysis.

RESULTS AND DISCUSSION

The vaccine was very well tolerated. No adverse reactions which could be attributed to immunization were reported by the 88 participants who received the Ty21a vaccine. Of the 82 placebo recipients, 3 noted a reaction. One child presented with headache, transient low grade fever, and a rash after the first dose; one child complained of malaise after all 3 doses; the third child reported a rash and a transient, low grade fever after the second dose. All reactions were mild and self-limiting.

Of the 170 children who completed the immunization regimen, paired serum samples were obtained from 160. One child was excluded from analysis for not following the protocol. The serum anti-*S. typhi* LPS antibody response is shown in Table 1. The seroconversion rate for vaccine recipients of all ages was 83% versus 14% for those who received the placebo ($P < 0.005$). There was no difference in the response rate for children aged 4, 5, or 6 (80-86%); $P > 0.5$). The seroconversion rate among the placebo recipients was higher than what one would have predicted. We have not been able to pinpoint the reason for this observance although several possibilities exist. Spread of the Ty21a strain from vaccine to placebo recipients is unlikely to be the cause since Ty21a is not excreted in the faces at the dosage given (Murphy et al., 1991). Alternatively, some children scheduled to receive placebo may have inadvertently been given a dose of vaccine. Interestingly, 50% of the placebo group who manifested a serological response attended the same day-care center. Finally, this finding could be attributed to the assay system used; specifically, assigning too low a cut-off value (0.15 O.D. units) to signify an immune response. However, increasing this value to 0.2 did not appreciably change the seroconversion rates of either group. Further increasing the cut-off value to 0.25 or 0.3 resulted in a nearly equal decline in seroconversion rates for both groups.

Table 1. Anti-*Salmonella typhi* lipopolysaccharide (LPS) antibody response following ingestion of vaccine or placebo

Age (Years)	Formulation	Seroconversion	P
4	Placebo	3/20 (15)	
	Vaccine	19/22 (86)	<0.01
5	Placebo	6/36 (16)	
	Vaccine	30/37 (80)	<0.01
6	Placebo	2/20 (10)	
	Vaccine	20/24 (83)	<0.01
Total	Placebo	11/76 (14)	
	Vaccine	69/83 (83)	<0.01

Fully 37% (63/170) of children had elevated baseline antibody levels (O.D. ≥ 0.75; based upon experience with North American volunteers who ingested *S. typhi* as part of vaccine challenge studies), suggesting prior exposure to *S. typhi* or another group D *Salmonella* (Levine et al., 1987). Of this subgroup, 89% (31/35) of those who received vaccine seroconverted, while only 39% (11/28) of the placebo recipients did so ($P < 0.005$). This finding indicates that vaccine take is not inhibited by prior exposure to *S. typhi*. One could, therefore, reasonably expect vaccination to boost background levels of immunity even in a region of high endemicity.

A significant rise in serum IgG anti-*S. typhi* LPS antibody was found to correlate well with protection against disease in field trials with school-aged children in Santiago, Chile (Levine et al., 1989). Therefore, seroconversion may serve as an accurate marker of vaccine take and predictor of immunity. Previous studies in which young children were immunized with Ty21a yielded low rates of seroconversion. Approximately 40% of Chilean children 6 to 9 years of age seroconverted while the immune response was meager (~20%) in children 2 to 3.5 years of age (Black et al., 1983). Vaccination of infants and toddlers did not induce a detectable humoral or cellular immune response (Murphy et al., 1991).

Several factors may account for the higher rate of seroconversion in the current study. Foremost is recognizing the critical role that vaccine formulation plays in immunogenicity and efficacy. In previous trials the Ty21a vaccine was administered by emptying the contents of a capsule into a buffer solution of milk and sodium bicarbonate. Studies in our laboratories have subsequently shown that while the Ty21a remains viable in such a buffer, there is pronounced bacteria clumping which requires at least 20 minutes of mixing to yield a homogenous suspension. In contrast, the current formulation gives a homogenous suspension with only minimal (0.5 to 1 min) mixing. Bacterial clumping could adversely affect gut transit time and ability to colonize the proximal ileum with a resultant decrease in immune response. Additionally, in the present study, great care was taken to insure that the vaccine was ingested under conditions which would facilitate passage through the stomach into the duodenum. The fact that a substantial proportion of the Thai children appeared to have prior exposure to *S. typhi* or antigenically related *Salmonella* may have resulted in a priming of the immune system, hence a more vigorous response. Genetic differences between the Thai and Chilean populations could also play a role.

The excellent safety and immunogenicity record of the Ty21a vaccine in 4 to 6 year old Thai children paves the way for studies in younger children. At present, a study of similar design is planned for 2 to 4 year old children. If similar results are obtained in this younger age group, it may prove possible to immunize at an age prior to when disease incidence peaks, thereby achieving the maximum public health benefit.

REFERENCES

Black, R.E., Levine, M.M., Young, C.R., Rooney, J., Levine, S., Clements, M.L., O'Donnell, S., Hughes, T.P., the Chilean Typhoid Committee, and Germanier, R., 1983, Immunogenicity of Ty21a attenuated *Salmonella typhi* given with sodium bicarbonate or in enteric-coated capsules, *Biol. Stand.* 53:9.

Institute of Medicine, 1986, Diseases of importance in developing countries, in: *New Vaccine Development: Establishing Priorities*, Vol. II, National Academy Press, Washington.

Levine, M.M., Herrington, D., Murphy, J.R., Morris, J.G., Losonsky, G., Tall, B., Lindberg, A.A., Svenson, S., Bagar, S., Edwards, M.F., and Stocker, B., 1987, Safety, infectivity, immunogenicity, and in vivo stability of two attenuated auxotrophic mutant strains of *Salmonella typhi*, *J. Clin. Invest.* 79:888.

Levine, M.M., Ferreccio, C., Black, R.E., Tacket, C.O., Germanier, R., and the Chilean Typhoid Committee, 1989, Progress in vaccines against typhoid fever, *Rev. Infect. Dis.* 11(Suppl 3):S552.

Levine, M.M., Ferreccio, C., Cryz, S., and Ortiz, E., 1990, Comparison of enteric-coated capsules and liquid formulation of Ty21a typhoid vaccine in randomised controlled field trial, *Lancet* 336:891.
Murphy, J.R., Bagar, S., Muñoz, C., Schlesinger, L., Ferreccio, C., Lindberg, A.A., Svenson, S., Losonsky, G., Koster, F., and Levine, M.M., 1987, Human immunity to *Salmonella typhi*: some characteristics of humoral and cellular immunity of individuals resident in typhoid endemic and typhoid free regions, *J. Infect. Dis.* 156:1005.
Murphy, J.R., Grez, L., Schlesinger, L., Ferreccio, C., Bagar, S., Muñoz, C., Wasserman, S., Losonsky, G., Olson, J.G., and Levine, M.M., 1991, Immunogenicity of *Salmonella typhi* Ty21a vaccine for young children, *Infect. Immun.* 59:4291.
Simanjuntak, C.H., Faleologo, F.P., Punjabi, N.H., Darmowigoto, R., Soeprawoto, Totosudirjo, H., Haryanto, P., Suprijanto, E., Witham, N.D., and Hoffman, S.L., 1991, Oral immunisation against typhoid fever in Indonesia with Ty21a vaccine, *Lancet* 338:1055.
Thisyakorn, U., Mansuwan, P., and Taylor, D.N., 1987, Typhoid and paratyphoid fever in 192 children in Thailand, *Am. J. Dis. Child.* 141:862.

VACCINES FOR PREVENTION OF ENTERIC BACTERIAL INFECTIONS CAUSED BY *SALMONELLAE*

Shousun C. Szu,[1] Douglas Watson,[1] Marina Hinojosa,[1] Rachel Schneerson,[1] John B. Robbins,[1] David N. Taylor,[2] and Andrew Trofa[2]

[1]National Institute of Child Health and Development
NIH, Bethesda, MD
[2]Division of Communicable Disease and Immunology
Walter Reed Army Institute of Research
Washington, DC

ABSTRACT

Two clinical trials of the Vi vaccine provided evidence that serum antibodies to the capsular polysaccharide of *Salmonella typhi* confer protective immunity to typhoid fever. The immunogenicity of the Vi has been improved by binding this polysaccharide to a protein. In laboratory animals, this conjugate had increased immunogenicity compared to the Vi alone and elicited booster responses. Because there are as yet no vaccines for non-typhoidal *Salmonellae*, we synthesized a non-toxic conjugate of the O-SP of *Salmonella typhimurium* O:4,12 which conferred protective immunity in the murine model at doses and routes of immunization which are clinically acceptable. The rationale for developing conjugate vaccines for prevention of other enteric bacterial infections is discussed.

INTRODUCTION

Capsular polysaccharides or lipopolysaccharides may be both virulence factors and protective antigens for *Salmonellae*. These saccharides confer virulence by "shielding" the organisms from the protective effects of serum complement. Their protective action is mediated by activation of complement, induced by serum antibodies. The capsular polysaccharide of *Salmonella typhi* (Vi) is a linear homopolymer of $\alpha(1\rightarrow4)$-GalpNacA variably O-acetylated at C-3 (Heyns and Kiessling, 1967). Two double-blinded, randomized and controlled field trials in areas with high attack rates of typhoid fever evaluated the efficacy of Vi vaccine in individuals ages 5 to 44 years (Acharya et al., 1987; Klugman et al., 1987). In both studies, the efficacy of one injection of Vi vaccine

Biology of Salmonella, Edited by F. Cabello *et al.*,
Plenum Press, New York, 1993

was about 70%. The use of Vi vaccine is limited by its immunologic properties of age-dependent and T-cell independent immunogenicity (Douglas et al., 1983). We have overcome these properties by developing methods to covalently bind Vi to proteins to form conjugates as has been done with other polysaccharides (Goebel and Avery, 1929). Our objective is to develop a Vi vaccine that may be injected concurrently with DTP as part of the routine vaccination in infancy (Sarnaik et al., 1990).

Non-typhoidal *Salmonellae* and *Shigellae* remain major causes of morbidity and mortality especially in infants and young children in developing countries. Although TAB vaccines were used for many years, no scientifically based data verified their effectiveness against non-typhoidal *Salmonellae*. *Salmonella typhimurium* (Group B) infection of mice has been considered as a useful model for typhoid fever. A systemic infection and death can be induced in mice either by feeding or by intraperitoneal injection of low inocula of *S. typhimurium*. The LPS is both a virulence factor and a protective antigen for *S. typhimurium*. Leive and co-workers showed that the virulence of *S. typhimurium* for mice was related to the comparatively low activation of the alternate complement pathway by its LPS (Liang-Takasaki et al., 1982). Both IgM and IgG serum antibodies to the O-specific polysaccharide (O-SP) of the LPS of *S. typhimurium* confer protective immunity to mice against this pathogen (Saxen et al., 1984; Colwell et al., 1984; Carlin et al., 1987). Investigators prepared conjugates of oligosaccharides of this O-SP bound to proteins and injected intraperitoneally in complete Freund's adjuvant, elicited protective immunity to *S. typhimurium* in mice (Svenson et al., 1979; Svenson and Lindberg, 1981). The dosage and immunization route of this conjugate were not suitable for clinical use.

We have reexamined this approach to induce protective immunity to non-typhoidal *Salmonellae* and adapted it for humans. The O-SP of Groups A, B and D *Salmonellae* share the same trisaccharide backbone (Factor 12 in the Kauffmann-White Scheme). The non-reducing termini have unique dideoxysugars that confer the group specificity to the repeating unit of the 0-SP. The group B specificity of *Salmonellae* (Factor 4) is conferred by abequose (3,6-dideoxy-α-D-Gal) 1→3 linked to the mannose on the backbone repeating unit (Wilkinson, 1977). Factor 5 is created by an O-acetyl linked to the C2 of abequose. Glucose may be linked α(1→4) to the Gal of the trisaccharide backbone creating Factor 12_2 or α(1→6) to the Gal creating Factor 1.

Unlike Vi of *S. typhi*, the LPS of *S. typhimurium* is toxic for use as a vaccine for humans. Removal of its lipid A results in a preparation containing the O-specific polysaccharide (O-SP) and the core; a non-immunogenic molecule probably due to its small size. We conjugated the O-SP of *S. typhimurium* (0:4,12) to tetanus toxoid (TT) by a method developed in our laboratory. Subcutaneous injections of 2.5 μg O-SP-TT conjugates in saline into young outbred mice elicited both IgM and IgG antibodies directed

Abe [O-acetyl, C_2] αGlu
1 1
↑ ↑
3 4,6
→2)-β-D-Manp(1→4)-β-L-Rhap(1→3)-α-D-Galp(1→

Factor 1. Group B 0-specific polysaccharide (0:1,4,5,12,12_2).

mostly against Factor 4. A representative conjugate, O-SP-TT_{477}, induced both active and passive protection in mice against lethal infection with *S. typhimurium* (Watson et al.,1992).

MATERIALS AND METHODS

Vi and Vi-protein conjugates. Vi was purified from *Citrobacter freundii* WR7011 and thiolated with cystamine in the presence of a water soluble carbodiimide (ethyl-3-(3-dimethylaminopropyl) carbodiimide, EDAC) (Szu et al., 1987). About 1-2% of the residues were thiolated. Reduction with 100 mM dithiothreitol produced Vi-cysteamine. Proteins were thiolated with the heterobifunctional reagent N-succinimidyl 3-(2-pyridyl-dithio) propionate (SPDP) (Carlsson et al., 1978). Vi-cysteamine was mixed with protein-SPDP and a conjugate was formed by disulfide exchange between Vi-SH and N-pyridyldithio. The conjugate was purified by gel filtration through CL-4B Sepharose. We used bovine serum albumin (BSA), diphtheria toxoid (DT), TT, cholera toxin (CT) or its B-subunit (CTB) and *Pseudomonas aeruginosa* recombinant exotoxin A (rEPA) as carrier proteins.

O-SP from S. typhimurium. LPS was purified from *S. typhimurium* strain TML (O:4,12) (Westphal and Jann, 1965). The O-SP was purified by heating the LPS in 1% acetic acid at 100°C for 90 minutes, ultracentrifugation and gel filtration through Sephadex G-50.

O-SP-tetanus toxoid conjugate. The conjugate was synthesized by a scheme described for *Haemophilus influenzae* (Schneerson et al., 1980). O-SP was activated by cyanogen bromide at pH 10.5 and bound to adipic acid dihydrazide (ADH). The O-SP-AH derivative was bound to TT in the presence of 50 mM EDAC and purified by gel filtration through Sepharose CL-4B.

Serology. Antigenicity of the polysaccharides, proteins and conjugates was determined by double immunodiffusion. Vi antibodies were measured by RIA and expressed in μg antibody/ml serum. Protein and LPS antibodies were assayed by ELISA using hyperimmune sera for protein antibodies and murine monoclonal IgM (5C7) and IgG (4E2) for *S. typhimurium* LPS antibodies as standards. Absorptions were performed by incubating sera with phenol-inactivated *Salmonellae* at 37°C for 1 hour. More than 95% of the 0-SP conjugate induced antibodies were removed by the homologous *S. typhimurium* strain TML. Serum absorptions were also carried out with the cross-reacting Group B, *S. typhimurium* LT (O:4,5,12), Group D, *S. enteritidis* (O:9,12) containing Factor 12 and the non-cross reacting Group C, *S. montevideo* (O:6,7),and the O-SP deficient *S. typhimurium* SL1764 (Ra mutant). The latter two strains share the same core region of the 0-SP used in the conjugate.

Immunization. Six week old female BALB/c or general purpose mice were injected subcutaneously biweekly, 3 times with 0.1 ml containing 25 μg of either the O-SP alone or the conjugated O-SP. Animals were bled 7-10 days after each injection. Juvenile rhesus monkeys were injected subcutaneously twice at one month intervals with 0.5 ml Vi or Vi-CT conjugate containing 25 μg Vi. Monkeys were bled before and 3 weeks after each injection.

Protection. One week after the third immunization, *S. typhimurium* W118, in mid-logarithmic growth phase, were injected intraperitoneally and the LD_{50} was determined 21 days later. The survivors were exsanguinated 14 days later (35 days post challenge) and their serum LPS antibodies assayed.

RESULTS AND DISCUSSION

Vi conjugate. The composition of the thiolated Vi, the SPDP-protein derivatives, and the characteristics of some of the conjugates are listed in Table 1. The ratio of protein to Vi ranged from 0.8 to 1.5. The toxicity of the CT in the conjugate was reduced 10^{3-4} fold. Molecular sizes of proteins after the SPDP-derivatization were unchanged, indicating there was no cross-linking of the protein as exemplified by Vi-TT (Figure 1). The Vi conjugates failed to enter a 7.5% polyacrylamide gel probably due to their large sizes.

Table 1. Yields and compositions of the Vi conjugates

Conjugate	Sulfhydryl/Vi wt ratio	SPDP/protein molar ratio	Conjugate protein/Vi wt ratio	% Yield
Vi-CT-II	0.5	4.5	1.5	18.0
Vi-CTB	1.5	3.8	0.82	6.4
Vi[a]-CTB	2.1	3.8	1.44	21.0

[a] Sulfhydryl groups were measured by the Ellman reaction with cysteimine as a reference
The Vi and proteins were measured by dye-binding methods. SPDP was measured by A_{343}.

Mice responded to the conjugate with higher levels of Vi antibodies than to the Vi alone (Table 2, left panel). Vi alone elicited a low response after the first injection and the antibody levels returned to pre-immune levels after the second injection. Injection of the Vi conjugates, in contrast, elicited higher levels of Vi antibodies after the first injection and a booster response after the second. Similar results were observed in juvenile monkeys (Table 3). No changes of Vi antibodies were observed in monkeys re-injected with the Vi alone. The majority of Vi antibodies induced by the Vi conjugate was of the IgG class, about 5% were of IgM and there were no detectable IgA antibodies (Figure 2). In a separate study (data not shown) we found that antibodies raised by the Vi alone as well as by the Vi conjugate failed to react with the de-O-acetylated Vi. This finding provided evidence that the conjugation procedure did not alter the structure of the Vi.

Comparable levels of CT antibodies were induced in mice by either Vi-CT conjugates or by the native CT (Table 2, right panel). Vi conjugates prepared with CT induced higher antitoxin levels than did CTB (Table 4). This suggests that the A subunit of CT is important in inducing antitoxin to both the homologous toxin and the cross-reactive LT.

S. typhimurium O-SP conjugate. The O-SP alone did not elicit antibodies in mice. This O-SP conjugated to TT ($O\text{-}SP\text{-}TT_{477}$) elicited IgM but not IgG antibodies after the first injection (Table 5). The second injection elicited a booster response of IgM antibodies (1.75 vs. 0.19 μg/ml, $p=0.001$) and low levels of IgG antibodies. Booster responses of both IgG and IgM antibodies were elicited by the third injection. There was a dose-related response to $O\text{-}SP\text{-}TT_{477}$: 12.5 μg of this conjugate elicited higher levels of both IgG and IgM antibodies ($p=0.01$) than the 2.5 μg dose.

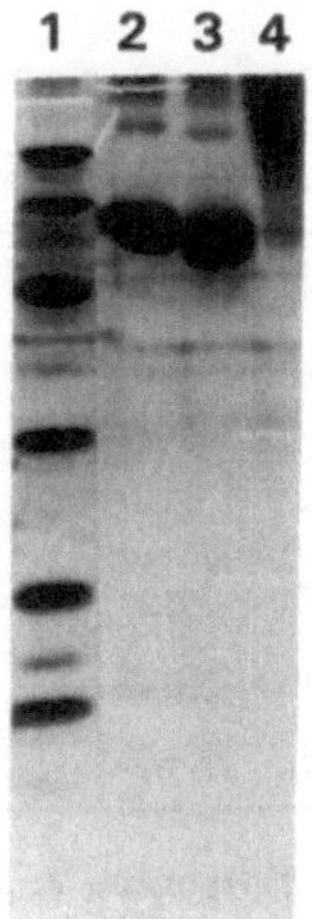

Figure 1. 7.5% SDS gel electrophoresis of tetanus toxoid (lane 2), SPDP-derivatized tetanus toxoid (lane 3), Vi-tetanus toxoid conjugate (lane 3) and standard (lane 1).

Table 2. Geometric mean serum antibody responses of BALB/c female mice injected with the Vi, CT, CTB or the Vi conjugates with either of these two proteins[a]

		Vi antibody (μg/ml) after injection:			CT antibody (10^4 U/ml) after injection:		
Immunogen	n	1	2	3	1	2	3
Vi	7	0.38[b]	0.08	0.09	0.01	<0.01	<0.01
CT	10	<0.01	<0.01	<0.01	9.5[c]	222.3[d]	978.4[e]
CTB	8	<0.01	<0.01	<0.01	7.9[c]	61.7[f]	232.7[d]
Vi-CT-11	10	2.12[g]	9.04[h]	6.28[h]	24.3[f]	927.0[e]	2,600.0[e]
Vi-CTB	10	2.73[g]	6.50[h]	3.26[h]	0.6[c]	31.3[f]	281.8[d]
Vi[s]-CTB	10	1.51[g]	0.76	1.89	1.6	33.5[f]	308.8[d]
Vi+CTB	8	0.50[b]	ND[j]	ND[j]	19.5[f]	ND[j]	ND[j]

[a] BALB/c mice were injected subcutaneously with 2.5 μg of the Vi alone, as a conjugate (Vi-CT-II, Vi-CTB, or Vi[s]-CTB), or physically mixed with 2.5 μg of CTB (Vi + CTB) at 10-day intervals. Ten mice from each group were exsanguinated after their last injection. Vi antibodies were measured by radioimmunoassay, and CT antibodies were measured by ELISA.

[b-l] Statistical significances are as follows: g versus b, P <0.0001; h versus i, P <0.01; f versus c, P <0.005; d and e versus f, P <0.001; and e versus d, P <0.01.

[j] ND, Not done.

Table 3. Geometric mean serum antibody responses of juvenile Rhesus monkeys immunized with the Vi, Vi-CT, or Vi-CTB[a]

		Vi antibody (μg/ml) after infection:				CT antibody (10^4 U/ml) after infection:			
Immunogen	n	None	1	2	3	None	1	2	3
Vi	6	0.07	0.23[b]	0.07[c]	ND	<0.01	<0.01	<0.01	ND
Vi-CT-11	8	0.08	1.05[d]	2.39[e]	ND	<0.01	0.16	2.04	ND
Vi-CTB	8	0.23	4.68[f]	6.58[g]	2.07[h]	<0.01	0.08	2.00	1.25
Vi[s]-CTB	8	0.16	1.71[i]	3.17[j]	0.76[h]	<0.01	0.16	4.28	2.65

[a] Monkeys were injected subcutaneously with 25 μg of the Vi alone or as a conjugate (Vi-CT-II, Vi-CTB, or Vi[s]-CTB), two or three times at 1-month intervals. The monkeys were bled before and 3 weeks after each injection. Vi antibodies were measured by radioimmunoassay, and CT antibodies were measured by ELISA and expressed in units. For post versus preimmunization of CT antibodies, P was <0.001 for all three conjugates. ND, not done.

[b-k] Statistical significances are as follows: b versus c, P = 0.001; e versus d, P = 0.01; g versus h, P = 0.02; j and i, P = 0.005; J versus k, P = 0.0001; f versus d, P = 0.02; e, g, and j versus c, P <0.001; and d, f, and i versus b, P <0.002.

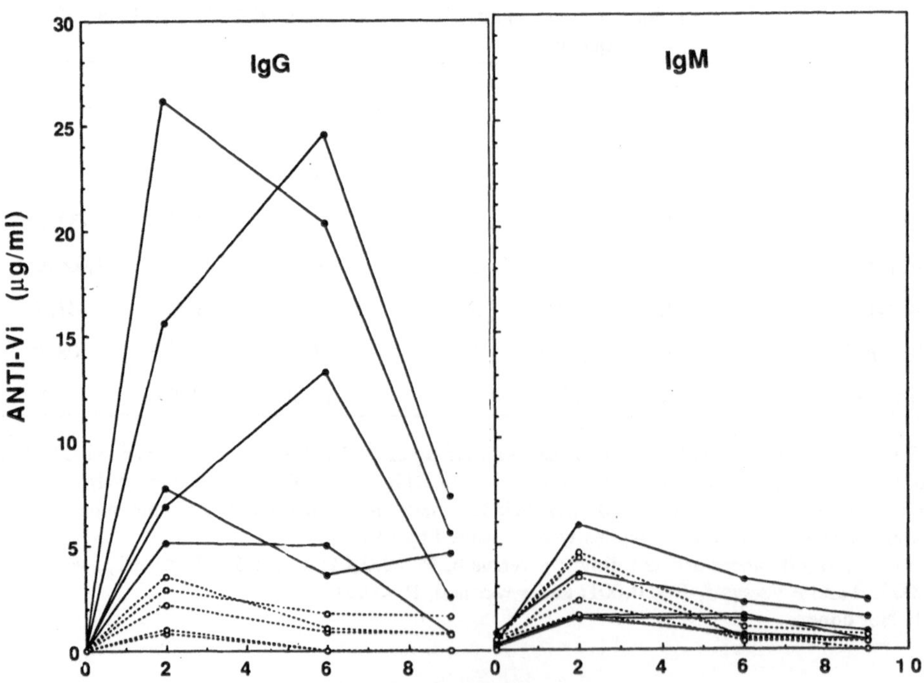

Figure 2. Serum antibody responses of juvenile Rhesus monkey injected with Vi or Vi conjugated to CTB.

Table 4. Geometric mean reciprocal serum neutralization titers against CT and the LT of *E. coli* in juvenile rhesus monkeys immunized with the Vi conjugates (n = 8 for each group)[a]

Conjugate	Titer after injection:					
	1		2		3	
	CT	LT	CT	LT	CT	LT
Vi-CT-11	0.39[b]	0.13[c]	76.1[d]	1.31[e]	ND	ND
Vi-CTB	0.003[f]	0.002[g]	6.75[h]	0.33[i]	16.0[o]	0.25[j]
Vi[a]-CTB	0.05[k]	0.03[l]	32.0[m]	2.38[n]	29.3[m]	2.83[n]

[a] The titers of preimmunization sera and those of monkeys injected with the Vi alone were 0. Juvenile rhesus monkey were injected with 5.0 μg of the Vi conjugates three times 3 weeks apart and bled 1 week before each injection and 3 weeks after the last injection. Antitoxins to CT and LT were measured by the CHO cell assay. ND, not done.
[b-n] Statistical significances are as follows: d and h versus b and f, P = 0.001; e versus c, and n versus l, P = 0.009; o and h versus f, P < 0.001.

Table 5. Serum antibodies of mice to Group B (O:4,12) after immunization with the O-specific polysaccharide of *Salmonella typhimurium* alone or conjugated to tetanus toxoid (O-SP-T_{477})

Immunogen	Dose (μg PS)	n	1st Infection IgG	1st Infection IgM	2nd Injection IgG	2nd Injection IgM	3rd Injection IgG	3rd Injection IgM	Responders IgG	Responders IgM
O-SP	2.5	10	ND[a]	ND	ND	ND	0	0	0/10	0/10
O-SP-TT_{477}	2.5	9	0	0.19	0.15	1.75	1.69	7.97	6/9	7/9
O-SP-TT_{477}	12.5	6	0.04	0.39	4.37	6.55	13.8	11.9	5/6	6/6
O-SP-TT_{477} + alum	2.5	8	0	0.29	0	0.37	0.43	1.49	6/8	7/8

[a]ND - Not done.
6-wk-old female pathogen-free CD-1 mice were injected 3 times subcutaneously biweekly with 0.1 ml of the above immunogens. Controls were injected with saline. The mice were bled before and 7 days after each subsequent injection. Antibody levels were determine by ELISA (METHODS AND MATERIALS) and are expressed as the geometric mean. Responders are defined as those mice with after the 3rd bleeding had greater than 0.4 μg IgG or 0.6 μg IgM.

Table 6. Specificity of serum antibodies in mice to the Group G LPS of *Salmonellae* (O:4,12) elicited by O-SP-TT_{477}

				Percent antibody adsorbed by cells			
Dosage			Antibody	Group B	Group D	Group C	"Rough"
(μg PS)	Ig	n	(μg/ml)	0:4,5,12	0:9,12	0:6,7	Ra LPS
2.5	IgG	5	6.09	53	17	6	14
12.5	IgG	5	30.7	46	21	22	19
2.5	IgM	8	6.54	37	0	0	0
12.5	IgM	5	15.7	50	2	6	6

Sera were from mice following their third immunization. Bacterial cells used for absorption were: Group B, *S. typhimurium* LT (0:4,5,12); Group D, *S. enteritidis* (0:9,12); Group C, *S. montevideo* (0:6,7); O-SP deficient *S. typhimurium* SL1764 (Ra mutant). The proportion of IgG and IgM antibodies removed by the absorption are presented as the arithmetic means.

Table 7. Protection against *Salmonella typhimurium* W118 in mice immunized with conjugate O-SP-TT_{477} and antibody levels in survivors

	Survivals and antibody levels[a]					
Challenge	Controls			Immunized		
dose[b]	Alive	IgG	IgM	Alive	IgG	IgM
10^2	8/8	14.7	5.18	ND	ND	ND
10^3	8/8	20.5	5.36	ND	ND	ND
10^4	5/8	76.4	4.07	9/9	118.0	11.8
10^5	1/8	1.7	1.45	9/9	65.24	18.2
10^6	0/8	-	-	9/9	41.3	14.3
10^7	ND	-	-	0	-	-

[a] Survival was recorded 21 days after challenge, at which time survivors were bled. Antibody levels are to serogroup B (O:4,12) LPS and are geometric means in μg/ml.
[b] Mice were challenged with indicated dose of *S. typhimurium* W118, 7 days after third dose of O-SP-TT_{477}.
6-week-old female CD-1 pathogen-free mice were immunized 3 times with 2.5 μg PS/dose of O-SP-TT_{477} (METHODS AND MATERIALS) and challenged intraperitoneally with 0.2 ml of *S. typhimurium* strain W118 7 days after the 3rd injection of O-SP-TT_{477}. Controls were injected with saline. Survival was recorded 21 days and the survivals bled 35 days after the challenge and their antibodies to the Group B LPS (0:4:12) assayed by ELISA.

The specificity of 0-SP-TT_{477} conjugate-induced antibodies is shown in Table 6. Absorption with *S. typhimurium* strain TML, from which the O-SP was prepared, removed >95% antibodies. The closely related Group B LPS, represented by *S. typhimurium* strain LT2 (O:4,5,12) removed ~50% of the antibodies. Group D *Salmonella*, whose LPS contains the same trisaccharide backbone (Factor 12) as that of group B, removed only ~20% of the IgG antibodies and none of the IgM. Absorption with either Group C or with the Ra mutant, which do not have O-SPs related to the Group B LPS but share the same core region, removed little or non-detectable antibodies. These data indicate that most of the antibodies elicited by the conjugate are specific to the Group B O-SP (Factor 4).

We examined the protective effect of active immunization of mice with this conjugate by challenge with *S. typhimurium* organisms (Table 7).

Three subcutaneous injections of 2.5 μg of O-SP-TT_{477} in saline conferred protection against lethal challenge by intraperitoneal injection of *S. typhimurium* strain W118. The LD_{50} was 160-fold greater than that of the unimmunized controls. There was a sharp end point of protection in the immunized group with 9/9 mice surviving the dose of 10^6 organisms but 0/9 surviving the dose of 10^7. Among the survivors, the antibody levels were higher in the immunized group receiving 10^5 organisms than in the non-immunized group, but these differences were not significant.

CONCLUSION

Typhoid fever and non-typhoidal *Salmonellae* cause serious and frequent diseases in many countries. The Vi polysaccharide elicits protective immunity against typhoid fever. We have modified the immunologic properties of this capsular polysaccharide in order to increase its efficacy and to allow its administration to infants along with their other routinely-scheduled bacterial vaccines. We have shown that Vi-protein conjugates elicit higher levels of antibodies than Vi alone and a booster response in laboratory animals. A clinical lot of Vi conjugate was prepared with *Pseudomonas aeruginosa* recombinant exoprotein A as the carrier. The safety and immunogenicity in adult volunteers of this conjugate is being compared to those of the Vi alone. An improved scheme of synthesizing Vi conjugates has been developed in our laboratory using adipic acid dihydrazide as a spacer. Vi conjugates synthesized by this method are more immunogenic in mice than those synthesized with SPDP. Further, we have found that fruit pectin, an α-D-(1-4)-galacturonic acid, when O-acetylated up to 95% at both the C2 and C3 positions, was more antigenic than the Vi polysaccharide. We plan to investigate this product as a possible immunogen to prevent typhoid fever.

Based upon our experience with other saccharide-protein conjugates, the O-SP vaccine described in this report, is likely to elicit protective levels of antibodies in infants. Further, conjugates may be given concurrently with DTP and *H. influenzae* type b conjugate to infants (Sarnaik, et al., 1990). The composition of our conjugates can be standardized so that the potency of new lots can be predicted by laboratory assays. We plan to evaluate the efficacy of our O-SP conjugates for prevention of osteomyelitis and other systemic infections in patients with sickle cell anemia who are at high risk for diseases caused by Group B *Salmonellae*. Recently, a new method of detoxifying LPS with hydrazine, an organic base, was developed in our laboratory. The base-detoxified LPS retained a larger portion of the core region. An improved O-SP conjugate using this new method will be investigated.

Whereas there is evidence for the protective immunity to *S. typhimurium* in mice conferred by serum 0-SP antibodies, there is no direct evidence for the host component that confers disease-acquired type-specific protection against shigellosis in humans (Cohen et al., 1988). There are similarities between the clinical signs, pathogenesis and protective immunity *Shigellae* and *Salmonellae* which we feel can be extended to vaccine development for shigellosis. Two findings provide information on this subject. The first is derived from studies of recruits in the Israeli army which show a correlation between serum IgG LPS antibodies and resistance to Shigellosis (Cohen et al., 1988; Cohen et al., 1991). The second is that convalescence from shigellosis confers type-specific protective immunity albeit incomplete and of limited duration. We propose that the O-SP of non-typhoidal *Salmonellae* and *Shigellae* are both virulence factors and protective antigens (Chu et al., 1991; Robbins et al., 1992). We have started a program to develop parenterally-injected O-SP conjugate vaccines, prepared from *Shigella dysenteriae* type 1, *S. flexneri* type 2a and *S. sonnei*. These O-SP conjugates have been shown to be safe and immunogenic in adult volunteers (Chu et al., 1991). Effectiveness trials with these O-SP conjugates are planned.

REFERENCES

Acharya, I.L., Lowe, C.U., Thapa, R., Gurubacharya, V.L., Shrestha, M.B., Cadoz, M., Schulz, D., Armand, J., Bryla, D. A., Trollfors, B., Cramton, T., Schneerson, R. and Robbins, J. B., 1987, Prevention of typhoid fever in Nepal with the capsular polysaccharide of *Salmonella typhi*, *N. Eng. J. Med.* 317:110.

Carlin, N.I., Svenson, S.B., and Lindberg, A.A., 1987, Role of monoclonal O-antigen antibody epitope specificity and isotype in protection against experimental mouse typhoid, *Microbial Path.* 2: 171.

Carlsson, J., Drevin, H., and Axen, R., 1978, Protein thiolation and reversible protein-protein conjugation.N-succinimidyl-3-(2-pyridyldithio)propionate, a new heterobifunctional reagent, *Eur. J. Biochem.* 173:723.

Chu, C., Liu, D., Watson, D.C., Szu, S.C., Bryla, D.A., Shiloach, D.J., Schneerson, R., and Robbins, J.B., 1991, Preparation, characterization and immunogenicity of conjugates composed of the O-specific polysaccharide of *Shigella dysenteriae* type 1 (Shiga's bacillus) bound to tetanus toxoid, *Infect. Immun.* 59:4450.

Cohen D., Green M.S., Block C., Rouach T., Ofek I., 1988, Serum antibodies to lipopolysaccharide and natural immunity to shigellosis in an Israeli military population, *J. Infect. Dis.* 157:1068.

Cohen, D., Green, M.S., Block, C., Slepon, R., and Ofek, I., 1991, A prospective study on the association between serum antibodies to lipopolysaccharide and attack rate of shigellosis, *J. Clin. Microbiol.* 29:386.

Colwell, D.E., Michalek, S.M., Briles, D.E., Jirillo, E., and McGhee, J.R., 1984, Monoclonal antibodies to *Salmonella* lipopolysaccharide: anti-O-polysaccharide antibodies protect C3H mice against challenge with virulent *Salmonella typhimurium*. *J. Immunol.* 133:950.

Douglas, R.M., Paton, J.C., Duncan, S.J., and Hansman, D.J., 1983, Antibody response to pneumococcal vaccination in children younger than five years of age, *J. Infect. Dis.* 148:131.

Goebel, W.F., and Avery, O.T., 1929, Chemo-immunological studies on conjugated carbohydrate protein. I. The synthesis of p-aminophenol-glucoside, p-aminophenol-galactoside and their coupling with serum globulin, *J. Exp. Med.* 34:521.

Heyns, K., and Kiessling, G., 1967, Strukturaufklarung des Vi-antigens aus *Citrobacter freundii (E. coli)* 396/38, *Carbohydr. Res.* 3:340.

Klugman, K.P., Koornhof, H.J., Gilbertson, I.T., Robbins, J.B., Schneerson, R., Schulz, D., Cadoz, M., Armand, J., Vaccine Advisory Committee, 1987, Protective activity of Vi capsular polysaccharide vaccine against typhoid fever, *Lancet* ii:1165.

Liang-Takasaki, C-J., Mäkelä, P.H. and Leive, L., 1982, Phagocytosis of bacteria by macrophages: changing the carbohydrate of lipopolysaccharide alters interaction with complement and macrophages, *J. Bact.* 128:1229.

Meadow, W.L., Schneider, H., and Beem, M.O., 1983, *Salmonella enteritidis* bacteremia in childhood. *J. Infect. Dis.* 152:185.

Roantree, R.J., 1967, *Salmonella* O antigens and virulence, *Ann. Rev. Microbiol.* 21:443.

Robbins, J.B., Chu, C., and Schneerson, R., 1992, Hypothesis for vaccine development: Protective immunity to enteric diseases caused by non-typhoidal *Salmonellae* and *Shigellae* is conferred by serum IgG antibodies to the O-specific polysaccharide of their LPS. *Clin. Infect. Dis.* 15:346.

Sarnaik, S., Kaplan, J., Schiffman, G., Bryla, D., Robbins, J.B., Schneerson, R., 1990, Studies on pneumococcus vacine alone or mixed with DTP and on pneumococcus type 6B and *Haemophilus influenzae* type b capsular polysacchaide-tetanus toxoid conjugates in 2- to 5-year-old children with sickle cell anemia, *Pediatr. Infect. Dis.* 9:181.

Saxen, H., Mäkelä, O., and Svenson, S.B., 1984, Isotype of protective anti-*Salmonella* antibodies in experimental mouse salmonellosis, *Infect. Immun.* 44:633.

Schneerson, R., Barrera, O., Sutton, A., Robbins, J.B., 1980, Preparation, characterization and immunogenicity of *Haemophilus influenzae* type b polysaccharide-protein conjugates. *J. Exp. Med.* 114:361.

Svenson, S.B., and Lindberg, A.A., 1981, Artificial *Salmonella* vaccine: *Salmonella typhimurium* O-antign-specific oligosaccharide-protein conjugates elicit protective antibodies in rabbits and mice, *Infect. Immun.* 32:490.

Svenson, S.B., Nurminen, M., and Lindberg, A.A.. 1979, Artificial *Salmonella* vaccine: O-antigen oligosaccharide-protein conjugates induce protection against infection with *Salmonella typhimurium*, *Infect. Immun.* 25:863.

Szu, S.C., Li, X., Schneerson, R., Vickers, J., and Robbins, J.B., 1989, Comparative immunogenicities of Vi polysaccharide-protein conjugates composed of cholera toxin or its B subunit as a carrier bound to high or lower molecular weight Vi, *Infect. Immun.* 57:3832.

Szu, S.C., Stone, A.L., Robbins, J.D., Schneerson, R., and Robbins, J.B., 1987, Vi capsular polysaccharide-protein conjugates for prevention of typhoid fever. *J. Exp. Med.* 166:1510.

Watson, D.C., Robbins, J.B., and Szu, S.C., 1992, Protection of mice against *Salmonella typhimurium* with O-specific polysaccharide-protein conjugate vaccine. *Infect. Immun.*, 60:4679.

Westphal, O., and Jann, K., 1965, Bacterial lipopolysaccharide extraction with phenol:water and further application of the procedure. *Meth. Carbohydr. Chem.* 5:83.

Wilkinson, S.G., 1977, Composition and structure of bacterial lipopolysaccharides. Ed. Sutherland. Surface carbohydrates of the procaryotic cell. Chapter 4. Academic Press, NY (1977).

DNA-BASED METHODS FOR DETECTION OF SALMONELLA ENTERICA

John E. Olsen[1], Søren Aabo[1], Lone Rossen[2],
Pernille D. Sørensen[2] and Ole F. Rasmussen[2]

[1]Deparment of Veterinary Microbiology, RVAU
Frederiksberg C., DK-1870 Denmark
[2]Biotechnological Institute
Lyngby, DK-2800 Denmark

INTRODUCTION

Salmonellosis is a world wide problem, and tests are performed routinely in order to prevent the spread of the disease through contaminated food. There has been an obvious need to develope rapid, specific and sensitive screening methods for use in the control of salmonellosis.

In recent years, DNA based detection methods (mixed- and single phase hybridization and polymerase chain reaction (PCR)) have become avaible methods used in direct detection of pathogenic bacteria in clinical-, food-, and environmental samples, and for rapid confirmation of traditionally cultured bacteria.

The present paper gives a brief review of reports dealing with DNA-based detection of *Salmonella enterica* and describes the development of a *Salmonella* specific PCR-assay based on the sequence of a *Salmonella* specific DNA fragment reported by Olsen and colleagues (1991).

HYBRIDIZATION METHODS FOR SALMONELLA DETECTION

An overview of key reports on DNA based methods for *Salmonella* detection is given in Table 1.

Three years after Moseley et al. (1981) first reported the detection of enterotoxigenic *Escherichia coli* by colony hybridization, Fitts et al. (1983) demonstrated randomly isolated DNA fragments that were specific for *Salmonella* and could be used to confirm the presence of salmonellae in pre-enrichment broths. The fragments had been selected from a DNA-libary based on the absence of hybridization to *E.coli*, and they were evaluated by hybridization to strains of 180 *Salmonella* serovars (Fitts 1985). The use of the fragments was commercialized for culture confirmation by the company GeneTrak (Flowers et al. 1987a), and the method received AOAC-approval, following a collaborative testing where the presence of *Salmonella* in enrichment broths was

Biology of Salmonella, Edited by F. Cabello *et al.*,
Plenum Press, New York, 1993

Table 1. Overview of DNA based methods for detection of *Salmonella.*

Reference	Assay characteristics		
	Probe molecule	Size(kb)	Company
Fitts et al. (1983)	Random cloned fragments of *S.* Typhimurium	3.6-6.0	GeneTrak
Gopo et al. (1988)	Random cloned fragment of *S.* Typhimurium	1.8	-
Wilson et al. (1990)	rRNA-derived oligo nucleotides	-	GeneTrak
Scholl et al. (1990)	Random cloned fragment of *S.* Enteritidis	1.6	-
Tsen et al. (1991)	Oligonucleotides based on *S.* Typhimurium sequence	-	-
Olsen et al. (1991)	Random cloned fragment of *S.* Typhimurium	2.3	-

detected equally well by DNA-hybridization and by traditional culture methods (Flowers et al. 1987b).

Gopo and colleagues (1988) introduced a biotin labelled polynucleotide probe that could by used to confirm the presence of *Salmonella* in broth cultures. The probe was common to all serovars tested and did not hybridize to non-*Salmonella* bacteria. The biotin-label used gave identical results to radiolabelling but with a an increased background signal.

Two more polynucleotide probes from the chromosome of *S.* Typhimurium has been reported to be *Salmonella* specific: Tsen and colleagues (1989, cit in Tsen et al. 1991) reported on the possible use for diagnostic purposes of a randomly cloned DNA fragment from *S.* Typhimurium. The sequence of this fragment has been published, and a number of oligo nucleotides derived from the sequence have been tested for *Salmonella* sensitivity and specificity (Tsen et al. 1991). Three oligo nucleotides have proved useful for *Salmonella* detection. The 2.3 kb *S.* Typhimurium fragment isolated by Olsen et al. (1990) has also been sequenced (Aabo et al. unpublished), and oligo nucleotides have been derived. The possible use of these oligo nucleotides as PCR primers is described below.

Finaly, Scholl et al. (1990) have isolated a 1.6 kb fragment from the chromosome of *S.* Enteritidis and used this to detect *Salmonella* in clinical samples using a novel sample processing technique.

A second generation, commercial *Salmonella* detection assay has been extensively tested on pure cultures and released by the company GeneTrak (Wilson et al. 1990). This assay is based on single phase hybridization of oligo nucleotides to rRNA genes and rRNA of the ribosomes, and the method confirms the presence of salmonellae in less than 48 hours. *S.* subsp. *bongori* (ssp. V) was not detected by the initial oligo nucleotides used, but is now covered by the assay. The assay has received AOAC approval based on a collaborative study (Curiale et al. 1990a) and in comparison with culture techniques and immunogenic detection, the DNA detection method was found to perform equally well both on pure cultures (Curiale et al. 1990b) and on innoculated and naturally contaminated food samples (Beumer et al. 1991).

DEVELOPMENT OF A SALMONELLA SPECIFIC PCR-ASSAY

The performance of the probe fragment published by Olsen et al. (1991) has been evaluated by hybridization to pure cultures of bacteria and to naturally contaminated samples (Olsen et al. 1991, Aabo et al. 1992b). The probe has not hybridized to 178 non-*Salmonella* strains of 52 species belonging to 23 genera of the family *Enterobacteriaceae*, so far tested, nor has it hybridized to the full indigenous flora of fecal- and food samples, and it is considered to be *Salmonella* specific.

As seen from Table 2, the fragment is well conserved in the genus, as the probe hybridizes to all 393 strains of 214 serovars of *Salmonella* tested. The probe reacted with all of the seven *Salmonella* subpecies, and the entire fragment seemed to be common to *Salmonella*, as two subfragments of 1.8 kb and 0.8 kb, similary, hybridized to all tested strains (Aabo et al. 1992b).

Table 2. Hybridization of the fragment isolated by Olsen et al. (1991) to pure cultures of *Salmonella*.

Salmonella subspecies	No. of serovars		No. of strains	
	No. tested	No. positiv	No. tested	No. Positive
S. supsp. *enterica*	142	142	308	308
S. supsp. *salamae*	26	26	39	39
S. subsp. *arizonae*	29	29	29	29
S. subsp. *houtenae*	15	15	15	15
S. supsp. *bongoria*	1	1	1	1
S. subsp. *indica*	1	1	1	1

Results based on Olsen et al. (1991) and Aabo et al. (1992).

The fragment has been sequenced (Aabo et al., unpublished), and eight oligonucleotides distributed over the fragment (Figure 1) were synthesized. The specificity and sensitivity of these oligo nucleotides were analysed by dot blot hybridizations to 15 *Salmonella* and 15 non-*Salmonella* strains. None of the oligo nucleotids hybridized to all of the *Salmonella* strains tested, but a *Salmonella* specific region was demonstrated (Aabo et al. 1992a). Part of this region, between the primers ST3 and ST7 (Figure 1) has been sequenced in 20 strains of 20 different serovars of *Salmonella*. Minor sequence variations were observed between strains but two primer sequences, ST11 (24mer) and ST15 (25mer) were conserved in all strains (Aabo et al. 1992a).

A PCR assay, employing these two primers and using an annealing temperature of 57°C specifically amplifies a 429 basepair product in strains of *Salmonella* (Aabo et al. 1992a). The assay has so far been evaluated by testing pure cultures of 146 *Salmonella* and 86 non-*Salmonella* strains. The results of these tests are shown in Table 3. The PCR-assay detected all but 2 of the 146 *Salmonella* strains tested, and none of the 86 non-*Salmonella* strains. Both false-negative strains belonged to mono phasic *S. arizonae* (ssp. IIIa) (Aabo et al. 1992a). Mono phasic *S. arizonae* are very rarely isolated

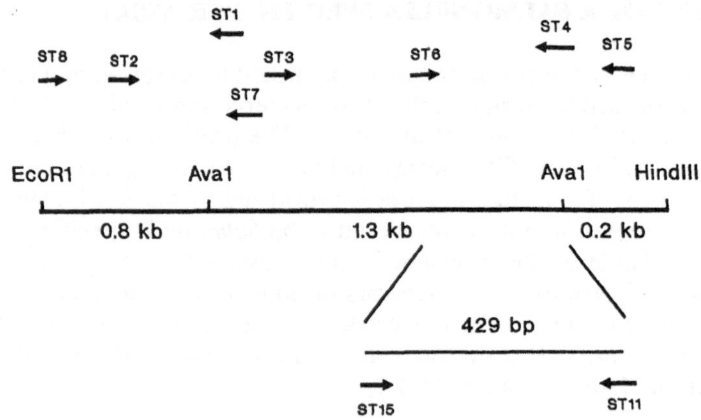

Figure 1. Schematic presentation of the 2.3 kb probe fragment of Olsen et al. (1991) showing key restriction sites, the position of the 8 oligo nucleotides originally tested and the two primers (ST11 and ST15) used for PCR-amplification.

from human cases of salmonellosis (Rowe and Hall, 1992), and the detection of these strains is less relevant for most purposes.

The assay has been preliminary tested for food quality assurance using an assay construction with a short 4 hours pre-enrichment in peptone water followed by enrichment in selinite broth for 16 hours, and *Salmonella* is detected by this method in app. 24 hours. The assay has a detection limit in the area of $5x10^3$ bacteria per ml. (Sørensen et al., unpublished) but work in ongoing with the dual aim of minimizing both the detection limit and the detection time. It is, however, not anticipated that the PCR assay can be used for detection of *Salmonella* in food samples without a preceding enrichment by culture procedures.

Table 3. Evaluation of a *Salmonella* specific PCR-assay by testing of pure cultures of bacteria.

Subspecies	No. tested		No. of Positive	
	Strains	Serovars	Strains	Serovars
S. subsp. *enterica*	95	69	95	69
S. subsp. *salamae*	23	21	23	21
S. subsp. *arizonae*	18	18	16	16
S. subsp. *houtenae*	8	8	8	8
S. subsp. *bongori*	1	1	1	1
S. subsp. *indica*	1	1	1	1
Non-*Salmonella*	86	-	0	-

Results based on Aabo et al. (1992a). The assay conditions were: 5 µl of an overnight broth culture was lysed at 94°C for 10 minutes in 50 µl 0.25% SDS 0.05 M NaOH. 5 µl of this was mixed with 100 µl PCR-buffer (Rossen et al. (1991) and was run for 35 cycles: 55°C 30", 72°C 1', 94°C 30". Finaly, the product was kept at 72°C for 10' before the PCR-product was analysed on a 1.5% agarose gel in TAE buffer.

REFERENCES

Aabo S., Rossen L., Sørensen P., Rasmussen O.F. and Olsen J.E. (1992a). Salmonella detection by polymerase chain reaction. Submitted (*Mol. Cell. Probes*).

Aabo S., Thomas A., Hall M.S., Smith H.W. and Olsen J.E. (1992b). Evaluation of a *Salmonella* specific DNA fragment by hybridization to pure cultures using radiolabelled and digoxygenin labelled probes. *APMIS* (in press).

Beumer R.R., Brinkman E. and Rombouts F.M. (1991). Enzyme-linked immunoassays for the detection of *Salmonella* spp.: a comparison with other methods. *Int. J. Food. Microbiol.* 12: 363-274.

Curiale K., Klatt M.J. and Mozola M.A. (1990a). Colorimetric deoxyribonucleic acid hybridization assay for rapid screening of *Salmonella* in food. *J. Assoc. Off. Anal. Chem.* 73: 248- 256.

Curiale K., Mciver D., Weathersby S. and Planer C. (1990b). Detection of salmonellae and other *Enterobacteriaceae* by commercial dexoyribonucleic acid hybridization and enzyme immunoassay kits. *J. Food Protech.* 53: 1037-1046.

Fitts R., Daimond M., Hamilton C. and Neri M. (1983). DNA-DNA hybridization assay for detection of *Salmonella* in foods. *Appl. Environ. Microbiol.* 46:1146-1151.

Fitts R. (1985). Development of a DNA-DNA hybridization test for the presence of *Salmonella* in foods. *J. Food Tech.* 39: 95-102.

Flowers R.S., Klatt M.J., Mozola M.A., Curiale M.S., Gabis D.A. and Siliker J.H. (1987a). DNA hybridization assay for detection of *Salmonella* in foods: collaborative study. *J. Assay. Off. Anal. Chem.* 70: 521-529.

Flowers R.S., Mozola M.A., Curiale M.S., Gabis D.A. and Siliker J.H. (1987b). Comparative study of a DNA hybridization method and the conventional culture procedure for detection of *Salmonella* in foods. *J. Food. Science* 52: 781-785.

Gopo J.M., Melis R., Filipska E., Meneveri R. and Filipska J. (1988). Development of a *Salmonella* specific biotynylated DNA probe for rapid routine identification of *Salmnoella*. *Mol. Cell. Probes.* 2: 271-279.

Moseley S.L., Hug I., Alim A.R.M.A., So M., Samadpour-Motalebi M. and Falkow S. (1981). Detection of enterotoxigenic *Escherichai coli* by colony hybridization. *J. Infect. Dis.* 142: 892-898.

Olsen J.E., Aabo S., Nielsen E.O and Nielsen B.B (1991). Isolation of a *Salmonella* specific DNA hybridization probe. *APMIS* 99: 114-120.

Rossen L., Holmstrøm K., Olsen J.E. and Rasmussen O.F. (1991). A rapid polymerase chain reaction based assay for the identification of *Listeria monocytogenes* in food samples. *Int. J. Food. Microbiol.* 14: 145-152.

Hall M.L.M. and Rowe B. (1992). *Salmonella arizonae* in the United Kingdom from 1966 to 1990. *Epidem. Infect.* 108: 59-65.

Scholl D.R., Kaufmann C., Jollick J.D., York C.K., Goodrum G.R. and Charache P. (1990). Clinical application of novel sample processing technology for the identification of salmonellae by using DNA probes. *J. Clin. Microbiol.* 28: 237-241.

Tsen H.-Y., Wang S.J., Roe B.A. and Green S.S. (1991). DNA sequence of a *Salmonella* specific DNA fragment and the use of oligonucleotide probes for *Salmonella* detection. *Appl. Microbiol. Biotechnol.* 35: 339-347.

Wilson S.G., Chan S., Deroo M., Vera-Garcia M., Johnson A, Lane D. and Halbert D.N. (1990). Development of a colorimetric, second generation nucleic acid hybridization method for detection of *Salmonella* in foods and a comparison with conventional culture procedure. *J. Food. Science.* 55: 1394-1398.

EFFICACY OF CARUMONAM TREATMENT FOR THE ERADICATION OF SYSTEMIC SALMONELLOSIS IN MICE

Letterio Bonina, Adriana Arena, Rosario Trifiletti, Pietro Mastroeni, and Daniela Iannello
Microbiology Institute, Medical School, Messina University
Piazza xx Settembre, 4
98100 Messina, Italy

Antibiotic therapy is an effective tool in the treatment of typhoid fever. Antibacterial agents that can efficiently penetrate within eucaryotic cells are likely to be more effective against intracellular bacterial parasites than molecules that can exert their activity only in the extracellular compartment. We reported that early treatment of *S. typhimurium* infected mice with different antibiotics led to the observation that the most effective antibacterial molecules are the ones that penetrate better into the phagocytic cells (Bonina et al 1990; Mastroeni et al 1984). Clearance of the bacteria from the RES was obtained with some but not all the antibiotics tested. Late administration of several antibiotics (after the suppression of bacterial exponential growth in the RES) has proved to be unable to eradicate the organisms from the RES and the reasons for these failures are largely unknown. In previous experiments we investigated the efficacy of short treatment with one monocyclic (aztreonam) and three bicyclic (ampicillin, cefazolin, ceftazidime) beta-lactam antibiotics, in controlling salmonella systemic infections. When the treatment was commenced early in the infection (exponential growth phase), we showed that the monobactam aztreonam was far more active than the other antibiotics both in innately susceptible BALB/c (*Ity* ^{S}) and resistant CBA (*Ity* ^{r}) mice (Bonina et al 1990).

The present study investigated the ability of the monobactam carumonam to eradicate a salmonella infection in BALB/c mice when the treatment is commenced in the later stages of the infection, after the establishment of the plateau. This experimental approach was chosen in the attempt to investigate the ability of the antibiotic treatment to eradicate a carrier state.

Biology of Salmonella, Edited by F. Cabello *et al.*,
Plenum Press, New York, 1993

S. typhimurium M525 has been described (Hormaeche 1979). Bacteria were grown overnight in stationary cultures in tryptic soy broth (Difco, Detroit, Michigan, USA) and snap frozen in liquid nitrogen. For inoculation, one vial was rapidly thawed, and appropriate dilutions in sterile PBS were made. The size of the inoculum was checked by pour plating in TSA (Difco, Michigan, USA). BALB/c mice were purchased from Charles River, Como, Italy. Animals of either sex, eight weeks old, were intravenously injected with $2x10^2$ *S. typhimurium* M525 in a lateral tail vein in 0.125 ml PBS. The monobactam carumonam (Bracco, Milan, Italy) was dissolved in sterile saline. The MIC of carumonam for *Salmonella typhimurium* M525 was 0.125 mg/l as determined by the broth dilution test (Washington, 1980). Carumonam was administered as a single daily ip dose. In all experiments, the treatment was started on day 8 and stopped on day 21 after bacterial challenge. Groups of four mice were killed by cervical dislocation before their daily dose of antibiotic. Livers and spleens were homogenised separately in distilled water and 0.1 ml aliquots of suitable dilutions were plated in duplicate on nutrient agar. Results are expressed as geometric mean $\log_{10}$ cfu (colony forming units) per whole organ $\pm$ standard deviation.

BALB/c mice were injected i.v. with *S. typhimurium* M525. The infection evolved as expected with the organisms growing in the spleen and liver at about 0.5 log a day. Bacterial growth in the RES was suppressed towards the end of the first week resulting in the expected plateau phase with counts that remained virtually constant throughout the experiment. The antibiotic therapy was commenced on day 8 and stopped on day 22. The table shows that by day 15 (one week of treatment) viable counts in spleens and livers of treated mice were significantly lower than in controls. By day 22 counts in treated mice showed a further decrease and were dramatically lower than in controls. At this time all the mice tested still carried low numbers of organisms in the RES and complete eradication was not achieved yet. Counts performed on days 29, 36 and 43 showed that no relapse of the infection occurred. Moreover no viable bacteria could be recovered in 2 out of 4 mice on day 29 and in 3 out of 4 mice on day 36, despite the treatment having been stopped on day 22. No bacteria were present either in the spleen or in the liver of any of the 4 mice sacrificed on day 43.

Thus, the administration of carumonam reduces dramatically the bacterial load in the RES also when the therapy is started later in the infection (after the establishment of the plateau). Surprisingly the eradication of the infection was obtained only long after the cessation of the therapy. The reasons for the superior efficacy of carumonam is not clear. It is likely that the *in vivo* pharmacokinetics of the monobactam positively influence its intracellular distribution and antibacterial activity. In fact, unlike other beta-lactam antibiotics, carumonam can efficiently penetrate into phagocytic cells by a dose dependent and energy-requiring process. The reasons for the marked efficiency of carumonam are still under investigation and the possible contribution of host factors in the complete clearance of the organisms from the RES will be evaluated.

TABLE 1
LOG_{10} COLONY FORMING UNITS OF SALMONELLA TYPHIMURIUM M525 IN LIVERS AND SPLEENS OF BALB/c MICE TREATED WITH CARUMONAM

Days after challenge	Untreated mice Livers	Untreated mice Spleens	Treated mice Livers	Treated mice Spleens
6	5.5±0.2	4.8±0.2		
8	5.2±0.2	4.9±0.2		
15	5.2±0.3	4.9±0.5	3.4±0.5	2.7±0.3
22	5.3±0.3	5.4±0.1	1.0±0.09	0.6±0.06
29	5.4±0.3	5.0±0.2	0.3±0.4	0.2±0.2
36	5.3±0.5	5.2±0.2	0.1±0.3	0.07±0.15
43	5.3±0.3	5.3±0.2	none	none

Data are expressed as the geometric mean ± standard deviation of viable counts/organ^{-1}.

REFERENCES

Bonina, L., Carbone, M., Matera, G., Teti, G., Joysey, S.H., Hormaeche, C.E., Mastroeni, P. 1990: Beta-lactam antibiotics (aztreonam, ampicillin, cefazolin, and ceftazidime) in the control and eradication of *Salmonella typhimurium* in naturally rresistant and susceptible mice. J. Antimicr. Chemother. 25: 813.

Hormaeche, C.E.,1979, The natural resistance of radiation chimeras to *S. typhimurium* C5. Immunology, 37:329.

Mastroeni, P., Bonina, L., Leonardi, M.S., Carbone, M., Berlinghieri, M.C.1984, Study of aztreonam intracellular kinetics and its relationships with different cell types. In Proceedings of the IV Mediterranean Congress of Chemotherapy, Rhodos, Greece, 345.

Washington, J.A., Sutter, V.L.,1980, Dilution susceptibility test: agar and macro-broth dilution procedures. In Manual of Clinical Microbiology, 3rd edn (Lennette, E.H., Balows, A., Hausler, W.J., Truant, J.P. Eds), pp 453-458. ASM, Washington, DC.

DESTRUCTION OF *SALMONELLA* BY OSMOTIC SHOCK AND LYSOZYME

Athina Chatzopoulou[1] and R.J. Miles[2]

[1] Department of Health, Eileen House, 80-94 Newington Causeway, London SE1 6EF, U.K.
[2] Microbial Physiology Research Group, Division of Biosphere Sciences, King's College London, Campden Hill Road, London W8 7AH, U.K.

ABSTRACT

Lysozyme is generally ineffective against Gram-negative bacterial cells, as the peptidoglycan layer of the cell wall is protected by a lipid-containing outer membrane. A number of procedures may disrupt the outer membrane sufficiently to allow access of lysozyme to peptidoglycan. These procedures generally involve the use of toxic chemicals or extreme physical conditions (e.g. of pH or temperature). However, we have developed procedures whereby osmotic and/or cold shock treatments may render Gram-negative cells, including *Salmonella* species, susceptible to lysozyme. These procedures are potentially applicable to the treatment of meats, poultry carcasses, eggs and vegetables. In addition, they are simple, relatively rapid and inexpensive, using crude egg white solution as the lysozyme source.

In the application of these procedures to food treatment we envisage a system whereby foods are either immersed in, or sprayed with a hypertonic saline solution and subsequently washed in water containing lysozyme/egg white. In laboratory experiments, the viable count of *Salmonella* cell suspensions were reduced by up to 99.999% when: (i) cells were incubated at 20-37 °C, for 10 min in nutrient media containing 0.8M NaCl and subsequently diluted (1:10) with a solution of 10ug/ml lysozyme (or 0.5mg/ml pasteurised egg white) ; or (ii) cells were incubated at 0°C for 1 min in 0.8M NaCl and similarly diluted. The extent of killing of *Salmonella* on artificially contaminated chicken skin and red meat were not as high as those for the same bacteria in suspension. However, cell kills of more than 90 % were reproducibly obtained.

Biology of Salmonella, Edited by F. Cabello *et al.*,
Plenum Press, New York, 1993

HUMAN VARIATIONS IN SUSCEPTIBILITY TO INFECTION WITH *S. TYPHI*: EVIDENCE FROM THE DISTRIBUTION OF INCUBATION PERIODS IN SINGLE-EXPOSURE EPIDEMICS

Jaap T. van Dissel and Ralph van Furth
Department of Infectious Diseases
University Hospital Leiden
P.O. BOX 9600
2300 RC Leiden, The Netherlands

INTRODUCTION

Studies in inbred strains of mice have led to the identification of an increasing number of genes that control an infection by *S. typhimurium*. Innate resistance seems to be regulated by a single, autosomal dominant, non-H-2 gene designated Ity (Plant and Glyn, 1976). Studies suggest that resident macrophages, exudate granulocytes and exudate macrophages, i.e., monocytes, are the effec-tor cells that express the Ity gene (van Dissel et al, 1987). In selected strains of mice, other resistance genes have been identified as well (O'Brien, 1986). In humans, only indirect evidence for the action of non-HLA-resistance genes is available; an approach used in the search for genetic determinants of susceptibility to infection with *S. typhi* was to study the modes of the epidemic curve in water- and foodborne infections (Naylor, 1983).

In common-vehicle, single-exposure epidemics, the main factors determining the in-cubation period are the infecting number and virulence of the micro-organisms, and the susceptibility of individual members of the exposed population (Frost, 1928; Sartwell, 1966). Variations of the latter would result in a multimodal distribution of incubation periods, defined as the interval between exposure to the micro-organisms and the onset of symptoms of disease caused by these bacteria. Obviously, such variations can only be expressed at intermediate attack rates: variations in susceptibility will not affect the distribution of the epidemic curve at very low attack rates -only susceptible individuals become ill which gives an unimodal epidemic curve- or at very high attack rates distribution of incubation periods is to narrow to differentiate between susceptible and resistant individuals.

In common-vehicle, single-exposure epidemics that occur in a homogeneous population, the distribution of onset dates around the mean should be skewed to the right

Biology of Salmonella, Edited by F. Cabello *et al.*,
Plenum Press, New York, 1993

(tail off towards the longer times) and comply to certain characteristics that depend on the infecting dose (Saaty, 1961; Shortley, 1965; Bailey, 1975). Thus, variability in susceptibility of individuals in the population exposed to *S. typhi* would be expected to cause a deviation of the observed epidemic curve from the theoretical distribution of incubation periods.

In an attempt to identify genetic determinants of susceptibility to infection with *S. typhi* in humans, we investigated *S. typhi* outbreaks for characteristics of the distribution of incubation periods.

MATERIAL AND METHODS

An extensive literature search was done in an effort to identify epidemics of typhoid fever reported in scientific journals from the start of this century. Only those articles were selected for the statistical analysis in which strong-though sometimes circumstantial-evidence of a single episode of exposure with a common vehicle was presented, along with a distribution of incubation periods, e.g., in a Table or comprehensive graph, and figures indicating the total population exposed and the number of persons that became ill. Characteristics of the distribution curve of dates of onset of the first symptoms of disease, i.e., mean, median, modes, skewness, variance, log-normality, were determined by the Statistical Package for Social Sciences (SPSS PC^{+}) program.

RESULTS

In 33 epidemics between 1908 and 1985, the dates of exposure and onset of symptoms of typhoid fever had been accurately recorded; 16 of these 33 were waterborne epidemics, the others were either foodborne (14) or milkborn (3).
In 32 epidemics including a total of 3461 persons with typhoid fever, on average 105 (median 58) persons became ill; one additional large epidemic involved 2021 persons. The population exposed was approximately known in 25 epidemics, allowing the calculation of attack rates that ranged from less than 1 percent to just over 90 percent (median 14.5 percent). These attack rates probably reflect the size of the infecting number of micro-organisms.

In most cases, the distributions of the incubation period took the form of a lognormal curve, with a range of 2 to 50 days. Skewness to the right was proportional to the attack rate and was more pronounced at the highest rates. Secundary cases rarely occur in the single-exposure epidemics, i.e., they constitute always less than five percent of all cases. In general, an extensive search led to the identification of an asymptomatic carrier that caused contamination of food or water.

The mean incubation period amounted to 16.3 days (range 7.2-22.7 days; median 18.1 days; Fig 1). In 12 epidemics with attack rates less than 20 percent, the mean incubation period was similar ($p<0.05$) and amounted to 20 days (SD 2.2 days). In epidemics with attack rates from 20 to 90 percent, the mean incubation period was inversely proportional to the attack rate and could satisfactory be described (r^2 0.85; $p<0.001$) by the equation:

$$\text{mean incubation period (days)} = 20.3 - 0.174 \times \text{attack rate (\%)}.$$

Five epidemics with intermediate attack rates showed a multimodal epidemic curve, with modes respectively at 6 to 10 and 16 to 21 days. Further analysis showed that the distribution of incubation periods in these epidemics reflects a summation of at least two lognormal distributions, each describing the incubation periods of a subgroup in the population exposed to *S. typhi*.

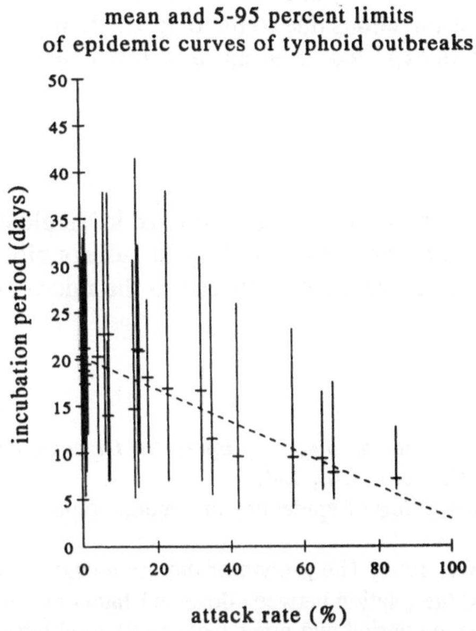

Figure 1. Interval between exposure to *S. typhi* and the onset of first symptoms of the disease, i.e., the incubation period, in relation to the attack rate, in 25 epidemics of typhoid fever. The mean incubation period, 5 and 95 percentile of the distributions of the incubation periods is given. For epidemics with attack rates less than about 20 percent, the incubation period is essentially constant, i.e., on average 20.3 days. At attack rates higher than 20 percent, the mean incubation period is inversely proportional to the attack rate.

DISCUSSION

Our data indicate that analysis of the distributions of the incubation periods can provide evidence for a variation in susceptibility to infection with *S. typhi* between individuals in the population exposed to the micro-organism. Such variation in susceptibility is evidenced by a multimodal distribution of onset dates in epidemics with intermediate attack rates, as well as other characteristics of the epidemic curves (van Dissel, 1992). The observed variation in susceptibility can, of course, be either innate or acquired, for instance by vaccination or prior infection. Because similar multimodal distribution curves have been observed in experimental *S. typhi* infections in human volunteers (Naylor, 1983), the variation in susceptibility probably is innate and results from genetic factors rather than acquired resistance.

At attack rates higher than 20 to 30 percent, the mean incubation period is inversely proportional to the attack rate. By contrast, for all epidemics with attack rates less than 20 percent, the mean incubation period was about 20 days, and the distributions of the incubations period similar. As suggested by Meynell and Meynell (1958), for infecting doses of micro-organisms much below the number required for an attack rate of 30 to 50 percent, the distributions of incubation periods take an identical form. Theoretically, it takes the form that is the same as if the infection was due to the outgrowth of only one micro-organism. When there is very little probability of symptoms due to an infection at all, i.e., reflected in low attack rates, there is a high probability that, if symptoms of infection do occur, they are due to the progeny of a single micro-organism in the infecting dose, and for infections due to the outgrowth of a single micro-organisms, the mean incubation period is expected to be more or less constant.

SUMMARY

Analysis of distributions of incubation periods in single-exposure outbreaks of typhoid fever can provide evidence for a variation in susceptibility to infection with *S. typhi* between individuals in the population exposed to the micro-organism.

REFERENCES

Bailey, N.T.J., 1975, *The Mathematical Theory of Infectious Diseases and Its Applications*. 2nd Ed. London, Charles Griffin & Co., Ltd.

Frost, W.H., 1928, Some conceptions of epidemics in general, reprinted in 1976, *Am. J. Epidem.* 103:141.

Meynell, G.G., Meynell, E.W., 1958, The growth of the microorganisms in vivo with particular reference to the relation between dose and latent period. *J. Hyg.* 56:323.

Naylor, G.R.E., 1983, Incubation period and other features of food-borne and water-borne outbreaks of typhoid fever in relation to pathogenesis and genetics of resistance. *Lancet* 1:864.

O'Brien, A.D., 1986, Influence of host genes on the resistance of inbred mice to lethal infection with *Salmonella typhimurium. Curr. Topics Microbiol. Immunol.* 24:37.

Plant, J., Glyn, A.A., 1976, Genetics of resistance to infection with *Salmonella typhimurium* in mice. *J. Infect. Dis.* 133:72.

Saaty, T.L., 1961, Some stochastic processes with absorbing barriers. *J. Roy. Statist. Soc. Ser B* 23:319.

Sartwell, P.E., 1966, The incubation period and the dynamics of infectious diseases. *Am. J. Epidem.* 83:204.

Shortley, G., 1965, A stochastic model for distributions of biological response times. *Biometrics* 21:562.

van Dissel, J.T., Stikkelbroeck, J.J.M., et al., 1987, *Salmonella typhimurium*-specific difference in rate of intracellular killing by resident peritoneal macrophages from *Salmonella*-resistant CBA and *Salmonella*-susceptible C57Bl/10 mice. *J. Immunol.* 138:4428.

van Dissel, J.T., 1992, submitted *Epidemiological Reviews*.

SEMISOLID MEDIA FOR DETECTION OF SALMONELLA SPP

Ildefonso Perales and Ana Audicana

Laboratorio de Salud Pública
Dirección de Salud de Vizcaya
Apdo. 6080. 48071 Bilbao, Spain

Salmonella isolation is a complex and multifactorial problem. Successful isolation depends on the behaviour of the different serotypes (more than 2.200 have been described), physiological state of the Salmonella cells, number of cells in the sample, type of sample, background flora, etc.

In general the easier situation is the isolation of Salmonella from faeces of a person with acute salmonellosis. In this case, direct plating on an appropiate culture medium will be enough to give isolated colonies. However, when the faeces are from a person with past salmonellosis or from an asymptomatic carrier, there is a need for an enrichment procedure that allows the detection of few cells of Salmonella in the sample.

When the isolation of Salmonella is required from environmental or food samples the difficulties are greater. In these cases is necessary to use highly selective media for inhibiting the background flora. However, as the Salmonella cells can be sublethally damaged by the environmental conditions or the processing procedures used with foods, these cells can be killed when added to selective culture media.

In order to achieve successful isolation of Salmonella from foods or environmental samples, most of the cultural methods include a secuential scheme with the following stages: 1) Pre-enrichment, 2) Selective Enrichment, 3) Selective and differential plating, 4) Identification. The final outcome depends on the combination of these four stages and the good or bad results of each step is influenced by the previous one. Each stage requires 18-24 hours, so that at least 4 days are necessary for finishing the analysis.

The procedures for the isolation of Salmonella from foods and enviromental samples are time-consuming and cumbersome. Also there is no method that allows the isolation of Salmonella with 100% efficiency; in particulor, plating media are not sufficiently selective. Most of the methods use liquid media for the pre-enrichment and enrichment, and solid media for the selective differentiation. However, the use of semisolid media based on a selective motility enrichment are an alternative for increasing the efficiency of recovery and to reduce the time necessary to achieve reliable results.

Semisolid media for the isolation of Salmonella spp. from different samples have been used for many years. Their use was increased since 1984 when Goossens et al. described a simplification of their procedure using Petri dishes. Some workers began testing their efficacy, and included them in the routine work. The method is based on the ability of Salmonella for migrating

Biology of Salmonella, Edited by F. Cabello *et al.*,
Plenum Press, New York, 1993

more rapidly than most of other bacteria, producing an enrichment and a differentation between the media that show a line of migration and the ones that not.

Semisolid media are most useful for the isolation of salmonellas from food and water samples where their isolation is otherwise particularly difficult.

At present the semisolid media most used are those derived from the Rappaport broth, made semisolid by the addition of agar. Semisolid Rappaport (SR) and modified semisolid Rappaport-Vassiliadis (MSRV) are used in different laboratories. MSRV is a very specific medium but SR is more productive and gives a better sensitivity-specificity ratio (Perales and Erkiaga,1991).

SR increases the number of positive samples compared with classical culture methods using liquid and solid media (Perales and Audicana,1989a; Perales and Erkiaga,1991). Moreover SR can substitute for both the enrichment broth and the plating medium, obtaining a reduction of one day in the time necessary for the isolation of Salmonella from foods and water (Perales and Audicana 1989b; Perales and Erkiaga,1991).

In addition, a procedure combining the use of semisolid media with membrane filtration of food samples and a latex agglutination technique, allowing the detection of Salmonella spp. from food and feed samples in 28 hours was developed in our laboratory. The method includes pre-enrichment of 25 g of sample in 225 ml of M-broth for 8 h at 35^0C. Subsequently upon sedimentation, 10 ml portions of the resuscitated macerates are filtered through Millipore HC membranes: some food suspensions require prior multi-enzyme digestion. Membranes are then inverted and transfered to the surface of two semisolid selective media: SR and a modified formula termed SRM44, i.e. the amount of tryptone was reduced to 2,67 g/l and proteose peptone (2,67 g/l) and yeast extract (1,06 g/l) were added, the concentration of magnesium chloride was increased to 21,26 g/l, the malachite green reduced to 0,021 g/l and the final pH was adjusted to 4,5 by the use of 1N HCl. The semisolid media are incubated for 14 h at 35^0C. If biomass migration is observed the most advanced portion is spotted onto the surface of tryptone soy yeast extract agar, incubated for 6 h at 42^0C and tested by Salmonella Latex Agglutination Reagent (Oxoid, UK).

This procedure allows the recovery of between 1 and 10 Salmonella cells, added to food samples, irrespective of their being fully vital or injured by heat or freezing. The productivity of the procedure was compared to that of a conventional technique (enrichment in selenite-cystine and Rappaport-Vassiliadis broths and plating onto bismuth sulfite, XLD and brilliant green agars) applied to 129 different naturally contaminated and artificially inoculated foods and feeds. Amongst the 76 samples that were found positive by any procedure SR/SRM44 detected 71 (93,4%) of the positives, and the conventional method 69 (90,7%).

REFERENCES

Goossens, H., Wauters, G., de Boeck, M., Janssens, M. and Butzler, J.P., 1984, Semisolid selective-motility enrichment medium for isolation of salmonellae from faecal specimens, J. Clin. Microbiol. 19:940-941.

Perales,I. and Audicana,A., 1989a, Evaluation of semisolid Rappaport medium for detection of salmonellae in meat products, J. Food Prot. 52:316-319.

Perales,I. and Audicana,A., 1989b, Semisolid media for isolation of Salmonella spp. from coastal waters, Appl. Environ. Microbiol. 55:3032-3033.

Perales,I. and Erkiaga,E., 1991, Comparison between semisolid Rappaport and modified semisolid Rappaport-Vassiliadis media for the isolation of Salmonella spp. from foods and feeds, Int. J. Food Microbiol. 14:51-58.

SALMONELLA FLAGELLIN - CARRIERS OF HETEROLOGOUS ANTIGENS AND IDENTIFICATION OF TWO NEW FLAGELLAR GENES

Gad Frankel[1], Sharon Moshitch,[1] David Zangen,[3] Adam Friedmann,[2] and Linda Doll[1]

[1]Department of Membrane Research and Biophysics
The Weizmann Institute of science, Rehovot 76100
[2]Department of Clinical Molecular Biology and [3]Pediatrics
Hadassah Medical Organization, Mount Scopus, Jerusalem 91240
Israel

INTRODUCTION

The bacterial flagellum consists of three distinct parts: the basal body, the hook and the filament (Macnab 1987). The filament, the main component of the flagellum, is composed of a repeating subunit of a single protein - flagellin. The cloning and sequencing of five flagellin genes, ca. 1500 bp, (Wi and Joys, 1985) determining non-cross reactive flagellar antigens showed complete identity for several hundred base-pairs between alleles at both the C- and N- terminal regions, but great diversity in the middle of the gene, which contains a "hypervariable" segment IV (of ca. 500 bp), with less than 30% amino acid homology for any pairwise comparison. Several lines of evidence indicate that region IV determines a part of the flagellin molecule exposed at the surface of the filament and determining antigen character.

In a recently developed method, Stocker and his colleagues (Newton et al., 1989) have used *Salmonella* flagellin to obtain expression of short heterologous amino acid sequences at

Biology of Salmonella, Edited by F. Cabello *et al.*.
Plenum Press, New York, 1993

the surface of the flagellar filament. This system consists of plasmid pLS408, which contains the *S. muenchen d* flagellin gene under its own promoter, and DNA fragments specifying the amino acid sequences which are inserted in-frame into segment IV of the cloned flagellin gene. In the present study, one of the five heterologous antigen insertions inhibited the migration of the bacteria through semi-solid medium while other two interfere with the delivery of the flagellin to the tip of the flagellar filament, resulted in flagellin accumulation in the bacteria as a cytoplasmic protein. In order to construct an alternative vector, we attempted cloning the gene encoding flagellar antigen *j*. The gene for this antigen is common in Indonesian isolates of *S. typhi* and have a deletion of 87 amino acids in the region IV (Frankel, et al., 1989). This study led us to identify and characterize what appear to be two new genes (located downstream to gene *fliC),* which are essential for biosynthesis of the flagella. According to the recommended nomenclature for flagellar genes (Iino et al., 1988), we have termed the two new genes *fliU* and *fliV* (Doll and Frankel, 1992).

RESULTS

We have used pLS408 as a vehicle for expression of several amino acid sequences, listed in Table 1, on the surface of the flagella. The resultant recombinant plasmids were used to transform the flagellin-deficient *S. dublin* strain SL5928. pLS408 complemented the induced mutation and restored the motile ability of the bacteria (Fig. 1). However, two of the five heterologous insertions (Table 1) abolished the ability of the cloned flagellin gene to restore motility and the bacteria remained immobilized, while another insert only inhibited motility (Fig. 1). Bacterial cell extracts of *S. dublin* strain SL5928 were analyzed by immunoblotting after electrophoretic separation in polyacrylamide gels (Fig. 2). No flagellin was observed associated with the untransformed strain (not shown), but flagellin was present in bacterial-cell extracts of SL5928 harboring plasmids, regardless of the their ability to complement the bacterial mutation. However, the amount of flagellin detected was lower in the immobilized plasmid-harboring strains compered with strains harboring "motility plasmids".

Since some of the inserts resulted in continuance immobilization of the receiver mutant strain, we attempted cloning the gene encoding flagellar antigen *j*. However, introduction of the cloned structural flagellin gene (pHJ103) to the flagellin-deficient *Salmonella* strain failed to complement the mutation, and the FliC flagellin protein was accumulated in the bacterial cytoplasm (not shown). However, if DNA sequences downstream to *fliC-j* were included, the resultant clone (pHJ104) restored the motile ability of the bacteria which again became flagellated. We have determined and characterized the DNA sequence of this region. It appears that it encodes to two new flagellar genes, which are essential for biosynthesis of the flagella (Doll and Frankel, 1992). According to the recommended nomenclature for flagellar genes, we termed the two new genes *fliU* and *fliV*.

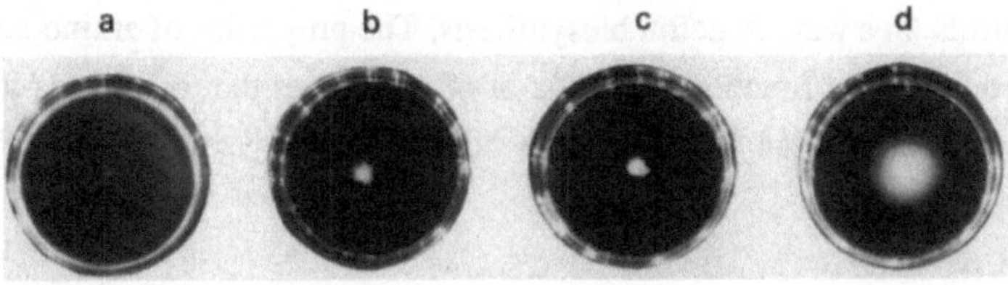

Figure 1. Results of motility assays preformed with the flagellin-deficient *S. dublin* mutant strain SL5928 (b), and with SL5928 harboring plasmids: pLS408 (a); pLS408-TCR (c); pLS408-HPQ (anti-sense) (d). The swarm zone diameter through the semi-solid medium reflects the motility of the strains.

Table 1. Amino acid sequences inserted in the flagellin hypervariable region and ability of the recombinant gene to complement *S. dublin* 5928 mutation.

Origin	Sequence	Motility.
Biotin binding site of avidin (sense strand)	RGEFTGTYITAVT	-
Biotin binding site of avidin (anti-sense strand)	GNSGDVGTGEFTT	+
HPQ-biotin analog (sense strand)	MLHPQPAA	+
HPQ-biotin analog (anti-sense strand)	RRPAAGAA	+/-
TCR of an autoimmune T-cell clone	YYCARERITTATDYW GQGTTLWSS	-

CONCLUSIONS

In the present study we have used plasmid pLS408 for expression of several amino acid sequences as part of the bacterial flagella. However, since some of the recombinants lost their ability to complement the flagellin-locus deletion of *S. dublin* SL5928, we have cloned gene *fliC-j* from *S. typhi* in order to use it as an alternative to gene *fliC-d*. Since *fliC-j* has a deletion in its hypervariable region we thought that it might tolerate what otherwise seems to be "problematic" insertions. As gene *fliC-j* itself could not complement the mutation of SL5928, we have added to the clone its downstream DNA region. We identified in this region what seems to be two new genes of the *Salmonella* flagellar regulon (termed *fliU* and *fliV*, encodes for proteins exhibiting molecular mass of 19 and 20 kDa). The fact that the amount of flagellin associated with immobilized recombinant-plasmid-harboring strains was lower compared with "motile constructs" suggests that the level of free cytoplasmic monomers controls, in a way, flagellin biosynthesis. The properties of amino acid sequences required for production of functional flagella or of sequences that disrupt its formation and the precise function of *fliU* and *fliV* is currently being studied by several laboratories.

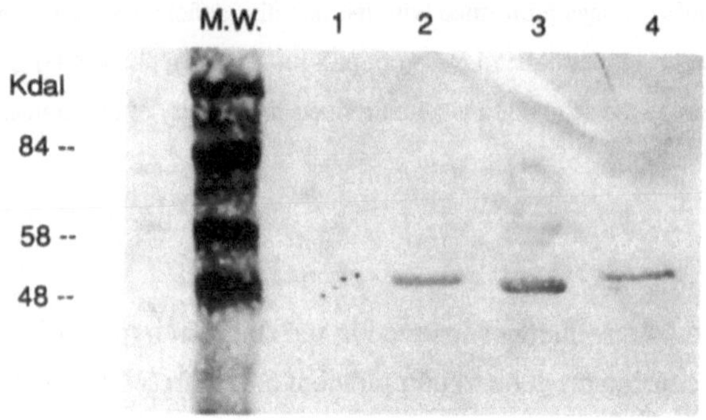

Figure 2. Western blotting using polyclonal antibodies against monomeric flagellin. Lane 1, pLS408-TCR-; lanes 2, pLS408-HPQ-; lane 3, pLS408- ; lane 4, pLS408-Biotin (sense)-transformed SL5982 cells.

REFERENCES

Doll, L., and Frankel, G., 1992, Cloning and DNA sequencing of two new genes, the products of which are essential for Salmonella flagellar biosynthesis, Gene, in press.

Frankel, G., Newton, S.M.C., Schoolnik, G.K., and Stocker, B.A.D., 1989, Intragenic recombination in a flagellin gene: characterization of the H1-j gene of Salmonella typhi, EMBO J., 10: 3149.

Iino, T., Komeda, Y., Kutsukake, K., Macnab, R.M., Matsumura, P., Parkinson, J.S., Simon, M.I., and Yamaguchi, S., 1988, New unified nomenclature for the flagellar genes of Escherichia coli and Salmonella typhimurium, Microbiol. Rev., 52: 533.

Macnab, R.M., 1987, Flagella. In: Neidhardt, F.C. (ed) Escherichia coli and Salmonella typhimurium. American Society for Microbiology, Washington DC, p. 70.

Newton, S.M.C., Jacob C.O., and Stocker, B.A.D., 1989, Immune response to cholera toxin epitope inserted in Salmonella flagellin, Science, 244: 70.

Wei, L.-N., and Joys, T.M., 1985, Covalent structure of three phase-1 flagellar filament of Salmonella, J. Mol. Biol., 186: 791.

PRELIMINARY CHARACTERIZATION OF TN*PHOA* MUTANTS OF SALMONELLA ENTERITIDIS WITH REDUCED INVASIVENESS *IN VIVO*

Homayoun Halavatkar and Paul A. Barrow

Agricultural and Food Research Council
Institute for Animal Health
Compton
Berkshire
RG16 ONN, United Kingdom

Several groups have studied the genetic basis of epithelial cell entry by *Salmonella* species using an *in vitro* system of cultured monolayers.[1,2,3] The aim of this study was to examine the molecular basis of *in vivo* invasivness as opposed to *in vitro* invasiveness of a phage type 4 strain of *S.enteritidis*. Using the day old chicken model the genetic basis for *in vivo* invasiveness was studied by identifying mutations which affect this phenotype. *In vivo* invasiveness was defined as the ability of organisms to penetrate the intestinal epithelial cells following oral inoculation and localise in the spleen within 24h. Approximately 500 Tn*phoA* insertion mutants of *S.enteritidis* S13 were examined. Seven Tn*phoA* mutants were identified as being less invasive by a low rate of isolation of the organisms from the spleen 24h after oral inoculation (13%-47% of chicks inoculated) compared to the parent strain and an invasive Tn*phoA* mutant (76%-87% respectively).

All the mutants were shown to be serum resistant, possessed smooth LPS, had similar growth rate *in vitro* compared to the parent strain and grew equally well on minimal medium and under anaerobic conditions. The efficiency of iron siderophore activity and iron uptake of the mutants were shown to be equivalent to the parent strain. Results of the uptake by the spleen following intramuscular inoculation indicated that the mutants which were identified as being less invasive by oral route localised and persisted in spleen as well as the parent strain. Southern blot analysis of the mutant strains showed that in each mutant the site of the Tn*phoA* insertion was different and each mutant had a single insertion.

The mutant strains were divided into four classes by their pattern of invasion for chicken kidney (CK) and vero cells. Strains of class 1 which were non-motile showed significantly reduced invasivness in both cell lines compared to the parent and other mutants. The class 2 mutant was almost as invasive as the parent strain in CK and vero cells. Strains of class 3 showed reduced invasion in both cell lines. Strains of class 4

Biology of Salmonella, Edited by F. Cabello *et al.*,
Plenum Press, New York, 1993

showed reduced invasion in CK cells but were almost as invasive as the parent strain in vero cells. This may suggest that the organism may express more than one invasion pathway.

The invasion of ileal wall by the parent strain and selected number of mutants belonging to each of the four classes identified *in vitro* was examined. The class 1 mutant which was significantly less invasive *in vitro* was almost as invasive as the parent strain and an invasive Tn*phoA* mutant at day 1 and 2 but showed reduced invasion at day 3. The invasion of ileal wall by the class 2 mutant was similar to the parent strain at day 1. However, the mutant showed slightly reduced invasion at days 2 and 3. One of the class 3 mutants showed reduced invasion of ileal wall over three days compared to the parent strain similar to that found in *in vitro* system. The other mutant behaved similarly except it appeared invasive at day 1. The class 4 mutants which were defective for entry in CK cells and not vero cells showed reduced invasion of ileal wall. These observations indicates that invasion *in vitro* is not an accurate indication of invasiveness *in vivo*.

REFERENCES

Galan, J.E. and Curtiss III, R., 1989, Cloning and molecular characterization of genes whose products allow *Salmonella typhimurium* to penetrate tissue culture cells, *Proc. Natl. Acad. Sci. USA* 86:6383-6387.

Lee, C.A. and Falkow, S., 1990, The ability of *Salmonella* to enter mammalian cells is affected by bacterial growth state, *Proc. Natl. Acad. Sci. USA* 87:4304-4308.

Stone, B.J., Garcia, C.M., Badger, J.L., Hassett, T., Smith, R.I.F., and Miller, V.L., 1992, Identification of novel loci affecting entry of *Salmonella enteritidis* into eukaryotic cells, *J. Bacteriol.* 174:3945-3952.

HYPO- AND HYPER- INVASIVE Tn*pho*A MUTANTS OF *SALMONELLA TYPHIMURIUM* (STRAIN TML)

Gillian R. Douce, Iqbal I. Amin, John Stephen, Gordan Dougan[1], and Stephen N. Chatfield[2]

Microbial Molecular Genetics and Cell Biology Group,
School of Biological Sciences,
University of Birmingham,
PO Box 363,
Birmingham B15 2TT, United Kingdom

Department of Biochemistry[1], and Medeva Group Research[2],
Imperial College of Science, Technology and Medicine,
Exhibition Road, London SW7 2AL, United Kingdom

Approximately 100 Tn*pho*A mutants of *Salmonella typhimurium* (Strain TML a clinical gastroenteritic strain of human origin) were generated. These were screened in the test already described (Douce *et al.*, 1991) for invasiveness. The latter was measured for each mutant as a percentage of that shown by their immediate parents on the same day.

Figure 1 shows the distribution of invasiveness of all the mutants tested in a once only analysis. A value of 100% means that the mutant has the same invasive properties as its parent. The skewed distribution means that the bulk of the mutations result in a diminution of invasiveness, suggesting that most Tn*pho*A insertions alter surface characteristics of the organism in a manner which affects attachment/invasion properties. Those groups of mutants which lay outside the two vertical hatched lines were re-analysed to yield 7 hypo-invasive and 4 hyper-invasive mutants (groups A and B respectively; Figure 2). Two of three rough mutants were half as invasive as their parents.

While the production of hypo-invasive mutants was expected the production of hyper-invasive mutants was not. Possible reasons for the latter phenotype include:1) inactivation of a surface expressed component which normally sterically inhibits receptor/ligand interaction; 2) transposon insertion into the C-terminal end of the exported protein which may affect the position and conformation of the protein in the membrane; 3) secondary effects on the expression of genes down stream normally regulated by the wild-type product; 4) modification of a gene involved in regulating responses to environmental stimuli. These possibilities are being investigated.

Biology of Salmonella, Edited by F. Cabello *et al.*,
Plenum Press, New York, 1993

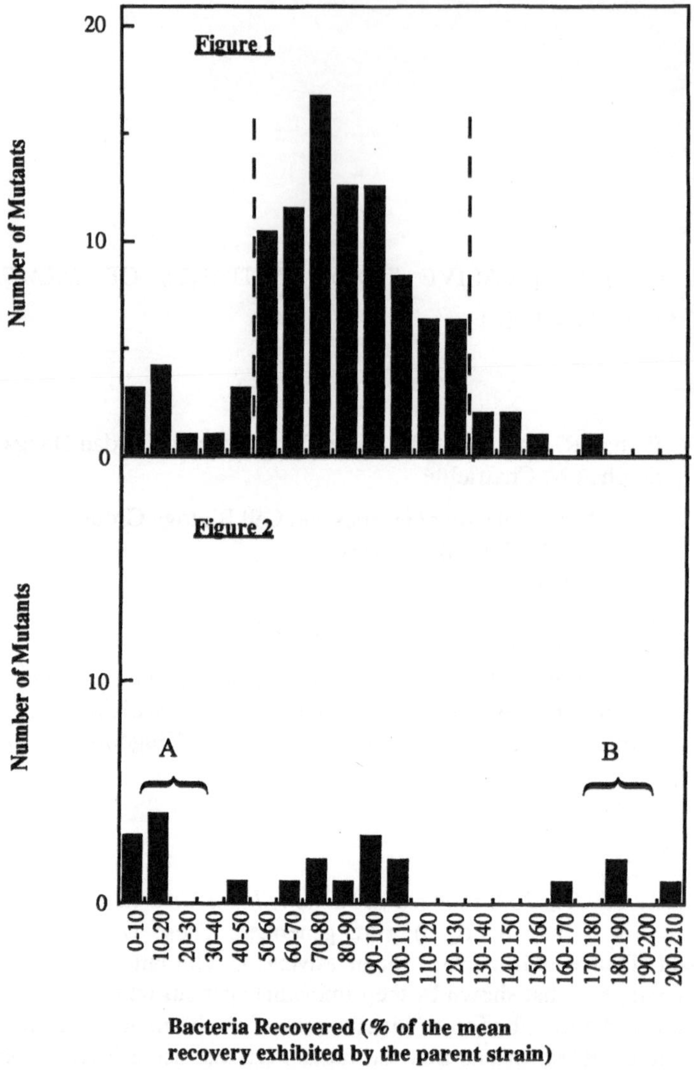

Figures 1 and 2: Bacterial invasiveness was measured as the number of gentamicin-resistant bacteria recovered as a percentage of the original inoculum. Invasiveness of mutants was expressed as a percentage of the invasiveness of their immediate parents, determined on the same day on the same multi-well tray. For this comparative analysis, the invasiveness of parents was normalised to 100%. Figure 2 represents a re-analysis of those mutants outside the vertical hatched lines.

REFERENCES

Douce, G. R., Amin, I. I., and Stephen, J., 1991, Invasion of HEp-2 cells by strains of *Salmonella typhimurium* of different virulence in relation to gastroenteritis, *J. Med .Microbiol.* 35: 349.

INVASION OF ILEAL MUCOSA BY STRAINS OF *SALMONELLA TYPHIMURIUM*: QUANTITATIVE STUDIES *IN VITRO*

Iqbal I. Amin, Gillian R. Douce, Michael P. Osborne [1], and John Stephen

Microbial Molecular Genetics and Cell Biology Group,
School of Biological Sciences, Physiology Department[1],
University of Birmingham,
PO Box 363,
Birmingham B15 2TT, United Kingdom

In the preceding contribution by Douce *et al.*, a series of Tn*pho*A mutants were described which were generated from *Salmonella typhimurium* strain TML and whose invasive phenotype was determined in a HEp-2 cell assay for invasiveness (Douce *et al.*, 1991). To date no quantitative system has been described in which invasiveness has been measured in functional gut *in vitro*. This contribution describes such a system.

In whole animal experiments gut invasion data are difficult to interpret since the distinction between cellular invasion of gut epithelia and tissue spread through the lamina propria and beyond is often hard to make or not taken into account in the first place. In addition, invasion data derived from studies in cultured cells is assumed to reflect the invasiveness of differentiated epithelia without formal proof that this is in fact the case.

Nearly every contemporary study on the invasiveness of enteric pathogens is based on the selection of non-invasive mutants derived from invasive wild-type organisms using tissue culture cells. The aim of this study was to develop an *in vitro* model using rabbit intestinal mucosa for quantitative studies of the initial attachment/invasion events in experimental gastroenteritic infections caused by *S. typhimurium*. A purpose built apparatus was used in which the bacterial challenge could be applied directly to the mucosal surface of rabbit ileum. This avoids the problems associated with conventional organ cultures in which invasion can (and does) occur *via* cut surfaces and submucosal routes.

The organ culture system was as described by Worton *et al.*, (1989) except that different solutions were used. To preserve the tissue *in vitro*, we sequentially varied the composition of the solutions used to bathe the mucosal and serosal surfaces of rabbit ileum. Tissues were incubated at 37°C and gassed with 95% O_2/ 5% CO_2 for variable periods following which they were fixed, embedded and sectioned for both light and electron microscopy. Previously it had been observed using Tissue culture media (Tcm) for both mucosal and serosal sides, that ileal tissue could be preserved up to about 2 h. However during this time a huge accumulation of fluid occurred in the lamina propria with degeneration of and loss of large tracts of epithelium (Worton *et al.*, 1989). This happened because fluid was continuously absorbed by villi and could not be drained away because of absence of a functional lymphatic

Biology of Salmonella, Edited by F. Cabello *et al.*,
Plenum Press, New York, 1993

drainage system and the low hydraulic conductivity of the tissue. Tissue culture media contain high [Na] and low [glucose] and sodium is known to play a pivotal role in the absorption of fluid by intestinal villi. Solution G (WHO rehydration solution; NaCl, 60 m*M*; $NaHCO_3$, 30 m*M*; KCl, 20 m*M*; and glucose, 111 m*M*) was used on the mucosal side since it contained lower [Na] and higher [glucose]. High [glucose] was used to see if it protected the tissue. The same solution - with 50 m*M* added mannitol (Solution M) - was used on the serosal side. Mannitol was used to increase the osmolarity on the mucosal side to pull absorbed fluid across the submucosa. Villi remained intact after 3 h of incubation; but by 4 h villi had an exploded appearance. Na^+ was then excluded from the mucosal side and replaced by choline$^+$ which is a non-absorbed cation used as substitute for Na^+ (Starkey *et al.*,1990).

Using Solution C (choline Cl, 60 m*M*; choline HCO_3, 30 m*M*; KCl, 20 m*M*; and glucose, 111 m*M*) in the mucosal side and Solution G on the serosal side villi remained intact up to 3 h with minimal fluid accumulation in the lamina propria but after 4 h the tissue degenerated. Moreover Solution C did not support the viability of *S. typhimurium* due, it was thought, to the absence of an available nitrogen source.

Attempts were therefore made to determine the minimum amount of Tcm (Minimum Essential Medium with Earle's salts without glutamine, to which glutamine was added to a final concentration of 2.0 m*M* and foetal calf serum to a final concentration of 10% (v/v)) required to supplement Solution C to extend the period for which tissue would remain viable. Using 10% Tcm (v/v) with Solution C (Mucosal medium) on the mucosal side and Solution G on the serosal side villi remained viable up to 4 h. Some fluid accumulated in the lamina propria which indicated that the villi were still functioning at this stage.

The mucosal medium supported the viability of *S. typhimurium*. This was important because organisms were to be suspended in and presented to the tissues in this medium in the invasion assay.

Invasion assays were carried out with 3 virulent strains of *S. typhimurium* (TML, W118, WAKE) and 4 avirulent strains (SL1027, LT-7, M206, Thax-1). The results, presented in the contribution from Stephen, Amin and Douce (this volume) show that the three virulent strains were distinguished from the four avirulent strains in terms of their invasiveness for gut epithelia.

REFERENCES

Douce, G. R., Amin, I. I.,and Stephen, J., 1991, Invasion of HEp-2 cells by strains of *Salmonella typhimurium* of different virulence in relation to gastroenteritis, *J. Med. Microbiol.* 35: 349.

Starkey, W. G., Candy, D. C. A., Collins, J., Spencer, A. J, Osborne, M. P. and Stephen, J., 1990, An *in vitro* model to study aspects of the pathophysiology of murine rotavirus-induced diarrhoea. *J. Pediatr. Gastroenterol. Nutr.* 10: 361.

Worton, K. J., Candy, D. C. A., Wallis, T. S., Clarke, G. J., Osborne, M. P., Haddon, S. J., and Stephen, J., 1989, Studies on early association of *Salmonella typhimurium* with intestinal mucosa *in vivo* and *in vitro*: relationship to virulence, *J. Med. Microbiol.* 29: 283.

MICROTUBULE-DEPENDENT VERSUS MICROFILAMENT-DEPENDENT INVASION MECHANISMS ENCODED BY *CITROBACTER FREUNDII*

Tobias A. Oelschlaeger and Dennis J. Kopecko

Department of Bacterial Immunology
Walter Reed Army Institute of Research
Washington, DC 20307-5100

Citrobacter freundii are common causative agents of UTI's and have been implicated in gastrointestinal diseases (Guarino et al., 1989). They are known to cause neonatal meningitis (Rae et al., 1991) and in isolated cases bacteraemia (Flegg and Mandal, 1989).

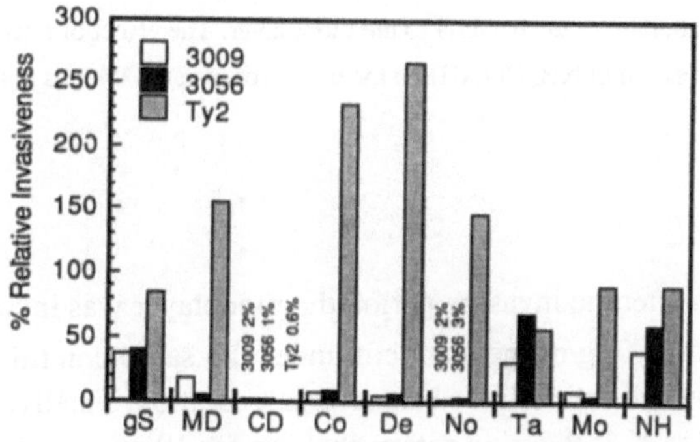

Figure 1 Effects of different inhibitors on the relative invasiveness of *Citrobacter freundii* 3009 and 3056 and *Salmonella typhi* Ty2 into the human embryonic intestinal cell line INT407. One hour before the addition of bacteria the INT407 cells were incubated with the inhibitors g-strophantin (gS, 250μM), monodansylcadaverine (MD, 250μM), cytochalasin D (CD, 2μM), colchicine (Co, 10μM), demecolcine (De, 1μM), nocodazole (10μM), taxol (Ta, 50μM), monensin (Mo, 40μM), and ammonium chloride (NH, 20mM). Without inhibitor the relative invasiveness is 100%.

Biology of Salmonella, Edited by F. Cabello *et al.*,
Plenum Press, New York, 1993

Two fresh UTI isolates of *Citrobacter freundii* (strains 3009 and 3056) were compared with the *Citrobacter freundii* lab strain 7004, *Salmonella typhi* Ty2 and the noninvasive *Escherichia coli* strain HB101 in their invasion ability in the presence or absence of biochemical inhibitors. Uptake was assessed using INT407 human embryonic intestinal epithelial cells, as well as human bladder epithelial cell lines T24 and 5637. The invasion assay was performed as described by Elsinghorst et al., 1989. To the confluent monolayer of epithelial cells in wells of a 24 well plate 50 bacteria per epithelial cell were added. Bacteria used for the invasion assay had reached A_{600}=0.4 - 0.6. After mixing for 1 min the bacteria were gently centrifuged at 200g for 5 min onto the monolayer. During the following 2 hrs incubation period (= invasion period) at 37°C in a 6%CO_2/94% air atmosphere the bacteria were allowed to enter the epithelial cells. To kill remaining

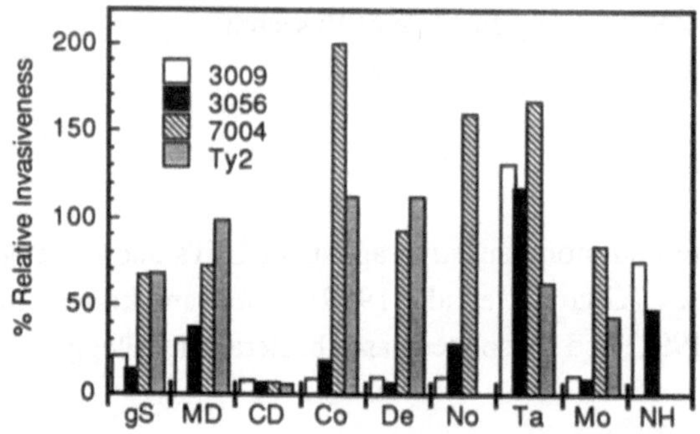

Figure 2 Effects on the uptake into cells of the human epithelial bladder cell line 5637 of *Citrobacter freundii* 3009, 3056, and 7004 and of *Salmonella typhi* Ty2. The inhibitors are listed in figure 1 and were added 1 hr prior to the addition of the bacteria to the monolayer. The effect of nocodazole (No) and NH_4Cl (NH) on Ty2 invasiveness and of NH_4Cl (NH) on the invasiveness of 7004 was not determined.

extracellular bacteria after the invasion period the monolayer was incubated for another 2 hrs in the presence of 100µg/ml gentamycin under the same conditions as before, after washing the epithelial cells with Earl's balanced salt solution. Finally, the monolayer was lysed with 0.1% Triton X-100 during rotary shaking for 10 min at room temperature and the intracellular bacteria were enumerated by plate count. To study the effect of inhibitors of eukaryotic cell function epithelial cells were incubated with inhibitors for 1hr before the addition of bacteria and the inhibitors were present during the invasion period. The concentrations of inhibitors were chosen to exhibite expected inhibitory properties but not to effect the viability of eucaryotic or procaryotic cells during the assay period.

The highest intracellular bacterial recovery (= % of inoculum surviving the gentamycin treatment) was achieved for all three *Citrobacter freundii* strains with both bladder cell lines, T24 and 5637. The recovery for 3009 was 12.8% and 27.6%, for 3056 9.3% and 8.9% and for 7004 1.7% and 0.2% with cell line T24 and 5637, respectively. With the intestinal INT407 cells, the recovery of 7004 was as low as with the noninvasive HB101 (0.02%). Therefore, invasion assays with 7004 in the presence of inhibitors were not performed with INT407 cells. Also, the UTI *Citrobacter freundii* strains 3009 and 3056 were less invasive for INT407 cells, reaching only 0.7% and 1.0% recovery, respectively. *Salmonella typhi* Ty2 showed the highest recovery with INT407 cells (10.2%) and showed somewhat reduced invasiveness with T24 (6.5%) and 5637 (8.5%) bladder cells.

Addition of the bacteriostatic antibiotics chloramphenicol or rifampicin to the bacteria at the beginning of the assay inhibited invasion >98%, implying that nascent protein synthesis is required for invasion ability of *Citrobacter freundii* 3009, 3056 and *Salmonella typhi* Ty2.

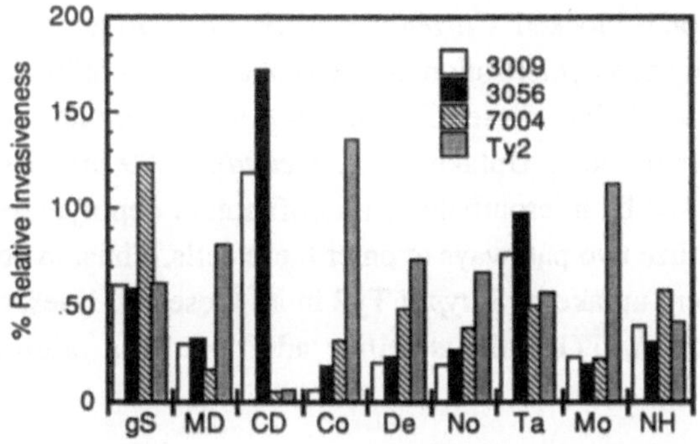

Figure 3. Relative invasiveness of *Citrobacter freundii* 3009, 3056 and 7004 and of *Salmonella typhi* Ty2 for the human epithelial bladder cell line T24 in the presence of different inhibitors. Before the addition of bacteria the epithelial cells were incubated for 1 hr with the inhibitor. The symbols and concentrations of the inhibitors are listed in the legend to figure 1.

Cytochalasin D, which depolymerizes microfilaments inhibited both *S. typhi* and *C. freundii* uptake by >93% in INT407 and 5637 cells (CD, Figure 1 and 2). In contrast, with T24 bladder cells, the uptake of the clinical *Citrobacter* isolates 3009 and 3056 was not reduced whereas 7004 and Ty2 invasion was decreased by ~95% (CD, Figure 3).

The inhibitors of coated-pit formation, g-strophantin and monodansylcadaverine, had no apparent effect on Ty2 and 7004 uptake into INT407, T24, and 5637 cells (However, strain 7004 invasion of T24 cells was reduced by 75% in the presence of monodansylcadaverine) but generally reduced *C. freundii* 3009 and 3056 invasion by 40% to 96% (gS and MD, Figure 1, 2, and 3).

Addition of monensin or ammonium chloride, which block endosome acidification, did not affect *Salmonella typhi* Ty2 invasion efficiency but markedly reduced the number of viable intracellular 3009 and 3056 *Citrobacter* in all three cell lines. The uptake of 7004 was only reduced into T24 cells. (Mo and NH, Figure 1, 2 and 3).

Depolymerization of microtubules by colchicine, demecolcine, or nocodazole led to a >70% reduction in *C. freundii* 3009 and 3056 invasion, whereas 7004 uptake was reduced only in T24 cells (by up to 68%) (Figure 3) but not in 5637 cells (Figure 2). Microtubules were not required for Ty2 uptake in all cell three cell lines (Co, De, and No, Figure 1, 2, and 3). Surprisingly, stabilization of microtubules in the presence of taxol did not reduce uptake of *Citrobacter freundii* in bladder cells. Ty2 invasion was decreased about two fold as was invasion of INT407 cells was for all four bacterial strains (Ta, Figure 1, 2, and 3).

Collectively these data suggest that *C. freundii* strains 3009 and 3056 synthesize a peptide surface ligand(s) of short half-life that triggers uptake into cultured epithelial cells. Inhibitors of coated-pit formation and of endosome acidification markedly reduced intracellular recovery of *Citrobacter freundii* but had no effect on control Ty2 invasion, suggesting the involvement of coated-pits in *Citrobacter* invasion and the activation of intracellular *Citrobacter* survival functions by acidification or a reduction of the uptake efficiency if acidification is blocked. *Citrobacter* 3009 and 3056 invasion of T24 cells was dramatically inhibited by depolymerization of microtubules and not by depolymerization of microfilaments, implying that these *Citrobacter* strains invade T24 cells using a microtubule-dependent process. Uptake of *C. freundii* 3009 and 3056 into 5637 and INT407 cells was blocked by microtubule or microfilament depolymerization suggesting that *Citrobacter* can utilize two pathways to enter these cells. Thus, in contrast to the strict microfilament-dependent uptake of *S. typhi* Ty2 in all these cell lines, the closely related *Citrobacter* isolates from UTI patients utilize additionally a microtubule-dependent invasion mechanism.

REFERENCES

Elsinghorst, E.A., L.S. Baron, and D.J. Kopecko 1989, Penetration of human intestinal epithelial cells by *Salmonella*: Molecular cloning and expression of *Salmonella typhi* invasion determinants in *Escherichia coli*. *Proc, Natl. Acad. Sci. USA* 86: 5173-5177

Flegg, P.J. and B.K. Mandal 1989, *Citrobacter freundii* bacteraemia presenting as typhoid fever, *J. Infect.* 18:171-173

Guarino, A., R. Giannella, and M.R. Thompson 1989, *Citrobacter freundii* produces an 18-amino-acid heat-stable enterotoxin identical to the 18-amino-acid *Escherichia coli* heat-stable enterotoxin (ST Ia). *Infect. Imm.* 57:649-652

Rae, C.E., A. Fazio, and J.P. Rosales 1991, Successful treatment of neonatal *Citrobacter freundii* meningitis with ceftriaxone. *DICP* 25:27-29

INTERACTION BETWEEN HUMAN CELL LINES AND *SALMONELLA TYPHIMURIUM*: EFFECTS OF PROTEIN SYNTHESIS, GROWTH PHASE, AND MULTIPLICITY OF INFECTION ON ADHESION AND INVASION

Johannes G. Kusters and Bernard A. M. van der Zeijst

Department of Bacteriology, Institute of Infectious Diseases
and Immunology, University of Utrecht, P.O. Box 80.165, 3508
TD Utrecht, The Netherlands

INTRODUCTION

Salmonella typhimurium is a widely spread Gram negative bacterium that is capable of causing disease in both man and animals. The first step in the infection process is the invasion of the epithelial cells of the gut. This invasion proceeds in two phases. First bacteria adhere to the host cells, then they penetrate these cells. In spite of their relevance to the pathogenesis of the bacterium, the adhesion and penetration mechanism(s) used by salmonella species are still obscure. Some studies claim that invasion of salmonella requires *de novo* protein synthesis and is dependent on the growth state of the bacterium, others deny these claims (B.B. Finlay et al., 1989; R.K. Ernst et al., 1990; and C.A. Lee and S. Falkow 1990). In order to clarify these contradictory data, monolayers of a cell line derived from human intestinal tissue (Int-407) were infected with a virulent strain of *S. typhimurium* (strain C52) and the kinetics of bacterial adhesion and invasion were studied.

RESULTS AND DISCUSSION

S. typhimurium strain C52 (P. Pardon et al., 1986) was obtained from dr. L. Norel (Inst. Pasteur, Paris, France). Henle Intestine 407 (Int-407) cells were obtained from Flow Laboratories Inc. Experiments were carried out essentially as described by B.B. Finlay et al., 1989. In our experimental approach we wanted to study the effects of the multiplicity of infection (m.o.i, i.e. the number of bacteria in the inoculum per cell), growth phase and the need for (ongoing) protein synthesis on the kinetics of both adhesion and invasion (defined as adhesion followed penetration). Confluent monolayers of cells were therefore overlaid with different amounts of bacteria in pre-warmed DMEM and placed in a CO_2 incubator for 5 to 180 min. Adhesion was determined as the number of viable bacteria that were adherent to glutardialdehyde

Biology of Salmonella, Edited by F. Cabello *et al.*,
Plenum Press, New York, 1993

180 min. Adhesion was determined as the number of viable bacteria that were adherent to glutardialdehyde fixed Int-407 cells incubated with these bacteria for the set time (glutardialdehyde fixation prevents invasion). Non-adherent bacteria were removed by five washes with phosphate buffered saline (PBS). Invasion was determined as the number of viable bacteria present in Int-407 cells after five washes with PBS to remove non-adherent bacteria followed by an 90 min. incubation of cells and bacteria in DMEM + 150 μg/ml Colistin. The latter treatment kills all extra-cellular bacteria while it does not affect the viability of intracellular bacteria.

Effect of moi on adhesion and invasion

To determine the effect of the m.o.i. we performed adhesion and invasion assays using various amounts of bacteria (derived from the logarithmic phase) per cell. When adhesion (and invasion) rates are expressed as the number of adhesive (or invasive) bacteria divided by the total number of (viable) bacteria, the adhesion rates ranged from 2% to 100%, and invasion rates ranged from 0.1% to 20% depending on the moi. Plotting the data as absolute numbers of bacteria that adhere to (or invade) the Int-407 cells reveals another feature. Adhesion (and invasion) show saturation kinetics at about 10 bacteria per cell. Scanning electronmicrographs (SEM) show that the adherent bacteria are more or less evenly distributed over the cells. At the saturation point the surface of the cells is certainly not completely covered by bacteria. Thus there probably is a limited number of receptors present on the surface of the cells.

Effect of growth phase on adhesion and invasion

The effect of the growth phase on adhesion and invasion was determined by infecting monolayers with exactly the same number of bacteria in either the logarithmic or the stationary growth phase. The same effects of the m.o.i. on the adhesion and invasion were observed as above were logarithmic cultures were used as inoculum. Thus no significant differences in the adhesion and invasion rates of were detected between bacteria in the logarithmic or the stationary phase.

Induced proteins

Bacteria were added to glutardialdehyde fixed cells and bacterial proteins were pulse labelled with ^{35}S-methionine. The bacteria were harvested and the radioactive (newly synthesized) proteins were separated by SDS-PAGE. Adhesion of *S. typhimurium* to host cells clearly induces *de novo* synthesis of certain bacterial proteins. These induced proteins may represent important virulence factors related to adhesion/penetration. Adhesion induced protein bands could even be visualized on silver-stained SDS-PAGE gels with total bacterial cell lysates, indicating synthesis of rather big quantities of induced protein.

Ongoing protein synthesis

To determine the role of protein synthesis in adhesion and invasion assays were performed either in the presence or absence of chloramphenicol. Chloramphenicol inhibits the bacterial protein synthesis but does not kill the bacteria. While the adhesion was not influenced by this treatment, invasion was completely abolished by the absence of bacterial protein synthesis.

To examine a need for ongoing protein synthesis, chloramphenicol was added at different times during the invasion assay. The observed kinetics show that ongoing protein synthesis is required for *S. typhimurium* invasion, since the invasion stops immediately after the addition of chloramphenicol.

General validation of the results

Similar results were obtained in adhesion and invasion assays of field isolates of *S. typhimurium* using monolayers of Int-407, HeLa, and HEp-2 cell lines (data not shown). We are now in the process of establishing the role of these induced proteins in either the adhesion or the invasion of *S. typhimurium*.

REFERENCES

Ernst, R.K., Dombroski, D.M., and Merrick, J.M., 1990. Anaerobiosis, type 1 fimbriae, and growth phase are factors that affect invasion of HEp-2 cells by Salmonella typhimurium. Infect. Immun. 58:2014.

Finlay, B.B., Heffron, F., and Falkow, S. 1989. Epithelial cell surfaces induce Salmonella proteins required for bacterial adherence and invasion. Science 243:940.

Lee, C.A., and Falkow, S., 1990. The ability of Salmonella to enter mammalian cells is affected by bacterial growth state. Proc. Natl. Acad. Sci. USA. 87:4304.

Pardon, P., Popoff, M.Y., Coynault, C., Marly, J, and Miras, I., 1986. Virulence-associated plasmids of Salmonella serotype typhimurium in experimental murine infection. Ann. Inst. Pasteur Microbiol., 137B: 47.

MOLECULAR CLONING AND EXPRESSION IN *Escherichia coli* OF THE *Salmonella typhi* GENE CLUSTER CODING FOR TYPE 1 FIMBRIAE

Gian Maria Rossolini[1], Patrizia Muscas[1], Alessandra Chiesurin[1], and Giuseppe Satta[2]

[1]Dipartimento di Biologia Molecolare.- Sezione di Microbiologia - Università di Siena, Siena, Italy
[2]Istituto di Microbiologia - Università Cattolica del S. Cuore, Rome, Italy

INTRODUCTION

Type 1 Fimbriae are proteinaceous fibrillar surface structures that have been found in several members of the family *Enterobacteriaceae* (Clegg and Gerlach, 1987). These structures are able to mediate microbial attachment to D-mannosyl residue-containing receptors present on various eukaryotic cells (Ofek et al., 1977; Ofek and Sharon, 1990). Bacteria carrying type 1 fimbriae are able to mediate a mannose-inhibitable (MI) hemagglutination of certain species of erythrocytes and of yeast cells (Paranchych and Frost, 1988).

Type 1 fimbriae have recently been demonstrated to play a key role in enterobacterial communicability (Bloch et al., 1992) and, owing to their adhesive properties, it has also been speculated that these structures play a role in microbial pathogenicity (Svanborg Edèn and Hansson, 1978; for reviews see: Finlay and Falkow, 1989; Orndorff and Bloch, 1990). The role of type 1 fimbriae in this sense, however, is still largely unclear and might vary in different bacteria (Clegg and Gerlach, 1987; Finlay and Falkow, 1988; Finlay and Falkow, 1989; Lockman and Curtiss III, 1992).

In an accompanying report we provide some evidence suggesting that type 1 fimbriae, which are apparently the only adhesin present in *Salmonella typhi* (Duguid et al., 1966; Halula and Stocker, 1987; Satta et al., in preparation), might play a role in typhoid fever pathogenesis at the point of microbial colonization and invasion of the intestinal epithelium (Satta et al., this volume). Here we report the molecular cloning of the *S. typhi* genes responsible for the production of type 1 fimbrial structures (*fim* genes) in *Escherichia coli*.

RESULTS

Molecular Cloning in *E. coli* of the *S. typhi fim* Genes Responsible for Production of Type 1 Fimbriae

To clone the *S. typhi fim* genes, a genomic library of the fimbriated *S. typhi* strain Sty4 was constructed in the cosmid vector cosKT1 and transduced in the *E. coli* strain HB101, which does not produce either type 1 fimbriae or any other fimbrial structure. Transductants were screened with a polyclonal antiserum raised against *S. typhi* type 1 fimbriae. Eight clones producing proteins recognized by the antiserum were found in this way.

The antibody-reactive transductants were subsequently assayed for both MI agglutinating

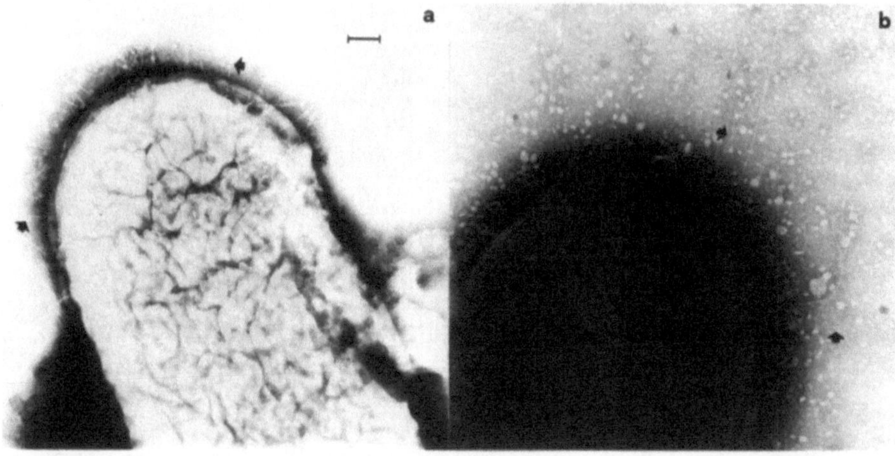

Figure 1. Electron micrographs of negatively stained *S. typhi* Sty4 (a) and of one of the antibody-reactive *E. coli* HB101 transductants producing *S. typhi* fimbriae (b). The arrows indicate fimbrial structures; the bar corresponds to 0.1 μm.

activity of guinea pig erythrocytes (GPE) and the presence of fimbrial structures at their surface, as determined by electron microscopy. This analysis showed that 6 of the antibody-reactive transductants were able to agglutinate GPE in a MI fashion and possessed peritrichous fimbrial structures morphologically identical to those of *S. typhi* (figure 1 and data not shown).

Mapping the *S. typhi fim* Genes within the Cloned DNA

Analysis of the DNA inserts carried by the cosmids present in fimbriated *E. coli* transductants allowed us to map the *fim* genes to a 17-kb *Bam*HI restriction fragment which, when subcloned in the plasmid pACYC184, was able to confer an agglutinating and fimbriated phenotype on the *E. coli* strain HB101 (figure 2). The restriction map of this DNA fragment was determined (figure 2). To confirm the origin of the cloned DNA, chromosomal DNA of the *S. typhi* strain Sty4 was digested with several restriction endonucleases (including *Bam*HI, *Bgl*II, *Cla*I, *Eco*RI, *Nde*I, *Sac*I, *Sal*I, *Stu*I, and *Xba*I) and restriction fragments were analyzed by Southern blot using the entire 17-kb *Bam*HI fragment as hybridization probe. In every case the hybridization profiles obtained were consistent with the assumption that the 17-kb *Bam*HI fragment corresponded to a single chromosomal fragment of Sty4 (data not shown).

Further subcloning showed that the smallest fragment still able to confer an agglutinating and fimbriated phenotype to *E. coli* HB101 was a 11-kb *Sac*I-*Hind*III restriction fragment. Smaller fragments were no longer able to direct fimbrial expression in the *E. coli* host (figure 2).

Expression of the *S. typhi fim* Genes in *E. coli* Strains Carrying a Complete Deletion of the Endogenous *fim* Genes

Since *E. coli* HB101 strains circulating in different laboratories may have retained all or part of the endogenous *fim* genes (Blomfield et al., 1991), this strain does not seem fully satisfactory as host for studying the expression of heterologous *fim* genes.

To study the expression of *S. typhi fim* genes in *E. coli*, we therefore transferred the recombinant plasmids containing the *S. typhi fim* genes into *E. coli* strains AAEC185 and AAEC072 which carry a defined and complete deletion of the endogenous *fim* genes (Blomfield et al., 1991). The transformants were assayed for the production of *S. typhi* fimbriae by hemagglutination and electron microscopy. Results of these experiments showed that the same *Salmonella* DNA fragments able to direct expression of fimbrial structures in *E. coli* HB101 were also able to do so in the two *E. coli* Δ*fim* strains (figure 2), thus confirming that all the *S. typhi* genes necessary for production of type 1 fimbrial structures were contained within the 11-kb *Sac*I-*Hind*III insert of plasmid pFS11HS.

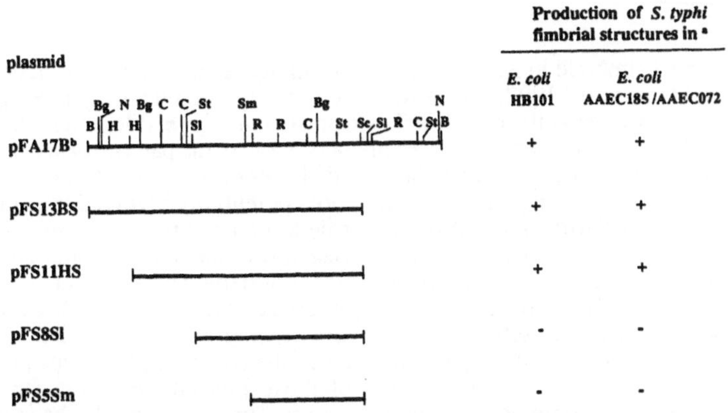

ª assayed by MI heamagglutination of GPE and electron microscopy.
ᵇ plasmid replicon was pACYC184 for pFA17B and pBluescript for all other plasmids.

Figure 2. Restriction endonuclease map of the 17-kb *Bam*HI chromosomal fragment from *S. typhi* Sty4 containing the *fim* gene cluster, and subcloning strategy. B, *Bam*HI; Bg, *Bgl*II; C, *Cla*I; H, *Hin*dIII; N, *Nde*I; R, *Eco*RI; Sc, *Sac*I; St, *Stu*I; Sl, *Sal*I; Sm, *Sma*I.

In Vitro Adhesion to Intestinal Epithelial Cells by *E. coli* Strains Producing *S. typhi* Type 1 Fimbriae

In *S. typhi* type 1 fimbriae can mediate adhesion to intestinal epithelial cells in vitro, being apparently the principal microbial structure exerting this function. The adhesion promoted by fimbriae is important for the invasion of epithelial cells (Satta et al., this volume). To verify if *S. typhi* type 1 fimbriae retain this adhesive function when produced in nonfimbriated *E. coli*, adhesion to the human intestinal epithelial cell line INT-407 was assayed using *E. coli* strains producing *S. typhi* type 1 fimbriae.

Results of these experiments showed that the *E. coli* strains HB101, AAEC185, and AAEC072 producing *S. typhi* fimbriae were able to adhere to the intestinal epithelial cells much more efficiently than the nonfimbriated parental strains, even though the adhesiveness was lower than that of fimbriated *S. typhi* (table 1). Unlike *S. typhi*, however, *E. coli* strains producing *S. typhi* fimbriae were not able to invade the epithelial cells (data not shown).

Table 1. Adhesion to the intestinal epithelial cell line INT-407 by *E. coli* strains producing *S. typhi* type 1 fimbriae and control strains.

strain	relevant phenotype	No of bacteria/cell[2]
E.coli HB101(pFS11HS)	Fim+(*Sty*)	8
E.coli AAEC185(pFS11HS)	Fim+(*Sty*)	9
E.coli AAEC072(pFS11HS)	Fim+(*Sty*)	8
E.coli HB101	Fim-	<1
E.coli AAEC185	Fim-	<1
E.coli AAEC072	Fim-	<1
S.typhi Sty4	Fim+	20

[1] For description of plasmids see figure 2.
[2] Mean value of three experiments; in each experiment 200 cells were subjected to microscopical examination.

DISCUSSION

The role of type 1 fimbriae in *Salmonella* pathogenicity is still unclear (Finlay and Falkow, 1988; Lockman and Curtiss III, 1992). Since we have recently obtained evidence suggesting that type 1 fimbriae, which are apparently the only adhesin present in *S. typhi* (Duguid et al., 1966; Halula and Stocker, 1987; Satta et al., in preparation), might play a role in the pathogenesis of typhoid fever (Satta et al., this volume), we began to study the genetic bases of these structures. The cloning and characterization of the genes coding for microbial factors potentially involved in pathogenicity, in fact, is a necessary step to clarify and confirm their role as pathogenicity determinants.

Production of recombinant *S. typhi* type 1 fimbriae allows the normally non-adhesive *E. coli* strains HB101 and AAEC185 to adhere to the intestinal epithelial cell line INT-407. Since these recombinant strains are not able to invade the same epithelial cells and are suitable as hosts for DNA libraries, they could represent a useful system for the cloning and characterization of invasins derived from *S. typhi* or other invasive organisms. In fact, the adhesive phenotype provided by the *S. typhi* fimbriae might facilitate the functioning of those invasins requiring stable microbial adhesion to the cell surface in order to exert their function. This is apparently the case for *S. typhi* invasin(s) (Satta et al., this volume), and it may be speculated that the higher invasion ability observed for *S. typhi* Ty2 as compared to that of *E. coli* HB101 carrying the *S. typhi invA-D* invasion determinants, reported by Elsinghorst et al. (1988), was due not only to the existence of multiple invasins in *S. typhi* but also to the fact that HB101 is not able to adhere efficiently to INT-407 cells, while adhesion is an important step in the invasion process of *S. typhi*.

EXPERIMENTAL PROCEDURES

Bacterial Strains and Genetic Vectors

The fimbriated *S. typhi* strain Sty4, selected for construction of the genomic library, was a clinical isolate. *E. coli* HB101 (Boyer and Roulland-Dussoix, 1969) was used as host for library transduction and for all recombinant DNA procedures. *E. coli* strains AAEC185 and AAEC072, carrying a complete deletion of the endogenous *fim* genes and used to study expression of *S. typhi fim* genes in *E. coli*, were kindly provided by Dr. Ian Blomfield (Blomfield et al. 1991). Hemagglutination assays with GPE to detect the presence of type 1 fimbriae were performed as described by Duguid et al. (1966). Electron microscopy for detection of fimbrial structures was performed as described by Korhonen et al. (1980).

The cosmid vector cosKT1 (Tartof and Hobbs, 1988) was used for construction of the *S. typhi* genomic library. Plasmid vectors pBluescript SK (Stratagene, La Jolla, Calif.), and pACYC184 (Chang et al., 1977) were used for subcloning.

Recombinant DNA Methodology

Basic recombinant DNA techniques were performed essentially as described by Sambrook et al. (1989). *S. typhi* high molecular weight chromosomal DNA was extracted as described by Frankel et al. (1989). Construction of the *S. typhi* genomic library in the cosmid vector cosKT1 was performed essentially as described by Tartof and Hobbs (1988). The ligation mixture was in vitro packaged in bacteriophage lambda transducing particles (DNA packaging kit, Boehringer GmbH, Mannheim, Germany) which were subsequently used to transduce the *E. coli* strain HB101. Immunoscreening of bacterial colonies with anti-fimbrial polyclonal antibodies was performed as described by Sambrook et al. (1989), the only exception being that transductants were first grown on Brain Heart Infusion (BHI) agar plates for 48 h at 37°C and subsequently lifted on nitrocellulose membranes (BA85, Scleicher & Schuell, Inc., Keene, N. H.). The preparation of rabbit polyclonal antibodies against *S. typhi* type 1 fimbriae has been described elsewhere (Satta et al., this volume). Anti-rabbit IgG coupled with horseradish peroxidase (Boehringer) were used to detect primary antibodies.

Adhesion and Invasion Assays

Adhesion to and invasion of epithelial cells by *S. typhi* and *E. coli* strains were assayed as described elsewhere (Satta et al., this volume). The human intestinal epithelial cell line INT-407 (Flow Laboratories, Inc., McLean, Va.) was used for adhesion and invasion assays.

ACKNOWLEDGMENTS

This work was supported in part by grant No. 90.00098.PF70 and No 92.01195.PF70 from the Italian National Research Council (C.N.R.) - Targeted Project "Biotecnologie e Biostrumentazione".

We are very grateful to Dr. Ian Blomfield and Colleagues for having kindly provided us with the *E. coli* strains carrying a complete deletion of the endogenous *fim* genes.

REFERENCES

Bloch, C.A., Stocker, B.A.D., and Orndorff, P.E., 1992, A key role for type 1 pili in enterobacterial communicability, *Mol. Microbiol.* 6:697.

Blomfield, I.C., McClain, M.S., and Eisenstein, B.I., 1991, Type 1 fimbriae mutants of *Escherichia coli* K12: characterization of recognized afimbriate strains and construction of new *fim* deletion mutants, *Mol. Micorbiol.* 5:1439.

Boyer, H.W., and Roulland-Dussoix, D., 1969, A complementation analysis of the restriction and modification of DNA in *Escherichia coli*, *J. Mol. Biol.* 41:459.

Chang, A.C.Y., and Cohen, S.N., 1977, Construction and characterization of amplifiable multicopy DNA cloning vehicles derived from the P15A cryptic miniplasmid, *J. Bacteriol.*, 134:1141.

Clegg, S., and Gerlach, G.F., 1987, Enterobacterial fimbriae, *J. Bacteriol.* 169:934.

Duguid, J.P., Anderson, E.S., and Campbell, I., 1966, Fimbriae and adhesive properties in salmonellae, *J. Pathol. Bacteriol.* 92:107.

Elsinghorst, E., Baron, L.S., and Kopecko, D.J., 1989, Penetration of human intestinal epithelial cells by *Salmonella*: molecular cloning and expression of *Salmonella typhi* invasion determinants in *Escherichia coli*, *Proc. Natl. Acad. Sci. USA* 86:5173.

Finlay, B.B., and Falkow, S., 1988, Virulence factors associated with *Salmonella* species, *Microbiol. Sci.* 5:324.

Finlay, B.B., and Falkow, S., 1989, Common themes in microbial pathogenicity, *Microbiol. Rev.* 53:210.

Frankel, G., Newton, S.M., Schoolnik, G.K., and Stocker, B.A.D., 1989, Intragenic recombination in a flagellin gene: characterization of the H1-j gene of *S. typhi*, *EMBO J.* 10:3149.

Halula, M.C., and Stocker, B.A.D., 1987, Distribution of the mannose-resistant hemagglutinin produced by *Salmonella* species, *Microbial Pathogenesis* 3:455.

Korhonen, T.K., Nurmiaho, E-L., Ranta, E., and Svanborg Edén, C., 1980, New method for isolation of immunologically pure pili from *Escherichia coli*, *Infec. Immun.* 27:569.

Lockman, H.A., and Curtiss III, R., 1992, Virulence of non-type 1-fimbriated and nonfimbriated nonflagellated *Salmonella typhimurium* mutants in murine typhoid fever, *Infect. Immun.* 60:491.

Ofek, I., Mirelman, D., and Sharon, N., 1977, Adherence of *Escherichia coli* to human mucosal cells mediated by mannose receptors, *Nature* 265:623.

Ofek, I., and Sharon, N., 1990, Adhesins as lectins: specificity and role in infection, *Curr. Topics Microbiol. Immunol.*

Orndorff, P.E., and Bloch, C.A., 1990, The role of type 1 pili in the pathogenesis of *Escherichia coli* infections: a short review and some new ideas, *Microbial Pathogenesis* 9:75.

Paranchych, W., and Frost, L.S., 1988, The physiology and biochemistry of pili, *Adv. Microb. Physiol.* 29:53.

Sambrook, J., Fritsch, E.F., and Maniatis, T., 1989, "Molecular Cloning: a Laboratory Manual", 2nd ed., Cold Spring Harbor Laboratory, Cold Spring Harbor, N.Y.

Svanborg Edèn, C., and Hansson, H.A., 1978, *Escherichia coli* pili as possible mediators of attachment to human urinary tract epithelial cells, *Infec. Immun.* 21,229.

Tartof, K.D., and Hobbs C.A., 1988, New cloning vectors and techniques for easy and rapid restriction mapping, *Gene* 67:169.

THE APPLICATION OF *SALMONELLA ENTERITIDIS aroA* VACCINES IN CHICKENS

Gerard L. Cooper[1], Lindsay M. Venables[1], Robin A.J. Nicholas[1],
Gavin A. Cullen[1] and Carlos E. Hormaeche[2]

[1] Bacteriology R+D Discipline,
Central Veterinary Laboratory,
Weybridge,
Surrey KT15 3NB, England

[2] University of Cambridge,
Department of Pathology,
Tennis Court Road,
Cambridge
CB2 1QP, England

INTRODUCTION

The incidence of *S. enteritidis* infection in humans has increased steadily since the mid-1980's and is now a significant public health problem in Britain (PHLS/SVS update, 1992). In 1990 *S. enteritidis* phage type 4 (PT4), associated primarily with infected chicken eggs and meat, accounted for *ca.* 35% of all *Salmonella* infections (PHLS/SVS update, 1991). In the chicken this serotype may cause invasive disease in broilers leading to increased mortality in some cases. In laying hens colonisation of the reproductive organs, with gross pathology detectable at *post-mortem,* may have no deleterious effect on egg production (Cooper *et al.*, 1989). Infection of the testes in male chickens (Bygrave and Gallagher, 1989) may result in venereal transmission within breeding flocks. It has been recognised for some time that *S. enteritidis* can be transmitted to progeny chickens by *in-ovo* infection (Snoeyenbos *et al.*, 1969). Unlike the successful eradication of *S. pullorum/gallinarum,* based on the serological identification of reactors with a stained *S. pullorum* antigen, attempts to eradicate this serotype from breeding stock by a policy based on bacteriological isolation and slaughter do not appear to have had an impact on the incidence of *S. enteritidis* infection in humans in Britain.

Live attenuated *Salmonella* vaccines based on mutation of the *aroA* gene are safe and protective in many species (Hoiseth and Stocker, 1981; Tackett *et al.*, 1992; Segall and Lindberg, 1991; Begg *et al.*, 1990). Such a vaccine may protect chickens against vertical transmission and lateral spread of *S. enteritidis* within flocks. We have developed models of infection to evaluate the potential efficacy of such vaccines for chickens.

PREPARATION AND EVALUATION OF *S. ENTERITIDIS aroA* VACCINES

The *aroA* mutation was introduced from *S. typhimurium* LT2 *aroA* 554::Tn*10* by transduction with phage P22 HT *int* (provided by G. Dougan). In all challenge experiments the virulent *S. enteritidis* strain 109, phage type 4, marked with a resistance to nalidixic acid was used.

VACCINE Se795*aroA*

Oral vaccination of newly-hatched White Leghorn chicks with *S. enteritidis* Se 795*aroA*, derived from the mouse-virulent parent strain Se795 (Hormaeche *et al.*, 1991), led to reduced intestinal shedding following an oral challenge. After an intravenous challenge the colonisation of livers and spleens was reduced in the absence of significant circulating IgG antibodies (Cooper *et al.*, 1990).

VACCINES Se267*aroA* AND LA5*aroA*

Vaccine candidates were then prepared from current *S. enteritidis* PT4 isolates from chickens. Strain LA5 was obtained from a broiler chicken and strain 267 was isolated from a laying hen, at post-mortem (Cooper *et al.*, 1989). In BALB/c mice vaccine strains 267 *aroA* and LA5 *aroA* were equally avirulent with an LD_{50} by intraperitoneal route of *ca.*10^7 CFU.

Newly-hatched chicks were vaccinated orally with 10^9 colony forming units (CFU) or 10^5 CFU repeated at weekly intervals until day 21. At eight-weeks of age, before challenge, seroconversion to LPS was detected in *ca.* 20% of the chickens immunised with 10^9 CFU of either vaccine strain. Seroconversion was not detected in those chickens vaccinated using the 10^5 x4-dose regimen. Oral vaccination with LA5*aroA* or 267*aroA*, using either regimen, led to a reduction in intestinal shedding when chickens were challenged orally. There was significant seroconversion in controls following oral challenge whereas antibody responses in vaccinated chickens indicated a prevention or limitation of invasion. There was also a reduction in colonisation of spleens and livers following an intravenous challenge.

These results suggested that, in the absence of circulating IgG antibodies, *aroA* vaccines given orally could induce a degree of cell-mediated immunity and induced protection at the gut level (Cooper *et al.*, 1992).

THE LAYING HEN

It is essential that a live vaccine, if it is to be of any practical use in the field, must protect the reproductive tract of laying hens. Newly-hatched hen chicks were vaccinated orally with 10^9 CFU of strain 267*aroA* and again at two weeks of age. A second group received booster doses at 16 and 18 weeks old. When challenged intravenously at 22-weeks-old there was a statistically significant reduction in the number of ovaries colonised in both vaccinated groups. Booster vaccination did not reduce the numbers of challenge strain in the ovary from the numbers found in chickens vaccinated with two doses only. However, colonisation of the caeca was completely prevented in the booster group suggesting increased immunity at the gut level which might prove desirable in the field.

LATERAL SPREAD

To prevent lateral spread and amplification of *S. enteritidis* PT4 infection within a flock young chickens should have sufficient immunity to prevent colonisation, either by the faecal-oral route or via the pericloacum. To simulate lateral spread we vaccinated newly-hatched chicks and introduced infected seeder chickens at three weeks of age. Colonisation of controls by the challenge strain resulted in heavy shedding which was, in some cases, at least as heavy as in the seeders. In contrast, oral vaccination prevented or significantly reduced colonisation by the challenge strain.

CROSS-PROTECTION

S. typhimurium shares the LPS O-12 antigen with that of *S. enteritidis* and remains the second most common serotype isolated from chickens. Newly-hatched chickens vaccinated orally with *S. enteritidis* Se 267*aroA* were challenged at eight weeks of age by intravenous or oral routes using a virulent *S. typhimurium* strain. There was no reduction in intestinal shedding or numbers of the *S. typhimurium* strain in vaccinated birds (Cooper *et al.*, submitted, 1992). Hormaeche *et al.* (1991) found that *S. typhimurium aroA* conferred very good protection against a homologous challenge but gave no protection against *S. enteritidis*.

SUMMARY

S. enteritidis aroA vaccines reduced intestinal shedding, reduced numbers of challenge organisms in livers, spleens ovaries and caeca and prevented or limited invasion after oral challenge. They reduced lateral spread from infected seeder birds but gave no protection against an oral or intravenous challenge with *S. typhimurium*.

REFERENCES

Begg, A.P., Walker, K.H., Love, D.N. and Mukkur, T.K.S., 1990, Evaluation of protection against experimental salmonellosis in sheep immunised with 1 or 2 doses of live aromatic-dependent *Salmonella typhimurium*, *Aus.Vet. J.* 67:294.

Bygrave, A.C. and Gallagher, J., 1989, Transmission of *Salmonella enteritidis* in poultry (letter), *Vet. Rec.* 124:571.

Cooper, G.L., Nicholas, R.A.J., and Bracewell, C.D., 1989, Serological and bacteriological investigations of chickens from flocks naturally infected with *Salmonella enteritidis, Vet. Rec.* 125:567.

Cooper, G.L., Nicholas, R.A.J., Cullen, G.A., and Hormaeche, C.E., 1990, Vaccination of chickens with a *Salmonella enteritidis aroA* live oral salmonella vaccine, *Microb. Pathog.* 9:255.

Cooper, G.L., Venables, L.M., Nicholas, R.A.J., Cullen, G.A., and Hormaeche, C.E., 1992, Vaccination of chickens with chicken-derived *Salmonella enteritidis* phage type 4 *aroA* live oral salmonella vaccines, *Vaccine* 10:247.

Cooper, G.L., Venables, L.M., Nicholas, R.A.J., Cullen, G.A. and Hormaeche, C.E., (submitted for publication 1992), Further studies on the application of *S. enteritidis aroA* vaccines in chickens.

Hoiseth, S.K. and Stocker, B.A.D., 1981, Aromatic dependent *Salmonella typhimurium* are non-virulent and effective as live vaccines, *Nature.*291:238.

Hormaeche, C.E., Joysey, H.S., Desilva, L., Izhar, M. and Stocker, B.A.D., 1991, Immunity conferred by Aro- salmonella live vaccines, *Microb. Pathog.* 10:149.

Public Health Laboratory Service and State Veterinary Service Update on *Salmonella* infection (edition 6). PHLS/SVS, Colindale, London, January 1991.

Public Health Laboratory Service and State Veterinary Service Update on *Salmonella* infection (edition 12). PHLS/SVS, Colindale, London, July 1992.

Segall, T. and Lindberg, A.A., 1991, *Salmonella dublin* experimental infection in calves: protection after oral immunisation with an auxotrophic *aroA* live vaccine, *J. Vet. Med.* 38:142.

Snoeyenbos, G.H., Smytser, C.F. and van Roekel, H., 1969, *Salmonella* infections of the ovary and peritoneum of chickens, *Avian Dis.* 13:668.

Tacket, C.O., Hone, D.M., Losonsky, G.A., Guers, L., Edelman, R. and Levine, M., 1992, Clinical acceptability and immunogenicity of CVD 908 *Salmonella typhi* vaccine strain, *Vaccine* 10:443.

EXPRESSION OF *BACILLUS ANTHRACIS* PROTECTIVE ANTIGEN IN *SALMONELLA TYPHIMURIUM*

Nick M. Coulson, and Richard W. Titball

Chemical and Biological Defence Establishment
Porton Down, Salisbury SP4 OJQ
United Kingdom

INTRODUCTION

Anthrax is an infectious disease known since antiquity. It is nearly universal in its geographic distribution and, although primarily a disease of herbivores, it can affect many species and it has important zoonotic implications. Anthrax is caused by *Bacillus anthracis*, a gram positive, non-motile, aerobic and facultatively anaerobic spore-forming organism. Two main virulence factors have been identified, one is a three-component protein exotoxin, termed the anthrax toxin, and the other a poly-D-glutamic acid capsule. These factors are carried on plasmids termed pXO1 and pXO2 respectively. Anthrax toxin is composed of three components, Protective Antigen (PA), Lethal Factor (LF) and Edema Factor (EF). PA is secreted as a 83 KDa protein which is cleaved in the serum, or on the cell surface, to release a 20 KDa N-terminal fragment. The remaining 63 KDa fragment binds to a specific cell surface receptor and can then bind either EF or LF. The combination of PA and LF is referred to as Lethal Toxin, and PA and EF is called Edema Toxin. This is illustrated in Figure 1. The toxins therefore fit the A/B model with the 63KDa C-terminal fragment of trypsin cleaved PA being equivalent to the receptor-binding region (B moiety) which can then bind and internalize either EF or LF (alternative A moieties). Edema Toxin causes oedema when injected intradermally in animals and EF is an adenylate cyclase. When injected intravenously in susceptible animals Lethal Toxin causes rapid death but it's mechanism of action is unknown although it is assumed to have an enzymatic action within the cytosol.

There is an effective parenterally administered Live Spore Vaccine (LVS) used in domesticated animals but which is obviously difficult to administer to susceptible wildlife, such as elephants in Africa. This LVS is not considered safe enough for human use in the west. The current human vaccines are made from filtered culture supernatant which consists mainly of PA. These vaccines require an initial four doses followed by annual boosters given by injection. Animal experiments have shown that a strong immune response to PA can be protective against challenge with even the most virulent strains of *B. anthracis* but the adjuvants used in the human vaccine do not appear to induce this level

Biology of Salmonella, Edited by F. Cabello
Plenum Press, New York, 1993

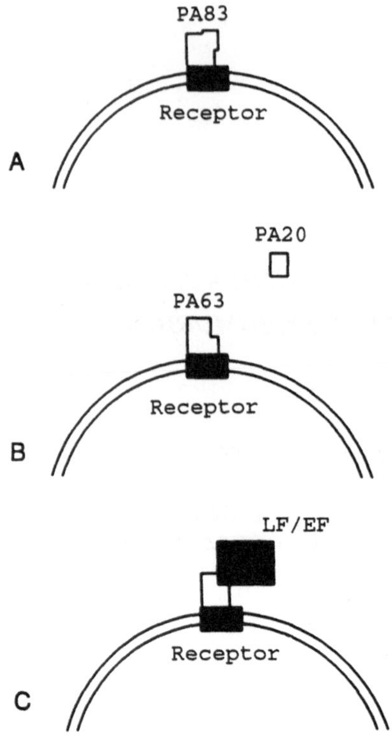

Figure 1. Schematic diagram of PA interaction with the cell. Adapted from Leppla (1991).
A. 83 KDa PA binds to cell specific receptor.
B. 20 KDa fragment cleaved off to expose LF/EF binding site on 63 KDa fragment.
C. LF or EF binds to 63 KDa fragment and is internalized by endocytosis.

of protection. Experimental work has also shown that after LVS use, protection can occur in the presence of only low anti-PA antibody titres suggesting that either the type of antibody is important or that cell-mediated immune responses are required.

A new human vaccine is required to improve the protection provided by the current vaccine preferably without the need for multiple injections. Although the LVS is suitable for domesticated animals, a single dose oral vaccine is the only practical approach for administration to wild herbivores. A vaccine vector which is orally active and will stimulate both protective antibodies and cell-mediated responses to PA is required.

During the search for a new effective oral typhoid vaccine, rationally attenuated Salmonella mutants have been developed. The experimental approach has used *Salmonella typhimurium* in a mouse model of the action of *S. typhi* in man. Several mutants have shown promise but the aromatic amino acid mutants (Hoiseth and Stocker, 1981) are now undergoing human trials. Early on in their development, it was realised that this system would have utility as a carrier of heterologous antigens to not only the mucosal immune system, but also the humoral and cell-mediated immune systems.

The *aroA⁻ S. typhimurium* vector was selected for initial studies. As it was known that the full length PA was only expressed at low levels from the *B. anthracis* promoter in *E. coli* an alternative strategy was needed. Therefore, as both the cell-receptor and LF/EF binding sites are located within the C-terminal 63 KDa fragment, it was decided to express this fragment from the *lac* promoter in a high copy number vector.

RESULTS

Plasmid pPA26 (Welkos et al., 1988) containing the open reading frame (ORF) of the PA gene (Figure 2) was moved via the r^-m^+ *S. typhimurium* LB 5010 into the *aroA*$^-$ *S. typhimurium* SL 3261. Bacteria were transformed by electroporation using a Biorad Gene Pulser at 1.25 kV, 25μF and 800 ohms. Expression of PA was barely detectable on Western blotting with either polyclonal or monoclonal antibodies (result not shown).

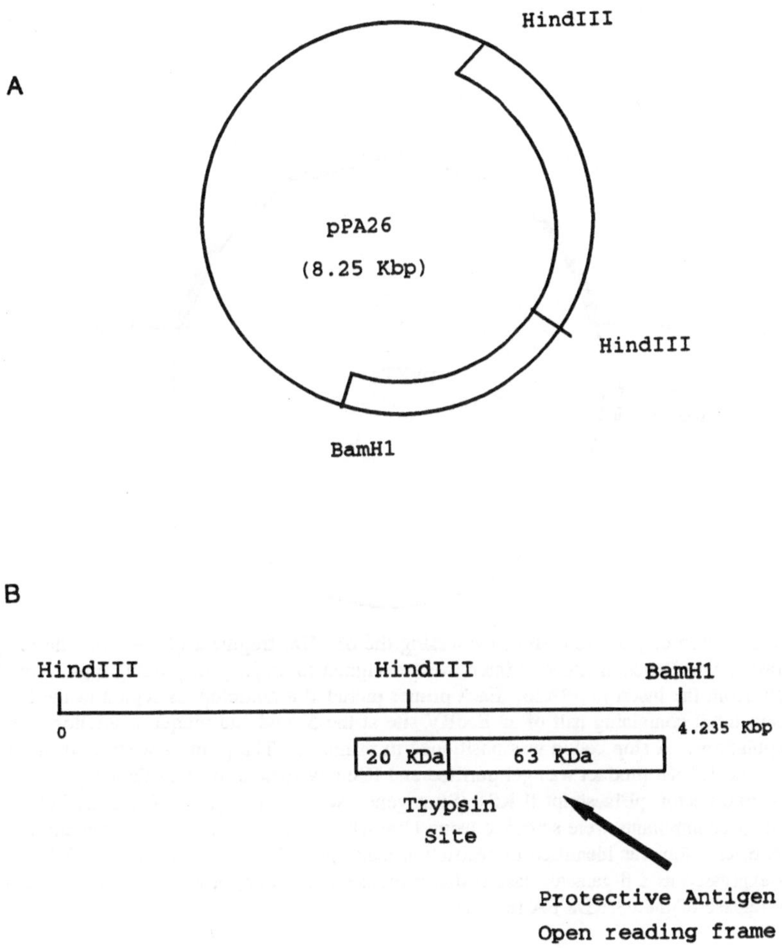

Figure 2. Details of plasmid pPA26

A. Plasmid pPA26 is a pBR322 based plasmid containing a 4.2 Kbp insert of *B. anthracis* DNA which encodes the full-length PA protein.

B. The position of the PA open reading frame is shown in relation to the 4.2 Kbp insert of pPA26. The mature protein is 83 KDa and consists of a 20 KDa N-terminal fragment which can be trypsin cleaved from a 63 KDa C-terminal fragment. The 63 KDa fragment contains both the cell and EF/LF binding regions.

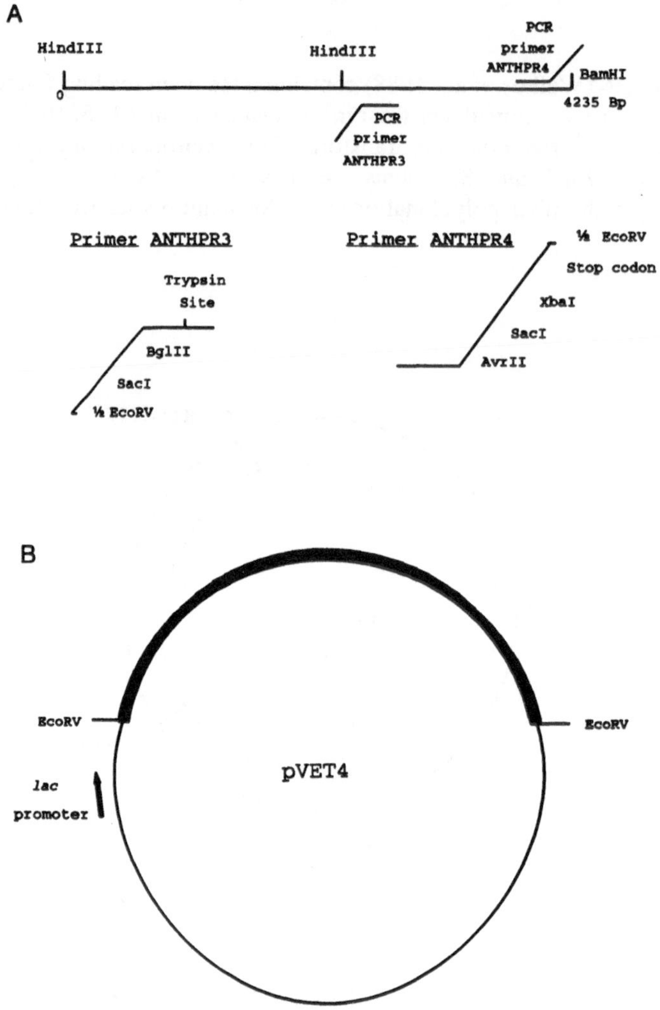

Figure 3. Construction of plasmid pVET4 expressing the 63 KDa fragment of PA from the *lac* promoter.
A. Polymerase Chain Reaction (PCR) primers were designed to amplify the coding region for the 63 KDa fragment from the insert in pPA26. Each primer included a homologous region to the PA gene of about 20 bases, a tail containing half of an EcoRV site at the 5' end and unique restriction sites to aid further manipulations. A stop codon was positioned in primer 2. The primers were used in a standard PCR reaction, the 1.7 Kb product was gel purified and Klenow treated to blunt the ends.
B. The phagemid vector, pBluescript II KS+ (Stratagene), was cut with EcoRV and the PCR product ligated into it. Recombinants were screened using blue/white selection and a clone containing the insert in the correct orientation was identified by restriction mapping. This was designated pVET4. The product was expressed as a β-galactosidase fusion protein incorporating a trypsin cleavage site which would allow release of the 63 KDa PA fragment.

As described in figure 3, plasmid pVET4 was constructed to express the 63KDa fragment of PA from the *lac* promoter in a high copy number vector. The plasmid expressed a fragment of the expected length in both *E. coli* and *S. typhimurium* SL 3261. This protein could only be visualised after western blotting and not on coomassie stained gels (see lane 3, figure 4). Shorter proteins can also be seen on the blot suggesting that the full length product is being degraded. In order to try and stabilise the full sized construct, it was decided to try and export the PA fragment into the periplasm using the signal sequence from the heat Labile toxin B-subunit (LT-B) of *E. coli.* A plasmid containing this sequence was obtained (Schodel and Will, 1989) and an appropriate fragment excised from pVET4 and cloned after the LT-B signal sequence. A clone expressing PA was designated pSIG13 and expressed a fragment of the expected length in *S. typhimurium*. Periplasmic and cytoplasmic preparations were made during expression of the protein in *E. coli* HB 101 and verified by the appropriate enzyme assays. The PA was found to be still within the cytoplasm (data not shown).

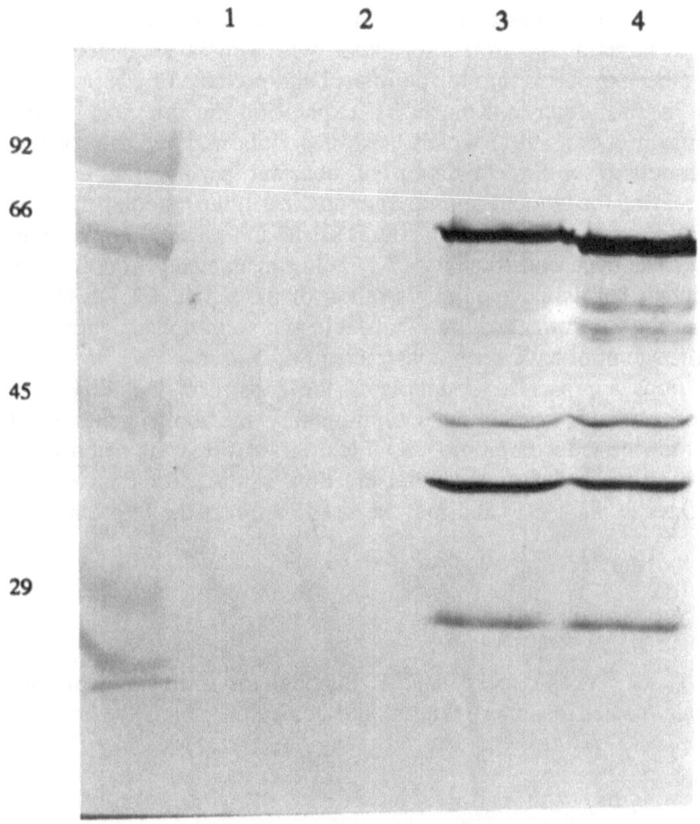

Figure 4. Expression of PA fragments in *aroA* attenuated *S. typhimurium* SL 3261
S. typhimurium SL 3261 *aroA* were transformed with either pPA26 (expressing PA from the *B. anthracis* promoter), pVET4 (expressing the 63 KDa PA fragment as a β-galactosidase fusion) or pSIG13 (with the PA fragment expressed after an *E. coli* signal sequence). Cultures were grown overnight in L-broth supplemented with ampicillin. Using OD measurements, equal numbers of cells (1-1.5 mls of culture) were spun and the pellet resuspended in 200 μl of 1xLB. The whole cell extracts were boiled for 3 mins. Aliquots of the extracts were run on 8% SDS-PAGE gels and blotted onto PVDF membrane. After blocking the membrane was probed with a mouse monoclonal antibody which binds to a site in the 63 KDa fragment. Binding was detected with a HRP linked anti-mouse secondary antibody and 3,3'-diaminobenzidine. Pre-stained molecular weight markers were run on the left hand side of the gel.

Lane 1 - *S. typhimurium* SL 3261
Lane 2 - *S. typhimurium* SL 3261/pPA26
Lane 3 - *S. typhimurium* SL 3261/pVET4
Lane 4 - *S. typhimurium* SL 3261/pSIG13

Groups of eight mice were vaccinated with *S. typhimurium* SL 3261 containing either pVET4, pSIG13 or pPA26. Mice were vaccinated orally on days 1 and 14 with approximately 10^{10} organisms. Anti-PA antibodies could not be detected in the sera from any of the groups. Preliminary data suggests that the plasmid was lost from the vaccine strains within a few days. This also occurs *in vitro* when selective pressure is not present.

DISCUSSION

The lack of detectable antibody response to these constructs is probably due to the small amount of antigen presented to the immune system. This could be a result of both the small amount of antigen produced and the instability of the constructs *in vivo*. Multiple intravenous immunisations may be more successful in eliciting a response.

The initial approach to increasing the amount of full-length product was unsuccessful. It was hoped that by targeting the protein for export with a signal sequence, the full-length protein might have been exported into the periplasm and protected from proteolytic action. This could have been further enhanced by expression in the *htrA* attenuated *S. typhimurium* (Johnson et al., 1991) which lacks a periplasmic serine protease which is thought to be associated with degradation of aberrant periplasmic proteins. Other approaches to increasing the amount of protein expressed in the cytoplasm include altering the promoter, mRNA secondary structure, codon usage and plasmid copy number. A *trc* promoter construct has been constructed and is being evaluated. Codon usage changes have been considered, but would require alteration of the whole 1.7 Kb ORF. There is some evidence that the C-terminal of PA is unstable and so various constructs containing natural and modified termini have been constructed.

There are various approaches to stabilising the constructs including chromosomal integration, use of plasmids containing complimentary metabolic genes or the use of anaerobic or macrophage induced promoters which only express the heterologous protein once the bacteria have reached certain cellular location. An anaerobic promoter, *nirB*, has been described (Oxer et al., 1991) and may be used to direct the expression of PA in a regulated manner.

ACKNOWLEDGMENTS

The authors thank B.A.D. Stocker and F. Schodel for strains and plasmids and G. Dougan and S. Fairweather for other material and advice.

REFERENCES

Hoiseth, S.K., and Stocker, B.A.D., 1981, Aromatic-dependent *Salmonella typhimurium* are non-virulent and effective as live vaccines, *Nature*. 291:238.

Johnson, K., Charles, I., Dougan, G., Pickard, D., O'Garoa, P., Costa, G., Ali, T., Miller, I. and Hormaeche, C., 1991, The role of a stress-response protein in *Salmonella typhimurium* virulence, *Mol Microbiol*. 5:401.

Leppla, S.H., 1991, The Anthrax Toxin Complex *in*: "Sourcebook of Bacterial Protein Toxins," J.E. Alouf, and J.H. Freer, ed., Academic Press, London.

Oxer, M.D., Bentley, C.M., Doyle, J.G., Peakman, T.C., Charles, I.G., and Makoff, A.J., 1991, High level heterologous expression in *E. coli* using the anaerobically-activated *nirB* promoter, *Nuc Acid Res*. 19:2889.

Schodel, F. and Will, H., 1989, Construction of a plasmid for expression of foreign epitopes as fusion proteins with subunit B of *Escherichia coli* heat-labile enterotoxin, *Infect Immun*. 57:1347.

Welkos, S.L., Lowe, J.R., Eden-McCutchan, F., Vodkin, M., Leppla, S.H. and Schmidt, J.J., 1988, Sequence and analysis of the DNA encoding protective antigen of *Bacillus anthracis*, *Gene*. 69:287.

SIV-SALMONELLA CONSTRUCTS AND THEIR POTENTIAL AS VACCINE CANDIDATES

Karen Strahan[1], Peter Kitchin[2] and Carlos Hormaeche[1]

[1]Dept. of Pathology, University of Cambridge, Tennis Court Road, Cambridge, CB2 1QP
[2]National Institute of Biological Standards and Control, Blanche Lane, Potters Bar, South Mimms, Herts., EN6 3QG

Live attenuated salmonellae are effective as carriers of antigens from other pathogens (combined live oral vaccines) and can elicit humoral, secretory and cellular immunity to the recombinant antigens.We have described a new class of live attenuated *Salmonella* vaccines harbouring lesions in *htrA*, a stress protein gene (Chatfield et al., 1992). These *Salmonella htrA* mutants are of reduced virulence when administered either orally or intravenously, and they are effective as live oral vaccines, conferring significant protection against oral challenge in BALB/c mice. The virulence and invasiveness of *Salmonella htrA* mutants was investigated in three models of increased susceptibility to salmonella infection. These included BALB/c mice, either given sublethal whole body irradiation (350 R) or administered rabbit anti-TNFα antiserum, and (CBA/N x BALB/C)F1 male mice which express the *xid* sex linked B cell defect of CBA/N mice and are more susceptible to salmonellae than female littermates. *S. typhimurium htra* mutants derived from virulent strains, C5046 (C5 *htrA*::Tn*phoA*) and BRD726 (SL1344 Δ*htrA*) were not more invasive in immunosuppressed mice than in normal controls in the three mouse models of defective immunity. These results indicate that susceptibility to *S. typhimurium htrA* vaccines is not enhanced by conditions of impaired resistance to infection.

SIV*gag*(p27), a viral structural protein, was expressed in a number of *Salmonella* vaccine strains. The plasmid pKA27, containing full-length p27 in pUC19, was introduced into *S. typhimurium* strains with *aroD* and *htrA* mutations and *S. dublin aroA*. Western blotting with anti-p27 McAb LH104-E showed that p27 was expressed in undegraded form by all transductants. The plasmid was retained *in vitro* without antibiotic selective pressure, but was lost when injected into mice and no antibody response to p27 was seen. Cellular responses to p27 and specific T-cell reponses, as determined by the production of interleukin-2, were also investigated. No specific T-cell response to p27 expressed in *Salmonella* was obtained.

Attempts to export the protein from the cell, in order to possibly reduce its toxicity and increase its immunogenicity were unsuccessful with both the pELB (Lei et al., 1987) and pLS408 (Newton et al., 1989) vectors. Attempts to increase the level of expression of p27 by replacing the *lac* promoter with the stronger *tac* promoter or the *in vivo* inducible *nirB* promoter were also unsuccessful. These results indicate that only moderate levels of expression of full length p27 can be tolerated by either *E. coli* or *Salmonella.*

An epitope(V2) from the envelope protein of SIV has been expressed in *E. coli* and *Salmonella* as a fusion to the B subunit of the heat-labile toxin (LT-B) of *E. coli*. LT-B is a potent immunogen and has attracted much attention as a potential vaccine. The immune response to foreign epitopes, which by themselves may be poorly immunogenic, might be improved when coupled to this carrier. This is currently under investigation with regard to the V2 epitope.

REFERENCES

Chatfield, S.N., Strahan, K., Pickard, D., Charles, I.G., Hormaeche, C.E., and Dougan, G. 1992. Evaluation of *Salmonella typhimurium* strains harbouring defined mutations in *htrA* and *aroA* in the murine salmonellosis model. Microb. Pathogen. 12: 145.

Lei, S.P., Lin, H.C., Wang, S.S., Callaway, J., and Wilcox, G. 1987. Characterisation of the *Erwinia carotovora pelB* gene and its product pectate lyase. J. Bacteriol. 169: 4379.

Newton., S.M.C., Jacob, C.O., and Stocker, B.A.D. 1989. Immune response to cholera toxin epitope inserted in *Salmonella* flagellin. Science 240: 1041.

Oxer, M.D., Bentley, C.M., Doyle, J.G., Peakman, T.C., Charles, I.G., and Makoff, A.J. 1991. High level heterologous expression in *E. coli* using the anaerobically-activated *nirB* promoter. Nuc. Ac. Res. 19:2889.

IN VIVO AND *IN VITRO* CHARACTERISATION OF ATTENUATED TN*PHO*A MUTANTS OF *SALMONELLA TYPHIMURIUM* C5

Mark Sydenham, Derek Pickard, Carlos Hormaeche#, Gordon Dougan* and Steven Chatfield

Vaccine Research Unit, Medeva Group Research, Imperial College of Science, Technology and Medicine, London SW7 2AY
#Department of Pathology, University of Cambridge, Cambridge, CB2 1QP
*Department of Biochemistry, Imperial College of Science, Technology and Medicine, London SW7 2AY

Defined mutations in such genes as *aro*A , *pur*A and *omp*R have been shown to attenuate *Salmonella* species, and strains harbouring such mutations make good, single dose oral vaccine strains (O'Callaghan *et al* 1988, Dorman *et al* 1989, Chatfield *et al* 1992). As an approach to identifying other virulence factors associated at or near the bacterial surface Tn*pho*A was used to mutagenise the mouse virulent strain of *S.typhimurium*, C5. The Tn*pho*A transposon is based on Tn*5* but carries the *Escherichia coli* alkaline phosphatase gene lacking the promoter and signal sequence (Manoil and Beckwith 1985). Functional activity of the alkaline phosphatase enzyme only occurs when secreted, thereby indicating insertion of the transposon into a gene whose product is secreted. Activity of the enzyme can be screened using a chromogenic substance such as X-gal, resulting in white colonies for non-functional activity and blue colonies for a functional fusion. Using this method a bank of mutants was obtained and screened for attenuation in BALB/c mice using the

Biology of Salmonella, Edited by F. Cabello *et al.*,
Plenum Press, New York, 1993

natural oral route of infection (Miller *et al* 1989). Of these, 15 were found to be attenuated, 9 having defects in LPS biosynthesis which were not used in this study. Of the remaining 6 mutants 2 have been shown to harbour a mutation in *htr*A, which encodes a stress protein and is thought to be important in survival within macrophages (Johnson *et al* 1991). A third mutant has now been shown to harbour a mutation in *osmZ*, which encodes a histone-like protein which may be involved in DNA supercoiling (Higgins *et al* 1988). The remaining 3 mutants were screened for their *in vivo* properties by both the oral and intravenous (i.v.) routes in BALB/c mice (Table 1).

Table 1. The *in vivo* properties of the attenuated *S. typhimurium* C5 Tn*phoA* mutants in BALB/c mice.

	Log LD50	
	i.v.	oral
C5	<1	5.0
BRD 439	5.41	9.17
BRD 441	2.94	9.14
BRD 873	5.22	9.73

As can be seen the levels of attenuation all differ, with mutants BRD439 and BRD873 being attenuated by more than 4 logs both i.v. and orally. Mutant BRD441 is only attenuated by 2 logs i.v. but 4 logs orally. The differing values may be a reflection of their ability to attach to and invade host tissues. The *in vitro* invasive properties of the mutants were characterised in a cell invasion assay using HEp-2 cells, a human larynx epidermal cell-line. Overnight stationary phase cultures of the mutant strains were incubated with HEp-2 monolayers for 3 hours, washed, treated with gentamycin for an hour to kill the extracellular bacteria, and the cells then lysed to release the intracellular bacteria. The percentage of bacteria invading the cells was then determined. All the strains show reduced invasion levels when compared to the wild-type parent strain, with the levels again differing between the three mutants. To determine whether these properties were a C5 strain restricted phenomenon or not, the attenuating lesion of BRD873 was transduced into the three other mouse virulent strains, *S.dublin,* SL1344, and HWSH, using P22 transduction. Whilst the *in vivo* properties of the new strains remain undetermined, the *in vitro* properties show that the introduction of the mutations results in reduced invasion levels in the HEp-2 cell assay when compared to the respective parent strains. The levels vary between the various backgrounds but all show at least a 75% reduction in invasion. Whether these reductions are reflected by a reduced ability of the mutants to bind to cells has not as yet been determined. The *in vitro* invasion levels are presented in Table 2.

Table 2. The *in vitro* levels of invasion of HEp-2 cells of Tn*phoA* mutants and their parent strains.

Strain	Invasion (% inoculum)	% wild type
C5	8.1%	100
BRD 439	2.57%	31.7
BRD 441	0.19%	2.34
BRD 873	0.73%	9.01
S.dublin	4.86%	100
BRD 874	0.11%	2.26
SL1344	13.8%	100
BRD 875	2.98%	21.59
HWSH	6.1%	100
BRD 876	0.71%	11.63

The insertion sites of BRD441 and BRD873 have now been sequenced and potential open reading frames determined. Comparison of sequence data has shown no homology with any published sequence data from invasion genes. Future work is to map these genes to the chromosome and to produce defined mutations to enable the characterisation of these "invasion" genes to be verified.

REFERENCES

Chatfield, S.N., Fairweather, N., Charles, I., Pickard, D., Levine, M.M., Hone, D., Posada, M., Strugnell, R.A., and Dougan, G., 1992, Construction of a genetically defined *Salmonella typhi* Ty2 *aro*A *aro*C mutant for the engineering of a candidate oral typhoid-tetanus vaccine, *Vaccine*. 10:53.

Dorman, C.J., Chatfield, S.N., Higgins, C.F., Hayward, C., and Dougan, G., 1989, Characterisation of porin and *omp*R mutants of a virulent strain of *Salmonella typhimurium*: *omp*R mutants are attenuated *in vivo*, *Infect. Immun.* 57:2136.

Higgins, C.F., Dorman, C.J., Stirling, D.A., Waddell, L., Booth, I.R., May, G., and Brewer, E., 1988, A physiological role of DNA supercoiling in the aromatic control of gene expression in *S.typhimurium* and *E.coli*, *Cell*, 52:569.

Johnson, K., Charles, I., Dougan, G., Pickard, D., O'Gaora, P., Costa, G., Ali, T., Miller, I., and Hormaeche, C., 1991, The role of a stress response protein in *Salmonella typhimurium* virulence, *Mol. Microbiol.* 5:401.

Manoil, C, and Beckwith, J., 1985, Tn*pho*A: A transposon probe for protein export signals, PNAS USA, 84:2833.

Miller, I.A., Maskell, D., Hormaeche, C., Johnson, K., Pickard, D., and Dougan, G., 1989, Isolation of orally attenuated *Salmonella typhimurium* following Tn*pho*A mutagenesis, *Infect. Immun.* 57:2758.

O'Callaghan, D., Maskell, D., Liew, F.Y., Easmon, C.S.F., and Dougan, G., 1988, Characterisation of aromatic- and purine-dependent *Salmonella typhimurium*; attenuation, persistence and ability to induce immunity in BALB/c mice, *Infect. Infect.*, 56:419.

IS VACCINATION WITH *S. TYPHI* Ty21A (VIVOTIF BERNA) SAFE IN IMMUNOCOMPROMIZED INDIVIDUALS?

Jaap T. van Dissel, Gabriella Goudriaan, Harry Guiot, and Ralph van Furth

Department of Infectious Diseases
University Hospital Leiden
P.O. Box 9600
2300 RC The Netherlands

INTRODUCTION

For protection against many infectious diseases including typhoid fever, travellers from industrialized countries who visit endemic regions in the developing world count on the efficacy of vaccination. Healthy persons can be effectively vaccinated against typhoid fever by the oral, live-attenuated *S. typhi* Ty21a, Vivotif Berna[tm] (Gilmaan et al, 1977; Woodruff et al, 1991). In immunosuppressed persons, however, it is assumed that there is an increased risk for adverse effects from vaccination with live vaccines including systemic infection with the attenuated vaccin strain. Therefore, *S. typhi* Ty21a is not administered to individuals immunocompromized due to, for instance, the use of high dose glucocorticosteroids, HIV-infection or hematologic malignancies.

S. typhi Ty21a possesses various mutations induced by random chemical mutagenesis, including the *galE* mutation that prevents the normal metabolism of galactose. In the presence of exogenous galactose, lipopolysaccharide is synthesized but accumulation of intermediate products results in osmotic lysis of the micro-organisms; in the absence of galactose, the bacteria cannot complete lipopolysaccharides and form a fragile outermembrane (Germanier and Furer, 1975).

Because of the vital mutations in the vaccin strain, it is assumed that in vivo, Ty21a does not persist in the tissues, irrespective of the immune status of the host. For obvious reasons, this supposition cannot be studied in immunosuppressed humans. This prompted us to investigate whether Ty21a can replicate in the tissues of immunosuppressed animals and cause a generalized, systemic infection.

Biology of Salmonella, Edited by F. Cabello *et al.*,
Plenum Press, New York, 1993

MATERIAL AND METHODS

Animals

Specific-pathogen free female Swiss mice, 20 to 25 g, were used.

Interventions to Render Animals Immunosuppressed

Hydrocortisone 15 mg s.c. was administered one day before experiments to cause a prolonged (≥ 14 days) reduction of the number of blood monocytes and lymphocytes; at the same time, this treatment results in an increase of peripheral blood granulocytes (Thompson and van Furth, 1970).

Whole-body irradiation with 8 Gy was given five days before experiments to cause a rapid and irreversible reduction in the number of blood granulocytes, monocytes and lymphocytes; the median survival time of irradiated animals is 14 days (van 't Wout et al, 1989).

Micro-organisms

The *galE* mutant of *S. typhi* Ty2, Ty21a (Vivotif Berna[tm]), was cultured overnight in Difco medium containing either 0.25% glucose [glu] or 0.25% galactose and 0.0025% glucose [gal.glu].

Infection of Normal and Immunosuppressed Animals

To investigate whether macrophages and granulocytes play an pivotal role in the eradication of *S. typhi* Ty21a, the course of infection was followed in normal and immunosuppressed mice after intravenous (i.v.) or intramuscular (i.m.) injection of live bacteria. Prior to injection, Ty21a were cultured overnight in either galactose and glucose [gal.glu] or glucose [glu] only. At various intervals after i.v. or i.m. (into thigh muscle) injection of bacteria, groups of four mice were killed. The liver, spleen and thigh muscle were removed and the number of viable bacteria per organ determined by tissue homogenization followed by a plate count of serially diluted homogenates.
Colony counts were used to calculate the geometric mean of viable *S. typhi* Ty21a per organ; in addition, rate constants of elimination or growth (dimension: hr^{-1}) were calculated according to the equation: rate constant = $[\ln N(t2) - \ln N(t1)]/t2-t1$, in which $N(t1)$ and $N(t2)$ are the number of viable bacteria at time t1 and t2 (hr).

RESULTS

During in vitro culture of Ty21a in Difco medium containing galactose but no glucose, the number of viable micro-organisms decrease, probably due to bacteriolysis. In medium containing galactose and glucose [gal.glu], however, a rapid outgrowth of Ty21a is observed at a rate identical to that in medium containing glucose [glu] only (Figure 1.). In a slide test, Ty21a cultured overnight in [gal.glu] agglutinated with polyvalent and specific (9 and 9,12) antiserum against O-side chains, whereas no agglutination was observed with Ty21a cultured overnight in [glu] (data not shown).

Following i.v. injection of 2×10^4 to 2×10^6 bacteria, the number of Ty21a [glu] and Ty21a [gal.glu] in the liver and spleen decreased in normal, and hydrocortisone-treated and irradiated mice (Table I). In irradiated mice, bacterial elimination was somewhat less rapid compared to normal and hydrocortisone-treated animals. Furthermore, in all mice, Ty21a [glu] were eliminated from the organs more rapidly ($p<0.01$) than bacteria cultured overnight with [gal.glu] (Table I).

In vitro experiments showed that fresh mouse serum was not bactericidal for Ty21a [gal.glu] or Ty21a [glu], indicating that the reduction of viable bacteria after i.v. injection was not due to serum sensitivity of Ty21a.

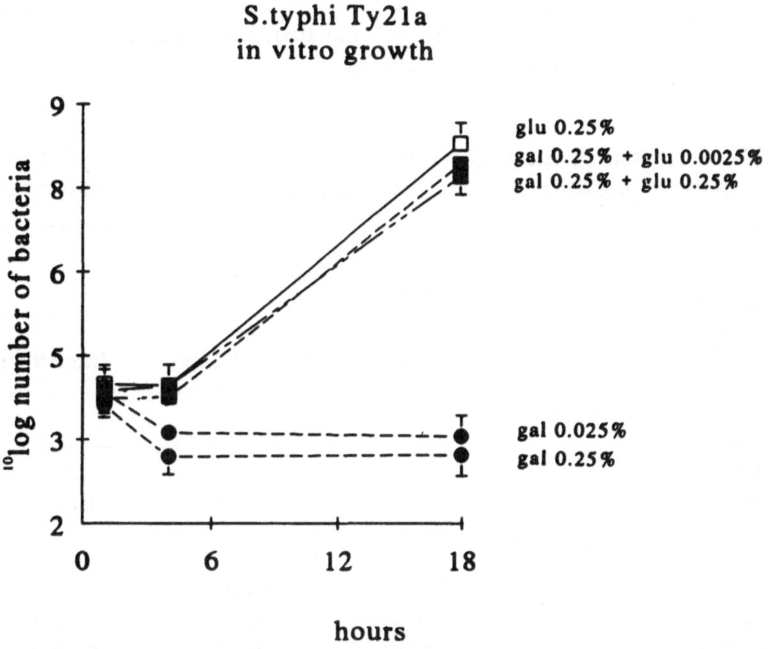

Figure 1. In vitro culture of the *galE* mutant of *S. typhi*, the Ty21a vaccin strain, in medium supplemented with galactose, glucose, or galactose and glucose (percentage indicated on the righthand-side of the graph). Minute amounts of exogenous glucose prevent bacteriolysis of Ty21a when cultured in the presence of galactose.

In normal mice, 1-2 $x10^6$ live Ty21a injected i.m. were eradicated from the thigh muscle in three to four days. In hydrocortisone-treated mice, bacterial elimination occurred even more rapid: no live Ty21a could be cultured from muscle homogenates two days after injection of the 1-2 $x10^6$ micro-organisms. In both normal and hydrocortisone-treated mice, Ty21a [glu] was eliminated much more rapidly from the thigh muscle than Ty21a [gal.glu].

By contrast, a marked outgrowth to more than 10^8 bacteria per tighmuscle characterized the course of infection after i.m. injection of 1-2 $x10^6$ live Ty21a into irradiated mice; the rate of bacterial outgrowth was identical for Ty21a [gal.glu] and Ty21a [glu] (Table I). In addition, bacteria could be recovered from the liver and spleen after i.m. injection of Ty21a [gal.glu] or Ty21a [glu] in irradiated mice but not in normal or hydrocortisone-treated animals, suggesting that after i.m. injection into whole-body irradiated, leukocytopenic mice a generalized infection occurs. These mice died 5 to 7 days after the i.m. injection, seemingly due to the infection with *S. typhi* Ty21a.

DISCUSSION

From the findings of this study it may be concluded that the elimination of *S. typhi* Ty21a from the tissues of mice depends on the integrity of the cellular defense mechanisms of the host: resident macrophages and especially exudate leukocytes are necessary to suppress the growth of Ty21a in the tissues of mice after i.v. or i.m. injection. This conclusion is supported by the finding that in irradiated, leukocytopenic mice, Ty21a can cause a systemic infection that can even be lethal.

Table 1. Rate constants (hr^{-1}) of elimination or growth of Ty21a in immunosuppressed mice after intravenous or intramuscular injection[1]

S. typhi Ty21a administered	Treatment of mice		
	none	hydrocortisone*	irradiation**
intravenous		spleen	
[glu]	-2.31	-2.12	-1.84
[gal.glu]	-1.51	-1.61	-1.22
intramuscular		muscle	
[glu]	-0.11	-0.23	+0.096
[gal.glu]	-0.07	-0.15	+0.092

[1] elimination rates (negative sign) or outgrowth rates (positive sign) were determined from geometric mean bacterial counts of spleen or muscle; *S. typhi* Ty21a was cultured overnight in Difco medium supplemented with either galactose and glucose [gal.glu], or glucose only [glu].
* hydrocortisone acetate 15 mg sc, one day before experiment.
** whole-body irradiation 8 Gy, five days before experiment.

Treatment of mice with hydrocortisone causes a prolonged reduction of the number of blood monocytes and lymphocytes but an increase in number of blood granulocytes, and results in an increased number of granulocytes in an inflammatory exudate compared to that in untreated mice (Silva et al., 1987). The more rapid elimination of i.m. injected Ty21a in hydrocortisone-treated mice compared to normal mice, therefore, is probably related to the higher number of exudate granulocytes in the hydrocortisone-treated mice that effectively deal with Ty21a.

In whole-body irradiated mice that lack circulating granulocytes and monocytes, Ty21a injected i.v. are rapidly cleared from the circulation and subsequently eliminated from the liver and spleen, presumably by resident macrophages that are not affected by the irradiation. However, irradiation prevents the increase in the number of blood granulocytes and blood monocytes and the influx of these cells into an inflammatory exudate in mice undergoing an acute inflammatory response (van 't Wout et al, 1989). In irradiated mice, the outgrowth of Ty21a after injection into muscle, i.e., tissue that lacks resident macrophages, indicates that exudate phagocytes play a pivotal role in eradication of Ty21a from the tissues.

S. typhi Ty21a cultured overnight in galactose and glucose complete the synthesis of the O-side chains; Ty21a grown in medium with glucose only do not complete the lipopolysaccharides. The combined in vitro and in vivo findings of this study indicate that completed lipopolysaccharide make Ty21a less vulnerable to the cellular defense mechanisms of the host: Ty21a [gal.glu] was eliminated less rapidly than Ty21a [glu] from muscle, liver and spleen. These results extend previous findings of others showing that, after ingestion by phagocytes, the intracellular killing of rough Gram-negative bacteria that lack complete lipopolysaccharide is more rapid than that of smooth bacteria of the same species (Weiss et al, 1980; Weiss et al, 1986).

In vitro, the presence of minute amounts of glucose prevents the bacteriolysis of *S. typhi* Ty21a cultured in medium with galactose. Obviously, in vivo, glucose is present in all tissues; thus it seems unlikely that the *galE* mutation alone causes attenuation of the bacteria (Silva et al, 1987). Indeed, recently it was shown that *galE* mutation does not cause attenuation of Ty2 in humans (Hone et al, 1988).

Although the precise nature of attenuation of Ty21a remains unclear, several large clinical trials and years of general use have proven the overall safety of the oral live-attenuated *S. typhi* Ty21a; no major complications of the Ty21a vaccin have been described in humans. Still, in immunocompromized persons, a parenteral, killed vaccin presents a safer alternative, when a protective immune reaction is anticipated in these individuals. Such a strategy is supported by our study showing that, at least in mice, eradication of *S. typhi* Ty21a depends on the functional activity of macrophages and granulocytes and is not due to an instrinsic property of the attenuated micro-organisms.

SUMMARY

To investigate whether S. typhi Ty21a can cause a systemic infection in immunosuppressed animals, mice were treated with hydrocortisone or received whole-body irradiation prior to injection of the vaccin strain. After i.v. injection, the bacteria were rapidly eliminated from the liver and spleen, presumably by phagocytosis and intracellular killing by resident macrophages; this elimination was not affected by the interventions to render mice immunosuppressed. Furthermore, the results indicate that after i.m. injection, eradication of the bacteria depends on an influx of exudate granulocytes and monocytes.

REFERENCES

Germanier, R. and Furer, E., 1975, Isolation and characterization of gal E mutant Ty 21a of *Salmonella typhi*: a candidate strain for a live, oral typhoid vaccine, *J. Infect. Dis.* 131:553.

Gilman, R.H., Hornick, R.B., et al., 1977, Evaluartion of a UPD-glucose-4-epimeraseless mutant of *Salmonella typhi* as a live oral vaccine, *J. Infect. Dis.* 136:717.

Hone, D.M., Attridge, S.R., et al., 1988, A *galE via* (Vi antigen-negative) mutant of *Salmonella typhi* Ty2 retains virulence in humans, *Infect. Immun.* 56:1326.

Silva, B.A., Gonzalez, C., et al., 1987, Genetic characteristics of the *Salmonella typhi* strain Ty21a vaccine, *J. Infect. Dis.* 155:1077.
Thompson, J. and van Furth, R., 1970, The effect of glucocorticosteroids on the kinetics of mononuclear phagocytes, *J. Exp. Med.* 131:429.
Weiss, J., Beckerdite-Quagliata, S., Elsbach, P., 1980, Resistance of Gram-negative bacteria to purified bactericidal leukocyte proteins. Relation to binding and bacterial lipopolysaccharide structure, *J. Clin. Invest.* 65:619.
Weiss, J., Hutzler, M., Kao, L., 1986, Environmental modification of lipopolysaccharide chain length alters the sensitivity of *Escherichia coli* to the neutrophil bactericidal/permeability-increasing protein, *Infect. Immun.* 51:594.
Woodruff, B.A., Pavia, A.T., et al., 1991, A new look at typhoid vaccination. Informationfor the practicing physician, *J. Am. Med. Assoc.* 265:756.
van 't Wout, J., Linde, I., et al., 1989, Effect of irradiation, cyclophosphamide, and etoposide (VP-16) on the number of peripheral blood and peritoneal leukocytes in mice under normal conditions and during acute inflammatory reaction, *Inflammation* 13:1.

MECHANISMS OF ACQUIRED IMMUNITY INDUCED BY A LIVE ATTENUATED *SALMONELLA ABORTUSOVIS* (RV6) VACCINE

Serge Bernard, Laurence Guilloteau, Dominique Buzoni-Gatel, Françoise Bernard, Michel Zygmunt, Guy Bézard and Frédéric Lantier

INRA
Laboratoire de Pathologie Infectieuse et Immunologie
37380 Nouzilly, France

INTRODUCTION

Salmonella abortusovis (SAO) is a pathogen specific for sheep that induces abortion. The role of the immunity in this disease was initially studied in a murine model. A live attenuated vaccine, the SAO Rv6 strain has been developed against ovine abortive salmonellosis. This vaccine efficiently controls the disease in domestic animal husbandry (Pardon et al., 1980; Pardon et al., 1990). In mice, this strain induces a specific and high immunity (Lantier et al., 1981; Lantier et al., 1983). These conditions lead the Rv6 strain as a convenient model for studying the active immunity against the causal agent.

In mouse, innate resistance or susceptibility to infection with several types of intracellular pathogens, including *Salmonella* (Plant and Glynn, 1974), *Leishmania* (Bradley, 1974) and *Mycobacteria* (Skamene et al., 1982) was shown to be controlled on the chromosome 1 by the *Ity/Lsh/Bcg* locus. In sheep, involvement of genetic control of susceptibility to *Salmonella* infection is postulated. The level of the infection in mice (Oswald et al., 1992) and sheep is characterized by inter-individual variations in the degree of bacterial control, and development of immunity (Lantier personal commination).

In mice, susceptibility to acute *S. typhimurium* desease was expressed within few hours after infection (O'Brien and Metcalf, 1982; Swanson and O'Brien, 1983). Bone marrow reconstitution experiments established that effector cells responsible for resistance against BCG (Gros et al., 1983), against *Leishmania* (Crocker et al., 1984), and against

Biology of Salmonella, Edited by F. Cabello *et al.*,
Plenum Press, New York, 1993

Salmonella (Hormaeche et al., 1990), were macrophages. Nevertheless, control of intracellular *Salmonella* infections of mononuclear phagocytes is associated with the development of host T-cell protective immunity (for review, see Maskell et al., 1987; Nauciel, 1990, Guilloteau et al., 1992). Most of the accumulated evidences has indicated that activated macrophages and T cells cooperate to mediate the destruction of the pathogen in the infected host (Murray et al., 1985; Nakato et al., 1990).

We have investigated the role of T cells in acquired immunity, with the Rv6 model of vaccination in mice, by adoptive transfer of primed spleen cells from vaccinated donor to naive recipient mice that were then challenged with a virulent strain of SAO.

RESULTS AND DISCUSSION

Passive Transfer of Acquired Immunity with Immune Spleen T Cells

The course of SAO infection was monitored in mice after passive transfer of unprimed or primed spleen cells from donor mice. Transfer of $1X10^8$ primed spleen cells, from one month vaccinated female BALB/cby mice (*Ity^S* strain) to syngeneic recipient mice, significantly reduced the number of *Salmonella* in the spleen of recipient mice challenged with the virulent strain of SAO, compared to naive mice. Groups of mice that received no cell or unprimed cells, developed a severe infection and died before Day 10. On Day 10, the number of bacteria in the spleen of mice transferred with primed cells had significantly decreased, compared to Day 3 or Day 6. Primed cells from vaccinated mice were able to transfer protection to naive mice even 6 months after vaccination (fig. 1).

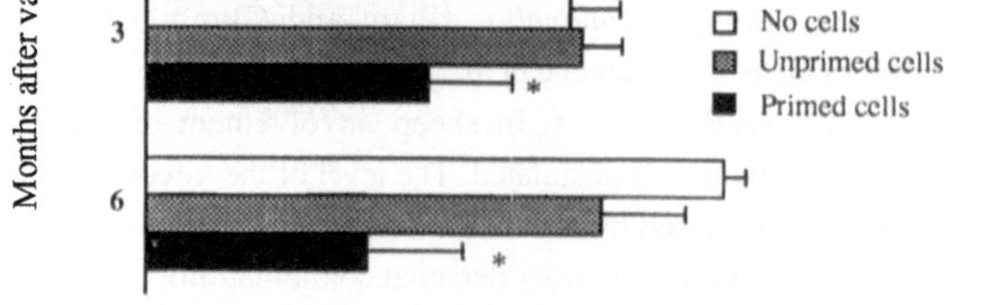

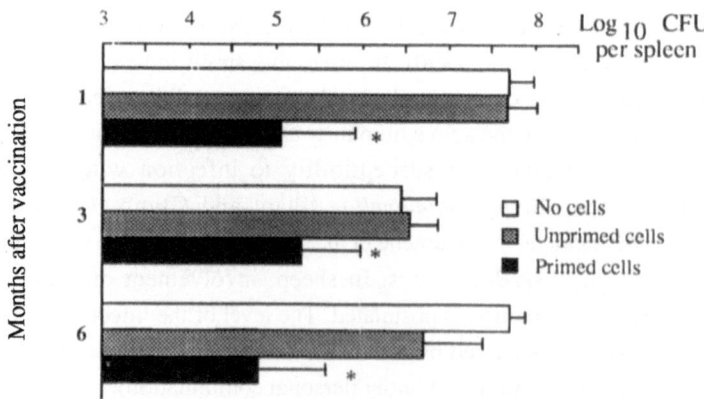

Figure 1. Passive transfer of immunity with spleen cells from mice vaccinated 1, 3 or 6 months previously with the SAO Rv6 strain. BALB/cby mice (8 weeks old) received $1X10^8$ cells from naive or primed SAO Rv6 mice ($5X10^7$ SAO Rv6 strain subcutaneously injected). At the time of transfer, no salmonella was detected in spleen and liver of vaccinated donor mice. Recipient mice, 24 h after the transfer, were challenged with $1X10^5$ viable virulent SAO 15/5 strain and were killed on Day 6 after the virulent challenge. Each data point is the mean number of *Salmonella* ± s.e. of at least 4 mice per group. Protection was evaluated by the enumeration of bacteria in the spleen. The statistical analysis was referred to naive mice : (*) :$P < 0.05$.

Effects of *In Vitro* Cell Depletion on the Adoptive Transfer of Protection with Immune Spleen T Cells

Treatment of primed spleen cells with normal rabbit serum or complement alone did not modify the protection conferred. On the contrary, treatment of cells with anti-Thy-1 serum with or without complement completely abrogated the protection measured on Day 6 after challenge, indicating that Thy-1 cells are mainly responsible of the transferred immunity. *In vitro* depletion of spleen T cell subsets with different monoclonal antibodies (Fig. 2) demonstrated the significant role of Lyt-2^+ cells in protection and to a lesser degree of L3T4^+ cells, although the difference between the mice which received anti-L3T4 treated cells and the one which received complement treated primed cells was not significant. Combined treatment of primed cells with anti-Lyt-2 and anti-L3T4 antibodies completely abrogated the ability of spleen cells to transfer cell mediated protection, demonstrating a probable active synergy between these two subsets of T cells. *In vitro* antibody treatment of primed cells did not modify the antibody response against challenge recipient mice (Fig. 2), suggesting there is no obvious relationship between an humoral antibody response from transferred B lymphocytes and protection, as already suggested (Eishentein and Sultzer, 1983). These results and the major role for L3T4^+ cell subset suggested in protection against *Salmonella thyphimurium* infection (Nauciel, 1990), could be an indication for a role of both L3T4^+ and Lyt-2^+ T cell subsets in transmission of passive protection against *Salmonella* (Guilloteau et al., 1992).

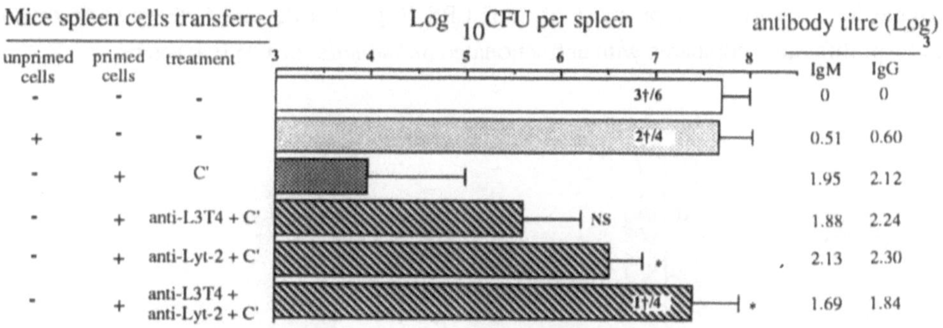

Figure. 2. Effects of *in vitro* treatment with anti-L3T4 anti-Lyt-2 monoclonal antibodies on the anti-SAO resistance adoptively transferred by primed cells. Spleen cells were incubated, with antibody and complement (C'). Naive mice were transfered with treated and untrated cells and challenged 24 h after in the conditions described figure 1. Bacterial counts were determined in the spleen on Day 6 after the challenge. Number of dead mice (†) was indicated in bars. The statistical analysis was referred to primed cells treated with complement : NS : not significant; (*) : $P < 0.05$. IgG and IgM anti-SAO antibody titres were determined by ELISA in pooled sera from each group of recipient mice on Day 6 after challenge.

Induction of Cell Lines Specific of SAO Cell Wall Fraction (CWF)

The central role of T cells to subsequent development of acquired immunity has been already claimed in murine salmonellosis (Collins, 1974; Chadler et al., 1986). More recently, adoptive transfer of *Salmonella* -specific T-cell lines from mice immunized with soluble antigen (spent medium antigen) from *S. typhimurium* protected naive syngeneic mice against a virulent challenge (Paul et al., 1985; Paul and Smith, 1988). To analyse the antigenic specificity of the protection obtained with T cells from immunized mice with the hypovirulent live SAO Rv6 strain, we decided to establish T cell lines.

Antigenic specificities of the T cells isolated from the spleen were performed with a proliferation assay. Whereas the dose response to Concanavalin A (0.1 to 10 µg/ml) of the spleen cells from primed and unprimed mice were similar, proliferation of spleen cells from vaccinated mice was significantly higher than the one of naive mice when stimulated by heat inactivated whole bacteria ($5X10^4$ to $5X10^7$ bacteria/ml). This stimulation was not only induced by the LPS fraction of the bacteria, since the response of the spleen cells from vaccinated mice to a purified LPS from *S. abortusequi* (SAE has the same O 4;12 serotype than SAO), gave the same response (or even less) than the one given by spleen cells from non immunized mice (data not shown).

Antigenic fractions from SAO bacteria were prepared as described for *Brucella* (Zygmunt et al., 1990). Briefly, washed heat inactivated bacteria were broken with glass beads in a Dyno-Mill apparatus, and first filtrated with hollow fiber cartridges of 0,1µm cutoff (Amicon). The cell wall fraction (CWF) in the retentate was collected and lyophilised. The ultrafiltrate was concentrated by a second ultrafiltration with hollow fiber cartridges of 100 kDa cutoff (Amicon). The cytoplasmic protein extracts (CEP H) in the retentate, and in the filtrate (CPE L) were collected and lyophilised. The spleen cell lymphocytes from mice vaccinated with Rv6 strain, were tested for their ability to proliferate in the presence of these antigenic fractions or in the presence of SAE LPS (Fig. 3). LPS and CPE gave no or non specific proliferation compared with unfractionated *Salmonella* or CWF antigens.

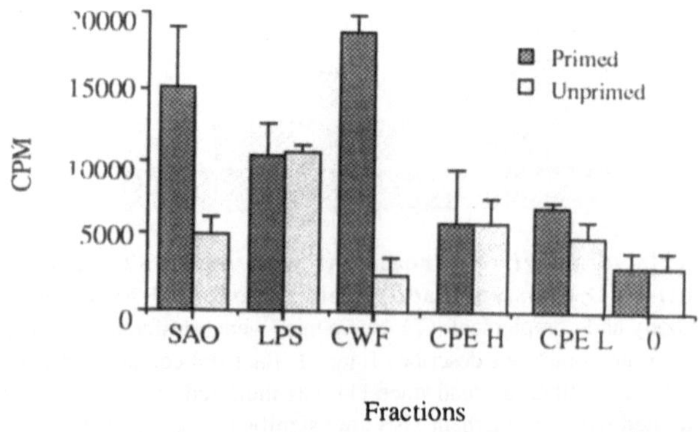

Figure 3. Mouse spleen cell proliferation assays with different antigenic preparations of SAO. Spleen cells from primed or unprimed mice ($2X10^5$ cells/well) have been cultured, for 4 days at 37°C, in 5% CO_2, in quadruplicate with mitogens or *Salmonella* antigens, in 96-well microtiter plates. Cells were pulsed with 1 µCi of {^{3} H} thymidine. The incorporation of radioactivity was measured 18 h later. . The cells were harvested onto glass filter (Scatron) and counted. SAO (10^6 CFU/ml), LPS from *Salmonella abortusequi* (0.25 µg/ml), cell wall fraction (CWF) and cytoplasmic protein extract (CPE H and CPE L) from SAO (2.5 µg/ml) were used as antigenic preparations. Results are expressed as means of quadruplicates, and standard deviation (bars).

Dose curve proliferation with CWF antigens showed that this fraction is able to generate a strong specific proliferation. High concentrations of CWF (>25 μg/ml) were toxic, whereas concentration between 2.5 and 25 μg/ml gave the best stimulation (fig. 4).

T cell lines were established from spleen cells obtained from mice vaccinated with SAO Rv6 strain. They were maintained and expanded by alternating weekly cycles of stimulation with CWF antigens and resting period without antigens (Vordermeier et al., 1991). After 4 cycles, the lines were assayed for Ag-specific proliferation responses (Fig. 5).

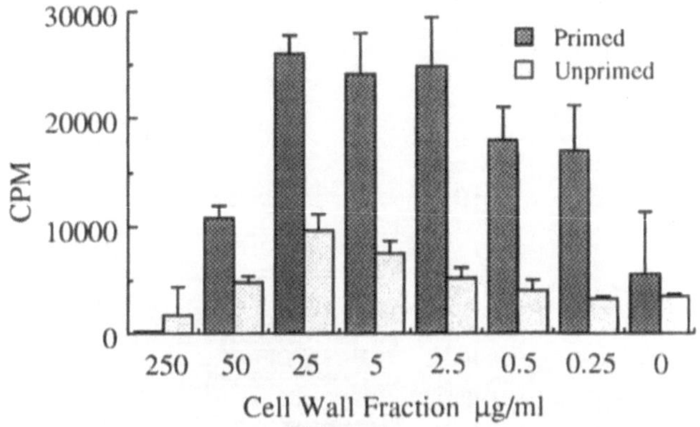

Figure 4. Dose curve proliferation of mice spleen cell with CWF antigens. Proliferation assay was done in the same conditions already described in legend of Fig. 3.

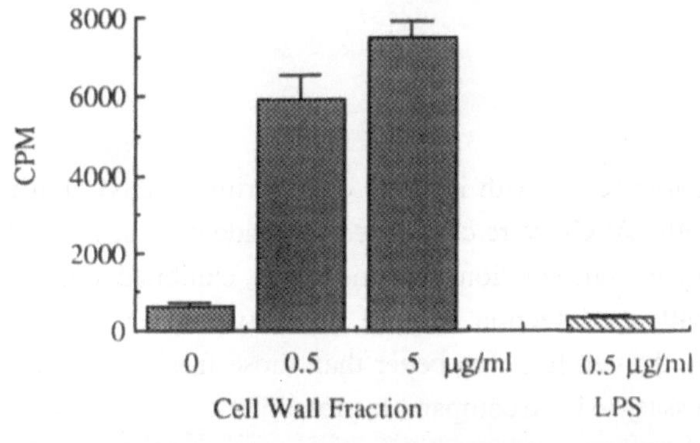

Figure 5. Cell line proliferation assays. T cell line was established from primed spleen cells from mice vaccinated with SAO Rv6 strain. They were maintained and expanded by alternating weekly cycles of stimulation with 4 μg/ml of CWF Ag, in the presence of $3X10^6$ irradiated spleen cells, and then resting without Ag.The cell line was checked for Ag-specific proliferation response by using $2X10^4$ cells/well and $3X10^5$ irradiated naive spleen cells/well (3000 rad). The final volume in each well of the 96-well microtiter plate, was 200 μl. General proliferation assay conditions were already described (Fig. 3).

The cell line responded in a dose-dependent manner to CWF, but not to homologous LPS. Twenty-four hour culture supernatant culture contained IL2 activity, whereas after 6 days of culture, with or without Ag, no anti-*Salmonella* antibody activity was found (data not shown). Immuno-enzymatic analysis after cyto-centrifugation revealed a majority of T cells. Unfortunately deeper study of the subset of these cells was not possible.

To assess the *in vivo* protective capacity of the *in vitro* maintained cells, these cells was passively transferred to naive mice. One day after the intravenous injection of $3X10^6$ cultured cells, the mice were challenged with the virulent 15-5 SAO strain. On the same

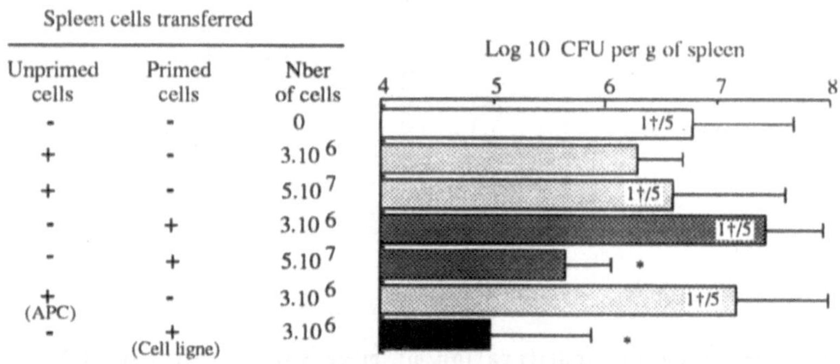

Figure 6. Passive transfer of immunity with specific CWF cell line established from primed mice spleen cells. Recipient BALB/cby mice (5) were injected intravenously with $3X10^6$ or $5X10^7$ viable spleen cells from naive or vaccinated donor mice ($5X10^7$ SAO Rv6 strain), $3X10^6$ cells from the CWF cell line or $3X10^6$ irradiated spleen cells (APC). Transferred mice were challenge 24 h later with $1X10^5$ viable virulent SAO 15/5 strain and killed on Day 6 after the virulent challenge. Protection was evaluated by the enumeration of bacteria in the spleen. The statistical analysis was referred to naive mice : (*) $P < 0.05$.

time, groups of mice transferred with immune cells (primed cells), non immune (unprimed cells), or irradiated cells (APC), were challenged in an identical fashion. The mice receiving the cell line or the higher concentration of primed cells, exhibited significant host protection ($P < 0.05$) to the challenge infection (Fig. 6). On Day 6 after the challenge, the passive protection induced by the cell line was better than those from vaccinated mice, since more cells (x 15) were necessary to have comparable protection.

To identify the antigen fractions that induced the proliferation, chromatography of SDS-solubilized CWF was performed on Sephacryl S-200 gel. Two main fractions (53 and 60), induced good specific proliferation of primed spleen cells (Fig.7).

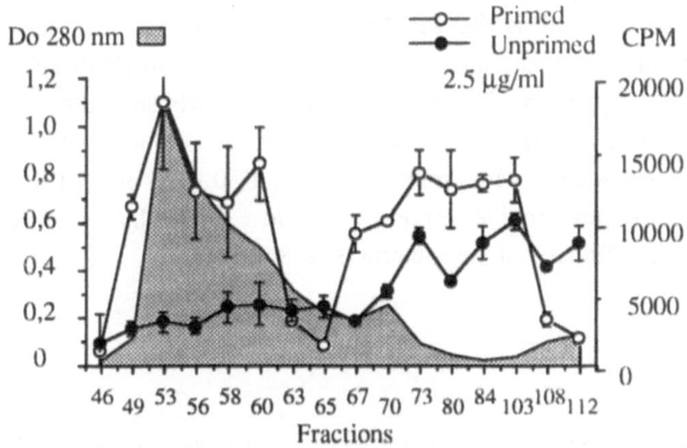

Figure 7. Mouse spleen cell proliferation assays with fractions of CWF. Chromatography on Sephacryl S-200 HR (Pharmacia) of CWF from SAO was done in 0.1 M NH_4CO_3, 0.1 % SDS buffer. Pool fractions of the column eluate were assessed for optical density (280 nm) and proliferative activity at 2.5 µg/ml. Proliferation assay of primed and unprimed mouse spleen cells, was done in the same conditions as already described (Fig. 3).

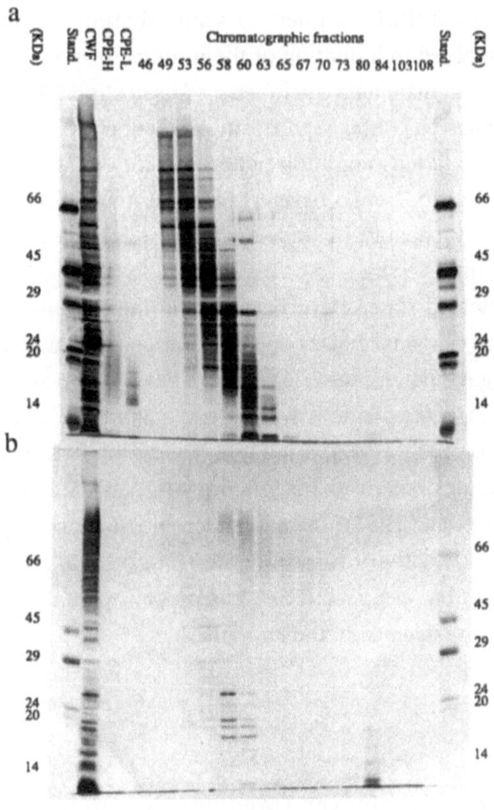

Figure 8. Pool fractions of the CWF column separation (Fig. 7) were analysed on a 11% polyacrylamide gel electrophoresis, in SDS denaturated conditions (SDS-PAGE) (Laemeli, 1970). Gels were silver stained either directly (Fig. 8a) for proteins, or after periodic oxidation (Fig. 8b) for oside residues and LPS (Dubray and Bézard, 1982).

The molecular weight range of the proteins that induced specific proliferation, analysed in polyacrylamide gel electrophoresis (SDS-PAGE) and silver stained, was between 20 and 100 KDa (Fig. 8a). Using silver staining after periodate oxidation (Dubray and Bézard, 1982) of S-200 fractions, a typical profile of smouth-LPS, regularly spaced bands ranging in 50 to 80 KDa were found (Fig. 8b). The fraction n° 53 was constituted with protein higher than 50 KDa, since the 60 fraction with protein lower than 50 KDa plus LPS. However, it was clear that non specific proliferative responses with low molecular weight components were mainly due to the LPS fraction of the bacteria (Fig. 8b).

CONCLUSIONS

These results are the first preliminary information that passive protection, against *Salmonella abortusovis* , can be transfer to naive mice with T cell lines. Induction of specific protective T cell clones anti-*Salmonella* were planed to characterize subsets of T cells involved in the active and passive protection of mice against SAO and to analyse the epitope(s) involved in the antigenic properties of the cell wall fraction (CWF). Nevertheless numerous parameters must be study to understand the induction of the protective immune response with the vaccinal strain in mice but also in sheep. Recirculation of this bacteria after vaccination with Rv6 SAO strain must be carefully studied to know dissemination of *Salmonella* Ags in body, and cell subsets involved in induction of the immune response. The route of inoculation is a capital parameter to vaccinate animals in term of protection. Previous reports suggested that the failure to T-cell lines to confer significant protection in mice was because the *in vitro* -maintained cells had an altered migration pattern (Dailey et al., 1985). In mice, with *S. thyphimurium* , antigen-specific T cells derived from peritoneal exudates and those derived from lymph nodes are only capable of transferring protection when the route of transfer is homologous (Paul and Smith, 1988).

Preliminary experiments done on vaccinated sheep have been shown that lymphocytes from lymph nodes gave strong proliferative responses after specific antigenic stimulation whereas blood and lymphatic lymphocytes gave no response (Data not shown). The absence of reaction of cells from efferent lymph can be explained by the low quantity of antigen presenting cells (APC) in this fluid (Bujdoso et al., 1989), although, a very low number of macrophages found in popliteal lymph samples, seem to be functionally sufficient in the regulation of immune response (De Martini et al., 1983). In fact, sheep blood lymphocytes (and may be lymphatic lymphocytes) need IL2 activity to proliferate (Emery et al., 1989). Investigations in this area must be undertaken on sheep with SAO as model, to study the T cell responses, in relationship with the genetic resistance of the animals.

ACKNOWLEDGMENTS

We would like to thank Dr Martin Vordermeyer (Medical Research Concil, Hammersmith Hospital, London, UK) and Dr Florence Velge (INRA, PPI,. Nouzilly, France), for their informations and assistances in the T cell line cultures.

REFERENCES

Bradley, D.J., 1974, Genetic control of natural resistance to Leishmania donovani, Nature. 250:354.

Budjoso, R., Hopkins, J., Dutia, B.M., Young, P., and McConnell, I., 1989, Characterization of sheep afferent lymph dendritic cells and their role in antigen carriage, J Exp Med. 170:1285.

Chadler, R., Sainis, K.B., and Lewis, N.F., 1986, Role of thymus-derivated lymphocytes in acquired immunity to salmonellòsis in mice, Microbiol Immunol. 30:1299.

Collins, F.M., 1974, Vaccines and cell-mediated immunity, Bacteriol Rev. 38:371.

Crocker, P.R., Blackwell, J.M., and Bradley, D.J., 1984, Expression of natural resistance gene LSH in resident liver macrophage, Infect Immun. 43:434.

Dailey, M.O., Gallatin, W.M., and Weissman, I.L., 1985, The in vivo behavior of T cell clones : altered migration due to the loss of lymphocyte surface homing receptor, J Mol Cell Immunol. 2:27.

De Martini, J.C., Fiscus, S.A., and Pearson, L.D., 1983, Macrophages in efferent lymph of sheep and their role in lectin-induced lymphocyte blastogenesis, Int Archs Allergy appl Immun. 72:110.

Dubray, G., and Bézard, G., 1982, A high sensitive periodic acid-silver stain for 1,2-diol groups of glycoproteins and polysaccharides in polyacrylamide gels, Annalyt Biochem. 119:325.

Eisenstein, T.K., and Sultzer, B.M., 1983, Immunity to Salmonella infection, Adv Exp Med Biol. 16:261.

Emery, D.L., Rothel, J.S., Kirkpatrick, A., and Maclaren, J.A., 1989, Generation, maintenance and reactivity of ovine T-lymphocyte clones derived from sheep immunized with pili from Bacteroides nodosus, Vet Immunol Immunopathol. 21:339.

Gros, P.E., Skamene, E., and Forget, A., 1983, Cellular mechanisms of genetically controlled host resistance to Mycobacterium bovis (BCG), J Immunol. 131:1966.

Guilloteau, L., Buzoni-Gatel, D., Bernard, F., and Lantier, F., 1992, Salmonella abortusovis infection in susceptible BALB/cby mice : importance of Lyt-2+ and L3T4+T cells in acquired immunity and granulomas formation, Microb Pathog. (Submitted)

Guilloteau, L., Buzoni-Gatel, D., Blaise, F., Bernard, F., and Pepin, M., 1991, Phenotypic analysis of splenic lymphocytes and immuno histochemical study of hepatic granulomas after a murine infection with Salmonella abortusovis, Immunology. 74:630.

Hormaeche, C.E., Mastroeni, P., Arena, A., Uddin, J., and Joysey, H.S., 1990, T cell do not mediate the initial suppression of a Salmonella infection in the RES, Immunology. 70:247.

Laemnli, U.K., 1970, Cleavage of structural proteins during the essemblage of the head of bacteriophage T4, Nature. 277:680.

Lantier, F., Pardon, P., and Marly, J., 1981, Vaccinal properties of Salmonella abortusovis mutants for streptomycin : screening with a murine model, Infect Immun. 34:492.

Lantier, F., Pardon, P., and Marly, J., 1983, Immunogenicity of a low-virulence vaccinal strain against Salmonella abortus ovis infection in mice, Infect Immun. 40:601.

Maskell, D.J., Hormaeche, C.E., Harrington, K.A., and Joysey, H.S., 1987, The initial suppression of bacterial growth in a Salmonella infection is mediated by localized rather than a systemic response, Microb Pathog. 2:295.

Murray, H.W., Spitalny, G.L., and Nathan, C.F., 1985, Activation of mouse macrophage in vitro and in vivo by interferon-gamma, J Immunol. 134:1619.

Nakato, Y., Onozuka, K., Terada, Y., Shinomiya, H., and Nakano, M., 1990, Protective effect of recombinant tumor necrosis factor-alpha in murine salmonellosis, J Immunol. 144:1935.

Nauciel, C., 1990, Role of CD4+ T cells and T-independant mchanisms in acquired resistance to Salmonella typhymurium infection, J Immunol. 145:1265.

O'Brien, A., and Metcalf, E.S., 1982, Control of early Salmonella typhimurium growth in innately Salmonella-resistant mice does not require fonctional T lymphocytes, J Immunol. 129:1349.

Oswald, I., Lantier, F., Moutier, R., Bertrand, M.F., and Skamene, E., 1992, Intraperitoneal infection with Salmonella abortusovis is partially controlled by gene closely linked with the Ity gene, Clin Exp Immunol. 87:373.

Pardon, P., Lantier, F , Marly, J., and Sanchis, R., 1980, Mise au point d'un vaccin contre la salmonellose abortive ovine, Bull Soc Vet Prat. 64:465.

Pardon, P., Sanchis, R., Marly, J., Lantier, F., Guilloteau, L., Buzoni-Gatel, D., Oswald, I., Pépin, M., Kaeffer, B., Berthon, P., and Popoff, M.Y., 1990, Experimental ovine salmonellosis (Salmonella abortusovis) : Pathogenesis and vaccination, Res Microbiol. 141:945.

Paul, C., Shalala, K., Warren, R., and Smith, R., 1985, Adoptive transfer of murine host protection to Salmonellosis with T-cell growth factor-dependent Salmonella-specific T-cell lines, Infect Immun. 48:40.

Paul, C., and Smith, R , 1988, Transfer of murine host protection by using interleukin-2-dependent T-lymphocyte lines. Infect Immun. 56:2189.

Plant, J.E., and Glynn, A.A., 1974, Natural resistance to Salmonella infection, delayed hypersentivity and Ir genes in different strains of mice, Nature. 248:345.
Skamene, E., Gros, P., Forget, A., Kongshavn, P.A.L., St Charles, , C., and Taylor, B.A., 1982, Genetic regulation of resistance to intracellular pathogens, Nature. 297:506.
Swanson, R.N., and O'Brien, A.D., 1983, Genetic control of the innate resistance of mice to Salmonella typhimurium : Ity gene is expressed in vivo by 24 hours after infection, J Immunol. 131:3014.
Vordermeier, H.M., Harris, D., Roman, E., Lathigra, R., Moreno, C., and Ivanyi, J., 1991, Identification of T cell stimulatory peptides from the 38-kDa protein of Mycobacterium tuberculosis, J Immunol. 147:1023.
Zygmunt, M.S., Martin, J.C., and Dubray, G., 1990, Analysis of immune response : Comparison of immunoblots after isoelectric focussing and sodium dodecyl sulfate polyacrylamide gel electrophoresis using cytoplasmic protein extract from brucella, FEMS Microbiol Letters. 70:263.

SYNTHESIS AND IMMUNOGENICITY OF A *SALMONELLA TYPHI* O-CHAIN-TETANUS TOXOID CONJUGATED VACCINE

Manoj Saxena and Jose L. Di Fabio

Bacterial Antigens and Antisera Section
Bureau of Biologics, Health and Welfare Canada
Ottawa, Ontario, Canada

INTRODUCTION

Typhoid fever continues to be a major cause of morbidity and mortality in the developing world. While the recently described oral Ty 21a and the parenteral purified Vi vaccines promise to be effective vaccines in the prevention of typhoid (Acharya et al., 1987; Levine et al., 1987), the need for more potent vaccines against typhoid fever cannot be overemphasised. The contribution of lipopolysaccharide 'O antigen' in the pathogenesis of *Salmonella typhi* (Mroczenski-Widley et al., 1989) highlights the potential of using lipopolysaccharide (LPS) or its components in the formulation of a vaccine against typhoid. This manuscript describes the synthesis of a *S. typhi* O-chain - tetanus toxoid conjugate vaccine and discusses the rationale and chemistry of the conjugation.

STRUCTURE OF *SALMONELLA TYPHI* LPS

S. typhi LPS like that of other gram negative bacteria is composed of three major portions, the Lipid A, the highly conserved core region and the serovar specific O-chain. Besides a host of other biological activities, LPS exhibits marked toxicity towards animals and human beings (Rapson, 1988). The toxicity of the LPS region is due to the Lipid A portion, while both core and O-chain are essentially non-toxic. In terms of immunity, however, it is the O-chain and to a lesser extent the core which appear to be critical. The toxicity of the LPS molecule which is a major impediment in its usage as an immunogen can therefore be abolished by the removal of the Lipid A component. The detoxification of the LPS relies on the acid lability of the KDO (2-keto-3-deoxy-D-manno-octulosonic acid) linkages. Mild treatment with acetic acid hydrolyses the KDO leaving a 'polysaccharide' region i.e. O-chain and core with the Lipid A separated. The Lipid A can then be easily removed by centrifugation.

Biology of Salmonella, Edited by F. Cabello *et al.*,
Plenum Press, New York, 1993

Removal of Lipid A while beneficial in terms of detoxification, results in decrease of immunogenicity. It becomes necessary therefore to conjugate the antigenic O-chain to a carrier protein for the development of a potent vaccine preparation.

CHEMISTRY OF CONJUGATION

A number of techniques are available for conjugating polysaccharide antigens to protein carriers (Dick and Beurret, 1989). In the present study the development of an O-polysaccharide-protein conjugate vaccine was essentially a two step process, first the oxidation of the polysaccharide and second, its chemical coupling to a protein carrier to form a stable conjugate. The use of periodate in the oxidation of polysaccharides is well established in carbohydrate chemistry (Seppala and Makela, 1989; Lugowski et al., 1990). In a simplistic representation, the reaction involves the oxidation of vicinal diols by periodate by cleavage of C-C bonds, generating in the

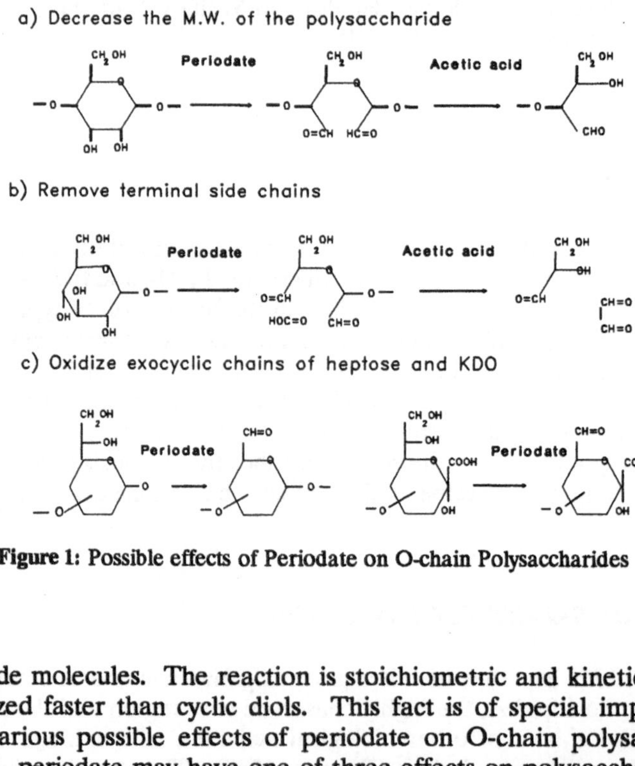

Figure 1: Possible effects of Periodate on O-chain Polysaccharides

process aldehyde molecules. The reaction is stoichiometric and kinetically, exocyclic diols are oxidized faster than cyclic diols. This fact is of special importance if we consider the various possible effects of periodate on O-chain polysaccharides. As shown in Fig. 1, periodate may have one of three effects on polysaccharides :
i) decrease the molecular weight of the O-chain, ii) remove terminal side chains and iii) oxidise exocyclic chains of heptoses and KDO. Thus, due to the kinetics of the reaction, it is possible to selectively oxidise the KDO and the heptoses leaving the rest of the molecule unaffected.

Oxidised O-chains can then be easily conjugated to amino groups by the method of Jennings and Lugowski (1981). The aldehyde on the sugar reacts with a free amino group on the protein to generate a Schiff's base which is further converted to a secondary amine by reduction with sodium cyanoborohydride, resulting in the formation of a polysaccharide- protein conjugate. The reaction known as **reductive amination** can be represented as

$$\text{Ps-CH=O} + \text{H}_2\text{N-Protein} \text{ ------------ } \text{Ps-CH}_2\text{-NH-Protein}$$

CONJUGATION OF *SALMONELLA TYPHI* O-CHAINS TO MONOMERIC TOXOID

Synthesis of Conjugate

S. typhi LPS (Sigma,USA) was treated by boiling with 1% acetic acid and Lipid A was removed by centrifugation. The resultant supernatant was chromatographed on Sephadex G-50 using 0.05 M pyridinium acetate, pH 5.4 as the eluent. The high molecular weight fractions (glycose positive) were collected, pooled and concentrated by lyophilization and represented the O-chains (Fig. 2). It is important to emphasise that the core region is always attached to the O-chain.

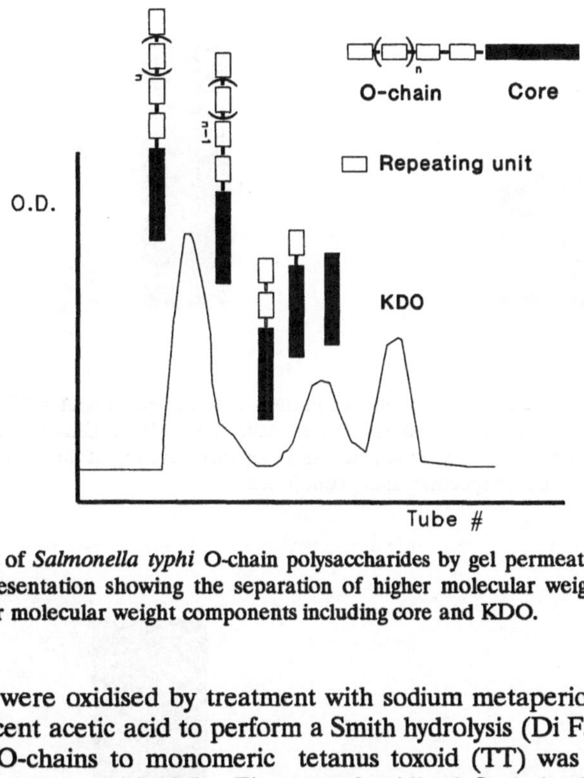

Figure 2. Isolation of *Salmonella typhi* O-chain polysaccharides by gel permeation chromatography. Diagrammatic representation showing the separation of higher molecular weight O-polysaccharide fractions from lower molecular weight components including core and KDO.

O-chains were oxidised by treatment with sodium metaperiodate followed by boiling in 1 per cent acetic acid to perform a Smith hydrolysis (Di Fabio et al., 1988). Conjugation of O-chains to monomeric tetanus toxoid (TT) was done as per the method of Lugowski et al. (1986). Five mg of oxidised O-chain dissolved in 0.1 N sodium bicarbonate, pH 8.1 (0.2 ml). Four mg of TT was added to it followed by 10 mg of sodium cyanoborohydride. The reaction mixture was left for 7 days at room temperature. At the end of the incubation period, the conjugate was purified by chromatography on Biogel A 0.5 with 0.01 M phosphate buffered saline (PBS), pH 7.0. Chemically, the composition of the conjugate was 67% protein and 33% polysaccharide. In terms of average molar composition, the conjugate thus consists of 4 O-chains attached to each tetanus toxoid molecule.

Mouse Immunogenicity studies

The immunogenicity of the conjugate was investigated by injecting Balb/c mice (18- 22 g each) with 5 or 20 µg of the conjugate in Freund's complete adjuvant. Animals were bled after three weeks, boosted at four weeks with the same concentration of conjugate in Freund's incomplete adjuvant, and bled again three weeks after boosting.

Development of immunity to the polysaccharide component of the conjugate i.e. *S. typhi* O- chain was quantified by measuring anti O-chain antibodies by Elisa (Fig. 3). Besides the generation of anti O-chain antibodies, there were a number of other important features of this data: a) there was a dose dependant effect, b) there was a booster effect and c) the conjugation process did not affect the immunogenicity of the toxoid component as significant levels of anti-TT antibodies were seen in all the immune sera (Fig.4).

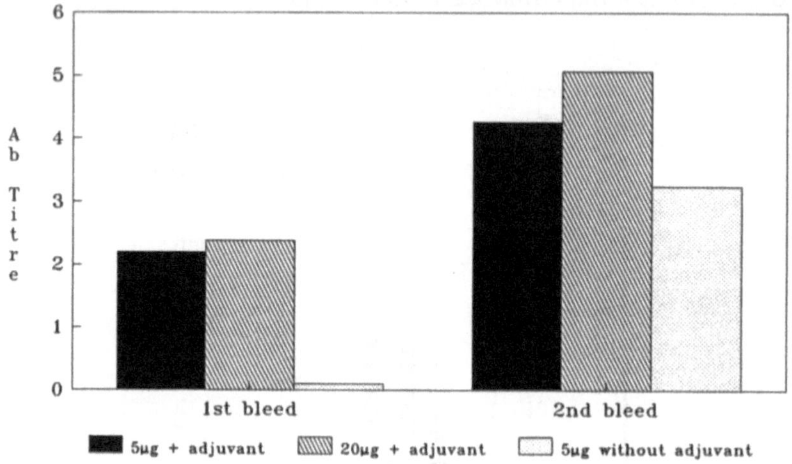

Figure 3. Serum anti- O-chain antibodies elicited in Balb/c mice injected with different doses of *S. typhi* O-chain tetanus toxoid conjugate. Antibodies were measured by indirect Elisa using a *S. typhi* O-chain-bovine serum albumin conjugate as the antigen. Antibody titres are log of the serum dilution that gave an OD of 1.0 at 450 nm under specified assay conditions.

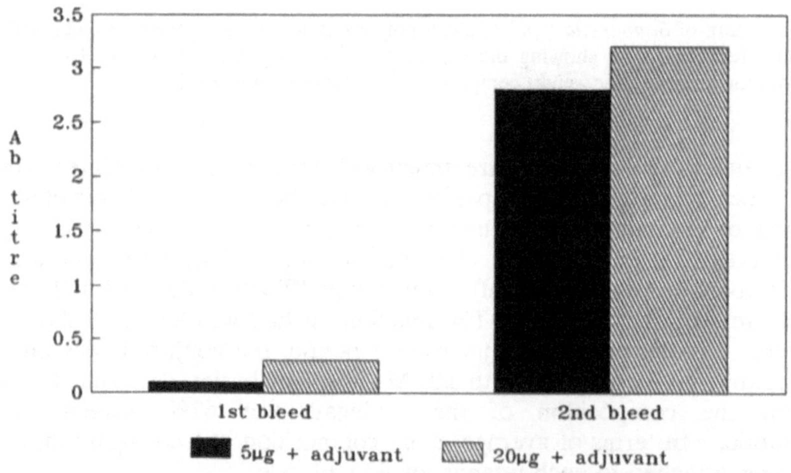

Figure 4. Serum anti-Tetanus Toxoid antibodies elicited in Balb/c mice immunized with 5 and 20 µg of *S. typhi* O-chain- toxoid conjugate. Antibody titres are log of serum dilution that gave an OD of 1.0 at 450 nm under specified conditions in an indirect Elisa using tetanus toxoid (plain) as the antigen.

DISCUSSION

Polysaccharide antigens are classically considered as T cell independent antigens (Grifiss et al., 1987; Lee, 1987) and thus are limited in their vaccine potential. However, the immunization data described here indicate the involvement of T-helper cells and the stimulation of memory cells. The conjugation process thus has changed the behaviour of the antigen into a T- dependent type. This is also evident from isotyping data of immune sera which showed primary sera to be predominant in IgM, while booster sera exhibited elevated IgG values. Similar alteration and enhancement of immunogenicity of polysaccharides or oligosaccharides by coupling to protein carriers has been reported for a host of bacterial pathogens (Beuvery et al., 1982; Anderson et al., 1983; Cryz et al., 1986). While the extent and exact role of LPS in the pathogenesis of *S. typhi* remains a subject of discussion (Isibasi et al, 1988; Mroczenski- Widley et al., 1989), antibodies to O-polysaccharide are expected to opsonize LPS and whole bacteria, thus providing an additional bactericidal effect on Gram-negative bacteria (Pennington and Menkes, 1981; Ziegler, 1988). Preliminary studies on the immune sera in this study showed significant bactericidal activity *in-vitro* towards *S. typhi* Ty2. Work is currently in progress to examine the immunogenicity of the conjugate with other adjuvants like aluminium hydroxide and to see their protective efficacy against *S. typhi* challenge in a mouse mucin model of typhoid.

ACKNOWLEDGEMENTS

MS acknowledges a visiting biotechnology fellowship from the Natural Sciences & Engineering Council of Canada.

REFERENCES

Acharya, V.L., Lowe, C.U., Thapa, R., Gurubacharya, V.L., Shrestha, M.B., Cadoz, M., Schulz, D., Armand, J., Bryla, D., Trollfors, B., Cramton, T., Schneerson, R. and Robbins, J.B., 1987, Prevention of typhoid fever in Nepal with the Vi capsular polysaccharide of *Salmonella typhi*. A preliminary report, *N. Engl. J. Med.*, 317: 1101.

Anderson, P.W., Pichichero, M.E., Insel, R.A., Betts, R., Eby, R. and Smith, D.H., 1983, Vaccines consisting of periodate cleaved oligosaccharides from the capsule of *Haemophilus influenzae* type b coupled to a protein carrier: structural and temporal requirements for priming in the human infant, *J.Immunol.*, 137:1181.

Beuvery, E.C., van Rossum, F. and Nagel, J., 1982, Comparison of the induction of immunoglobulin M and G antibodies in mice with purified pneumococcal type 3 and meningococcal group C polysaccharides and their protein conjugates, *Infect. Immun.*, 37:15.

Cryz, S.J., Sadoff, J.C., Furer, E. and Germanier, R., 1986, *Pseudomnas aeruginosa* polysaccharide-tetanus toxoid conjugate vaccine : safety and immunogenicity in humans, J. Infect. Dis., 154:682.

Dick, W.E. and Beurret, M., 1989, Glycoconjugates of bacterial carbohydrate antigens. A survey and consideration of design and preparation factors, *in*: "Conjugate Vaccines", J.M.Cruse and R.E.Lewis, Jr., ed., Karger, Basel.

Di Fabio, J.L., Perry, M.B. and Brisson, J.R., 1988, Structure of the antigenic O-polysaccharide produced by *Salmonella eimsbuttel*, *Biochem. Cell. Biol.*, 66:107.

Griffiss, J.M., Apicella, M.A., Greenwood, B. and Makela, P.H., 1987, Vaccines against encapsulated bacteria : a global agenda, *Rev. Infect. Dis.*, 9:176.

Isibasi, A., Ortiz, V., Vargas, M., Paniagua, J., Gonzalez, C., Moreno, J. and Kumate, J., 1988, Protection against *Salmonella typhi* infection in mice after immunization with outer membrane proteins isolated from *Salmonella typhi* 9,12,d,Vi, *Infect. Immun.*, 56:2953.

Jennings, H.J. and Lugowski, C., 1981, Immunochemistry of Group A, B and C Meningococcal polysaccharide-tetanus conjugates, *J.Immunol.*, 127:1011.

Lee, C.J., 1987, Bacterial capsular polysaccharides: biochemistry, immunity and vaccine, *Mol. Immunol.*, 24:1005

Levine, M. M., Ferreccio, C., Black, R.E., Germanier, R., Chilean Typhoid Committee., 1987, Large scale field trial of Ty21a live oral typhoid vaccine in enteric coated capsule formulation, *Lancet* 1:1049.

Lugowski,C., Kulakowska, M. and Romanowska, E.,1986, Characterization and diagnostic application of a lipopolysaccharide core oligosaccharide protein conjugate, *J.Immunol. Methods.*, 95:187.

Lugowski,C., Kulakowska, M. and Romanowska, E., 1991, Saccharide-protein covalent conjugates : immunochemical characterization of *Citrobacter* O36 oligosaccharide-tetanus toxoid conjugates, *FEMS Microbiol. Letts.*, 76:1.

Mroczenski-Wildey, M.J., Di Fabio, J.L. and Cabello, F.C., 1989, Invasion and lysis of Hela cell monolayers by *Salmonella typhi* : the role of lipopolysaccharide, *Microb. Pathogenesis*, 6:143.

Pennington, J.E. and and Menkes, E., 1981, Type-specific vs. cross protective vaccination for Gram-negative pneumonia, *J. Infect. Dis.*, 144:599.

Rapson, N.T., 1988, Biological and clinical effects of bacterial lipopolysaccharide, *Curr. Opin. Infect. Dis.*, 1:919.

Seppala, I. and Makela,O., 1989, Antigenicity of dextran-protein conjugates in mice. Effect of molecular weight of the carbohydrate and comparison of two modes of coupling, *J. Immunol.* 4:1259.

Ziegler, E.J., 1988, Protective antibody to endotoxin core: the emperor's new clothes?, *J.Infect. Dis.*, 158, 286.

USE OF *SALMONELLA SPP.* CARRIER STRAINS TO DELIVER *BORDETELLA PERTUSSIS* ANTIGENS IN MICE USING THE ORAL ROUTE

Carlos A. Guzmàn,[1] Kenneth N. Timmis,[2] and Mark J. Walker[3]

[1]Institute of Microbiology, University of Genoa, Italy
[2]GBF – National Research Centre for Biotechnology, Braunschweig, Germany
[3]Biology Department, University of Wollongong, Australia

INTRODUCTION

Bordetella pertussis is the etiological agent of whooping cough. This disease is particularly severe in young children and may lead to neurological disorders and death (Walker, 1988). The incidence of whooping cough has been reduced largely by mass immunization with a heat–killed whole cell vaccine routinely administered in combination with diphtheria and tetanus toxoids (Fine *et al.*, 1987). The benefits associated with this vaccine clearly outweigh the risks of rare but severe adverse effects, but nevertheless public concern about safety and immunogenicity has resulted in decreased vaccination rates accompanied by an increase in disease incidence (Miller *et al.*, 1982). Therefore, non-toxic and highly immunogenic vaccines are urgently needed to restore confidence in vaccination reducing the incidence of this important childhood disease.

Detoxified pertussis toxin (PT) and filamentous hemagglutinin (FHA) are considered prime candidates for inclusion in defined acellular vaccines against *B. pertussis* (Kimura *et al.*, 1990; Olander *et al.*, 1990). Traditionally, anti–pertussis vaccination has employed an intramuscular route. However, conventional vaccines administered parenterally are unable to elicit mucosal antibody responses (Shahin *et al.*, 1990; Thomas *et al.*, 1989) which may protect the host from both colonization and disease. An alternative to this approach is to stimulate mucosal and systemic immune responses by oral vaccination with *aro*A mutants of *Salmonella spp.* expressing recombinant antigens as immunoglobulin (Ig) A responses are typical of oral immunization (Brown *et al.*, 1987; Czerkinsky *et al.*, 1987; Guzmàn *et al.*, 1991a; Hoiseth *et al.*, 1981).

In the present work we describe the systemic and mucosal immune responses elicited following oral immunization of mice using *S. typhimurium aroA* mutants expressing recombinant FHA and PT S1 subunit (rPT–S1).

Biology of Salmonella, Edited by F. Cabello *et al.*,
Plenum Press, New York, 1993

MATERIALS AND METHODS

Bacterial Strains, Plasmids and Media

The bacterial strains used in this work were: *B. pertussis* TohamaI (Sato *et al.*, 1972); *S. typhimurium aroA* SL3261 (Hoiseth *et al.*, 1981). The plasmids used in this work were: pTX42 (Locht *et al.*, 1986); pCG26 (Guzmàn *et al.*, 1991b); pUC18 (Yanisch–Perron *et al.*, 1985); and pJLA506 (M. Walker), a derivative of the expression plasmid pJLA503 (Schrauder *et al.*, 1987) with a modified multiple cloning site.

S. typhimurium strains were grown in Luria broth or on Luria agar plates (Sambrook *et al.*, 1989). *B. pertussis* was grown on Bordet Gengou agar base (Difco) supplement with 1% glycerol and 15% (vol/vol) defibrinated horse blood. Ampicillin was used at 100 μg/ml.

DNA Manipulations

The DNA manipulations were performed as previously described (Sambrook *et al.*, 1989). The polymerase chain reaction (PCR) primers S1–5'*Nde*I (5'–GGCATATGCTGG GCACTCGGGCAATT–3') and S1–3'*Xba*I (5'–CCAGGTCTAGAACGAATA–3') were used for cloning the S1 subunit cistron as a *Nde*I/*Xba*I fragment.

Tissue Culture Methods and *In Vitro* Adhesion and Invasion Assays

Adhesion and invasion assays using the human embryonic intestinal cell line Intestine 407 (ATCC CCL–6) were performed as previously described (Guzman *et al.*, 1991a).

Mouse Immunization and Immunological Techniques

Six– to seven–week–old female BALB/c mice were immunized in groups of 5 with one dose on day 0 and boosted with an identical dose 30 and 40 days later as shown in table 1. The animals were sacrificed 10 days after the last booster. Immunization was performed and samples were collected as previously described (Guzmàn *et al.*, 1991a). For the determination of subclass–specific antibodies against FHA and PT in serum and lung washes, enzyme–linked immunosorbent assays (ELISA) were performed as previously described (Guzmàn *et al.*, 1991a), using sera or lung washes from control groups (C and G) as the blank for the ELISA readings.

Table 1. Mouse immunization protocols.

Group	Strain	Dose	Route	Viability
A	*B. pertussis* Tohama	10^6	i.p.[1]	Heat killed
B	SL3261	10^6	i.p.	Live
C	SL3261 (pCG26)	10^6	i.p.	Heat killed
D	SL3261 (pCG26)	10^6	i.p.	Live
E	SL3261 (pMW151)	10^6	i.p.	Heat killed
F	SL3261 (pMW151)	10^6	i.p.	Live
G	SL3261	10^9	Oral	Live
H	SL3261 (pCG26)	10^9	Oral	Heat killed
I	SL3261 (pCG26)	10^9	Oral	Live
J	SL3261 (pMW151)	10^9	Oral	Heat killed
K	SL3261 (pMW151)	10^9	Oral	Live

1: i.p. mean intraperitoneal immunization.

RESULTS AND DISCUSSION

Expression of Recombinant FHA and rPT–S1 in *S. typhimurium*

FHA was efficiently expressed in the attenuated *S. typhimurium aro*A mutant SL3261 at levels higher than those found in wild type *B. pertussis* when the gene *fhaB* was cloned in the vector pJLACG1 (Guzmàn *et al.*, 1991b). The upstream signals of the gene were replaced by the lambda tandem P_R and P_L promoters (under the control of the temperature sensitive cl repressor protein) and the translation initiation region engineered to optimize translational efficiency. The sequence encoding for the first 15 amino acids of FHA was replaced by a linker designed to optimize mRNA structure.

The S1 subunit of PT operon, originating from plasmid pTX42, was amplify by PCR including the leader sequence. The 5' primer contains the ATG start codon in the *Nde*I restriction site, whereas the 3' primer contains the TAG stop codon. The amplified fragment was cloned firstly as a *Nde*I/*Xba*I fragment into the pUC18 vector and subsequently, as a *Nde*I/*Eco*RI fragment, into the expression vector pJLA506 under the control of the lambda tandem P_R and P_L promoters (pMW151). Recombinant PT–S1 was expressed in SL3261 strain after induction at 37°C or 42°C at levels significantly higher than those found for *B. pertussis* (data not shown).

In Vitro Adhesiveness and Invasiveness of Vaccinal Strains

The presence of plasmid pCG26 and pMW151, and expression of the recombinant proteins in SL3261 strain did not affect *in vitro* invasiveness of the tested strains, which retained the ability to adhere to and invade the Intestine–407 cell line (table 2).

Table 2. *In vitro* adhesiveness and invasiveness of strains used in vaccination protocols.

Strain	Adhesiveness[1]	Invasiveness[2]
SL3261	15 ± 5	3 x 10^3
SL3261(pCG26)	13.2 ± 5.1	3.2 x 10^3
SL3261 (pMW151)	15.2 ± 3.3	5 x 10^3

[1] Mean number of bacteria per Intestine–407 cell ± standard deviation.
[2] CFU of viable intracellular bacteria per coverslip.

Assessment of Specific Anti–FHA and Anti–PT Antibody Responses in Vaccinated Mice

Following oral immunization of mice with strains expressing either recombinant FHA or rPT–S1, specific antibodies were elicited consisting mainly of serum IgG, and both IgG and IgA in lung washes (figures 1). The antibody levels were significantly higher than those obtained in mice orally immunized with killed bacteria. Significative levels of IgA were detected in lung only after peroral immunization. The IgG present may diffuse from capillary vesels to the lower respiratory tract.

Utilization of live oral vaccines expressing *B. pertussis* antigens, which stimulate both a systemic and lung responses, may provide an attractive alternative to purified component vaccines to protect both against infection and disease, favouring erradication of a disease for which there is only a human reservoir. The oral route may have higher acceptance rates than parenteral administration; oral immunization may also reduce or eliminate the most frequent local side effects associated with whooping cough vaccines. This delivery system would be cost effective to produce and easily administered, both particularly useful traits when contemplating massive immunization programs, especially in developing countries.

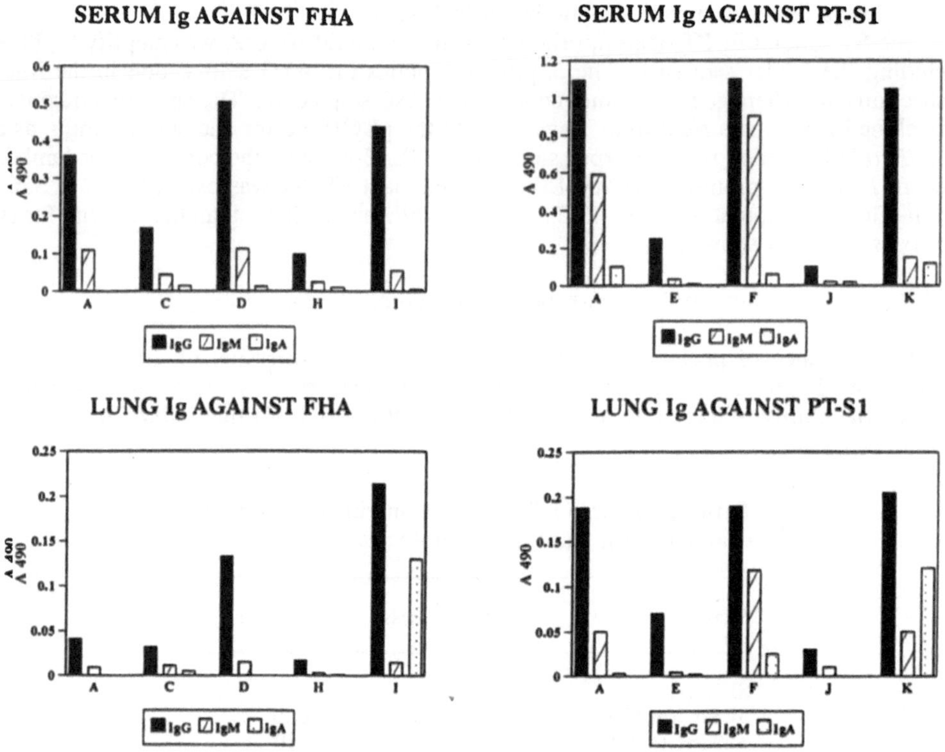

Figure 1. Levels of specific antibodies in sera and lungs after immunization of mice.

REFERENCES

Brown, A., Hormaeche, C.E., Demarco de Hormaeche, R., Winther, M., Dougan, G., Maskell, D.J., and Stocker, B.A.D., 1987, An attenuated *aroA Salmonella typhimurium* vaccine elicits humoral and cellular immunity to cloned beta-galactosidase in mice, J. Infect. Dis. 155:86.

Czerkinsky, C., Prince, S.J., Michalek, S.M., Jackson, S., Russell, M.W., Moldoveanu, Z., McGhee, J.R., and Mestecky, J., 1987, IgA antibody-producing cells in peripheral blood after antigen ingestion: evidence for a common mucosal immune system in humans, Proc. Natl. Acad. Sci. USA. 84:2449.

Fine, P.E.M., and Clarkson, J.A., 1987, Reflections in the efficacy of pertussis vaccines, Rev. Infect. Dis. 9:866.

Guzmàn, C.A., Brownlie, R.M., Kadurugamuwa, J., Walker, M.J., and Timmis, K.N., 1991a, Antibody responses in the lungs of mice following oral immunization with *Salmonella typhimurium aroA* and invasive *Eschèrichia coli* strains expressing the filamentous hemagglutinin of *Bordetella pertussis*, Infect. Immun. 59:4391.

Guzmàn, C.A., Walker, M.J., Rhode, M., and Timmis, K.N., 1991b, Direct expression of *Bordetella pertussis* filamentous hemagglutinin in *Escherichia coli* and *Salmonella typhimurium aroA*, Infect. Immun. 59:3787.

Hoiseth, S.K., and Stocker, B.A.D., 1981, Aromatic-dependent *Salmonella typhimurium* are non-virulent and effective as live vaccines, Nature (London). 291:238.

Kimura, A., Mountzouros, K. T., Relman, D.A., Falkow, S., and Crowell, J.L., 1990, *Bordetella pertussis* filamentous hemagglutinin : evaluation as a protective antigen and colonization factor in a mouse respiratory infection model, Infect. Immun. 58:7.

Locht, C., Barstad, P.A., Coligan, J.E., Mayer, L., Munoz, J.J., Smith, S.G., and Keith, J.M., 1986, Molecular cloning of pertussis toxin genes, Nucl. Acids Res. 14:3251.

Miller, D.L., Alderslade, R., and Ross, E.M., 1982, Whooping cough and whooping cough vaccine: the risks and benefits debate, Epidemiol. Rev. 4:1.

Olander, R.M., Muotiala, A., Karvonen, M., Kuronen, T., and Runeberg-Nyman, K., 1990, Serum antibody response to *B. pertussis* Tn5 mutants, purified PT and FHA in two different mouse strains and passive protection in the murine intranasal infection model, Microbial. Pathogen. 8:37.

Sambrook J., Fritsch, E.F., and Maniatis, T., 1989, "Molecular Cloning. A Laboratory Manual. Second Edition", Cold Spring Harbor Laboratory Press, New York.

Sato, Y., and Arai, H., 1972, Leukocytosis-promoting factor of *Bordetella pertussis*. I. Purification and characterization, Infect. Immun. 6:899.

Schrauder, B., Blocker, H., Frank, H., and McCarthy, J.E.G., 1987, Inducible expression vectors incorporating the *Escherichia coli atpE* translational initiation region, Gene. 52:279.

Shahin, R.D., Brennan, M.J., Li, -Z.M., Meade, B.D., and Manclark, C.R., 1990, Characterization of the protective capacity and immunogenicity of the 69 kD outer membrane protein of *Bordetella pertussis*, J. Exp. Med. 171:63.

Thomas, M.G., Ashworth, L.A., Miller, E., and Lambert, H.P., 1989, Serum IgG, IgA, and IgM responses to pertussis toxin, filamentous hemagglutinin, and agglutinogens 2 and 3 after infection with *Bordetella pertussis* and immunization with whole-cell vaccine, J. Infect. Dis. 160:838.

Walker, E., 1988, Clinical aspects of pertussis, *in*: "Pathogenesis and Immunity in Pertussis," A. C. Wardlaw, and R. Parton, ed., John Wiley and Sons, Chichester.

Yanisch-Perron, C., Vieira, J., and Messing, J., 1985, Improved M13 phage cloning vectors and host strains: nucleotide sequences of the M13mp18 and pUC19 vectors, Gene. 33:103.

T-CELL RESPONSES IN MICE IMMUNISED WITH LIVE aroA SALMONELLA VACCINES

Bernardo Villarreal, Pietro Mastroeni, Raquel Demarco de Hormaeche and Carlos E. Hormaeche

Department of Pathology, Tennis Court Road, Cambridge CB2 1QP, UK

The search for of improved salmonella vaccines has seen the development of live attenuated salmonellae with lesions in genes of the aromatic pathway (Aro vaccines). These can be given orally, cause few side reactions and confer excellent protection in animals and are being tested in humans. They can also be used to deliver recombinant antigens from other pathogens as combined vaccines.

The mechanisms of immunity to salmonellae remain unclear. Both humoral and cellular mechanisms are important in protection, and the superior efficacy of live *vs.* killed vaccines is probably due to the ability of the former to elicit cell mediated immunity (CMI). However, for unknown reasons not all live vaccines are effective. The nature of the antigens triggering the protective cellular response are unknown, and evidence has been presented for the involvement of both protein and polysaccharide (LPS) antigens in protective cellular immunity. A better understanding of CMI in salmonellosis is important for rational vaccine design.

We have studied CMI to salmonella antigens by cellular proliferation as measured by ^{3}-H-methyl-thymidine (TdR) incorporation and T-cell specific interleukin (IL-2/IL-4) production. Innately susceptible (*Ity*^{S}) BALB/c mice were vaccinated i.v. with *S. typhimurium* SL3261 *aroA* (Hoiseth and Stocker, 1981) which confers solid protection against virulent challenge (Hormaeche *et al.*, 1991). Two months later the vaccine had been cleared from the RES, whole spleen cells or T-cell enriched (nylon wool non-adherent) cell suspensions were prepared and tested against purified Westphal LPS

Biology of Salmonella, Edited by F. Cabello *et al.*,
Plenum Press, New York, 1993

and a whole cell soluble extract from the virulent wild type *S. typhimurium* C5 (C5SE).

The mitogenicity of LPS for mouse B cells was countered by polymixin B (PB) and/or detoxification of the antigens by mild alkaline hydrolysis. PB blocks LPS by binding to lipid A. Mild alkaline hydrolysis removes ester linked fatty acids from lipid A while conserving the O-specificity of the polysaccharide antigen. Alkali treatment of the whole cell extract (C5SENaOH) caused extensive protein degradation.

Spleen cells were cultured at 1 and $2x10^6$/ml in 200 µl/well in round bottom 96-well plates. Antigens were added to a final concentration of 2 µg/ml. Samples of supernatant taken after two days were frozen for later measurement of interleukins. TdR incorporation was measured after three days' culture.

LPS was, as expected, mitogenic for naive total spleen cells (TdR incorporation), whereas alkali treated LPS or LPS in the presence of PB, C5SE, C5SENaOH with or without PB were not mitogenic. Immune cells incorporated TdR in the presence of both LPS and C5SE and it was not possible to distinguish between the responses to these antigens. Addition of polymixin partially reduced this response. Alkaline hydrolysis of the antigens slightly decreased the proliferative response of immune cells to LPS, but surprisingly increased the response to the whole cell extract to a level above the reponse to treated LPS. This was shown for both total spleen cells and for T-cell enriched cells.

However, the observed proliferative responses of immune cells to purified LPS did not appear to represent specific T-cell responses. Measurement of IL-2/IL-4 production using the IL-2/IL-4 dependent CTLL cell line (MTT method) showed that the LPS did not trigger significant production of IL-2/IL-4. In contrast, the protein-rich whole cell extract produced a clear IL-2/IL-4 response, which was greater when using the alkaline treated (degraded) extract.

In conclusion, although immune cells react to purified LPS with a proliferative response, (TdR incorporation in both total spleen cells and T-cell enriched cells), LPS does not trigger T-cell activation as measured by IL-2/IL-4 production. A whole cell extract before and after alkaline treatment triggers both TdR incorporation and IL-2 /IL-4 production, responses to the alkali treated antigen being higher than to the non-treated antigen. The results suggest that the salmonella antigens involved in cell mediated immunity may be protein rather than LPS. The system will permit analysis of T-cell responses to salmonella antigens as well as to antigens expressed in salmonellae.

REFERENCES

Hoiseth, S. K., and Stocker, B. A. D., 1981, Aromatic-dependent *Salmonella typhimurium* are non-virulent and effective as live vaccines, *Nature* 291:238.

Hormaeche, C. E., Joysey, H. S., Desilva, L., Izhar, M., and Stocker, B. A. D., 1991, Immunity conferred by *aro*$^-$ salmonella live vaccines,. *Microb. Pathogen.* 10:149.

ROLE OF T-CELLS, TNFα AND IFNγ IN IMMUNITY TO ORAL CHALLENGE WITH VIRULENT SALMONELLAE IN MICE VACCINATED WITH *ARO*⁻ LIVE ATTENUATED SALMONELLA VACCINES

Pietro Mastroeni, Bernardo Villarreal-Ramos and
Carlos E. Hormaeche

Division of Microbiology and Parasitology
Department of Pathology, University of Cambridge
Tennis Court Road, Cambridge CB2 1QP, UK

Typhoid fever is still a widespread disease that affects many parts of the world and more effective vaccines are needed (Robbins *et al.*, 1990). Unfortunately the rational development of salmonella vaccines has been hampered by our incomplete understanding of the basic immunological mechanisms operating in experimental typhoid and in the human disease.

Live vaccines are generally believed to confer better protection than killed ones. Nevertheless not all live attenuated vaccine strains are protective and the reasons for these failures are largely unknown (Hormaeche *et al.*, 1991; Smith *et al.*, 1984). In fact the basic mechanisms of immunity and the immunological correlates of protection induced by live attenuated vaccine strains have not been elucidated yet. The relative importance of cellular *vs.* humoral mechanisms and the role of T-cells and cytokines in immunity induced by live attenuated salmonella vaccines are not clear and are still a matter of some dispute.

The new generation of live attenuated aromatic dependent oral vaccines are safe and effective in animals (Hormaeche, 1991) and are being tested in humans, with promising results (Tacket *et al.*, this volume). *S. typhimurium* SL3261 is a transposon generated, *aroA* derivative of the virulent *S. typhimurium* 1344 strain (Hoiseth *et al.*, 1981). In our mouse model the vaccine confers solid, long term and species specific protection against oral

Biology of Salmonella, Edited by F. Cabello *et al.*.
Plenum Press, New York, 1993

challenge with virulent organisms (Hormaeche *et al.*, 1991). Cross protection against unrelated organisms vanishes as soon as the vaccine strain is cleared from the organs of the reticuloendothelial system (RES). Thereafter protection is serotype specific and presumably involves the specific recall of immunity.

In the present study we investigated the requirement for T-cells, IFNγ and TNFα in the specific recall of immunity to oral challenge with virulent organisms in innately susceptible BALB/c mice immunised the live attenuated *S. typhimurium* SL3261 vaccine strain. BALB/c mice were vaccinated intravenously with *ca.* 10^6 organisms of the attenuated vaccine strain and challenged orally at least two months later with the virulent *S. typhimurium* C5. Naive mice were included in each experiment. In naive mice the infection progressed rapidly with bacterial counts in the RES increasing at about 1 log a day; no mice survived beyond day 8. Conversely immune mice efficently suppressed bacterial growth in the RES and no deaths occurred. *In vivo* administration of anti-CD4 and anti-CD8 monoclonal antibodies dramatically exacerbated the course of the secondary infection in immunised mice. On day 8 of the secondary infection, bacterial counts in the organs of the immunised T-depleted mice were much higher than in immune controls and no mice survived beyond day 15. Thus depletion of $CD4^+$ and $CD8^+$ T-cells severely impairs long term immunity to oral challenge with virulent organisms in mice vaccinated with *S. typhimurium* SL3261.

The effect of selective depletion of either $CD4^+$ or $CD8^+$ T-cells was significant although less marked than depletion of both T-cell subsets. Selective depletion of $CD4^+$ cells significantly impaired resistance both eight and fourteen weeks after vaccination as determined by estimation of bacterial numbers in organ homogenates. Depletion of $CD8^+$ cells alone had less effect on host resistance eight weeks than fourteen weeks after immunisation.

Administration of anti-IFNγ or anti-TNFα antibodies also exacerbated a secondary infection in vaccinated mice. IFNγ was detectable in sera of both controls and T-cell depleted mice on day 8 of the secondary infection as well as in sera of anti-TNFα treated mice on day 6 of infection. Thus IFNγ seems to be necessary (its neutralization exacerbates the course of the secondary infection) but not sufficient (it is still detectable in sera of T-depleted or anti-TNFα treated mice) for the enhanced resistance observed in immune mice during a secondary oral infection with virulent organisms.

On day 8 of the secondary infection, histological examination of spleen, liver and mesenteric lymph nodes of immune mice showed isolated granulomatous lesions. Hepatosplenomegaly with minute grossly visible focal lesions was observed in T-cell depleted immune mice. A widespread mononuclear cell infiltrate together with patchy necrosis was present. The histopathology in anti-IFNγ treated mice was qualitatively similar to that seen in T-cell depleted mice. In contrast, no cellular infiltration was observed in the organs of the anti-TNFα treated mice, in which the lesions appeared similar or worse than those seen in naive mice, with severe necrosis. Splenomegaly was not as marked as in T-cell depleted mice.

The present results indicate that T-cells, IFNγ and TNFα are important in the specific recall of immunity to virulent salmonellae conferred by immunization with live vaccines. T-cells, IFNγ and TNFα all seem to be crucial for the expression of the bactericidal activity of the RES. TNFα seems to be also involved in the recruitment of mononuclear cells and granuloma formation during a secondary oral infection with a virulent *Salmonella* strain.

REFERENCES

Hoiseth, S.K., and Stocker B.A.D., 1981, Aromatic dependent *Salmonella typhimurium* are non virulent and effective as live vaccines, *Nature*, 291: 39.

Hormaeche, C.E., Joysey, H.S., Desilva, L., Izhar, M., and Stocker, B.A.D., 1991, Immunity conferred by Aro^- salmonella live vaccines, *Microb. Pathogen.* 10: 149.

Hormaeche, C.E., 1991, Live attenuated salmonella vaccines and their potential as oral combined vaccines carrying heterologous antigens, *J. Immunol. Methods.* 142: 113.

Robbins, A.A., 1990, Progress towards vaccines we need and we do not have, *Lancet* 335: 1436.

Smith, B.P., Reina-Guerra, M., Hoiseth, S.K., Stocker, B.A.D., Habasha, F., Johnson, E., and Merrit, F., 1984, Aromatic dependent *Salmonella typhimurium* as modified live vaccines for calves, *Am. J. Vet. Res.* 45: 59.

REDUCED VIRULENCE OF *S. TYPHIMURIUM* WITH MUTATIONS IN GLOBAL REGULATORY GENES

J.A. Harrison[1], D. Pickard[2], A. Khan[1], S.N. Chatfield[2], C.J. Dorman[3], G. Dougan[4], and C.E. Hormaeche[1]

[1]Department of Pathology, University of Cambridge, Tennis Court Road, Cambridge CB2 1QP, UK.
[2]Medeva Group Research and
[4]Department of Biochemistry, Imperial College, London SW7 2AY, UK.
[3]Department of Biochemistry, Medical Sciences Institute, University of Dundee, Dundee DD1 4HN, UK

Some proteins which play a role in pathogenicity are expressed on the surface of the *Salmonella* cell or are located within the periplasm, *e.g.* flagella, exotoxins, fimbrial adhesins, haemolysins and outer membrane proteins.

We are characterising a bank of Tn*phoA* mutants of the mouse virulent *S. typhimurium* C5 (Miller *et al.*, 1989) which show reduced virulence in mice. Tn*phoA* (Manoil and Beckwith, 1985) is a reporter transposon which allows the isolation of mutants with insertions in genes coding for secreted proteins. Tn*phoA* is a Tn*5* derivative carrying the *E. coli* alkaline phosphatase *phoA* gene minus its natural promoter and signal sequence; mutants harbouring insertions in genes coding for secreted proteins can be identified as blue colonies on the chromogenic substrate X-P. A mutant of reduced virulence designated C5060 from the bank described by Miller *et al.* (1989) was characterised in this study.

The gene from C5060 harbouring the Tn*phoA* insertion was sequenced and found to be *osmZ (hns)*. Hns is a 15.6 kDa histone-like DNA binding protein which exists predominantly as a dimer. It is known to affect DNA supercoiling, colanic acid synthesis, glycine betaine transport, the ratio of OmpC to OmpF, β-glucoside uptake and expression of Type 1 fimbriae. Mutations in *osmZ* can affect supercoiling in different osmolar conditions; using pUC18 as a reporter plasmid a greater relaxation of the DNA was seen in C5060 than in the parent C5 in both LB and LB + 0.4 M NaCl. The greatest relaxation of plasmid DNA occurred in the high salt conditions for both strains. C5060 was more mucoid than the wild type, also typical of *osmZ* mutants with altered colanic acid synthesis.

However, Hns is an intracellular protein, and not secreted, and there is no apparent reason for a Tn*phoA* mutant to appear blue on X-P. Tn*phoA* had inserted into *osmZ* such that the open reading frames of *phoA* and *osmZ* were in opposite directions. Computer analysis of the sequence failed to detect another gene which could be driving expression of PhoA, and non-denaturing PAGE analysis showed that C5060 did not in fact produce any PhoA. Conversely, a *phoN*⁻ mutant of C5060 (*i.e.* lacking the native acid phosphatase of *Salmonella*) was white on X-P, suggesting that the *osmZ* mutation of C5060 in some way altered cell physiology.

The virulence of the *osmZ* mutant was tested in BALB/c mice. The LD_{50} increased approximately 1000-fold by both the i/v and oral routes. The *osmZ*::Tn*phoA* lesion was transduced into *S. typhimurium* SL1344 and *S. enteritidis* Se795; the i/v LD_{50} of the mutants were increased by approximately 10^4. The C5060 mutant increased more slowly than the wild-type C5 in the RES. Restoration of *osmZ* by co-transduction using a neighbouring marker (*oppB*255::Tn*10*) showed that whereas three transductants regained the non-mucoid phenotype of the wild type, only one regained virulence.

The results suggest that *osmZ* is important for virulence in *Salmonella*, but the attenuation of some of the mutants studied may be due to lesions secondary to the *osmZ* mutation.

REFERENCES

Manoil, C. and Beckwith, J., 1985, Tn*phoA*: A transposon probe for protein export signals *P.N.A.S.* 82: 8129.

Miller, I., Maskell, D., Hormaeche, C.E., Johnson, K., Pickard, D. and Dougan, G., 1989, Isolation of orally attenuated *Salmonella typhimurium* following Tn*phoA* mutagenesis. *Infect. Immun.* 57:2758.

INDEX